OXFORD

endorsed by edexcel

GCSE Maths

Foundation

Dave Capewell
Formerly Westfield School, Sheffield

Geoff Fowler
Mathematics Strategy Manager, Birmingham

Peter Mullarkey
Netherhall School, Maryport, Cumbria

Katherine Pate
Mathematics Publishing Consultant

OXFORD
UNIVERSITY PRESS

Great Clarendon Street, Oxford OX2 6DP

Oxford University Press is a department of the University of Oxford.
It furthers the University's objective of excellence in research, scholarship, and education by publishing worldwide in

Oxford New York

Auckland Cape Town Dar es Salaam Hong Kong Karachi
Kuala Lumpur Madrid Melbourne Mexico City Nairobi
New Delhi Shanghai Taipei Toronto

With offices in

Argentina Austria Brazil Chile Czech Republic France Greece
Guatemala Hungary Italy Japan South Korea Poland Portugal
Singapore Switzerland Thailand Turkey Ukraine Vietnam

First published 2006

British Library Cataloguing in Publication Data

Data available

ISBN 978 019 915087 8

10 9 8 7 6 5 4 3

Typeset by MCS Publishing Services Ltd, Salisbury, Wiltshire

Printed by Printplus, China

Acknowledgements

The publisher would like to thank Edexcel for their kind permission to reproduce past exam questions. Edexcel Ltd. accepts no responsibility whatsoever for the accuracy and method of working in the answers given.

This high quality material is endorsed by Edexcel and has been through a rigorous quality assurance programme to ensure that it is a suitable companion to the specification for both learners and teachers.
This does not mean that its contents will be used verbatim when setting examinations nor is it to be read as being the official specification – a copy of which is available at www.edexcel.org.uk

The Publisher would also like to thank Peter Hind for his authoritative guidance in preparing this book.

The image on the cover is reproduced courtesy of Photodisc.

The Publisher would like to thank the following for permission to reproduce photographs:

Figurative artwork is by Peter Donnelly.
P14 Oxford University Press; **p68** Oxford University Press; **p74** all Oxford University Press; **p87** both Oxford University Press; **p117** Justin Kase/Alamy; **p118** Tim Graham/Alamy; **p136** Derek Holloway; **p152** Photodisc/Oxford University Press; **p155** Photodisc/Oxford University Press; **p163** Oxford University Press; **p164** Ingram/Oxford University Press; **p209** Peter Adams Photography/Alamy; **p246** Hemera/Oxford University Press; **p251** Sami Sarkis/Alamy; **p260** Oxford University Press; **p274** Jack Sullivan/Alamy; **p282** Oxford University Press; **p311** Science Photo Library; **p320** Photodisc/Oxford University Press; **p355** Elizabeth A Whiting/Corbis; **p364** Oxford University Press; **p391** Araldo de Luca/Corbis.

Technical artwork is by MCS Publishing Services Ltd.

About this book

This book has been specifically written to help you get your best possible grade in your Edexcel GCSE Mathematics examinations. It is designed for students who have achieved level 4 or below at Key Stage 3 and are looking to progress to a grade E at Foundation level of GCSE.

The authors are experienced teachers and examiners who have an excellent understanding of the Edexcel two-tier specification and so are well qualified to help you successfully meet your objectives.

The book is made up of units which are based on the Edexcel specification, and provide coverage of the National Curriculum strands at Key Stage 4.

The units are:

The last unit is labelled the **+ section**, and is designed to provide you with access to grades D and C. This material is linked to the topics in the main units.

Problem solving is integrated throughout the material as suggested in the National Curriculum.

How to use this book

This book is made up of units of work that are colour-coded: Algebra (green), Data (blue), Number (orange) and Shape, space and measures (pink).

Each unit starts with an overview of the content, so that you know exactly what you are expected to learn.

This unit will show you how to

- Understand place value
- Order positive and negative numbers and represent them as positions and transitions on a number line
- Multiply and divide by powers of 10

The first page of a unit also provides prior knowledge questions to help you revise before you start – then you can apply your knowledge later in the unit. The answers to these questions are given at the back of the book.

Before you start ...

You should be able to answer these questions.

	Review
1 Write the number three hundred and four in figures.	Key stage 3
2 Calculate 7×10.	Key stage 3

Inside each unit, the content develops in double page spreads that all follow the same structure.

Each spread starts with a list of the learning outcomes and a summary of the keywords:

N1.1 Place value

This spread will show you how to:

- Understand place value and order positive numbers
- Multiply and divide by powers of 10

Keywords
Digit
Order
Place value

Key points are highlighted in the text so you can see the facts you need to learn:

- **You can find the area of any rectangle by using the formula:**
  Area of rectangle = length × width

width
length

Examples showing the key skills and techniques you need to develop are shown in boxes. Also margin notes show tips and reminders you may find useful:

Example

Simplify.

a $4m + 2p + 3m$ **b** $2x + 5y + 3x + y$

a $4m + 2p + 3m = 4m + 3m + 2p$
$= 7m + 2p$

b $2x + 5y + 3x + y = 2x + 3x + y + 5y$
$= 5x + 6y$

Rearrange the terms to collect like terms together.

$y = 1y$

Each exercise is carefully graded, set at three levels of difficulty:

- The first few questions are mainly repetitive to give you confidence, and simplify the content of the spread
- The questions in the middle of the exercise consolidate the topic, focusing on the main techniques of the spread
- Later questions extend the content of the spread – some of these questions may be problem-solving in nature, and may involve different approaches.

At the end of the unit is an exam review page so that you can revise the learning of the unit before moving on. The key Edexcel objectives are identified:

D3

Exam review

Key objectives

- Understand and use the probability scale
- Understand and use estimates or measures of probability from theoretical models (including equally likely outcomes)

Summary questions, including past exam questions, are provided to help you check your understanding of the key concepts covered and your ability to apply the key techniques.

In the **+ section**, you will find a range of topics that you will be familiar with at an easier level. References to the **+ section** are made throughout the book, where you can skip ahead and extend the topic to grades D and C.

An Edexcel formula page is provided near the end of the book so you can see what information you will be given in the exam.

Contents

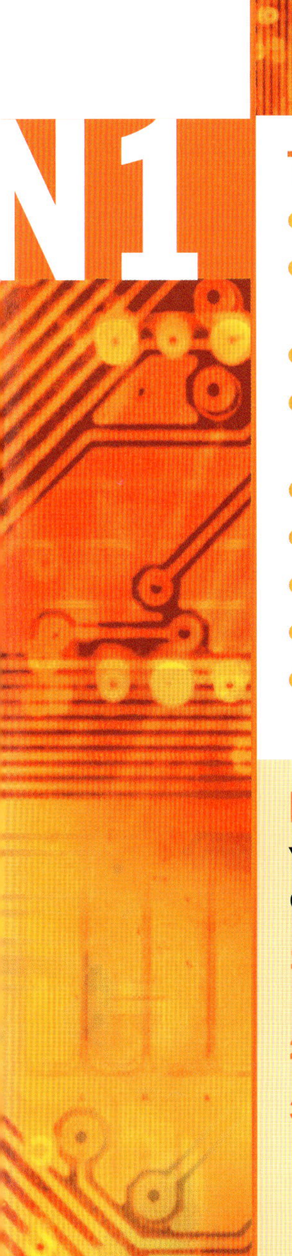

Properties of number

This unit will show you how to

- Understand place value
- Order positive and negative numbers and represent them as positions and transitions on a number line
- Multiply and divide by powers of 10
- Read measurements and information from scales, dials and timetables
- Order temperatures and position them on a number line
- Calculate changes in temperature
- Add, subtract and multiply with negative numbers
- Understand the terms 'factor' and 'multiple'
- Find the common factors and common multiples of two numbers

Before you start ...

You should be able to answer these questions.

Review

1 Write the number three hundred and four in figures. — Key stage 3

2 Calculate 7×10. — Key stage 3

3 How long is the journey from Seaton to Defford? — Key stage 3

Station	Arr./Dep. Time
Ayton	09:00
St Bees	09:20
Seaton	09:35
Defford	09:55
Extrop	10:15

4 What is the temperature on this thermometer? — Key stage 3

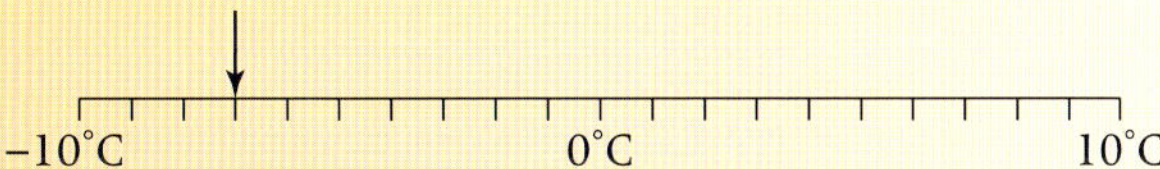

5 What is $5 - 17$? — Key stage 3

6 What is $40 \div 8$? — Key stage 3

N1.1 Place value

This spread will show you how to:
- Understand place value and order positive numbers
- Multiply and divide by powers of 10

Keywords
Digit
Order
Place value

- The value of each **digit** in a number depends upon its place in the number. This is called its **place value**.

In the number 4207

Thousands	Hundreds	Tens	Units
4	2	0	7

The digit 4 stands for 4 thousands

The digit 2 stands for 2 hundreds

The digit 7 stands for 7 units

You write this number in words as four thousand, two hundred and seven.

You can compare numbers using a place value table. Compare the place values of the first non-zero digits. If they are the same, compare the second digits. Keep comparing digits until you find a difference.

Example

Put these numbers in **order** from lowest to highest.
17 6300 5993 2330 2426 12 540

Put the numbers in a place value table.

Ten thousands	Thousands	Hundreds	Tens	Units
			1	7
	6	3	0	0
	5	9	9	3
	2	3	3	0
	2	4	2	6
1	2	5	4	0

The digit **1** stands for 1 ten.
The digit **6** stands for 6 thousands.
The digit **5** stands for 5 thousands.
The digit **2** stands for 2 thousands.
The digit **2** stands for 2 thousands.
The digit **1** stands for 1 ten thousand.

Look at the first non-zero digits.
You can now put all the numbers in order except for 2330 and 2426.
For these two numbers you need to look at the second digit

2330 2426

In order the numbers are 17, 2330, 2426, 5993, 6300, 12 540.

You can use a place value table to multiply and divide by 10 or 100.

- To multiply a number by 10 all the digits move one place to the left.

$37 \times 10 = 370$

Hundreds	Tens	Units	
	3	7	× 10
3	7	0	

The 0 holds the digits in place.

- To divide a number by 100 all the digits move two places to the right.

$4850 \div 100 = 48.5$

Thousands	Hundreds	Tens	Units	•	tenths	
4	8	5	0	•		÷ 100
		4	8	•	5	

Exercise N1.1

1 Write each of these numbers in figures.

a eighty-seven

b one hundred and forty-three

c four hundred and six

d four hundred and sixty

e two thousand and fifty-three

f eight thousand, five hundred and three

g eight thousand, five hundred and thirty

h thirty-four thousand, six hundred and forty

i thirty thousand, four hundred and sixty-four

j two hundred and six thousand, five hundred and three

2 These are the areas of eight countries in square kilometres.
Write each number in words.

a Algeria 2 381 741   **b** Chile 756 950

c France 543 965   **d** India 3 166 829

e China 9 596 960   **f** United Kingdom 244 100

g USA 9 368 900   **h** Australia 7 682 300

DID YOU KNOW?

Russia is the largest country in the world with an area of 17 075 000 km^2.

3 Put each list of numbers in order, starting with the smallest.

a 56   34   9   112   178   89   139

b 2372   1784   2386   1990   3233   3022

c 40 500   45 045   4555   4005   40 545   44 054

d 240 440   204 044   24 445   245 004   42 024   242 404

4 Draw a place value table to calculate each of these.
The first one has been started for you.

a 29×10

Thousands	Hundreds	Tens	Units	•	tenths	hundredths
		2	9	•		
				•		

b 132×10 **c** $590 \div 10$ **d** 17×100

e $6400 \div 10$ **f** $73 \div 10$ **g** $345 \div 100$

5 Calculate each of these.

a 12×10 **b** 4×100 **c** $320 \div 10$ **d** $4600 \div 100$

e 30×10 **f** 4.6×10 **g** $230 \div 100$ **h** $659 \div 10$

i 34×1000 **j** 3.56×100 **k** $23.6 \div 10$ **l** 0.345×100

N1.2 Reading scales

This spread will show you how to:

- Represent numbers as positions and transitions on a number line
- Read measurements and information from scales, dials and timetables

Keywords
Estimate
Number line
Scale
Timetable

You use a ruler to measure lengths.
You use a weighing scale to measure weight or mass.
You use a measuring jug to measure a volume of liquid.

- **Every number can be represented as a position on a number line.**

Most of the **scales** you read are number lines.

Example

Write the reading shown on each scale.

a

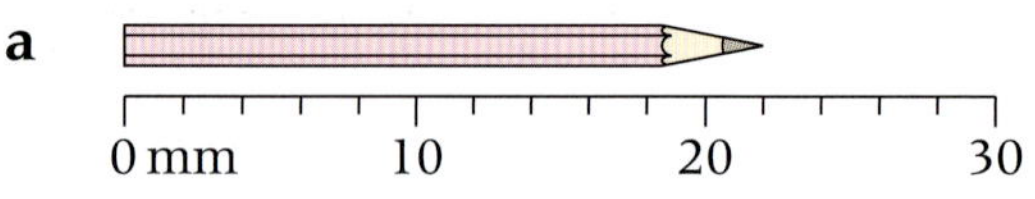

b

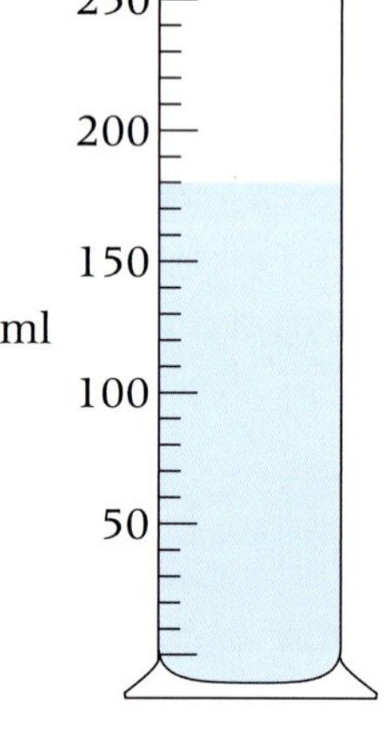

a The length of the pencil is between 20 and 30 mm.
There are 5 spaces between 20 and 30 mm.
So 5 spaces represent 10 mm.
Each space represents $10 \div 5 = 2$ mm.
The pencil is 22 mm long.

b The reading is between 150 and 200 ml.
There are 5 spaces between 150 and 200 ml.
So 5 spaces represent 50 ml.
Each space represents $50 \div 5 = 10$ ml.
The reading is 180 ml.

- You can **estimate** a measurement from a scale.

You need to be able to read **timetables** for buses and trains.

Example

Station	Depart	Depart	Depart
Penrith	08:00	09:00	10:00
Kendal	08:20	09:20	10:20
Lancaster	08:35	09:35	10:35
Preston	08:55	09:55	10:55
Warrington	09:15	10:15	11:15

How long does it take the 09:20 train from Kendal to get to Warrington?

The train leaves Kendal at 09:20 and arrives at Warrington at 10:15.
The journey time is:

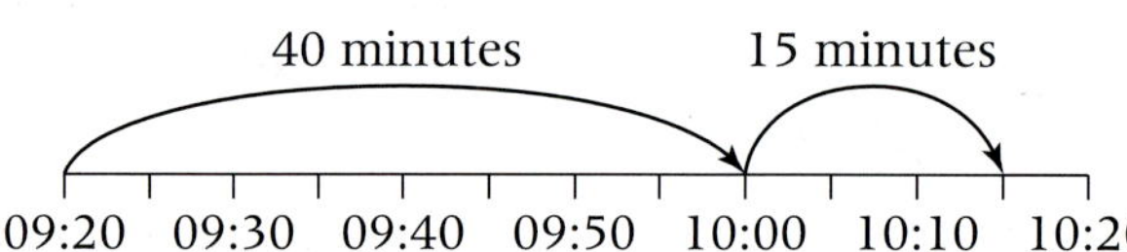

From 09:20 to 10:00 = 40 minutes
From 10:00 to 10:15 = 15 minutes
Total time = 55 minutes

Exercise N1.2

1 Write the number each of the arrows is pointing to.

a

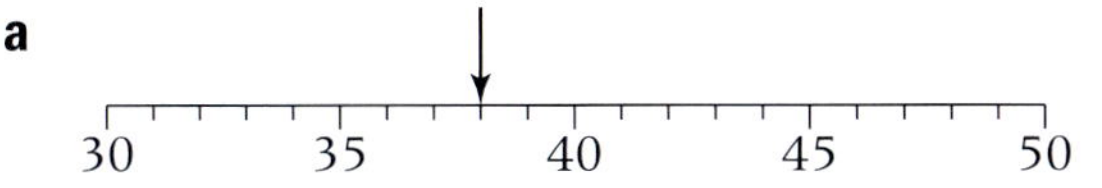

b

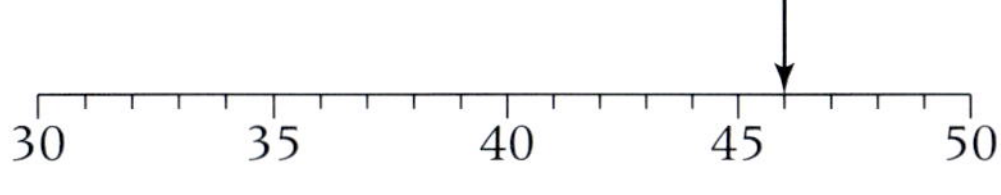

c

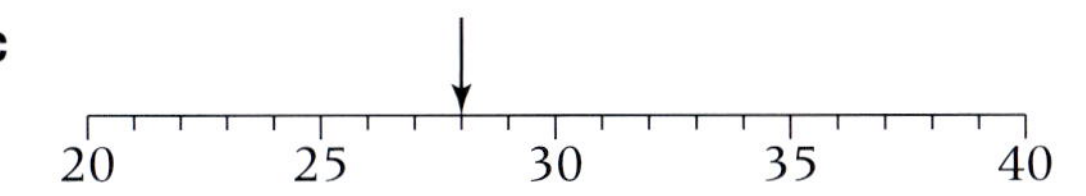

d

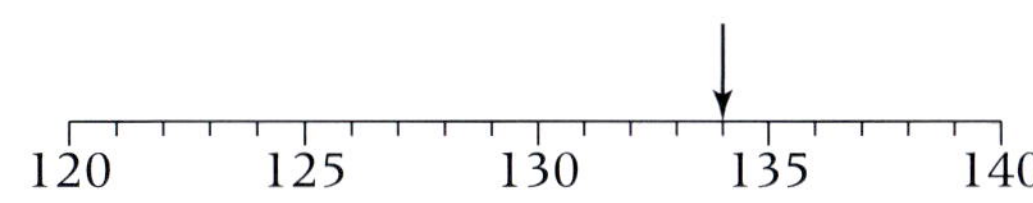

e

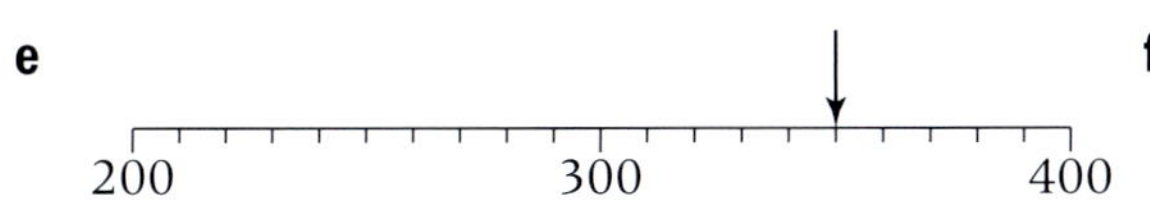

f

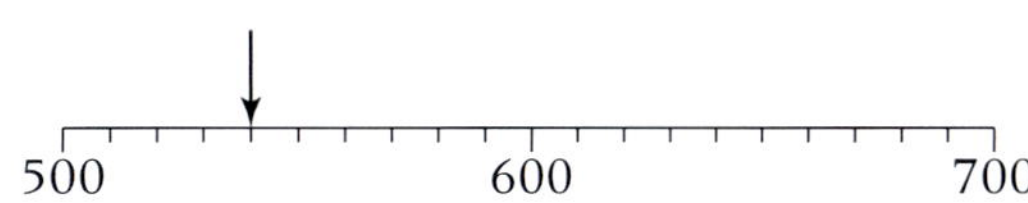

2 Write the reading shown on each scale.

a

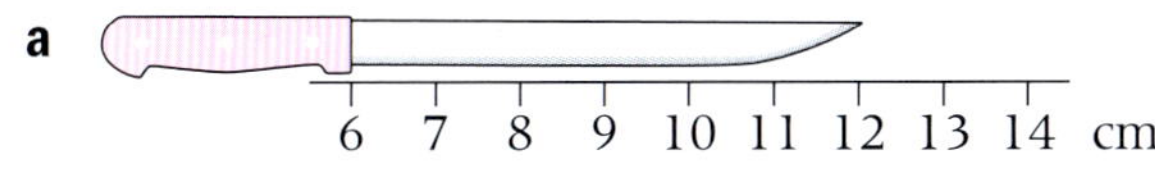

b

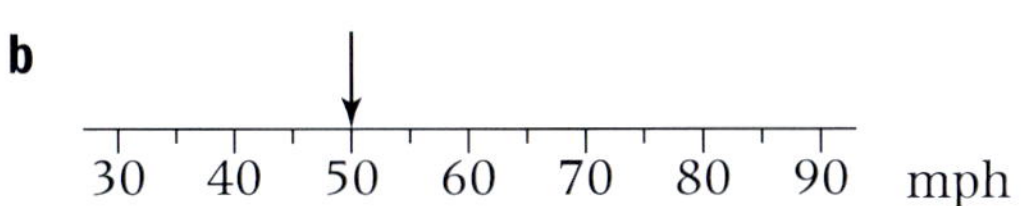

c

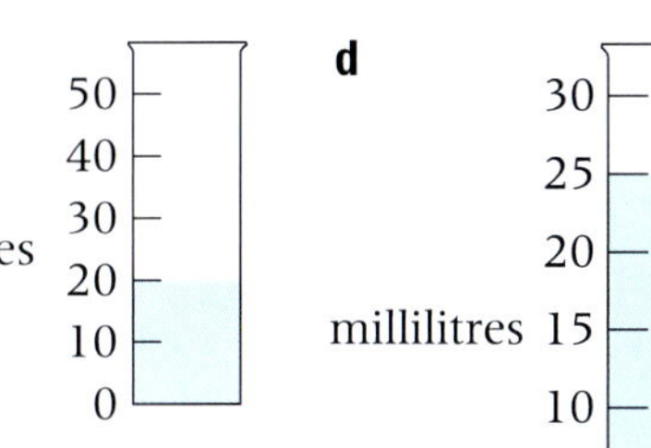

d

30
25
20
millilitres 15
10
5
0

3 Write the readings on each scale.

a

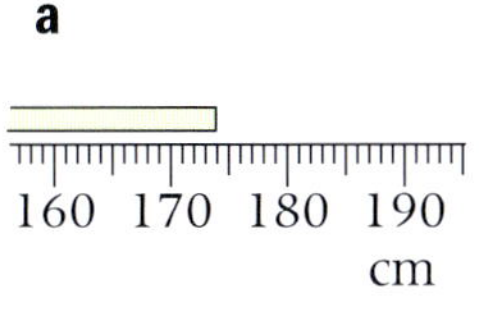

b

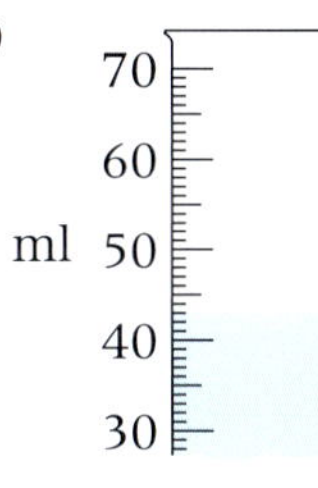

c

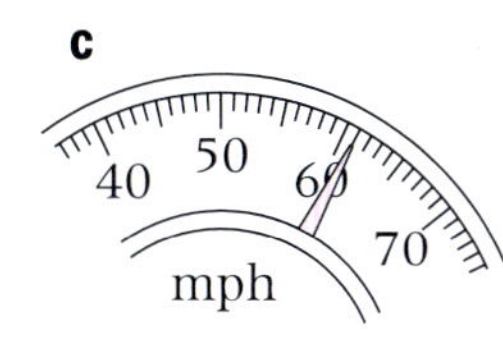

d

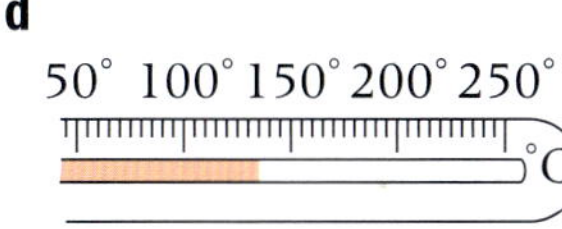

4

Lancaster–Keswick bus timetable			
Lancaster	12:30	14:20	17:25
Kendal	13:35	15:30	18:35
Windermere	14:03	15:58	19:03
Ambleside	14:18	16:18	19:23
Grasmere	14:36	16:36	19:41
Keswick	14:56	16:56	20:01

a What time does the 12:30 bus from Lancaster arrive in Keswick?

b A bus arrives in Grasmere at 19:41.
At what time did it leave Kendal?

c How many minutes does it take the 13:35 bus from Kendal to get to Ambleside?

d Karen catches the 17:25 bus from Lancaster. How long does it take her to travel to Windermere?

Temperature

This spread will show you how to:

- Order temperatures and position them on a number line
- Calculate changes in temperature

Keywords

Degrees Celsius
Negative
Number line
Temperature

- You use **negative** numbers for **temperatures** below 0°C.

These readings show the temperatures in Sydney and Moscow.

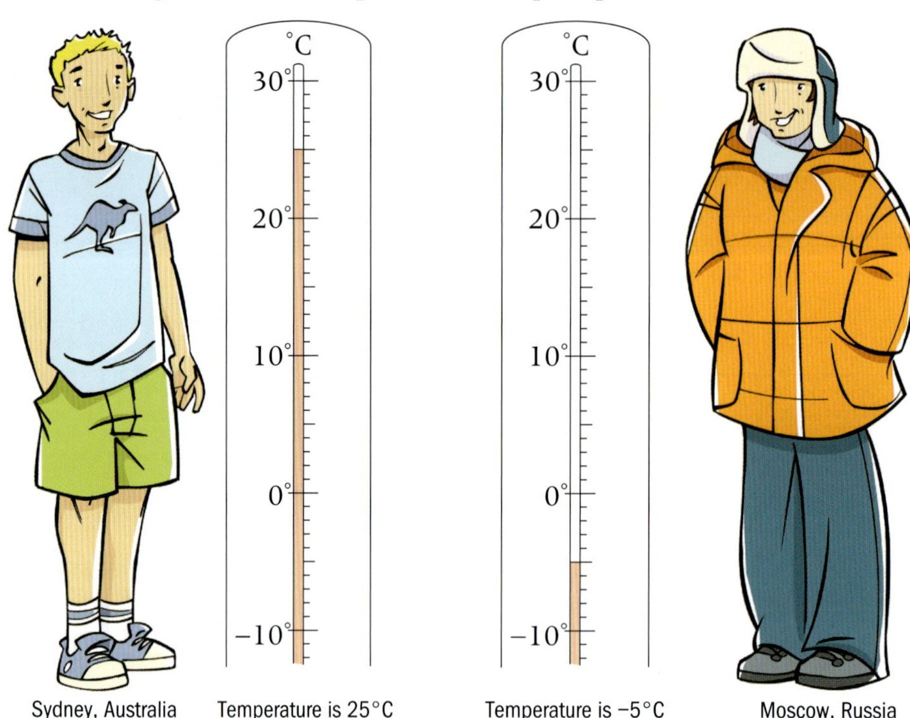

Sydney, Australia Temperature is 25°C Temperature is −5°C Moscow, Russia

The temperature scale is a **number line**.

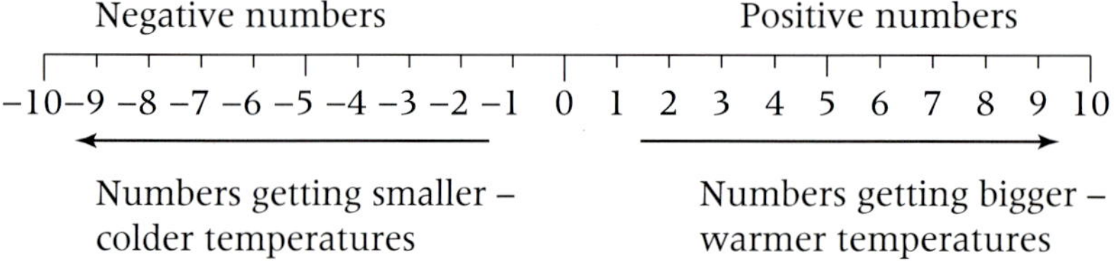

Example

The temperature in London is 13 °C. During the night the temperature falls by 18°. What is the night-time temperature in London?

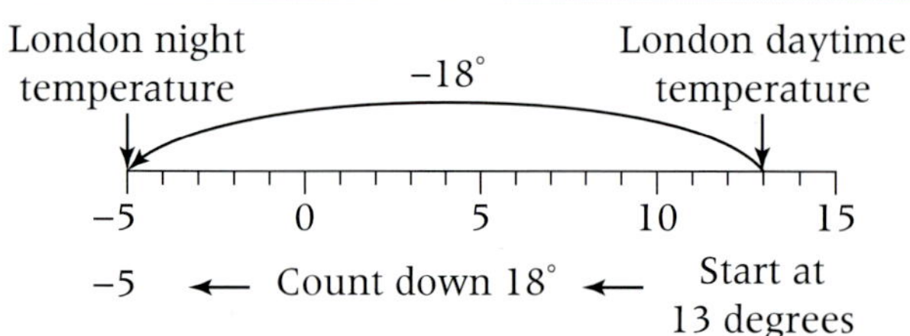

The night-time temperature in London is −5°C.

Exercise N1.3

1 Write the temperature on each of these thermometers.

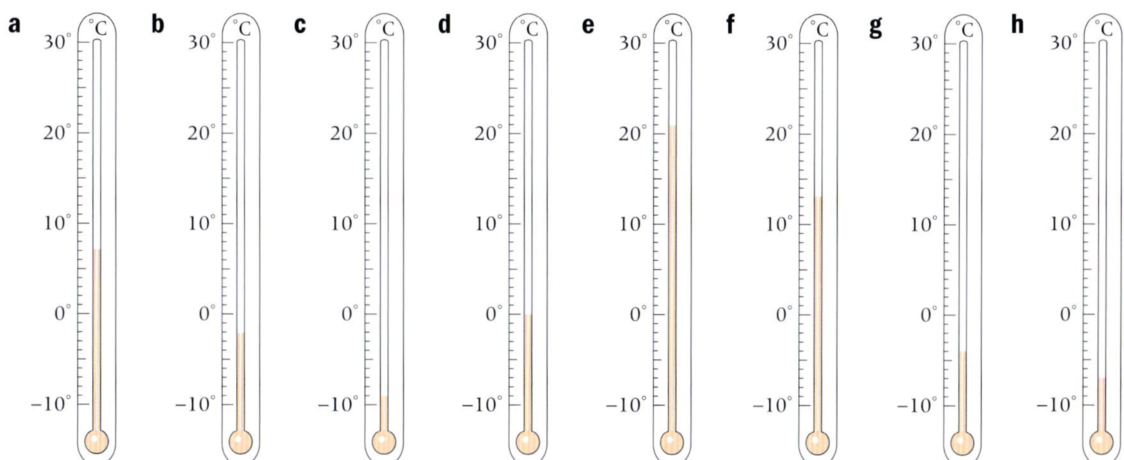

2 Write the hottest and coldest temperatures in each of these lists.
All the temperatures are in degrees Celsius.

a	−3	−6	0	7	−2	12
b	0	−12	−15	−8	−3	−17
c	12	21	−2	14	9	3
d	−12	15	19	−31	28	8

3 What temperature is

a 5 degrees higher than 7 °C

b 10 degrees higher than 2 °C

c 4 degrees higher than −1 °C

d 12 degrees higher than −15 °C

e 4 degrees lower than 12 °C

f 5 degrees lower than 2 °C

g 8 degrees lower than −3 °C

h 15 degrees lower than 7 °C?

4 This table shows the highest and lowest temperatures, in °C, in several cities around the world.

City	Perth	Cape Town	Copenhagen	Calgary	Manchester
Highest summer temperature	36	40	21	16	28
Lowest winter temperature	12	15	-12	-25	-2

a Which city had the highest summer temperature?

b Which city had the lowest winter temperature?

c Which city had the biggest difference in temperature between summer and winter?

d Which city had the smallest difference in temperature between summer and winter?

N1.4 Calculating with negative numbers

This spread will show you how to:

- Order negative numbers and position them on a number line
- Add, subtract and multiply with negative numbers

Keywords
Add
Multiply
Negative number
Number line
Subtract

- **You can order negative numbers using a number line.**

Example

Place these numbers in order, starting with the smallest.

−3 −4 2 −5 −2 4

Using a number line

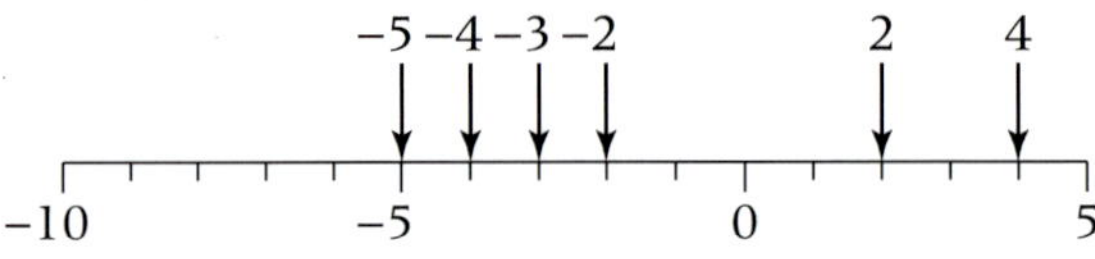

−5 is further away from zero than −4, so it is smaller.

The correct order is −5, −4, −3, −2, 2, 4.

You can use a number line to **add** and **subtract negative numbers**. There are two rules to remember.

- **Adding a negative number is the same as subtracting a positive number.**
- **Subtracting a negative number is the same as adding a positive number.**

Example

Calculate

a $8 + -3$ **b** $-8 - -3$

a Start at 8 and subtract 3 (move to the left).

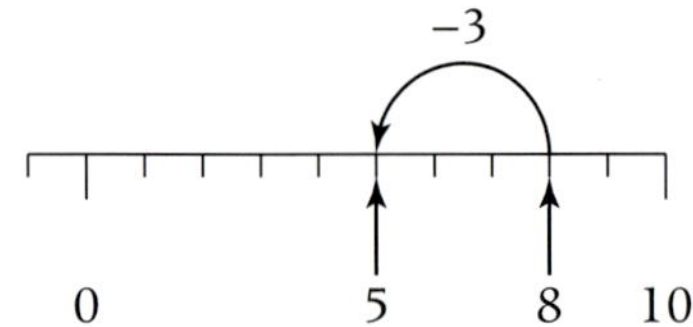

$8 + -3 = 8 - 3 = 5$

b Start at −8 and add 3 (move to the right).

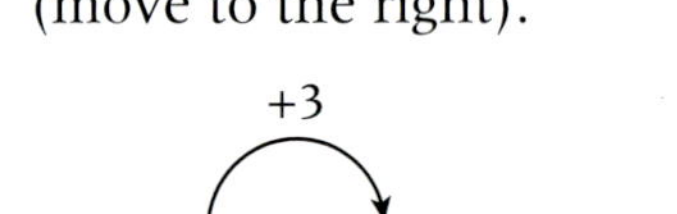

$-8 - -3 = -8 + 3 = -5$

You can **multiply** with negative numbers.
There are two rules to remember.

- **Negative number × positive number = negative number**

$-3 \times 4 = -12$

- **Negative number × negative number = positive number**

$-3 \times -4 = 12$

Exercise N1.4

1 Copy and complete this table, working out the new temperature in each case.

Start temperature (°C)	Change in temperature	New temperature (°C)
14	−11	
7	−10	
13	−6	
−5	+11	
−8	−5	

2 Calculate these.

a $3 + 5$ **b** $15 - 12$ **c** $7 - 10$ **d** $4 - 13$ **e** $14 - 17$

f $12 - 9$ **g** $-4 + 8$ **h** $-3 + 12$ **i** $-15 + 11$ **j** $-3 + 7$

k $-8 - 2$ **l** $-3 + 5$ **m** $-2 - 6$ **n** $-9 - 7$ **o** $-3 + 8$

p $-21 + 14$ **q** $21 - 14$ **r** $-21 - 14$ **s** $-13 - 23$ **t** $-16 + 34$

3 Calculate these.

a $16 + -3$ **b** $8 + -6$ **c** $13 + -5$ **d** $-7 + -3$ **e** $-6 + -7$

f $-4 + -8$ **g** $-8 + -2$ **h** $12 - -7$ **i** $5 - -11$ **j** $-3 - -6$

k $-9 - -2$ **l** $-4 - -2$ **m** $-6 - -10$ **n** $-3 + -2$ **o** $-3 - -2$

p $15 + -11$ **q** $17 - -13$ **r** $14 + -13$ **s** $15 - -11$ **t** $-11 + -14$

4 In these pyramids the brick that sits directly above two bricks is the sum of those two bricks. For example:

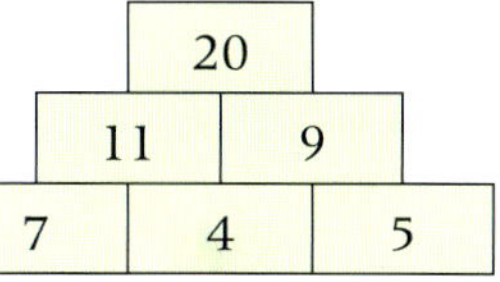

Copy and complete these pyramids.

a

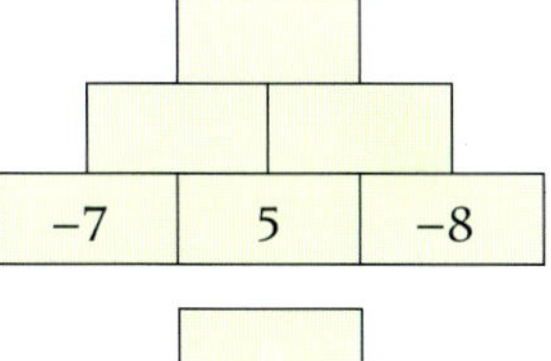

b

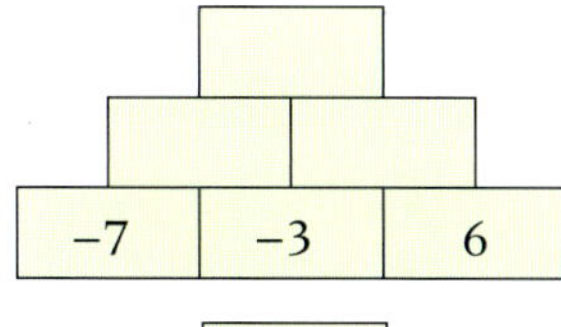

c

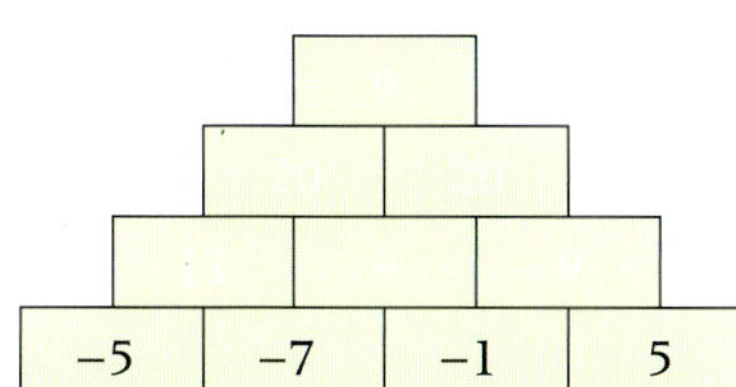

d

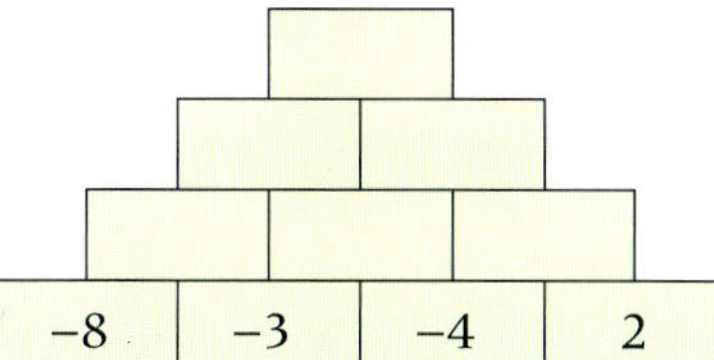

5 Calculate these.

a 3×5 **b** -3×4 **c** -2×10 **d** -3×8 **e** -4×5

f -3×-2 **g** -4×-2 **h** -6×-3 **i** -5×2 **j** -7×-3

N1.5 Factors and multiples

This spread will show you how to:

- Understand the terms 'factor' and 'multiple'
- Find the common factors and common multiples of two numbers

Keywords
Common factor
Common multiple
Factor
Multiple
Product

Any number can be written as the product of two **factors**.

$20 = 2 \times 10$ so 2 and 10 are factors of 20.

- **The factors of a number are those numbers that divide into it exactly, leaving no remainder.**

You can often write a number as the **product** of two factors.

Product means multiply together.

24 can be written as 4×6 or 3×8 or 2×12 or 1×24.

The factor pairs are 1×24, 2×12, 3×8 and 4×6.

You can write the factors in a list: 1, 2, 3, 4, 6, 8, 12, 24.

- **A common factor is a factor that is common to two different numbers.**

Example

Write the common factors of 30 and 42.

The factors of 30 are 1 2 3 5 6 10 15 30

The factors of 42 are 1 2 3 6 7 14 21 42

The common factors of 30 and 42 are 1, 2, 3 and 6.

You can use factor pairs:
1×30
2×15
3×10
5×6

The first five **multiples** of 20 are 20, 40, 60, 80 and 100.

$1 \times 20 = 20$ $2 \times 20 = 40$ $3 \times 20 = 60$
$4 \times 20 = 80$ $5 \times 20 = 100$

- **A common multiple is a multiple that is common to two different numbers.**

Example

Write the first three common multiples of 8 and 12.

The multiples of 8 are 8 16 24 32 40 48 56 64 72 80 ...

The multiples of 12 are 12 24 36 48 60 72 84 ...

The first three common multiples of 8 and 12 are 24, 48 and 72.

Exercise N1.5

1 This factor diagram shows all the factor pairs of 12.

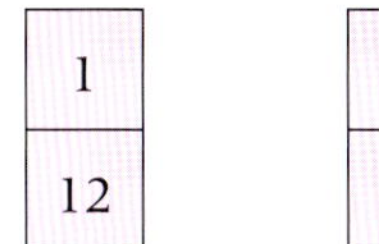

1	2	3
12	6	4

$1 \times 12 = 12$ $2 \times 6 = 12$ $3 \times 4 = 12$

The factors of 12 are 1, 2, 3, 4, 6 and 12.

Copy and complete these factor diagrams.

a 18  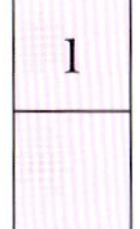

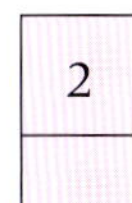

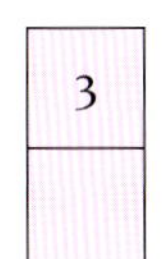

b 20  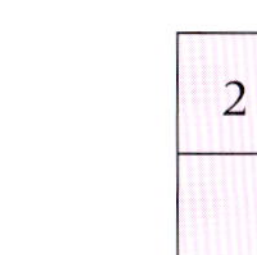

c 30 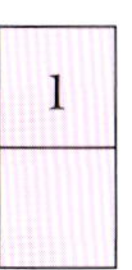

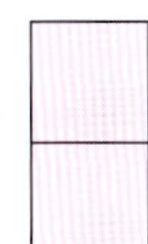

d 14 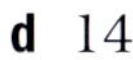 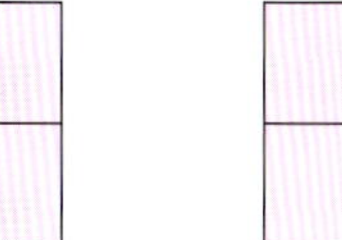

2 Draw factor diagrams for

a 15 **b** 22 **c** 28 **d** 36 **e** 40

3 Write all the factor pairs for each number.

a 8 **b** 16 **c** 23 **d** 34 **e** 10
f 26 **g** 42 **h** 48 **i** 39 **j** 44

4 Write the first three multiples of each number.

a 7 **b** 9 **c** 12 **d** 15 **e** 17
f 25 **g** 30 **h** 32 **i** 45 **j** 50

5 Find the common factors of

a 10 and 20 **b** 12 and 15 **c** 20 and 25
d 8 and 20 **e** 21 and 28 **f** 30 and 40
g 12 and 28 **h** 9 and 36 **i** 24 and 30

6 Find the first two common multiples of

a 6 and 10 **b** 9 and 12 **c** 4 and 6
d 10 and 15 **e** 14 and 21 **f** 20 and 30

7 **a** Write a multiple of 20 that is bigger than 200.
b Write a multiple of 15 that is between 100 and 140.
c Write a multiple of 6 that is bigger than 70 but less than 100.

N1 Exam review

Key objectives

- Understand and use positive and negative numbers, both as positions and translations on a number line
- Use the concepts and vocabulary of factor, multiple and common factor
- Interpret scales on a range of measuring instruments, including those for time and mass
- Add, subtract, multiply and divide integers and then any number
- Multiply or divide any number by powers of 10

1 **a** Write the number marked with an arrow. (1)

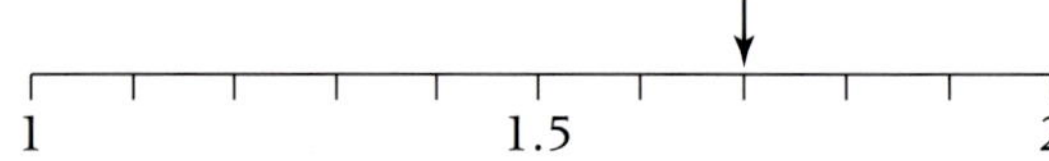

b **i** Calculate 8.85×10.

ii Copy the number line and mark the position of your answer to part **i**.

86 87 88 89 90 91 92

(2)

2 Here is a map of the British Isles. The temperatures in some places one night last winter are shown on the map.

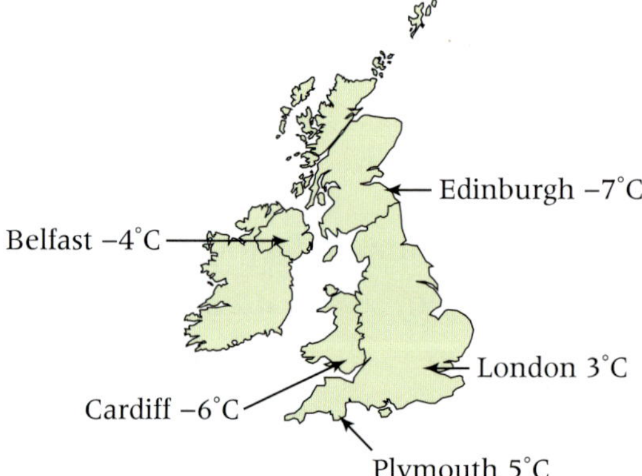

a **i** Write the names of the two places that had the biggest difference in temperature.

ii Work out the difference in temperature between these two places. (3)

b Two pairs of places have a difference in temperature of 2 °C.
Write the names of these places. (2)

(Edexcel Ltd., 2003)

S1 Measures, length and area

This unit will show you how to

- Make sensible estimates of measurements
- Convert measurements from one unit to another
- Know rough metric equivalents to imperial units
- Find the perimeter of a shape by counting squares and measuring
- Find the area of shapes by counting squares, measuring and using the formula for the area of a rectangle

Before you start ...

You should be able to answer these questions.

Review

1 Evaluate each of these. — Key stage 3

a $4.2 + 3.6$ **b** $3.8 + 5.6$
c $2.5 + 3$

2 Work out each of these. — Unit N1

a 40×10 **b** 14×1000
c 3.1×10 **d** 13.4×100
e 6.3×1000 **f** $400 \div 10$
g $6000 \div 100$ **h** $430 \div 100$
i $640 \div 1000$ **j** $3.1 \div 10$

3 Evaluate each of these. — Unit N1

a $4 \times 4\frac{1}{2}$ **b** $6 \times 2\frac{1}{2}$
c 8×1.75 **d** $40 \times 5 \div 8$
e $3 \times 8 \div 5$

4 Find the perimeter and area of this shape. — Key stage 3

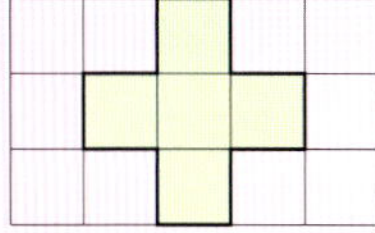

5 Measure this line — Key stage 3

a in millimetres
b in centimetres.

S1.1 Measurement

This spread will show you how to:

- Make sensible estimates of measurements

Keywords
Capacity
Length
Mass
Measure
Metric

Metric units are based on the decimal system.

- **You can measure length and distance using metric units.**

millimetre (mm)

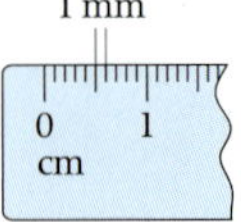

1 mm is $\frac{1}{10}$ of 1 cm
10 mm = 1 cm

centimetre (cm)

The thickness of your little finger is about 1 cm.

metre (m)

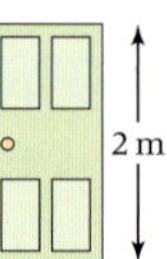

The height of a door is about 2 m.

kilometre (km)

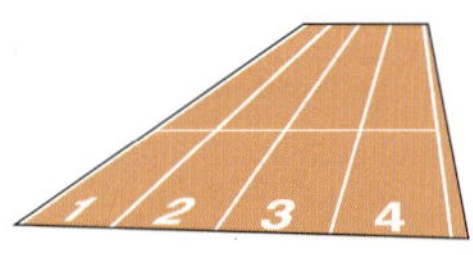

$2\frac{1}{2}$ times round the running track is about 1 km.

You use a ruler to measure short lengths.
You use a tape measure or trundle wheel to measure longer lengths.

Example

Give a sensible metric unit to measure

a the distance travelled on a car journey
b the mass of an apple
c the amount of water in a bath.

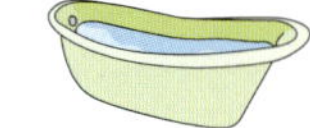

a kilometres (km)
b grams (g)
c litres

- **You can measure mass using metric units.**

Mass is linked to weight.

milligrams (mg)
1 mg is $\frac{1}{1000}$ of a gram
1000 mg = 1 g

gram (g)

A peanut weighs about 1 g.

kilogram (kg)

A bag of sugar weighs 1 kg.

tonne (t)

A small car weighs about 1 tonne.

- **You can measure capacity or volume using metric units.**

Capacity is the amount of liquid a container holds.

millilitre (ml)

A teaspoon holds about 5 ml.

centilitre (cl)

A can of drink holds 33 cl.

litre (l)

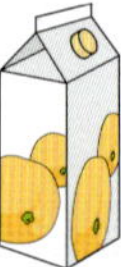

A carton of fruit juice holds 1 litre.

Exercise S1.1

1 Four metric units for measuring distance are

metre kilometre centimetre millimetre

a Write them in order of size starting with the smallest.

b Write the correct abbreviations next to your answers.

2 Which metric unit of length would you use to measure

a the length of a swimming pool **b** the thickness of a coin

c the distance from London to Paris **d** the height of a house?

3 Which of these measurements could be 2.5 cm?

a height of a room **b** height of a table

c diameter of a coin **d** length of a book

For questions 4–6 choose the most suitable answer.

4 The length of a car is about

a 30 cm **b** 300 mm **c** 3 m **d** 0.5 m

5 The length of a playing card is about

a 9 mm **b** 0.9 m **c** 90 cm **d** 9 cm

6 The width of a playing card is about

a 60 mm **b** 60 cm **c** 0.6 m **d** 6 m

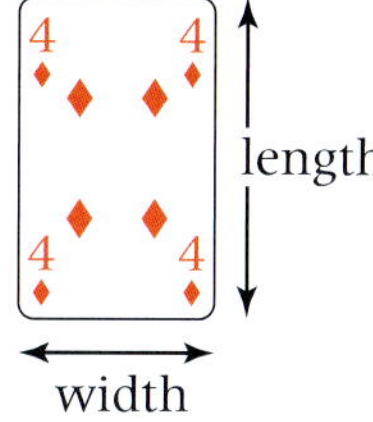

7 Which metric unit of mass would you use to weigh

a a person **b** a bus

c a banana **d** a piece of paper?

For questions 8–10, choose the most suitable answer.

8 The weight restriction for baggage on an aircraft could be

a 15 g **b** 15 t **c** 15 mg **d** 15 kg

9 A tub of margarine could weigh

a 500 kg **b** 500 g **c** 500 mg **d** 500 t

10 The weight of flour used to make pastry could be

a 200 mg **b** 200 g **c** 200 kg **d** 200 t

11 Three metric units for measuring capacity are

centilitre litre millilitre

a Write them in order of size, starting with the smallest.

b Write the appropriate abbreviation next to your answer.

S1.2 Metric measures

This spread will show you how to:

- Convert measurements from one unit to another

Keywords
Capacity
Convert
Length
Mass
Metric

- **Metric units are based on the decimal system.**

You can **convert** between units in the metric system, by multiplying or dividing by 10, 100, 1000, ...

Length

10 mm = 1 cm

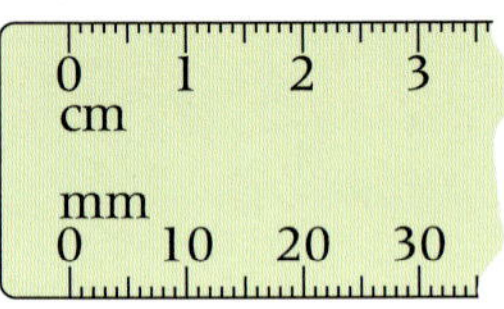

100 cm = 1 m

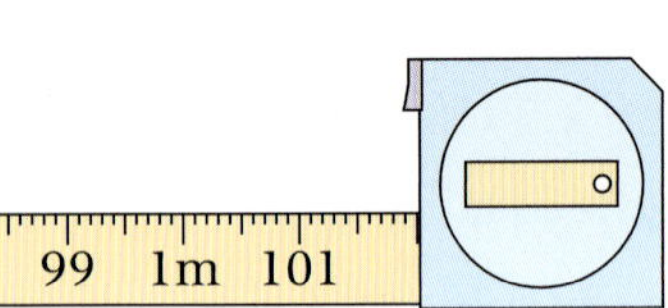

1000 m = 1 km

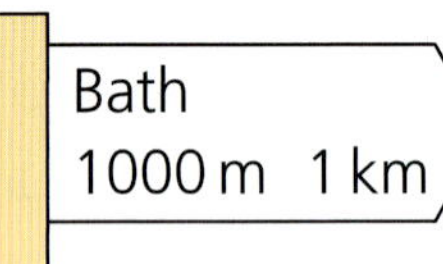

mm = millimetre
cm = centimetre
m = metre
km = kilometre

Mass

1000 mg = 1 g
1000 g = 1 kg
1000 kg = 1 t

Capacity

1000 ml = 1 litre
100 cl = 1 litre

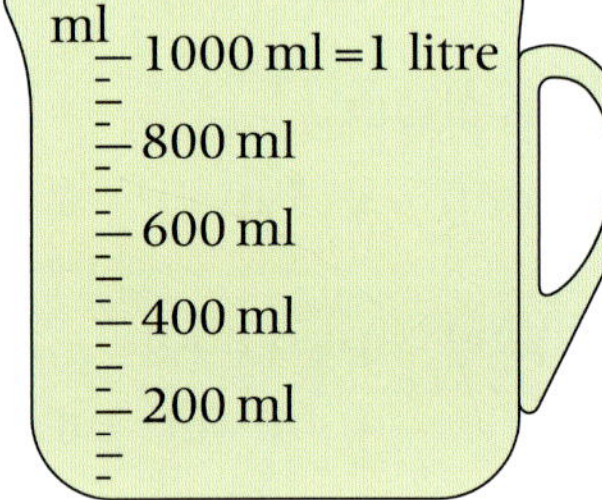

mg = milligrams
g = gram
kg = kilogram
t = tonnes

ml = millilitre
cl = centilitre
l = litre

Example

The height of a ceiling is 2.4 m.
Change 2.4 m to

a centimetres **b** millimetres.

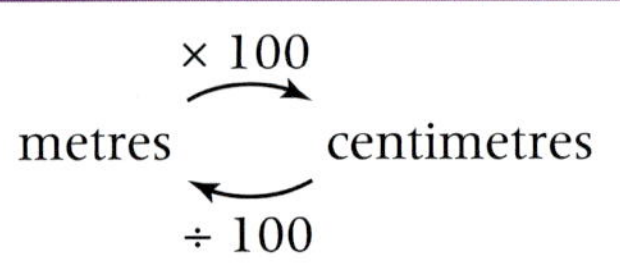

a 1 m = 100 cm
2.4 m = 2.4 × 100 cm = 240 cm

b 1 cm = 10 mm
240 cm = 240 × 10 mm = 2400 mm

Example

A bottle of wine holds 75 cl.
Change 75 cl to

a litres **b** millilitres.

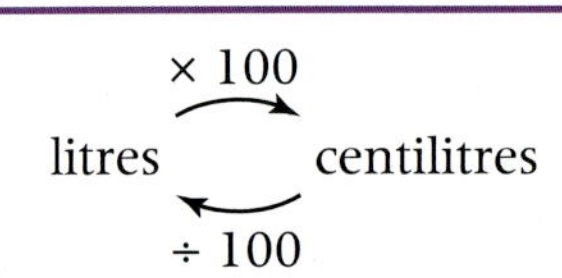

a 1 litre = 100 cl
75 cl = 75 ÷ 100 litre = 0.75 litre

b 1 litre = 1000 ml
0.75 litre = 0.75 × 1000 = 750 ml

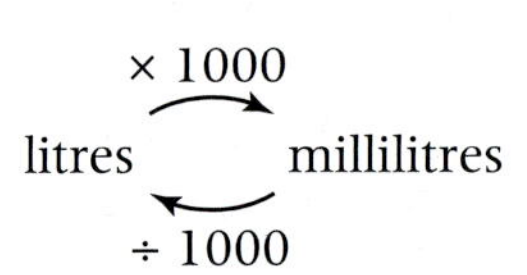

Exercise S1.2

1 Convert these measurements to millimetres.

a 5 cm **b** 8 cm **c** 15 cm **d** 6 cm 7 mm **e** 19 cm 3 mm
f 4.5 cm **g** 4.3 cm **h** 10.6 cm **i** 80 cm **j** 1 m

2 Convert these measurements to centimetres.

a 60 mm **b** 85 mm **c** 240 mm **d** 63 mm **e** 4 mm
f 4 m **g** 10 m **h** 3.5 m **i** 1.6 m **j** 1.63 m

3 Convert these measurements to metres.

a 400 cm **b** 450 cm **c** 475 cm **d** 470 cm **e** 50 cm
f 1 km **g** 4 km **h** 0.5 km **i** 3.5 km **j** 18 km

4 Convert these weights to kilograms.

a 8000 g **b** 7000 g **c** 7500 g **d** 500 g **e** 200 g
f 1 t **g** 1.5 t **h** 3.5 t **i** 10 t **j** 100 t

5 Convert these weights to grams.

a 1 kg **b** 4 kg **c** 0.5 kg **d** 4.5 kg **e** 3 kg 500 g
f 2 kg 400 g **g** 2.4 kg **h** 1000 mg **i** 500 mg **j** 2500 mg

6 Convert these capacities to litres.

a 1000 ml **b** 3000 ml **c** 500 ml **d** 4500 ml **e** 4750 ml
f 100 cl **g** 200 cl **h** 50 cl **i** 250 cl **j** 70 cl

7 Write these lengths in order of size, smallest first.
2.11 m 212 cm 2011 mm 209 cm

8 Write these heights in order of size, smallest first.
173 cm 1.7 m 1.75 m 176 cm 171 cm

9 A lorry can carry a maximum load of five tonnes.
Crates are made up, each weighing 625 kg.
How many crates can the lorry take?

10 A glass holds 200 ml. How many glasses can
I pour from a 1-litre bottle of lemonade?

11 One ream (500 sheets) of A4 paper weighs 0.5 kg.
Calculate the weight of one sheet of paper in grams.

12 Sarah buys a 0.5 kg bag of rice. Each portion of rice is 150 g.
How many complete portions can she get from the bag?

S1.3 Metric and imperial measures

This spread will show you how to:

- Know rough metric equivalents to imperial units

Keywords
Capacity
Equivalents
Imperial units
Length
Mass
Metric units

Most people in the world use **metric units**.
Some people still use **imperial units**.
It is useful to know the metric **equivalents** of imperial units.

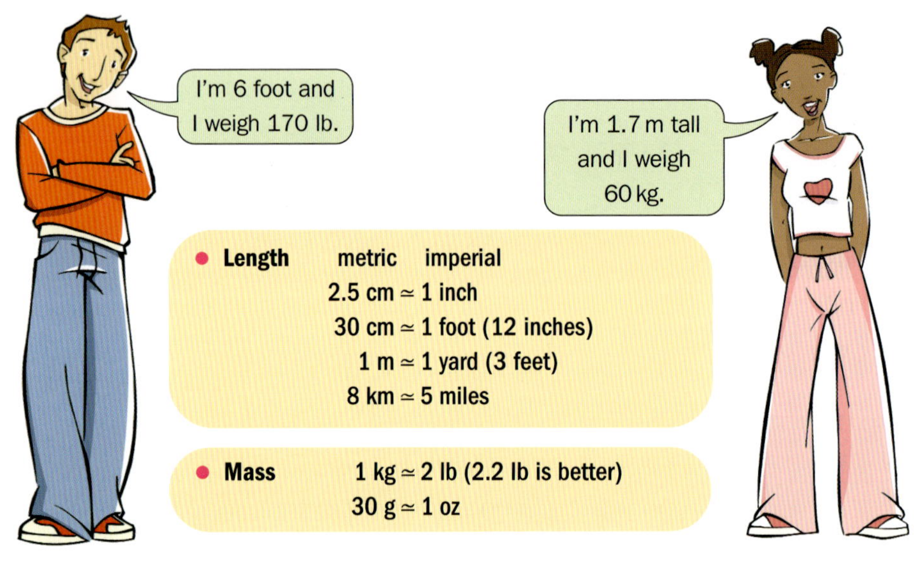

≃ means approximately equal to.

lb is pounds.

oz is ounces.

- **Capacity** metric imperial
  - 600 ml ≃ 1 pint
  - 1 litre ≃ 1.75 pints
  - $4\frac{1}{2}$ litres ≃ 1 gallon

A litre is more than a pint.

Example

In 2005 all Ireland's road signs were changed from imperial to metric units.
Change 50 miles to kilometres.

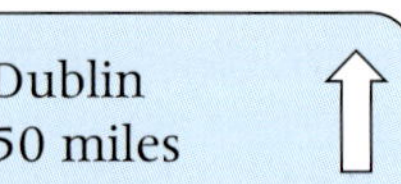

×10 (5 miles = 8 km ; 50 miles = 80 km) ×10

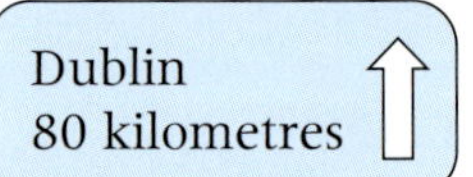

Example

I put 8 gallons of petrol in my car.
Approximately how many litres is that?

1 gallon ≃ $4\frac{1}{2}$ litres
8 gallons ≃ $4\frac{1}{2} \times 8$ litres
= 36 litres

Exercise S1.3

1 Convert these distances to miles.

a Madrid 8 km
b Valencia 240 km
c Benidorm 96 km
d Barcelona 120 km
e Granada 32 km
f Malaga 304 km
g Alicante 104 km
h Bilbao 68 km

2 Convert these distances to kilometres.

a Leeds 20 miles
b Sheffield 40 miles
c York 100 miles
d Manchester 70 miles
e Liverpool 250 miles
f London 45 miles
g Nottingham 35 miles
h Birmingham 55 miles

3 Convert these speeds to miles per hour.

a 64 km/h **b** 24 km/h **c** 16 km/h **d** 48 km/h **e** 80 km/h

km/h = kilometres per hour

4 Convert these speeds to kilometres per hour.

a 30 mph **b** 50 mph **c** 70 mph **d** 40 mph **e** 20 mph

5 Use 1 kg ≃ 2.2 lb, to convert these weights to pounds.

a 2 kg **b** 10 kg **c** 8 kg **d** 60 kg **e** 50 kg

6 Use 1 kg ≃ 2 lb, to convert these weights to kilograms.

a 4 lb **b** 60 lb **c** 100 lb **d** 25 lb **e** 11 lb

7 Use 1 litre ≃ 1.75 pints, to convert these capacities to pints.

a
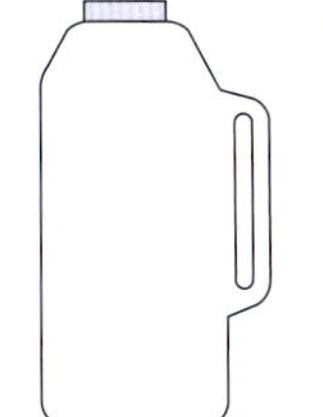
4 litres

b
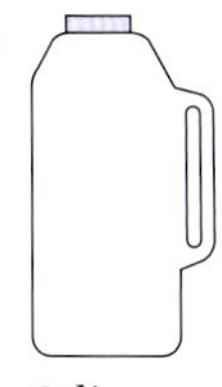
3 litres

c
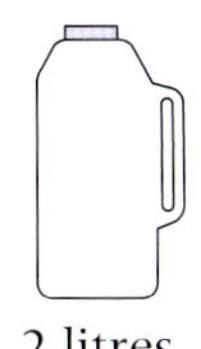
2 litres

d
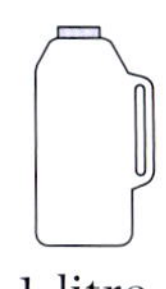
1 litre

e
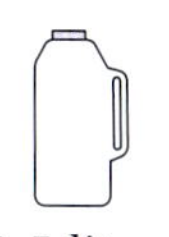
0.5 litres

8 Use 1 gallon ≃ $4\frac{1}{2}$ litres, to convert these capacities to litres.

a 2 gallons **b** 8 gallons **c** 0.5 gallons
d 2.5 gallons **e** 40 gallons **f** 500 gallons

S1.4 Perimeter

This spread will show you how to:

- Find the perimeter of a shape by counting squares and measuring

Keywords

Length
Perimeter
Rectangle
Unit
Width

- The **perimeter** is the distance all round a shape.
- You measure perimeter in units of **length**, for example centimetre (cm), metre (m).

Example

Find the perimeter of each shape. State the **units** of your answers.

a

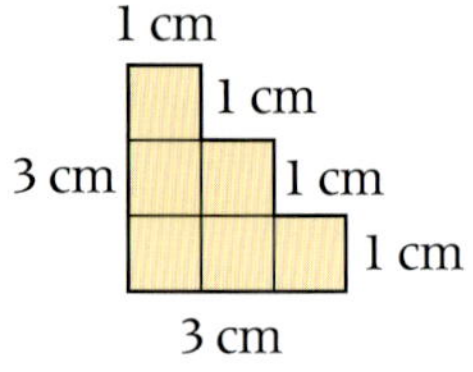

b

1 cm
1.4 cm

---

a Perimeter $= 1 + 1 + 1 + 1 + 1 + 1 + 3 + 3$
$= 12$ cm

b Perimeter $= 1 + 1.4 + 1 + 1.4 + 1 + 1.4 + 1 + 1.4$
$= 9.6$ cm

You can find the perimeter of a **rectangle** by counting cm square lengths.

Perimeter $= 3 + 2 + 3 + 2$
$= 10$ cm

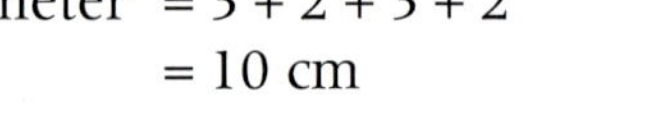

You can find the perimeter without a centimetre grid.

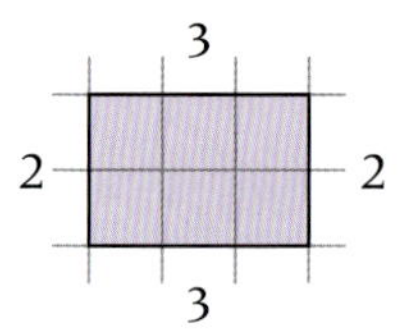

- For any rectangle
  Perimeter = length + width + length + width
  = 2 × length + 2 × width
  Perimeter $= l + w + l + w$
  $= 2l + 2w$

width (w)
length (l)

Example

The length of a rectangle is 5.8 cm.
The perimeter of the rectangle is 19.4 cm.
Calculate the width of the rectangle.

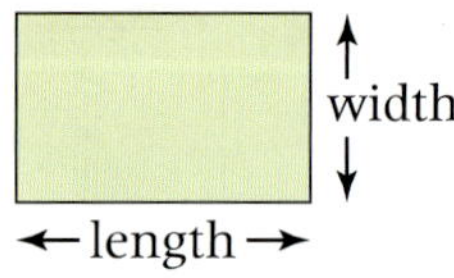

---

width + 5.8 + width + 5.8 = 19.4 cm
2 × width + 11.6 = 19.4
2 × width = 19.4 − 11.6
= 7.8

$$\text{width} = \frac{7.8}{2}$$

= 3.9

So width = 3.9 cm

Check:
5.8
5.8
3.9
+3.9
19.4

Exercise S1.4

1 Find the perimeter of each shape. Each square represents 1 cm.

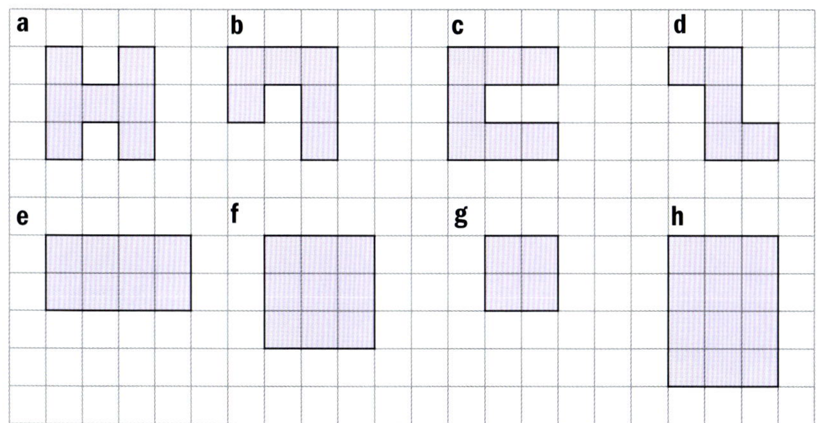

2 Calculate the perimeters of these shapes. State the units of your answers.

a 3 cm, 7 cm **b** 5 m, 10 m **c** 15 mm, 5 mm **d** square, 8 cm

e 10 m, 8 m, 6 m **f** 6 cm, 6 cm, 6 cm **g** 5 cm, 8 cm, 6 cm, 3 cm **h** 6 cm, regular hexagon

A regular hexagon has 6 equal sides.

3 A rectangular field is 80 m long and 35 m wide.
Calculate the perimeter of the field. State the units of your answer.

4 The perimeter of each rectangle is 24 cm. Calculate the unknown lengths.

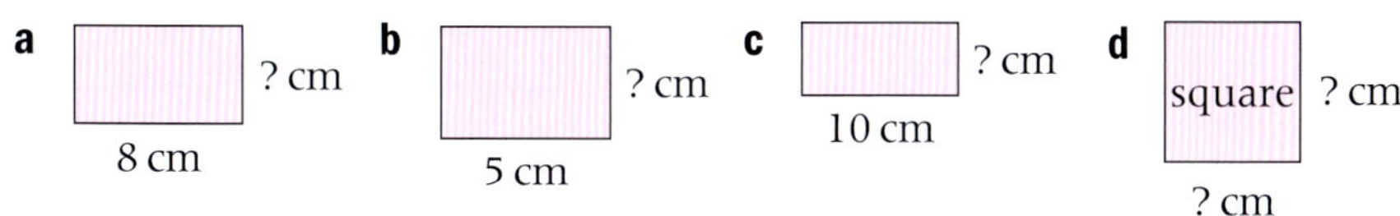

5 The perimeter of each rectangle is 36 cm. Calculate the unknown lengths.

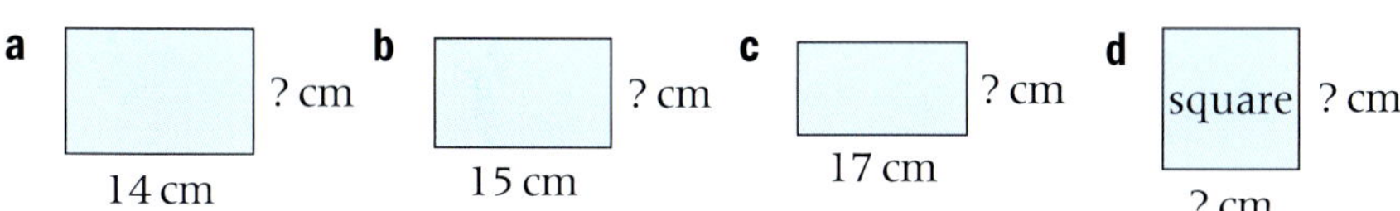

Area

This spread will show you how to:

- Find the area of shapes by counting squares
- Find the area of a rectangle using the formula

Keywords
Area
Rectangle
Square centimetre (cm^2)
Square metre (m^2)

- **Area is the amount of surface a shape covers.**

You can find the area of a shape by counting the number of squares on a centimetre grid.

Each square is equal to an area of 1 cm^2.

The area of the circle is about 12 cm^2.

1 **square metre** = 1 m^2
1 square millimetre = 1 mm^2

You can find the area of a **rectangle** by counting squares.

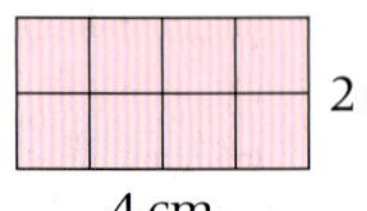
2 cm
4 cm

Area = 4×2
= 8 cm^2

There are 2 rows of 4 squares.

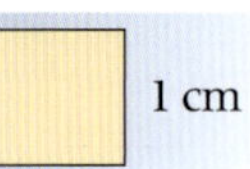
1 cm
1 cm

is 1 **square centimetre** or 1 cm^2.

- **You can find the area of any rectangle by using the formula:**
  Area of rectangle = length × width

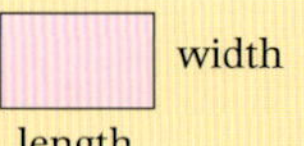
width
length

This formula works for all rectangles.

Example

Calculate the area of this rectangle.
State the units of your answer.

8 cm
12 cm

Area = $12 \times 8 = 96$ cm^2

Example

I buy 12 square paving slabs. Each slab measures 1 metre by 1 metre.
I want to make a rectangle using all 12 slabs.

a What sizes of rectangles can I make?
b Calculate the area of each rectangle.

a

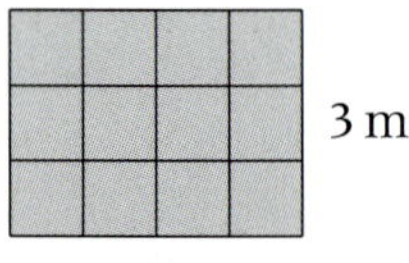
3 m
4 m

b Area = $3 \times 4 = 12$ m^2

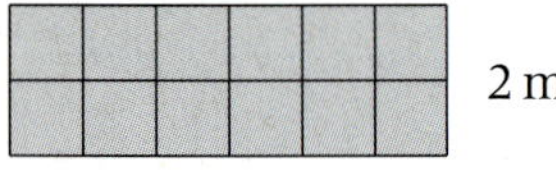
2 m
6 m

Area = $2 \times 6 = 12$ m^2

1 m
12 m

Area = $1 \times 12 = 12$ m^2

Factors of 12 are

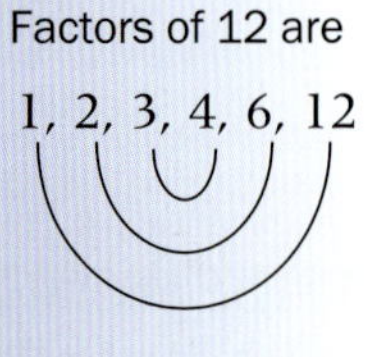

Exercise S1.5

1 Which is larger, one square centimetre or one square metre?

2 Find the area of each shape. Each square represents 1 cm^2.

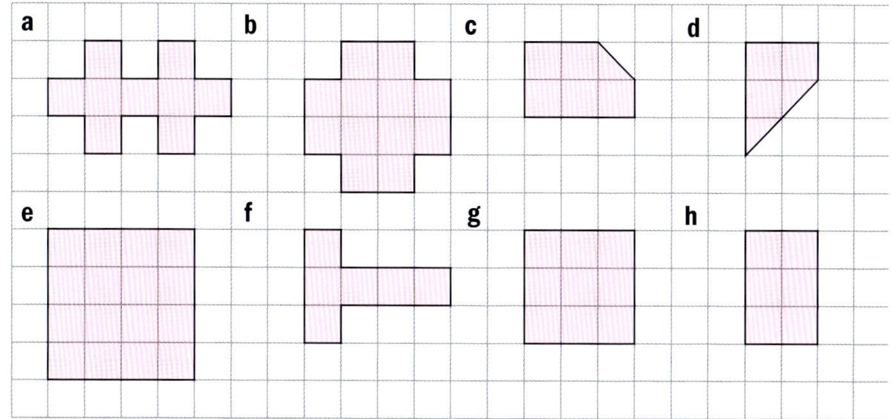

3 Calculate the areas of these rectangles. Remember to give the units of your answers.

a 3 cm, 5 cm

b 3 m, 9 m

c 10 cm, 20 cm

d square, 9 cm

e 1.5 m, 6 m

f 3 m, 4.5 m

g 15 cm, 12 cm

h square, 4.5 cm

4 Calculate the missing lengths. Give the units of your answers.

a Area 48 cm^2, 6 cm, ? cm

b Area 60 m^2, 5 m, ? m

c

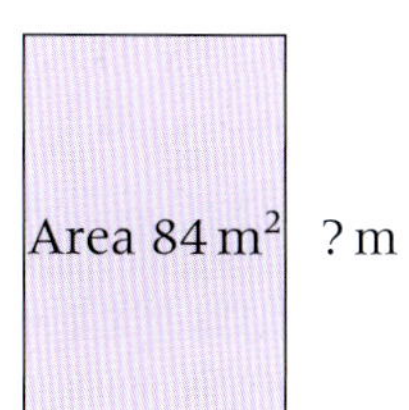

d

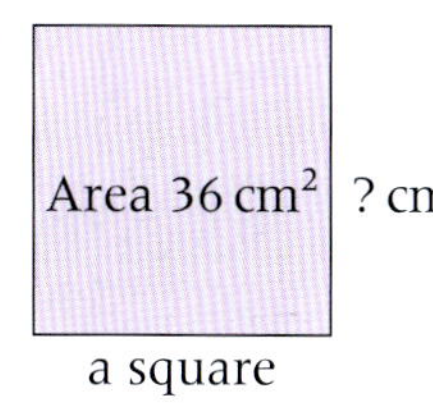

e

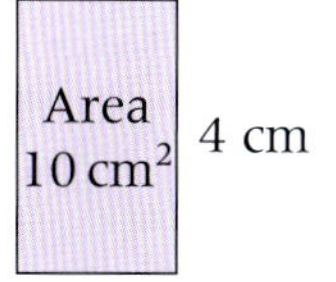

f

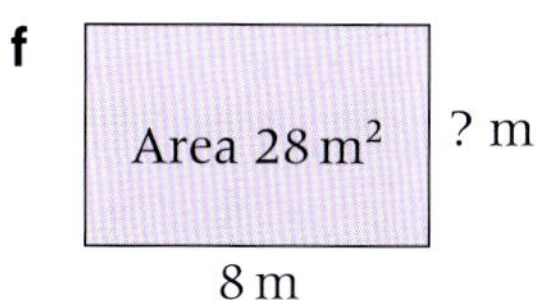

g

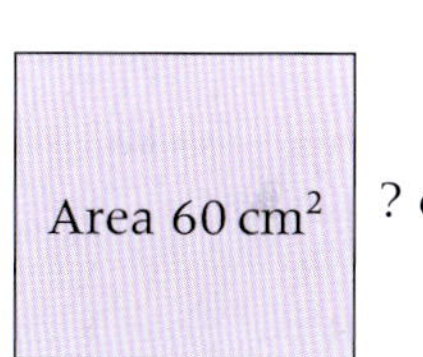

h

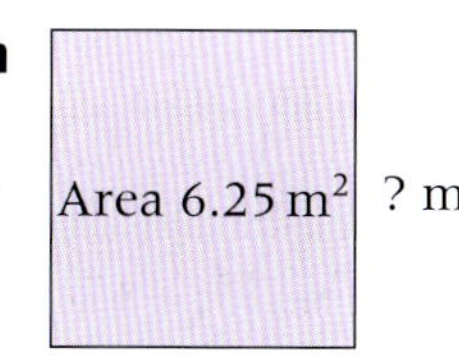

S1 Exam review

Key objectives

- Convert measurements from one unit to another
- Know rough metric equivalents of pounds, feet, miles, pints and gallons
- Make sensible estimates of a range of measures in everyday settings
- Find areas of rectangles, recalling the formula, understanding the connection to counting squares and how it extends this approach

1 The grid shows a shape made of centimetre squares.

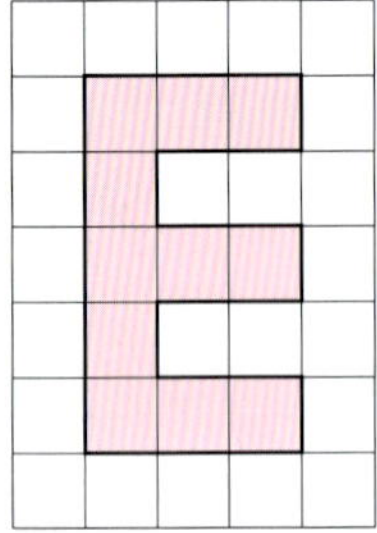

a Find the perimeter of the shaded shape. (1)
b Find the area of the shaded shape. (2)

2 Copy and complete this table.
Write a sensible unit for each measurement.

	Metric	Imperial
The weight of a turkey	....	pounds
The volume of water in a swimming pool	....	gallons
The width of this page	centimetres	....

(3)

(Edexcel Ltd., 2003)

A1 Expressions

This unit will show you how to

- Understand and use the vocabulary of algebra
- Use letters to represent numbers in algebraic expressions
- Use the four rules of arithmetic
- Simplify expressions by collecting like terms
- Rearrange algebraic expressions involving different letters
- Substitute positive and negative numbers into an expression, including those involving division, and work out its value

Before you start ...

You should be able to answer these questions.

Review

1 Work out these. — Unit N1

a $4 + 3 - 2$ **b** $6 - 2 + 5$

2 Work out these. — Key stage 3

a 3^2 **b** 2^2 **c** 4^2

3 There are 5 CDs in a packet. — Key stage 3
How many CDs are there in 3 packets?

4 Work out these. — Unit N1

a $4 - -3$ **b** $2 + -3$

c $-3 + 5$ **d** $-4 - -1$

5 Work out these. — Unit N1

a -3×2 **b** 4×-2

c $6 \div -3$ **d** $-8 \div -2$

A1.1 Letter symbols

This spread will show you how to:

- Use letters to represent numbers in algebraic expressions

Keywords
Expression

- **You can use letters to represent numbers.**

There are 6 eggs in a box.

In **2** boxes there are $6 \times \mathbf{2} = 12$ eggs.
In **3** boxes there are $6 \times \mathbf{3} = 18$ eggs.
In n boxes there are $6 \times n = 6n$ eggs.

In algebra $6n = 6 \times n$.
You do not write the $\times$ sign.

Example

There are 12 sweets in one packet.

a How many sweets are there in 4 packets?
b How many sweets are there in x packets?

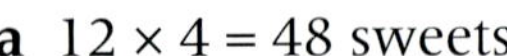

a $12 \times 4 = 48$ sweets
b $12 \times x = 12x$ sweets

- **An expression is a collection of letters and numbers with no = sign.**

Example

One apple costs 20 pence.

a Work out the cost of 3 apples.
b Write an expression for the cost of y apples.

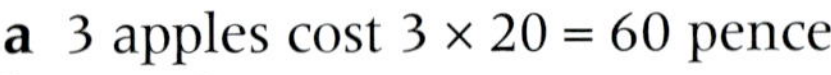

a 3 apples cost $3 \times 20 = 60$ pence
b y apples cost $y \times 20 = 20y$ pence

$y \times 20 = 20 \times y$
Write numbers before letters.

You can write expressions to represent real situations.

Example

There are m pens in one box.

a How many pens are there in 3 boxes?
b Write an expression for the number of pens in 3 boxes plus 5 extra pens.
c 6 pens are taken out of a box.
Write an expression for the number of pens left in the box.

a $3 \times m = 3m$ pens
b $3m + 5$
c $m - 6$

$3m$ in 3 boxes, plus 5 more.

Exercise A1.1

1 Stamps cost 30 pence each.

a How much do 8 stamps cost? **b** How much do n stamps cost?

2 Chews cost 20 pence each.

a How much do 10 chews cost? **b** How much do x chews cost?

3 Pencils cost 8 pence each.

a How much do 8 pencils cost? **b** How much do m pencils cost?

4 There are 12 biscuits in a packet.
How many biscuits are there in

a 5 packets **b** 3 packets

c x packets **d** n packets?

5 Work out the cost of

a 2 kg of potatoes **b** 3 kg of carrots

c 10 kg of potatoes **d** 4 kg of carrots.

Write an expression for the cost of

e n kg of potatoes **f** x kg of carrots

g y kg of potatoes **h** p kg of carrots.

6 Daniel has x DVDs.

a Lisa has twice as many DVDs as Daniel.
Write an expression in terms of x for the number of DVDs Lisa has.

b Charlie has 4 more DVDs than Daniel.
Write an expression in terms of x for the number of DVDs Charlie has.

c Sareeta has 3 fewer DVDs than Lisa.
Write an expression in terms of x for the number of DVDs Sareeta has.

7 Match each expression in box **A** with an expression in box **B**.

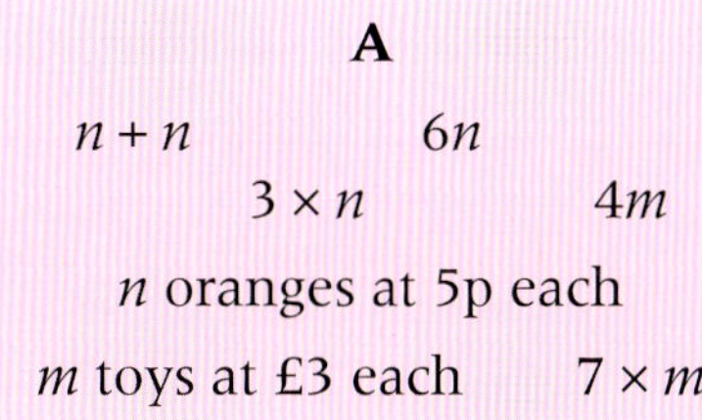

A

$n + n$ $6n$

$3 \times n$ $4m$

n oranges at 5p each

m toys at £3 each $7 \times m$

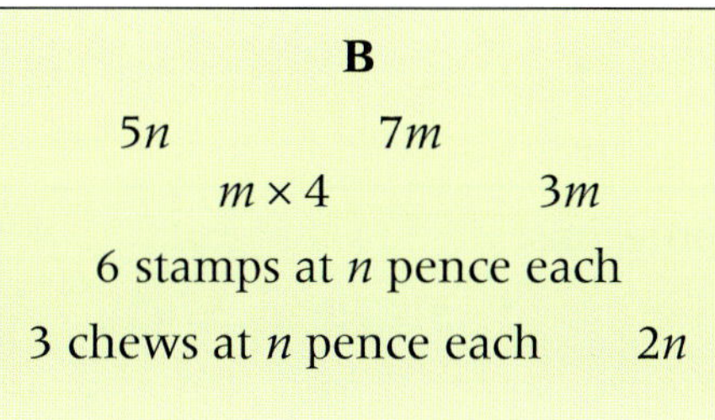

B

$5n$ $7m$

$m \times 4$ $3m$

6 stamps at n pence each

3 chews at n pence each $2n$

A1.2 Simplifying expressions

This spread will show you how to:

- Use the four rules of arithmetic
- Simplify expressions by collecting like terms

Keywords
Simplify
Squared
Term

- **You can add and subtract in algebra, just like you do with numbers.**

$m + m + m + m = 4$ lots of $m = 4 \times m = 4m$

$2 + 2 + 2 + 2$
$= 4$ lots of 2
$= 4 \times 2 = 8$

Addition

$4m \quad + 2m \quad = 6m$

$m + m + m + m \quad + m + m \quad = m + m + m + m + m + m$

Subtraction

$4m \quad - m \quad = 3m$

$m + m + m + m \quad - m \quad = m + m + m$

- **You can simplify multiplications and divisions in algebra.**

This topic is extended to expanding and factorising on page 372.

Multiplication

$3 \times a = 3a$

$x \times y = xy$

$y \times y = y^2$

You say 'y **squared**'.
$3 \times 3 = 3^2 (= 9)$

Division

$x \div 2 = \dfrac{x}{2} \qquad 4a \div 7 = \dfrac{4a}{7}$

Example

Simplify these expressions.

a $n + n + n$ **b** $c + c + c + c + c$ **c** $7x + 4x$
d $9y - y$ **e** $4p + 8p - 3p$

a $n + n + n = 3n$
b $c + c + c + c + c = 5c$
c $7x + 4x = 11x$
d $9y - y = 9y - 1y$
$= 8y$
e $4p + 8p - 3p = 12p - 3p$
$= 9p$

- **A term is an individual part of an expression.**

$4x - 2x$ is an expression.
$4x$ and $2x$ are terms.

To make calculations easier you can write terms in a different order.

Keep each term with its sign.

Example

Simplify these expressions.

a $3q - 5q + 2q$ **b** $t - 4t + 8t$

a $3q - 5q + 2q = 3q + 2q - 5q$
$= 5q - 5q$
$= 0$

b $t - 4t + 8t = t + 8t - 4t$
$= 9t - 4t$
$= 5t$

Exercise A1.2

1 Write each statement as a single amount. Use a letter to represent each item.
For example, 6 apples + 2 apples can be written as $6a + 2a = 8a$.

a 8 apples + 7 apples
b 6 boys + 5 boys
c 3 CDs + 5 CDs
d 8 chairs and 13 chairs
e 9 toys and 23 toys
f 18 books + 8 books
g 13 chocolates – 8 chocolates
h 6 cakes – 6 cakes
i 15 sweets – 7 sweets
j 13 bananas – 6 bananas

2 Simplify these expressions.

a $m + m + m$
b $n + n + n + n + n + n + n$
c $y + y + y + y + y + y$
d $z + z + z - z + z$

3 Collect the terms and simplify each expression.

a $6a + 7a$
b $5n + 12n$
c $6t + 9t$
d $3x + 16x$
e $9r - 5r$
f $6f - 2f$
g $12g - 5g$
h $15r - 7r$
i $16n + 4n + 3n$
j $6c + 8c + 3c$

4 Write each of these as a single term.

a $4 \times m$
b $p \times 3$
c $r \times s$
d $12 \times q$
e $5 \times g$
f $c \times 4$
g $d \times 8$
h $j \times k$
i $n \times n$
j $e \times 15$
k $t \times t$
l $6 \times m \times n$

5 Simplify each expression.

a $8n + 3n + 4n$
b $3m + 2m + 7m$
c $8p + 6p + 3p$
d $5q + 7q + 6q$
e $12x - 7x - 4x$
f $8w - w - 3w$
g $4a + 6a - 3a$
h $12b - 3b - 4b$
i $3j - 4j + 2j$
j $k - 5k + 6k$

6 Simplify these divisions.

The first one has been done for you.

a $d \div 4 = \frac{d}{4}$
b $x \div 3$
c $y \div 7$
d $t \div 9$
e $2a \div 3$
f $3n \div 4$
g $5p \div 7$
h $2v \div 4$

7 Use these four terms and three signs to make the different totals.

$3x$, $5x$, $2x$, $7x$	+, +, –

a $13x$
b $3x$
c $7x$
d $11x$

A1.3 Collecting terms

This spread will show you how to:

- Simplify expressions by collecting like terms
- Rearrange algebraic expressions involving different letters

Keywords
Like terms

- **An expression can include terms with different letters.**

Example

Bags of sweets come in two sizes.

There are n sweets in a small bag There are p sweets in a large bag

a Rupal buys 1 large bag and 1 small bag.
How many sweets does he buy altogether?

b Maisie buys 2 small bags and 3 large bags.
How many sweets does she buy altogether?

a 1 small bag + 1 large bag
n sweets + p sweets
$n + p$ sweets

b 2 small bags + 3 large bags
$2 \times n$ sweets + $3 \times p$ sweets
$2n + 3p$ sweets

- **Terms with the same letter are called like terms.**
  You can simplify expressions by collecting like terms.

Example

Simplify.

a $4m + 2p + 3m$ **b** $2x + 5y + 3x + y$

a $4m + 2p + 3m = 4m + 3m + 2p$
$= 7m + 2p$

b $2x + 5y + 3x + y = 2x + 3x + y + 5y$
$= 5x + 6y$

Rearrange the terms to collect like terms together.

$y = 1y$

Example

Simplify these expressions.

a $2e + 5f + 6e - 2f$ **b** $6u - v - 3u + 4v$ **c** $3m + 2n - m - 4n$

a $2e + 5f + 6e - 2f$
$= 2e + 6e + 5f - 2f$
$= 8e + 3f$

b $6u - v - 3u + 4v$
$= 6u - 3u + 4v - v$
$= 3u + 3v$

c $3m + 2n - m - 4n$
$= 3m - m + 2n - 4n$
$= 2m + -2n$
$= 2m - 2n$

Keep each term with its sign.

Exercise A1.3

1 Bags of peanuts come in two sizes. There are x peanuts in a small bag. There are y peanuts in a large bag.

a Sebastian buys 1 small bag and 1 large bag. How many peanuts does he buy?

b Adam buys 3 large bags. How many peanuts does he buy?

c Gabi buys 2 small bags and 1 large bag. Write an expression for the number of peanuts she buys.

d Sushma buys 4 small bags. Write an expression for the number of peanuts she buys.

e Kofi buys 3 small and 2 large bags. Write an expression for the number of peanuts he buys.

2 Simplify each expression by collecting like terms together.

a $a + b + a + b$ **b** $c + d + c + c + c$

c $3e + 2f + 4e + 3f$ **d** $5g + 7h + 2g + h$

e $3i + j + 4i + 5j$ **f** $3u + 5v + 2v + u$

3 Simplify these.

a $4a + 2b - 3a + b$ **b** $6x + 4y - 2x + 2y$

c $5m + 3n + 2m - n$ **d** $8s + 3t + s - 2t$

e $3p + 2q - p + 3q$ **f** $2c + 3d + 3c - 2d$

4 Simplify these by collecting like terms.

a $4a - 6b - 3a + 8b$ **b** $3c - 4d - 2c + 5d$

c $7u - 4v - 2u + 5v$ **d** $6x - 5y - 5x + 11y$

e $12m - 5n - 3m + 10n$ **f** $9p - 4q - 3p + 6q$

5 Simplify these.

a $5e - 3f - 2e + 2f$ **b** $4g - 6h - 3g + 4h$

c $j - 3k + 4j + 2k$ **d** $2r - 4s - r + s$

e $7t + 3u - 4t - 5u$ **f** $9v - 5w + 3v - 2w$

6 Make sets of three matching expressions, using one expression from each box in each set.

A	B	C
$3x + 5y - x + 2y$	$2x + 5y$	$7y - 3x + 5x - 2y$
$2x - 4y + 3x + 2y$	$3y + 7x + 4y - 5x$	$4x + 4y - 3x$
$2x + 4y - x$	$5x - 2y$	$7y + 2x$
$2y + 3x - x + 3y$	$3x + 6y - 2x - 2y$	$2x - 4y + 2y + 3x$

A1.4 Substitution

This spread will show you how to:

- Substitute numbers into an expression and work out its value

Keyword

Substitute

You can replace letters with number values.

This is called **substituting**.

- **You can substitute numbers into an expression and work out the value of the expression.**

Example

If $c = 4$, work out

a $5c$ **b** $2c + 1$ **c** $3c - 2$

a $5c = 5 \times c$ Write the multiplication sign.
$= 5 \times 4$ Write 4 instead of c.
$= 20$ Follow the order of operations.

b $2c + 1 = 2 \times c + 1$
$= 2 \times 4 + 1$
$= 8 + 1 = 9$

c $3c - 2 = 3 \times c - 2$
$= 3 \times 4 - 2$
$= 12 - 2 = 10$

- **You can substitute values into expressions with more than one letter.**

Example

Work out the value of each expression when $x = 2$ and $y = 1$.

a $x + 3$ **b** $y + 4$ **c** $x + y$
d $2x - y$ **e** $3y - 2x$ **f** x^2

a $x + 3 = 2 + 3$
$= 5$

b $y + 4 = 1 + 4$
$= 5$

c $x + y = 2 + 1$
$= 3$

d $2x - y = 2 \times x - y$
$= 2 \times 2 - 1$
$= 4 - 1 = 3$

e $3y - 2x = 3 \times y - 2 \times x$
$= 3 \times 1 - 2 \times 2$
$= 3 - 4 = -1$

f $x^2 = x \times x$
$= 2 \times 2 = 4$

Exercise A1.4

1 If $a = 3$, work out the value of these.

a $2a$ **b** $a + 1$ **c** $4a$ **d** $a - 2$

e $2a + 3$ **f** $3a - 4$ **g** $4a + 2$ **h** $4a - 5$

2 If $c = 4$ and $d = 2$, calculate the value of these.

a $3c$ **b** $2d$ **c** $d + c$ **d** cd

e $2c + 4d$ **f** $3c - 2d$ **g** $3d - c + 2$ **h** $4c + d - 5$

3 Find the value of each expression when $m = 2$ and $n = 5$.

a $m + n$ **b** $n - m$ **c** $2m + n$ **d** $m - n$

e m^2 **f** $n - 4m$ **g** $3m + n - 5$ **h** $2m - 3n + 1$

4 Work out the value of each expression when $x = 3$, $y = 5$ and $z = 2$.

a $4x + 3$ **b** $2z + y$ **c** $3y - z$ **d** $x + y + z$

e $2x - y + 3z$ **f** $3z - 2x + y$ **g** $4x^2$

5 Work out the value of these when $a = 3$, $b = 5$, $c = 4$ and $d = 6$.

a a^2 **b** $2a^2$ **c** b^2 **d** $2b^2$

e c^2 **f** $2c^2$ **g** d^2 **h** $2d^2$

6 A taxi company calculates its fares using

$$F = 2 + n$$

where F is the fare in pounds and n is the number of miles.

a What is the fare for a journey of 2 miles?

b What is the fare for a journey of 6 miles?

7 A rent-a-car company charges for its cars using

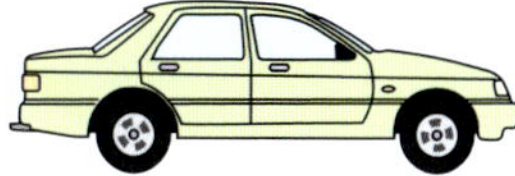

$$C = 30 + 20n$$

where C is the cost in £s and n is the number of days hired.

a What is the charge for 3 days hire?

b What is the charge for 14 days hire?

c If the charge is £210, for how many days was the car hired?

8 Work out the value of each letter, then read the coded word.

$m = 3$ $n = 8$ $p = 5$ $q = -2$

$H = 4n + 2q$ $L = 6m + 3p$ $O = m^2 + 7$ $A = 2q + 10$

$T = pq$ $C = 2n - 3p$ $E = np + 2m$

1 28 16 1 16 33 6 −10 46

A1.5 More substituting

This spread will show you how to:

- Substitute numbers into an expression and work out its value

Keywords
Negative

- **You can substitute negative values into expressions.**
  You use the rules for calculating with negative numbers.

Example

When $x = 2$ and $y = -3$, work out the value of

a $x - 3$ **b** $y + 5$ **c** $x - y$

a $x - 3 = 2 - 3$
$= -1$

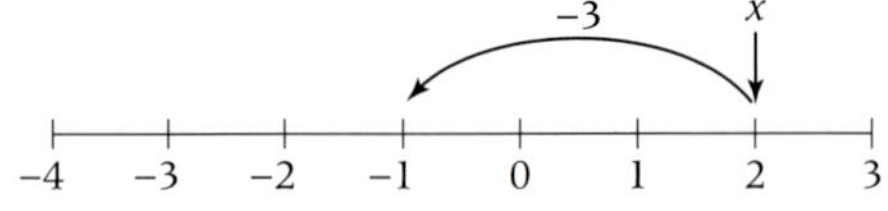

b $y + 5 = -3 + 5$
$= +2$

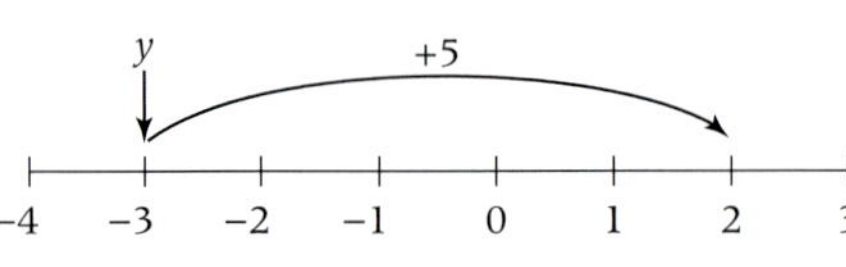

c $x - y = 2 - -3$
$= 2 + 3$
$= 5$

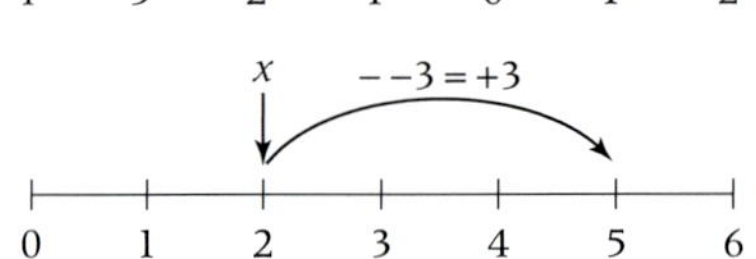

'Subtract −3' is the same as 'add +3'.

Example

Find the value of these expressions when $p = -2$ and $q = 4$.

a $3p - q$ **b** p^2

a $3p - q = 3 \times -2 - 4$
$= -6 - 4$
$= -10$

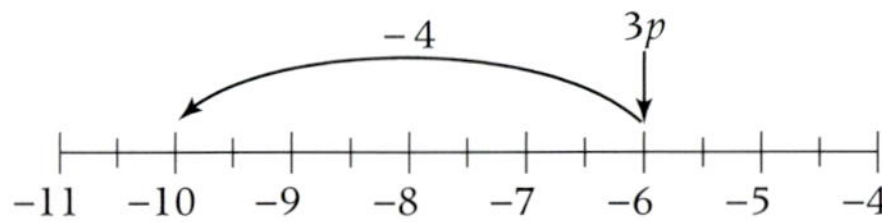

b $p^2 = p \times p$
$= -2 \times -2$
$= 4$

Negative × negative = positive

- **You can substitute values into expressions involving division.**

Example

Work out the value of each expression when $a = 3$ and $b = -2$.

a $\frac{4a}{6}$ **b** $\frac{a - b}{5}$

a $\frac{4a}{6} = \frac{4 \times a}{6}$
$= \frac{4 \times 3}{6} = \frac{12}{6} = 2$

b $\frac{a - b}{5} = \frac{3 - -2}{5}$
$= \frac{3 + 2}{5} = \frac{5}{5} = 1$

Exercise A1.5

1 Use this number line to work out the answers to these.

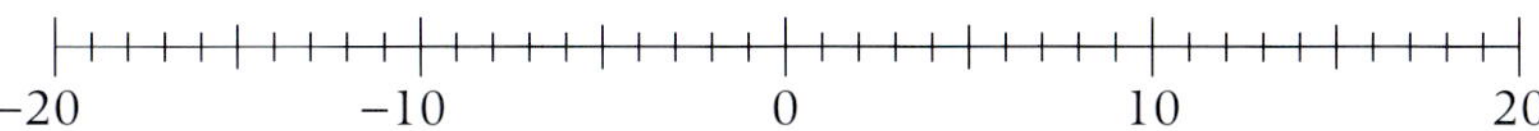

a $-7+3$ **b** $4-8$ **c** $-4+11$ **d** $-3-7$

e $-6+4$ **f** $6-18$ **g** $4-6-7$ **h** $-3-8+5$

2 If $x=-5$, work out the value of these.

a $3x$ **b** $x+2$ **c** $2x-1$ **d** $1-x$

e x^2 **f** $3-2x$ **g** $2x^2$ **h** $4x+15$

3 Work out the value of these expressions when $c=3$ and $d=-2$.

a $3c+1$ **b** $3d$ **c** d^2 **d** $2d+5$

e c^2 **f** $2d+4c$ **g** $2c-d$ **h** $d-c$

4 If $x=-3$, $y=2$ and $z=4$, work out the value of these.

a $x+y$ **b** y^2-5 **c** x^2+2 **d** $2y+z$

e $3y+2x$ **f** z^2-2y **g** $3z+2x$ **h** $3x+2y-z$

5 Work out the value of each of these expressions when $e=4$, $f=2$, $g=5$ and $h=-3$.

a $\frac{3e}{g}$ **b** $\frac{8g}{10}$ **c** $\frac{6f}{3}$ **d** $\frac{2g+3}{7}$

e $\frac{6g}{f}$ **f** $\frac{3h+4g}{8}$ **g** $\frac{5e}{fg}$ **h** $\frac{(2g+6)}{(e-f)}$

6 Calculate the value of each expression when $r=2$, $s=4$ and $t=-3$.

a $\frac{s}{2}$ **b** $\frac{6r}{3}$ **c** $\frac{t}{3}$ **d** $\frac{s+r}{3}$

e $\frac{t-5}{2}$ **f** $\frac{s\times r}{3}$ **g** $\frac{3r}{t}$ **h** $\frac{st}{r}$

7 You can use this rule to convert temperatures in °C to °F

$$F=\frac{9C}{5}+32$$

where F represents the temperature in °F and C represents the temperature in °C.

a Substitute the temperature 15 °C into the rule to convert it to °F.

b Convert −5 °C to °F.

A1 Exam review

Key objectives

- Distinguish the different roles played by letter symbols in algebra, knowing that letter symbols represent definite unknown numbers in equations
- Substitute numbers into a formula
- Understand that the transformation of algebraic expressions obeys and generalises the rules of arithmetic
- Manipulate algebraic expressions by collecting like terms

1 **a** Simplify

$n + 2m - 2n + m$. (2)

b Given that $n = -1$ and $m = 2$, work out the value of the expression in part **a**. (2)

2 Eggs are sold in boxes.

A small box holds 6 eggs.

Hina buys x small boxes of eggs.

Write down, in terms of x, the total number of eggs in these small boxes. (1)

(Edexcel Ltd., 2004)

N2

Whole number calculations

This unit will show you how to

- Round positive whole numbers and decimals
- Estimate the answers to problems
- Use a range of mental methods for addition and subtraction of whole numbers
- Use a range of written methods for addition and subtraction of whole numbers
- Use a range of mental methods for multiplication and division of whole numbers
- Use a range of written methods for multiplication and division of whole numbers

Before you start ...

You should be able to answer these questions.

	Review
1 Round 2916 **a** to the nearest 1000 **b** to the nearest 100 **c** to the nearest 10.	Key stage 3
2 Calculate each of these. **a** 95 − 47 **b** 76 + 29	Unit N1
3 Calculate each of these. **a** 576 + 239 **b** 914 − 685	Unit N1
4 Calculate each of these. **a** 5×6 **b** $48 \div 8$ **c** 12×6 **d** $282 \div 2$ **e** 12×14 **f** $91 \div 7$	Key stage 3

N2.1 Rounding

This spread will show you how to:

- Round positive whole numbers to the nearest 10, 100 or 1000 and decimals to the nearest whole number
- Estimate the answer to problems

Keywords
Approximate
Estimate
Rounding
Whole number

Rounding gives you the **approximate** size of a number.
When rounding numbers to a given degree of accuracy, look at the next digit.
If it is 5 or more, the number is rounded up. Otherwise it is rounded down.

You can round a number to the nearest 10, 100 or 1000.

You can round a decimal to the nearest whole number.

Example

a Round 4639 to the nearest 100.
b Round 235.723 to the nearest whole number.

a Look at the **tens** digit.

Thousands	Hundreds	Tens	Units
4	6	3	9

4639

4500 4600 4650 4700 4800

The **tens** digit is **3**, so the number is rounded **down** to 4600.

You can say that 4639 ≈ 4600 (to the nearest 100).

b Look at the **tenths** digit.

Hundreds	Tens	Units	•	tenths	hundredths	thousandths
2	3	5	•	7	2	3

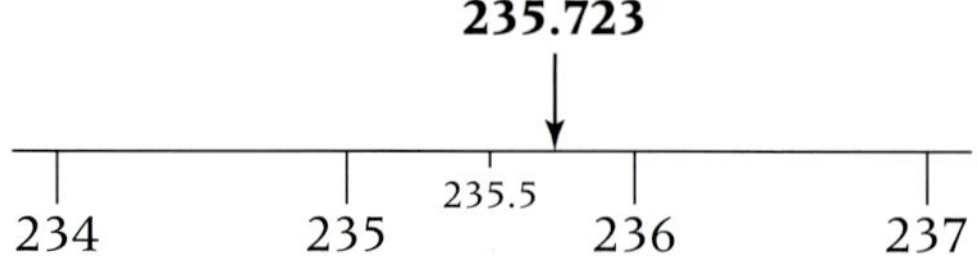

The **tenths** digit is **7**, so the number is rounded **up** to 236.

You can say that 235.723 ≈ 236 (to the nearest whole number).

This topic is extended to rounding decimals on page 362.

- **You can estimate the answer to a calculation by rounding the numbers.**

Example

John brings 324 tin cans to the recycling centre.
Habib brings 387 tin cans to the recycling centre.
About how many tin cans do the two boys bring to the recycling centre?

You can round each number to the nearest 100 to make a good estimate.

324 ≈ 300 (nearest 100)
387 ≈ 400 (nearest 100)

The boys bring about 300 + 400 = 700 tin cans to the recycling centre.

You have to decide whether to round the numbers to the nearest 10, 100 or 1000.

Exercise N2.1

1 Round each of these numbers to the nearest 10.

a 48 **b** 89 **c** 483 **d** 792 **e** 2638 **f** 6193

2 Round each of these numbers to the nearest 100.

a 343 **b** 484 **c** 882 **d** 2732 **e** 5678 **f** 16 491

You could use a number line sketch to help you.

3 Round each of these numbers to the nearest 1000.

a 3448 **b** 2895 **c** 4683 **d** 36 927 **e** 62 532 **f** 261 932

4 Round each of these numbers to the nearest

i 1000 **ii** 100 **iii** 10.

a 3472 **b** 81 382 **c** 1236.4 **d** 283.4 **e** 13 998 **f** 9999

5 Round each of these numbers to the nearest whole number.

a 4.8 **b** 3.9 **c** 11.6 **d** 25.074 **e** 16.286 **f** 435.972

6 Here are some statistics about eight footballers.

Name	Goals scored in a season	Passes completed in a match	Maximum shot speed (km/h)	Time to run 100 m (s)	Distance run in a match (m)
Peter Beattie	3	23	97.2	10.2	1623
Thierry Angel	26	52	89.6	10.95	12 453
Frank Schmiechal	1	38	69.2	13.7	11 276
Rudi Baros	18	29	93.9	11.03	13 789
Jan van Nistelroo	12	68	92.3	12.37	8673
Milan Dickov	5	15	78.3	16.4	9374
Paul Henry	21	33	83.7	11.06	9898
Phil Lampard	19	46	91.7	12.6	10 372

For each player round

a the distance run in a match to the nearest 1000 m

b the maximum shot speed to the nearest 10 km/h

c the time to run 100 m to the nearest 1 second.

7 Use rounding to estimate the answer to each of these calculations.

a 43 + 189 **b** 1563 + 28 **c** 1602 + 2654

N2.2 Mental addition and subtraction

This spread will show you how to:

- Use a range of mental methods for addition and subtraction of whole numbers

Keywords
Compensation
Mental methods
Partitioning

Partitioning

Split the numbers into parts (hundreds, tens and units).
Then add or subtract the parts.

Strategies to help you work out additions and subtractions in your head are called **mental methods**.

Example

Calculate **a** $37 + 78$ **b** $95 - 43$

a $37 + 78 = 30 + 7 + 70 + 8$
$= 70 + 30 + 15$
$= 100 + 15$
$= 115$

b $95 - 43 = 95 - 40 - 3$
$= 55 - 3$
$= 52$

Compensation

When the number you are adding or subtracting is nearly a multiple of 10 or a multiple of 100, round the number up or down.

Example

Calculate **a** $63 + 29$ **b** $174 - 39$

a $63 + 29 = 63 + 30 - 1$
$= 93 - 1$
$= 92$

b $174 - 39 = 174 - 40 + 1$
$= 134 + 1$
$= 135$

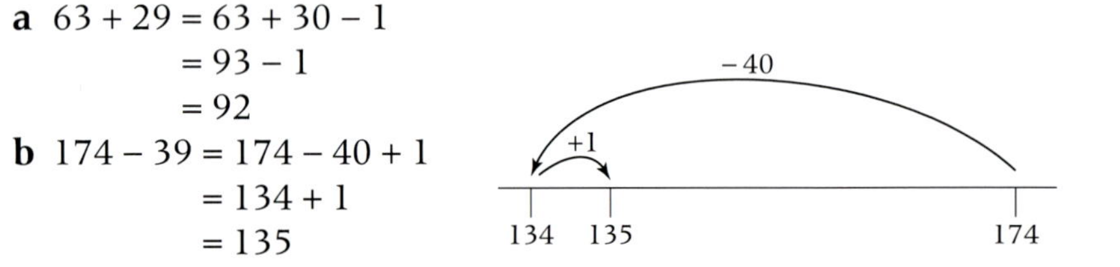

Remember to add or subtract to finish your calculation.

You can also count up from the smallest number to the largest number.

This is called shopkeeper's subtraction. Use a number line to help you.

Example

Gerta earns £408 a week. She spends £187 a week on her mortgage. How much money does she have left?

£408 − £187
Count up from 187 to the next 100 (that is, 200).
Count up from 200 to the next 100 below 408 (that is, 400).
Count up from 400 to 408.

Add together the counts you have made (that is, 13 + 200 + 8).

£408 − £187 = 13 + 200 + 8 = £221

Exercise N2.2

1 Write the answers to each of these calculations.

a 17 + 13 **b** 17 − 8 **c** 16 + 17 **d** 18 − 9 **e** 19 + 19 **f** 27 − 14

2 Find the missing number in each of these calculations.

a 63 + ? = 100 **b** ? + 43 = 100 **c** 31 + ? = 100 **d** 53 + ? = 100

3 Use a mental method for each of these calculations.

a 27 + 43 **b** 48 + 12 **c** 27 + 23 **d** 26 + 52 **e** 34 + 23 **f** 61 + 28

4 Use a mental method for each of these calculations.
Write the method you have used.

a 25 + 36 **b** 46 + 31 **c** 72 − 19 **d** 48 + 49 **e** 56 + 38 **f** 123 + 39

5 Use a mental method of calculation to solve each of these problems.

a Liam spends 43p on a can of cola and 39p on a packet of crisps. How much money has he spent?

b Brad scores 29 points in a game of rugby. In the next game he scores 23 points. What is his total score of points for the two games?

c Laura looks after 29 horses. Emma looks after 37 horses. How many horses do the girls look after altogether?

d Thomas owns 56 CDs. He sells 39 of his CDs. How many CDs does he have left?

6 Use a mental method to calculate these.

a 134 + 29 **b** 245 + 85 **c** 313 + 63

d 278 + 75 **e** 378 + 208 **f** 512 − 369

7 **a** Using each of the digits 1, 2, 3, 4 and 5 only once, what is the largest answer for an addition calculation that you can make?

☐☐☐ + ☐☐

b Using each of the digits 1, 2, 3, 4 and 5 only once, what is the smallest answer for a subtraction calculation that you can make?

☐☐☐ − ☐☐

8 Use a mental method for each of these calculations.
Write the method you have used.

a Decrease 132 by 64.

b How many less than 156 is 79?

c 234 + ☐ = 507. What is the value of ☐?

d 638 subtract 199.

e What is the total of 259 and 325?

N2.3 Written addition and subtraction

This spread will show you how to:

- Use a range of written methods for addition and subtraction of whole numbers
- Estimate the answers to problems

Keywords
Addition
Subtraction
Written method

- There is a standard **written method** for addition.

To calculate 3518 + 765:

Set out the calculation in columns: 3 5 1 8

Add the units: 7 6 5

8 + 5 = 13 = 10 + 3

carry 1 ten

Add the tens:

10 + 60 + 10 = 80

Add the hundreds:

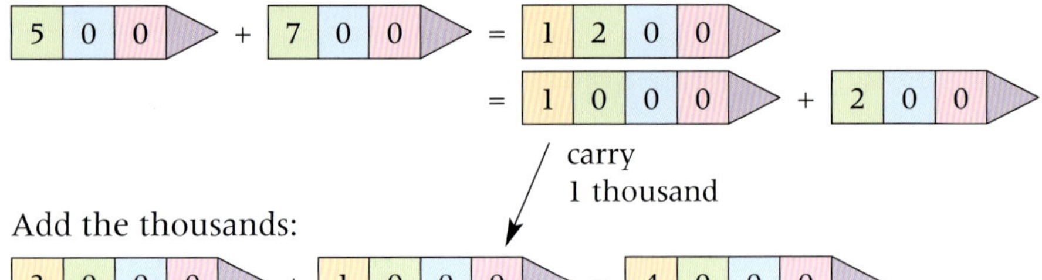

Add the thousands:

3000 + 1000 = 4000

So 3518 + 765 = 4000 + 200 + 80 + 3.

This is based on the **partitioning** mental method.

You should always estimate the answer.

3518 + 765
≈ 3500 + 800
= 4300

You could set out the calculation like this:

$$\begin{array}{r} 3\,5\,1\,8 \\ +\ \ \ 7\,6\,5 \\ \hline 4\,2\,8\,3 \\ {\scriptstyle 1}\ \ \ {\scriptstyle 1}\ \ \end{array}$$

- There is a standard written method for subtraction that is based upon the partitioning mental method.

Example

Calculate 3518 − 765.

	$^{2}\not{3}$	$^{14}\not{5}$	$^{1}1$	8	Set out the calculation in columns
−		7	6	5	
	2	7	5	3	Subtract the units 8 − 5 = **3**
					Subtract the tens **10 − 60 is not possible!**
					Move 1 hundred into the tens column 110 − 60 = **50**
					Subtract the hundreds **400 − 700 is not possible!**
					Move 1 thousand into the hundreds column 1400 − 700 = **700**
					Subtract the thousands 2000 − 0 = **2000**

Exercise N2.3

1 Use a written method to work out these calculations.

a 234 + 133 **b** 342 + 445 **c** 437 + 261 **d** 487 + 311

2 Use a written method to work out these calculations.

a 356 − 234 **b** 769 − 436 **c** 786 − 472 **d** 167 − 35

3 Use a written method to work out these.

a 432 + 549 **b** 537 + 249 **c** 345 + 273

d 487 + 432 **e** 472 + 715 **f** 654 + 533

In questions 3–6, you may need to carry.

4 Use a written method for each of these additions.

a 368 + 725 **b** 793 + 432 **c** 527 + 395

d 438 + 287 **e** 383 + 769 **f** 783 + 978

5 Use a written method for each of these subtractions.

a 562 − 417 **b** 837 − 618 **c** 748 − 563 **d** 865 − 472

6 Use a written method for each of these subtractions.

a 584 − 378 **b** 738 − 568 **c** 483 − 277 **d** 407 − 321

7 Use a written method for each of these calculations.

a 265 + 846 **b** 371 + 686 **c** 831 − 488

d 1345 + 874 **e** 2576 − 784 **f** 4762 − 1875

8 Yvonne is planning a holiday. She has a maximum budget of £1800. There are three parts to her holiday: transport, accommodation and spending money. She can choose various options for each part of her holiday:

Transport
Flights £378
Train £289
Car £278

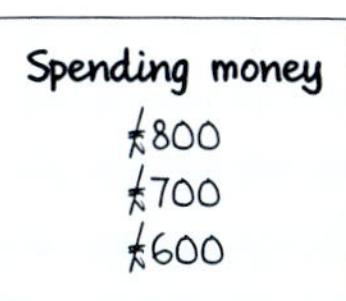

Accommodation
Villa £637
Hotel £892
Camping £278

Spending money
£800
£700
£600

For example car + villa + spending money
£278 + £637 + £800 = £1715

Write three choices she could make which would be within her budget of £1800.

9 A battery costs 83p.
Saima has £5. She buys as many batteries as she can.
Work out the amount of change Saima should get from £5.

N2.4 Mental multiplication and division

This spread will show you how to:

- Use a range of mental methods for multiplication and division of whole numbers

Keywords
Divisibility
Factor
Mental method
Multiple
Partitioning

There are lots of strategies you can use to help you work out multiplications and divisions in your head.

Partitioning
Split the numbers into parts (hundreds, tens and units).
Then multiply or divide each part separately.

Example

Calculate **a** 16×7 **b** $342 \div 3$

a $16 = 10 + 6$

$16 \times 7 = (10 \times 7) + (6 \times 7)$
$= 70 + 42$
$= 112$

b $342 = 300 + 42$

$342 \div 3 = (300 \div 3) + (42 \div 3)$
$= 100 + 14$
$= 114$

Remember
To multiply a number by 10, all the digits move one place to the left of the decimal point, for example, $10 \times 7 = 70$.

Compensation
Use this method when the number you are multiplying by is nearly a **multiple** of 10 or a multiple of 100.

Example

Calculate 14×9.

$9 = 10 - 1$

$14 \times 9 = (14 \times 10) - (14 \times 1)$
$= 140 - 14$
$= 126$

Here are some other methods you can use.

1 $23 \times 20 = 23 \times 2 \times 10$

Think of multiply by 20 as multiply by 2, followed by multiply by 10

$23 \times 2 = 46$
$46 \times 10 = 460$

$23 \times 20 = 460$

2 $84 \div 6 = 84 \div 2 \div 3$

$84 \div 2 = 42$
$42 \div 3 = 14$

$84 \div 6 = 14$

$\mathbf{6} = \mathbf{2} \times \mathbf{3}$ so $84 \div \mathbf{6}$ is the same as $84 \div (\mathbf{2} \times \mathbf{3})$. This means you can divide by 2 and then by 3.

Doubling and halving
Double one of the numbers and halve the other before you multiply.

$12 \times 15 = 6 \times 30$
$= 180$

Doubling
Double both of the numbers before you divide.

$80 \div 5 = 160 \div 10$
$= 16$

Exercise N2.4

1 Calculate these.

a 4×7 **b** $120 \div 10$ **c** 2×34 **d** $78 \div 2$
e 5×6 **f** $340 \div 10$ **g** 93×2 **h** 5×9

2 Use the mental method of partitioning to calculate these. Show the method you have used.

a 6×11 **b** 13×6 **c** 7×13 **d** 11×9 **e** 16×6 **f** 12×6

3 Use the mental method of partitioning to calculate these. Show the method you have used.

a $36 \div 3$ **b** $48 \div 4$ **c** $88 \div 8$ **d** $69 \div 3$ **e** $65 \div 5$ **f** $48 \div 3$

4 Use the mental method of compensation to calculate each of these. Show the method you have used.

a 7×9 **b** 12×9 **c** 13×11 **d** 11×19 **e** 14×11 **f** 21×6

5 Use the mental method of factors to calculate each of these. Show the method you have used.

a 13×20 **b** $48 \div 6$ **c** 23×4 **d** 7×8 **e** $132 \div 6$ **f** 17×4

6 Use the mental method of halving and doubling to calculate each of these. Show the method you have used.

a 4×27 **b** 3×14 **c** 33×4 **d** 14×15

7 Use an appropriate mental method to calculate each of these.

a 7×12 **b** 8×9 **c** $54 \div 6$ **d** 11×16 **e** 12×12 **f** $124 \div 4$

8 Use an appropriate mental method to solve each of these problems.

a Harry buys 13 CDs at £9 each.
How much does he spend?

b Kelvin shares £76 between his four grandchildren.
How much does each child receive?

c Hope runs every 20 metres in six seconds.
How long will it take her to run 400 metres?

d Pens are packed into boxes of six.
How many boxes will be needed for 78 pens?

e Trent pays for nine bars of chocolate.
Each bar of chocolate costs 26p.
How much money does he spend?

N2.5 Written multiplication and division

This spread will show you how to:

- Use a range of written methods for multiplication and division of whole numbers
- Estimate the answers to problems

Keywords
Dividend
Divisor
Estimate
Grid method
Standard method
Whole number

You can use the **grid method** to multiply whole numbers.

Example

Calculate 312×42

$42 \times 312 \approx 40 \times 300$
$= 12\,000$

You should always **estimate** the answer first.

×	300	10	2
40	$40 \times 300 = 12\,000$	$40 \times 10 = 400$	$40 \times 2 = 80$
2	$2 \times 300 = 600$	$2 \times 10 = 20$	$2 \times 2 = 4$

The calculation is split into four simpler calculations.

$42 \times 312 = 12\,000 + 400 + 80 + 600 + 20 + 4 = 13\,104$

You can use written methods for multiplication and division to solve problems.

Example

Ali buys 16 chocolate bars. Each bar costs 14 p. How much does he spend?

Estimate: $16 \times 14 \approx 20 \times 10 = 200$ pence

```
              16
            × 14
16 × 10      160
16 × 4       +64
             224
```

Ali spends 16×14 pence = 224 pence
= £2.24.

This is the **standard method** for multiplication.

Example

Karen shares £85 between her five children. How much will each child get?

Estimate: $85 \div 5 \approx 100 \div 5 = £20$

```
5)85
 −50     5 × 10
  35
 −25     5 × 5
  10
 −10     5 × 2
   0        17
```

$85 \div 5 = \mathbf{17}$

So each child receives £17.

This is the informal 'chunking' method for division.

You subtract 5 (the **divisor**) a total of 17 times from 85 (the **dividend**).

Exercise N2.5

1 Copy and complete these multiplications. Use the same method with parts **c** and **d**.

a 14×8

$\times$	**8**
10	$10 \times 8 = 80$
4	$4 \times 8 =$

$14 \times 8 = 80 +$ ____
$=$

b 12×9

$\times$	**9**
10	$=$
2	$=$

$12 \times 9 =$ ____ $+$ ____
$=$

c 14×9 **d** 17×8

2 Copy and complete these multiplications using the grid method. Remember to do a mental approximation first.

a 12×13

$\times$	**10**	**3**
10	$10 \times 10 = 100$	$10 \times 3 = 30$
2	$2 \times 10 =$	$2 \times 3 =$

$12 \times 13 = 100 + 30 +$ ____ $+$ ____
$=$

b 15×16

$\times$	**10**	**6**
10	$10 \times 10 =$	$10 \times 6 =$
5	$=$	$=$

$15 \times 16 =$ ____ $+$ ____ $+$ ____ $+$ ____
$=$

c 15×18

$\times$	**10**	**8**
10	$=$	$=$
5	$=$	$=$

$15 \times 18 =$ ____ $+$ ____ $+$ ____ $+$ ____
$=$

3 Use an appropriate method of calculation to work out each of these.

a 13×124 **b** 17×183 **c** 13×167 **d** 23×143 **e** 62×158
f 83×176 **g** 36×254 **h** 53×271 **i** 83×512

4 Use an appropriate method of calculation to work out each of these.

a $138 \div 6$ **b** $176 \div 8$ **c** $140 \div 5$ **d** $217 \div 7$
e $297 \div 9$ **f** $448 \div 8$ **g** $112 \div 8$ **h** $126 \div 7$

5 Use an appropriate method to solve each of these problems.

a Ian uses 19 text messages. Each message costs 7p.
What is the total cost of the text messages?

b Samira buys 17 fruit bushes at a cost of £7 per bush.
What is the total cost of the 17 fruit bushes?

c Mikka has 14 packets of sweets. Each packet contains 14 sweets.
How many sweets does he have altogether?

d Albert buys 13 chocolate bars. Each bar costs 16 pence.
How much does he spend?

N2 Exam review

Key objectives

- Develop a range of strategies for mental calculation
- Use standard column procedures for addition, subtraction, multiplication and division of integers
- Make mental estimates of the answers to calculations

1 Work out each of these.

a $27 + 52$ (1)

b $483 - 76$ (1)

c 3×25 (1)

d $156 \div 12$ (1)

2 54 327 people watched a concert.

a Write 54 327 to the nearest thousand. (1)

b Write down the value of the 5 in the number 54 327. (1)

(Edexcel Ltd., 2003)

A2 Simple equations

This unit will show you how to

- Add subtract, multiply and divide any number
- Understand and use inverse operations
- Use letters to represent numbers in algebra
- Understand and use the vocabulary of algebra
- Solve equations with whole number coefficients, including those involving multiplication and division
- Write equations for word problems
- Use the balance method to solve equations
- Check a solution is correct by substituting it back into the equation

Before you start ...

You should be able to answer these questions.

	Review
1 Copy and complete. **a** $9 + \square = 15$ **b** $\square - 4 = 11$ **c** $18 - \square = 7$ **d** $12 = \square + 9$	Unit N2
2 Copy and complete. **a** $5 \times \square = 20$ **b** $\square \times 6 = 36$ **c** $40 = \square \times 5$ **d** $15 = \square \times 3$	Unit N2
3 Copy and complete. **a** $4 \times 3 = 12$ $12 \div \square = 4$ $12 \div \square = 3$ **b** $7 \times 5 = 35$ $35 \div \square = 7$ $35 \div \square = 5$	Unit N2
4 Write two division facts for each of these multiplication facts. **a** $6 \times 3 = 18$ **b** $7 \times 4 = 28$ **c** $9 \times 4 = 36$ **d** $5 \times 6 = 30$	Unit N2
5 When $x = 3$ and $y = 4$, find the value of **a** $x + y$ **b** $x - y$ **c** $x \times y$ **d** $\frac{3x}{y}$	Unit A1

A2.1 Function machines

This spread will show you how to:

- Add subtract, multiply and divide any number
- Use inverse operations

Keywords
Input
Inverse
Operation
Output

- **You can write calculations using function machines.**

A function machine has

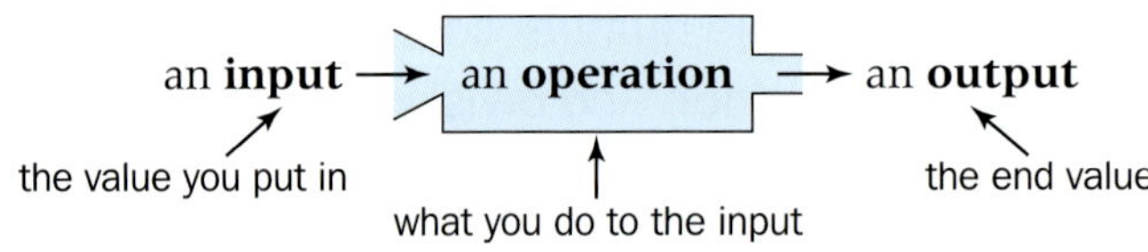

Example

Work out the outputs for these function machines.

a 2 → [+5] →

b 12 → [÷3] →

a 2 → [+5] → 7

b 12 → [÷3] → 4

- **Every operation has an inverse operation.**
  The inverse operation 'undoes' the operation.

- **You can work backwards through a function machine using inverse operations.**

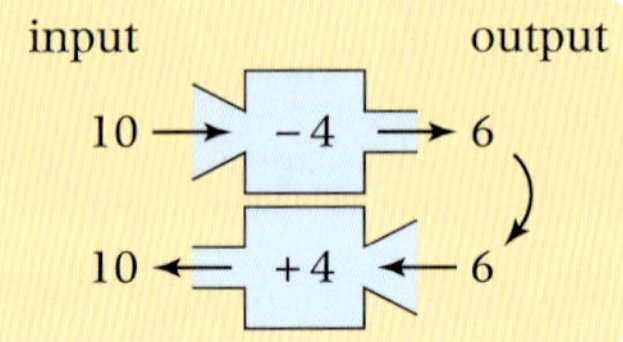

The inverse of 'add' is 'subtract'
The inverse of 'multiply' is 'divide'.

Example

Draw the inverse machines for these function machines.

a 3 → [×2] → 6 b 5 → [+13] → 18 c 10 → [÷2] → 5

a 3 ← [÷2] ← 6 b 5 ← [−13] ← 18 c 10 ← [×2] ← 5

- **You can use function machines to solve 'think of a number' problems.**

Example

I think of a number and multiply it by 4.
My answer is 20.

What is my number?

My number → [×4] → 20

5 ← [÷4] ← 20

My number is 5.

Exercise A2.1

1 Work out the outputs for these function machines.

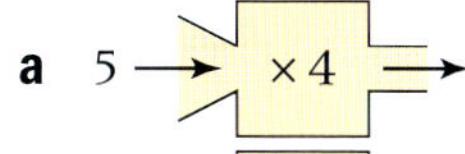

a 5 → ×4 → **b** 8 → ×2 → **c** 9 → +7 →

d 15 → −8 → **e** 20 → ×3 → **f** 18 → ÷3 →

g 14 → −6 → **h** 19 → +12 → **i** 27 → ÷9 →

2 Copy each machine and write the outputs for the inputs given.

a

input	machine	output
8		___
5		___
3	→ +7 →	___
0		___
−1		___

b

input	machine	output
4		___
2		___
1	→ ×3 →	___
0		___
−1		___

3 Copy and complete these function machines.

a 7 → ×? → 21 **b** 5 → ×? → 35 **c** 4 → ×? → 24

d 7 → −? → 2 **e** 6 → +? → 11 **f** 4 → +? → 17

g 12 → ÷? → 4 **h** 24 → ÷? → 6 **i** 12 → ×? → 60

4 Write the inverse operation for each of these operations.

a ×2 **b** +4 **c** −3 **d** ÷6

e −7 **f** ×5 **g** ÷2 **h** +11

5 Copy and complete the tables for these function machines.

a → ÷2 →

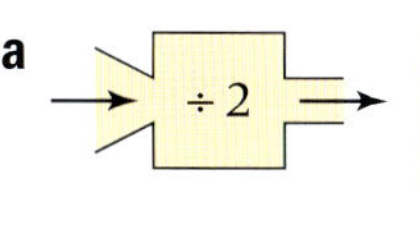

Input	Output
6	
10	
24	
	15

b → −4 →

Input	Output
5	
13	
	11
20	

6 Use function machines to solve these 'think of a number problems':

a I think of a number and add 12.
The answer is 25. What number did I think of?

b I think of a number and divide it by 4.
The answer is 5. What number did I think of?

A2.2 Solving equations

This spread will show you how to:

- Use letters to represent numbers in algebra
- Understand and use the vocabulary of algebra
- Set up and solve simple equations

Keywords
Equation
Expression
Solution
Solve

- **An expression is made up of terms, containing letter symbols and numbers.**

$3x+2$ $\quad x-4$ $\quad 2x$ $\quad 4x+1$
are all expressions.

- **An equation includes letter and number terms and an equals sign.**

$x+2=5$ is an equation.

In an equation, the letter symbol represents one number value. You can work out the value of a letter in an equation using number facts.

For example, $x+2=5$ and $3+2=5$.
So $x=3$ is the **solution** of the equation.
Working out the value of the unknown is called solving the equation.

- **You can use function machines to solve equations.**

Example

Solve these equations.

a $x+5=12$ **b** $y-6=10$

a

$x \rightarrow +5 \rightarrow 12$

$7 \leftarrow -5 \leftarrow 12$

$x=7$

b

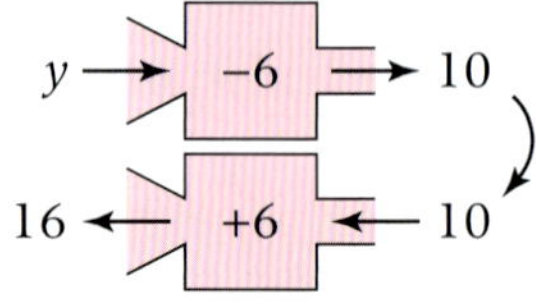

$y=16$

Draw the function machine for the equation.

Draw the inverse function machine.

- **You can write an equation for a word problem and solve it.**

Example

Tony had some pairs of socks.
For Christmas he got 4 more pairs of socks.
He counted all his socks and found he had 11 pairs altogether.
How many pairs of socks did he have to start with?

n pairs + 4 pairs = 11 pairs $\quad n+4=11$

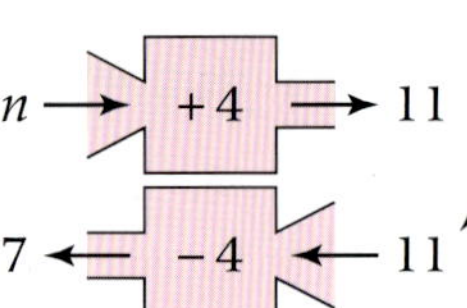

$n=7$
Tony had 7 pairs of socks to start with.

Use the letter n to represent the number of pairs of socks Tony started with.

Write an equation to represent the problem.

Use function machines to solve the equation.

Exercise A2.2

1 Use number facts to work out the missing numbers in these calculations.

a $6 + \square = 10$ **b** $2 + \square = 7$ **c** $5 + \square = 8$

d $3 + \square = 12$ **e** $6 + \square = 18$ **f** $7 = \square + 3$

g $9 = 4 + \square$ **h** $14 = \square + 6$ **i** $19 = 13 + \square$

2 Use number facts to work out the missing numbers in these calculations.

a $8 - \square = 2$ **b** $12 - \square = 6$ **c** $10 - \square = 3$

d $13 - \square = 2$ **e** $15 - \square = 8$ **f** $8 = 10 - \square$

g $5 = \square - 10$ **h** $7 = \square - 5$ **i** $20 = \square - 6$

3 Copy and complete these function machines to solve these equations.

a $x + 3 = 6$

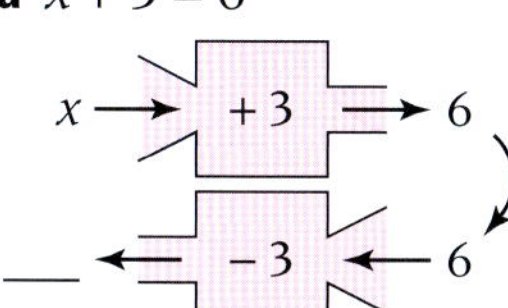

b $a - 8 = 5$

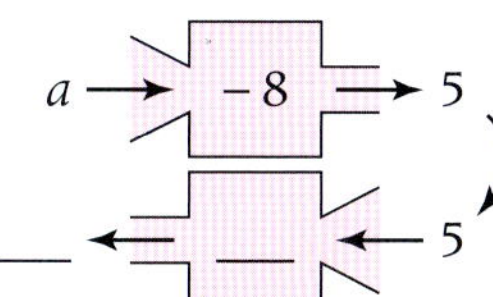

c $b + 4 = 3$

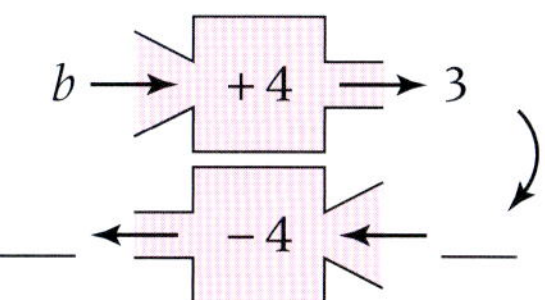

4 For each function machine, draw the inverse function machine and find the value of the unknown.

a $a \rightarrow +8 \rightarrow 15$ **b** $b \rightarrow +6 \rightarrow 23$ **c** $c \rightarrow +7 \rightarrow 12$

d $d \rightarrow +11 \rightarrow 20$ **e** $e \rightarrow +15 \rightarrow 48$ **f** $f \rightarrow +13 \rightarrow 21$

g $g \rightarrow +27 \rightarrow 40$ **h** $h \rightarrow +40 \rightarrow 100$ **i** $i \rightarrow -5 \rightarrow 3$

j $j \rightarrow -8 \rightarrow 4$ **k** $k \rightarrow -3 \rightarrow 9$ **l** $l \rightarrow -13 \rightarrow 9$

m $m \rightarrow -25 \rightarrow 6$ **n** $n \rightarrow -10 \rightarrow 90$ **o** $o \rightarrow -15 \rightarrow 15$

5 Use function machines to solve these equations.

a $x + 3 = 7$ **b** $y + 6 = 10$ **c** $n - 1 = 6$

d $m + 9 = 14$ **e** $r - 3 = 10$ **f** $9 + v = 0$

g $y - 2 = 7$ **h** $x - 4 = 3$ **i** $z - 6 = 2$

j $m - 5 = 9$ **k** $z + 2 = 15$ **l** $p - 7 = 9$

m $q - 4 = 8$ **n** $n + 4 = 11$ **o** $a + 16 = 27$

A2.3 Solving multiplication and division equations

This spread will show you how to:

- Solve equations with whole numbers, including those involving multiplication and division

Keywords
Equation

- **You can use function machines to solve equations involving multiplication and division.**

The equation $3x = 12$ means 'x multiplied by $3 = 12$'.

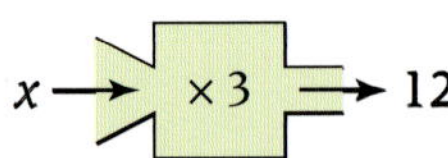

Solve the equation using an inverse machine.

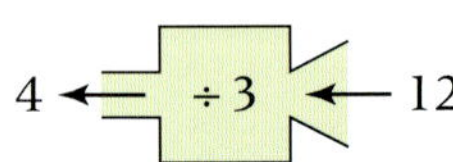

$x = 4$

The equation $\frac{y}{4} = 6$ means 'y divided by $4 = 6$'.

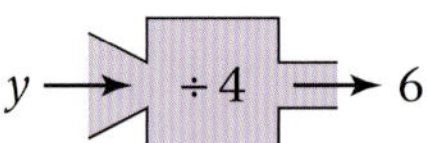

Solve the equation using an inverse machine.

$24 \leftarrow \boxed{\times 4} \leftarrow 6$

$y = 24$

Example

Solve these equations using number facts.

a $3s = 12$ **b** $\frac{t}{2} = 5$

a $3 \times s = 12$ and $3 \times 4 = 12$
So $s = 4$

b $t \div 2 = 5$ and $10 \div 2 = 5$
So $t = 10$

Example

Solve these equations using function machines.

a $6m = 30$ **b** $\frac{n}{7} = 3$

a

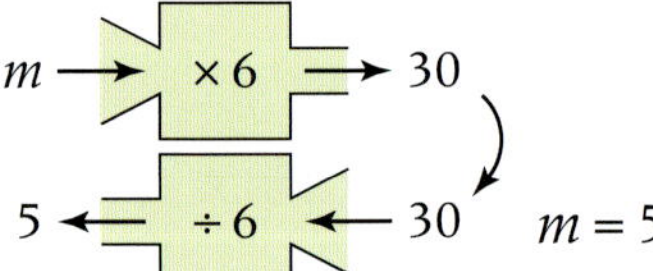

$m = 5$

b

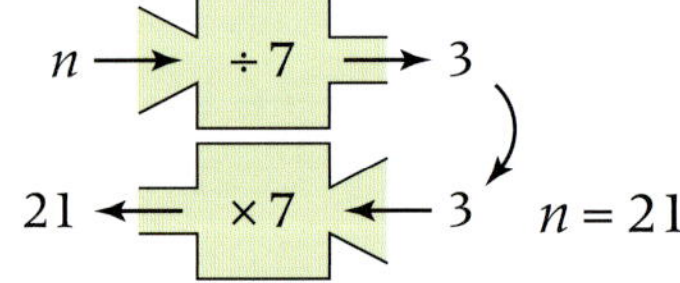

$n = 21$

Example

A pizza is cut into m slices. Ali and Kate share the pizza equally between them.

a Write an expression for the number of slices they have each.
b Ali and Kate have 6 slices each. Use your answer to part **a** to write an equation.
c Solve your equation to find the value of m.

a $\frac{m}{2}$ **b** $\frac{m}{2} = 6$ **c**

$m \rightarrow \boxed{\div 2} \rightarrow 6$

$12 \leftarrow \boxed{\times 2} \leftarrow 6$

$m = 12$

$\frac{m}{2}$ means 'm divided by 2'.

Exercise A2.3

1 Use number facts to find the missing numbers.

a $3 \times \square = 15$ **b** $4 \times \square = 20$ **c** $5 \times \square = 25$ **d** $2 \times \square = 16$

e $8 \times \square = 32$ **f** $4 \times \square = 24$ **g** $6 \times \square = 30$ **h** $\square \times 3 = 12$

i $\square \times 5 = 30$ **j** $\square \times 8 = 24$ **k** $\square \div 2 = 4$ **l** $\square \div 5 = 5$

m $\square \div 3 = 5$ **n** $\square \div 4 = 6$ **o** $6 \times \square = 18$ **p** $18 \div \square = 6$

q $20 \div \square = 5$ **r** $\square \times 7 = 35$ **s** $\square \div 7 = 7$ **t** $\square \times \square = 36$

2 Copy and complete these function machines and their inverses to find the value of the unknowns.

a

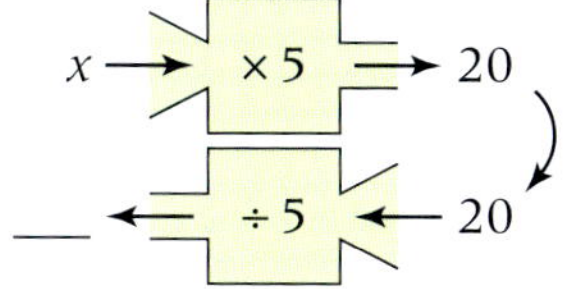

b

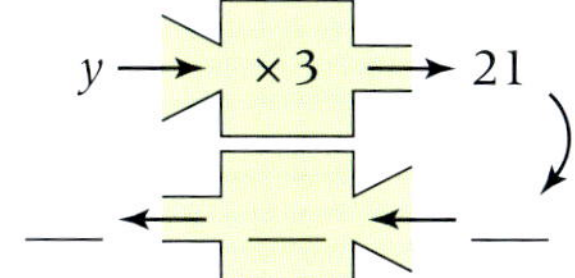

c

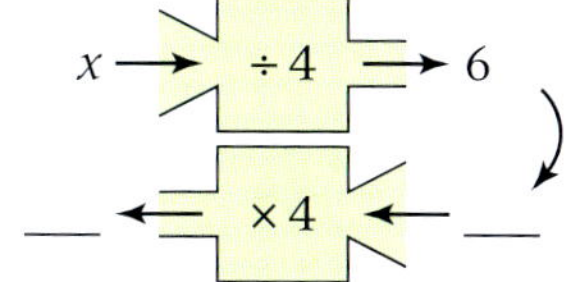

3 Match each function machine in set A with an equation in set B.

Set A

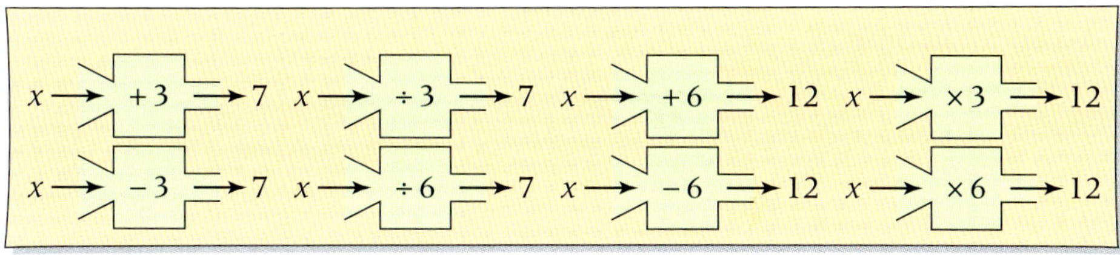

Set B

$3x=12$ $6x=12$ $x-6=12$ $x+6=12$

$\frac{x}{3}=7$ $x+3=7$ $x-3=7$ $\frac{x}{6}=7$

4 Draw function machines for these equations and use inverse machines to solve them.

a $4x = 20$ **b** $3y = 15$ **c** $5z = 30$ **d** $2r = 12$

e $4s = 28$ **f** $3t = 18$ **g** $2v = 22$ **h** $2w = 18$

i $\frac{x}{3} = 5$ **j** $\frac{y}{2} = 24$ **k** $\frac{z}{5} = 5$ **l** $\frac{r}{3} = 8$

5 Maisie has 5 boxes of pencils.
Each box contains n pencils.

a Write an expression for the number of pencils Maisie has.

b Maisie counts the pencils. She has 40 in total.
Use your expression from part **a** to write an equation.

c Solve your equation to find n, the number of pencils in a box.

A2.4 The balance method

This spread will show you how to:

- Use the balance method to solve equations

Keywords
Balance
Inverse

In this set of scales, the two sides **balance**.

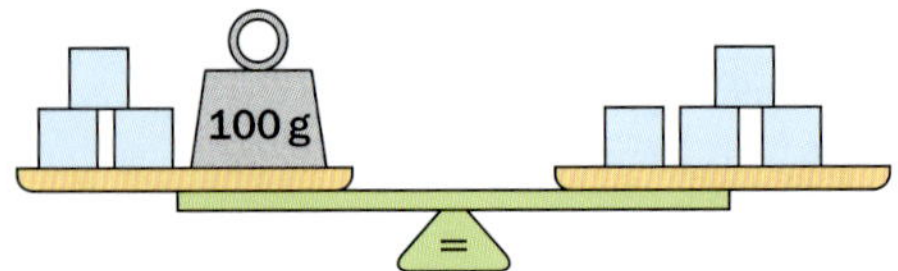

Add 2 boxes to each side.

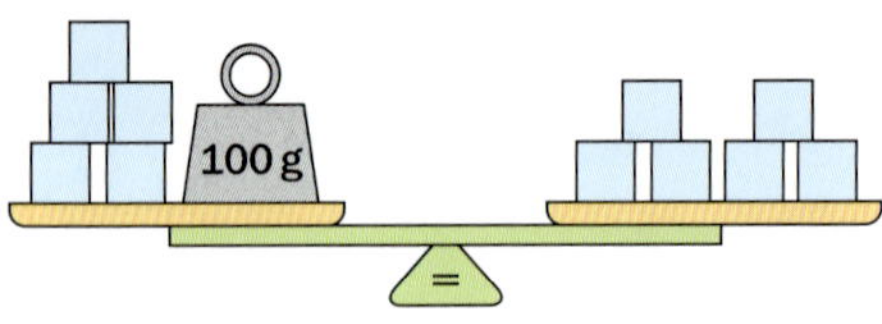

Still balanced.

Take 3 boxes from each side.

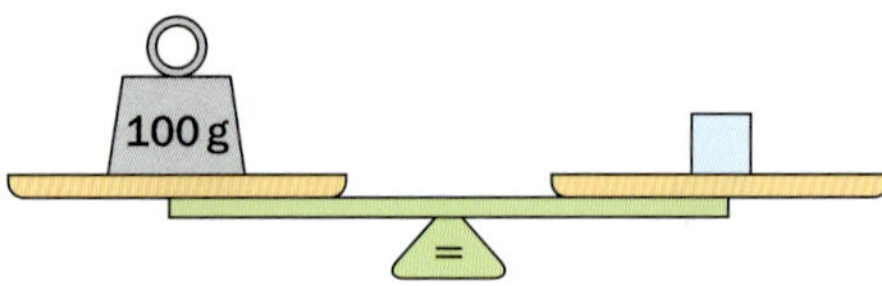

Still balanced. 1 box = 100 g

If you add or subtract (take away) the **same amount** from both sides, the scales still balance.

- **The two sides of an equation balance.**

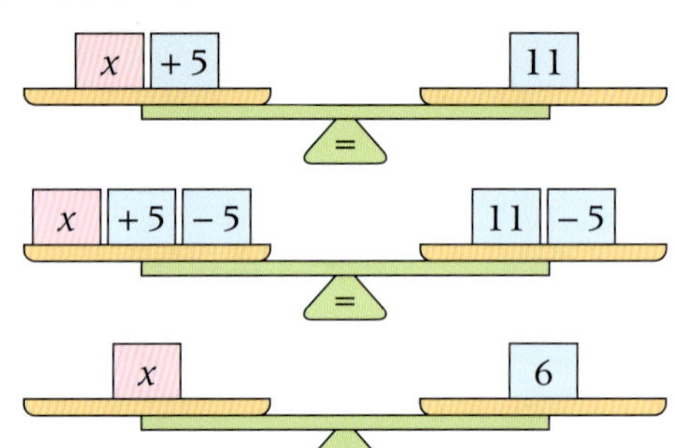

Consider the equation
$x + 5 = 11$

Subtract 5 from each side
$x + 5 - 5 = 11 - 5$

The solution is $x = 6$.

The **inverse** of +5 is −5.
Use the inverse operation to get x on its own.

- **You can use the balance method to solve an equation. You do the same to each side to keep the equation balanced.**

Example

Use the balance method to solve these equations.

a $m + 4 = 15$ **b** $n - 3 = 8$

a
$$m + 4 = 15$$
$$m + 4 - 4 = 15 - 4$$
$$m = 11$$

b
$$n - 3 = 8$$
$$n - 3 + 3 = 8 + 3$$
$$n = 11$$

In **a** the inverse of +4 is −4. Subtract 4 from both sides.

In **b** the inverse of −3 is +3. Add 3 to both sides.

Exercise A2.4

1 For each of these diagrams work out the weight of one box.

a

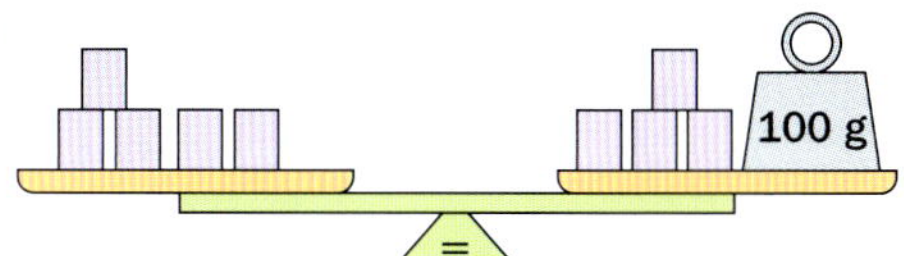

b

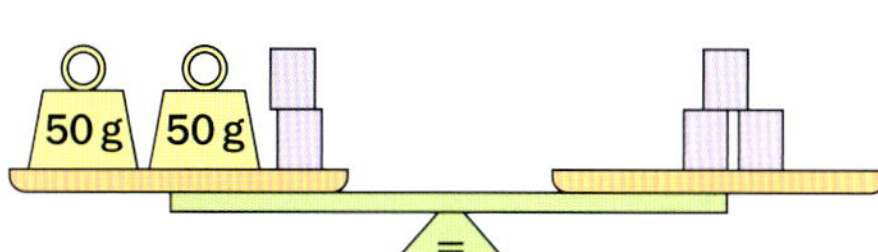

c

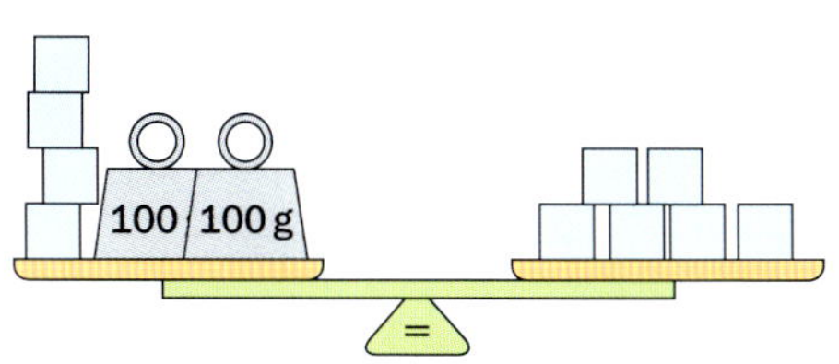

d

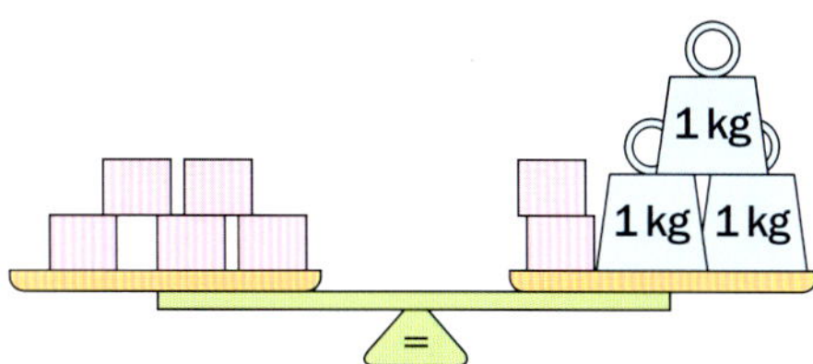

2 Copy and complete these balances to solve the equations.

a $x + 6 = 19$

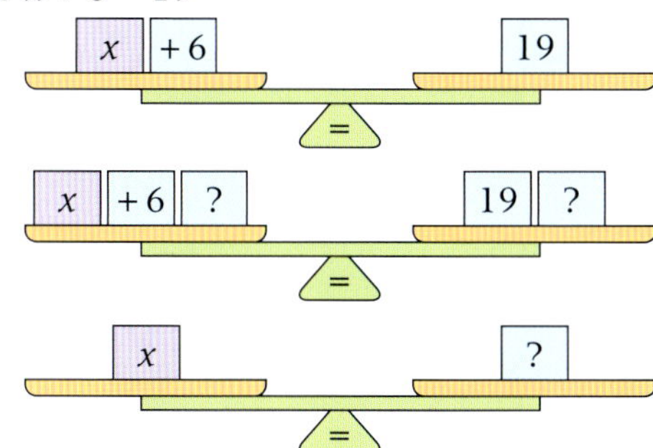

b $x - 4 = 8$

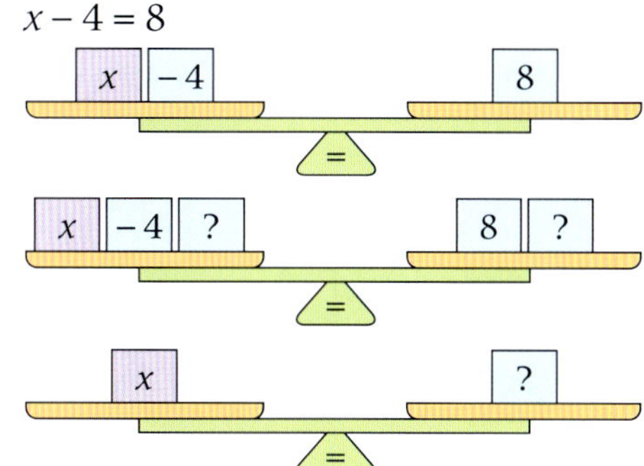

3 Copy and complete to solve these equations.

a $m + 7 = 13$ The inverse of +7 is ___

Subtract 7 from both sides:

$m + 7$ ___ $= 13$ ______

$m =$ ________

b $n - 9 = 17$ The inverse of −9 is ___

Add ____ to both sides:

$n - 9$ ____ $= 17$ ____

$n =$ _______

4 Match each equation in set A to its solution in set B.

Set A:

$13 - x = 10$ $x - 4 = -3$ $x + 5 = 7$

$x + 17 = 21$ $x + 8 = 14$ $x + 6 = 15$

$x - 7 = 1$ $x - 3 = 4$ $22 + x = 27$

Set B:

$x = 1$	$x = 2$	$x = 3$
$x = 4$	$x = 5$	$x = 6$
$x = 7$	$x = 8$	$x = 9$

5 Solve these equations using the balance method.

a $x + 5 = 8$ **b** $x + 3 = 14$ **c** $x + 8 = 13$

d $x + 3 = 18$ **e** $x + 9 = 0$ **f** $5 + x = 11$

A2.5 More solving equations

This spread will show you how to:

- Use the balance method to solve equations
- Check a solution is correct by substituting it back into the equation

Keyword

Balance
Substituting

These scales are **balanced**.

3 boxes weigh 300 g.

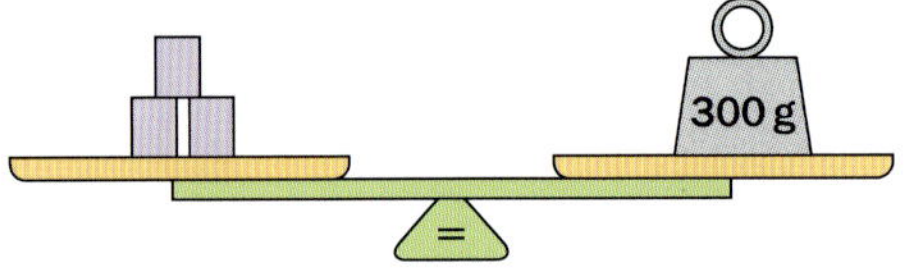

Double (×2) both sides.

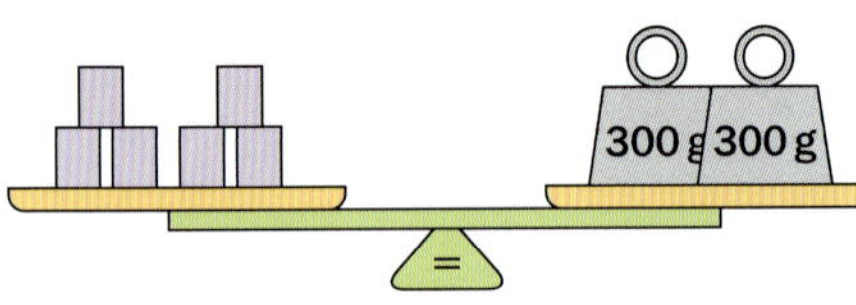

The scales still balance.
6 boxes weigh 600 g.

Divide both sides by 3.

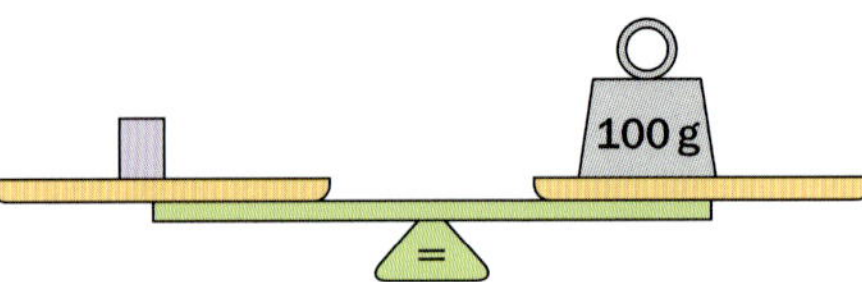

The scales still balance.
1 box weighs 100 g.

- **You can solve equations using the balance method.**
  To keep an equation balanced
  – You can add or subtract the same amount from both sides
  – You can multiply or divide both sides by the same number.

Example

Solve these equations using the balance method.

a $3x = 15$ **b** $\frac{x}{4} = 5$

---

a $3x = 15$

Divide both sides by 3:

$3x \div 3 = 15 \div 3$

$x = 5$

b $\frac{x}{4} = 5$

Multiply both sides by 4:

$\frac{x}{\not{4}} \times \not{4} = 5 \times 4$

$x = 20$

x is multiplied by 3. The inverse of ×3 is ÷3.

$\frac{x}{4} = x \div 4$. The inverse of ÷4 is ×4.

- **You can check that a solution is correct by substituting it back into the equation.**

Example

Daisy solves the equation $\frac{x}{7} = 4$.

Her solution is $x = 28$. Is she correct?

Substitute $x = 28$ into $\frac{x}{7}$: $\frac{28}{7} = 28 \div 7 = 4$

So $x = 28$ is correct

The correct value of x will give the answer 4.

Exercise A2.5

1 For each of these diagrams work out the weight of one box.

a

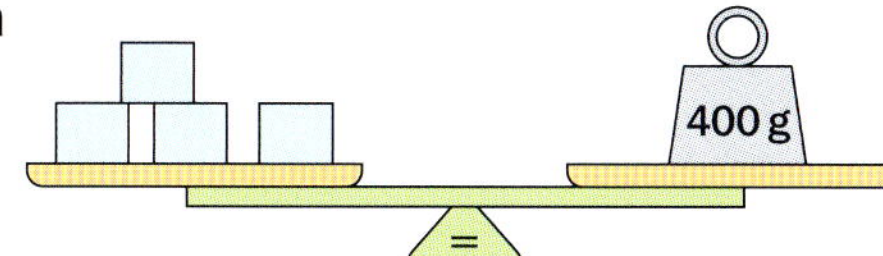

b

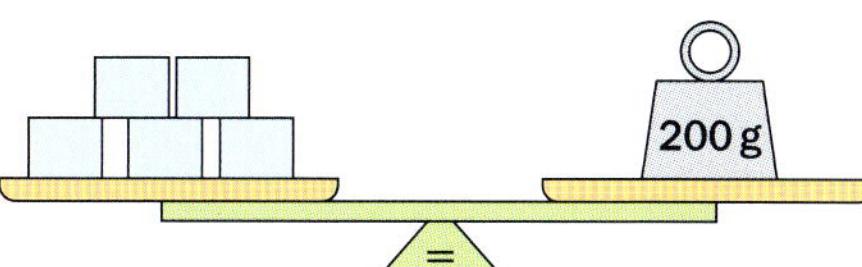

c

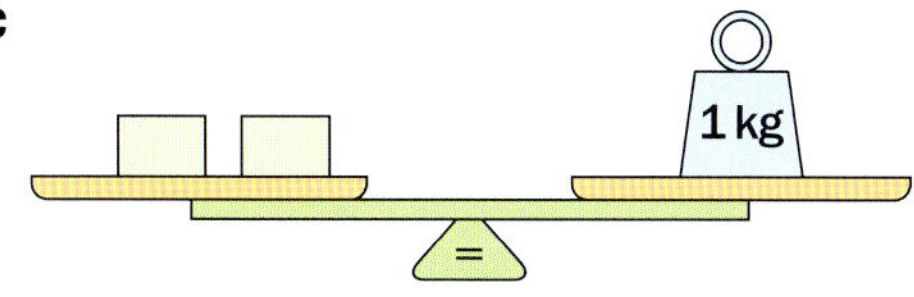

d

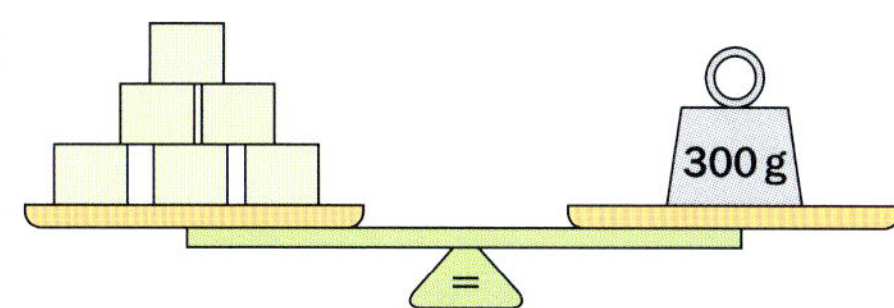

2 Copy and complete to solve these equations.

a $6m = 24$ The inverse of ×6 is ____

Divide both sides by ___:

$6m \div \_\_\_ = 24 \div \_\_\_$

$m = \_\_\_\_$

b $\frac{n}{5} = 11$ The inverse of ÷5 is ____

Multiply both sides by ___:

$\frac{n}{5} \times \_\_\_\_ = 11 \times \_\_\_\_$

$n = \_\_\_\_$

3 Solve these equations using the balance method.

a $5x = 15$ **b** $3x = 21$ **c** $6x = 18$ **d** $4x = 36$

e $28 = 4x$ **f** $7x = 28$ **g** $50 = 25x$ **h** $4x = 10$

4 Use the balance method to solve these equations.

a $\frac{s}{5} = 5$ **b** $\frac{t}{12} = 3$ **c** $\frac{u}{2} = 4$ **d** $\frac{v}{7} = 3$

e $\frac{v}{5} = 9$ **f** $\frac{w}{3} = 8$ **g** $10 = \frac{x}{5}$ **h** $3 = \frac{y}{9}$

5 Tom and Anya both solve the equation $7x = 56$.

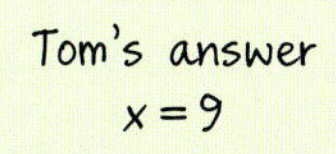

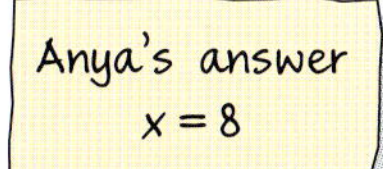

Who is correct? Explain how you worked it out.

6 Solve these equations.
Check your answers using substitution.

a $a + 7 = 11$ **b** $g - 7 = 8$ **c** $\frac{c}{4} = 9$ **d** $d - 5 = -2$

e $3e = 21$ **f** $9 + f = 5$ **g** $4g = 0$ **h** $\frac{h}{4} = 2.5$

A2 Exam review

Key objectives

- Set up simple equations
- Substitute numbers into a formula
- Solve linear equations, with integer coefficients, in which the unknown appears on either side
- Solve simple equations by using inverse operations or by transforming both sides in the same way

1 Solve each equation.

a $y - 1 = 5$ (1)

b $2x = 8$ (1)

c $z + 2 = 3z - 2$ (1)

2 Here is a table for a two-stage number machine.

It multiplies by 2 then subtracts 1.

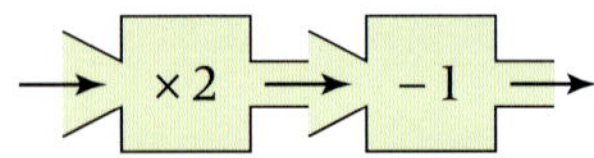

Input	Output
1	1
2	3
3	
5	
	15

Copy the table and fill in the missing numbers. (3)

(Edexcel Ltd., 2003)

D1 Collecting data

This unit will show you how to

- Understand the difference between primary and secondary data
- Identify which primary data is needed and in what format, including grouped data
- Collect discrete data and grouped discrete data using a data collection sheet
- Understand and use frequency tables to show data
- Collect data using various methods, including observation, controlled experiment, questionnaires and surveys
- Use random sampling methods, taking steps to minimise bias
- Calculate total frequency from a discrete frequency table
- Use two-way tables for discrete data

Before you start ...

You should be able to answer these questions.

Review

1 Put these numbers in order of size, smallest first.

a 37, 42, 17, 6, 30, 19, 26, 29

b 118, 135, 106, 121, 130, 115

c 156, 145, 154, 165, 166, 155, 144

Unit N1

2 Calculate each of these.

a $63 + 58$ **b** $48 + 96$

c $73 + 95 + 84$ **d** $138 + 275$

e $63 + 5$ **f** $38 - 15$

g $96 - 47$ **h** $136 - 54$

i $258 - 69$ **j** $432 - 166$

Unit N2

3 The table shows the average times of sunset and sunrise for 6 months of the year.

	Sunrise	Sunset
January	07:53	16:20
March	06:06	18:17
May	05:00	20:45
July	04:58	21:04
September	06:35	19:00
November	07:23	15:58

What time does the sun

a rise in July **b** set in May?

Key stage 3

D1.1 Data collection sheets

This spread will show you how to:

- Use data collection sheets for discrete data
- Understand and use frequency tables to show data

Keywords
Data
Data collection sheet
Frequency table
Tally chart

Before this headline could be written, information called **data** was collected.

Examiners tip
The topics in this unit are useful for your Data coursework.

- You can collect data using a **data collection sheet**. (This one is a **tally chart**.)

Colour	Tally	Frequency
Red	~~IIII~~ ~~IIII~~ III	13
Blue	~~IIII~~ II	7
White	~~IIII~~ ~~IIII~~ I	11

~~IIII~~ = 5

- The data can also be shown using a **frequency table**.

Colour	Frequency
Red	13
Blue	7
White	11

7 people chose Blue

Example

The students in a Year 10 class were asked to name their favourite season. Use the data-collection sheet to tally the data. Calculate the frequencies.

Season	Tally	Frequency
Spring		
Summer		
Autumn		
Winter		

Winter Summer Spring Spring Winter Autumn Summer
Spring Autumn Summer Winter Winter Summer Summer
Spring Summer Summer Summer Spring Spring

Season	Tally	Frequency
Spring	~~IIII~~ I	6
Summer	~~IIII~~ III	8
Autumn	II	2
Winter	IIII	4

Exercise D1.1

1 Tickets to see an Irish band called Ceol cost £5, £10, £15 or £20. These price tickets are sold one morning

These values are all in pounds (£).

15	10	10	5	5	5	5	5	10	20
20	5	5	5	5	10	5	10	5	15
15	20	20	5	5	10	10	10	10	5
5	5	10	5	10	15	20	20	20	20

a Copy and complete the tally chart.

b How many £5 tickets were sold?

c How many tickets were sold altogether?

Price	Tally	Number of tickets
£5		
£10		
£15		
£20		

2 The vowels in a paragraph on the front page of a newspaper are

a	e	i	a	e	o	u	e	i	a
e	o	a	i	e	u	e	o	i	o
u	o	u	i	e	e	o	i	a	e
a	o	e	i	o	e	e	i	o	u
a	a	e	e	o	i	u	e	o	e

a Copy and complete the data collection sheet to show the vowels in the paragraph.

b Which vowel occurred the most?

Vowel	Tally	Frequency
a		
e		
i		
o		
u		

3 A tetrahedron dice is numbered 1, 2, 3, 4.

a Copy and complete the data collection sheet to show these scores.

4	2	1	3	1	3	4	2
3	3	2	2	2	3	4	4
2	4	1	1	2	4	1	2
1	2	1	3	2	1	4	3
2	3	1	4	4	2	3	1

b State the number of times a 3 was rolled.

Score	Tally	Frequency
1		
2		
3		
4		

4 A class of students are asked to give the month of their birthday.

Mar	May	Apr	Jun	Sep	Mar	Jun	Feb
Sep	Dec	Nov	Nov	Apr	Mar	Jul	Aug
Aug	Mar	Apr	Feb	Jan	Sep	Jun	Jun
Sep	Nov	Oct	Jan	Jun	Dec	Oct	Feb

a Draw and complete a tally chart to show this information.

b Calculate the number of students in the class.

c In which month were most students born?

D1.2 Observation, controlled experiment and sampling

This spread will show you how to:

- Collect data using various methods including observation and controlled experiment
- Calculate total frequency from a discrete frequency table
- Use random sampling methods, taking steps to minimise bias

Keywords
Biased
Controlled experiment
Data collection sheet
Observation
Random sample

- You can collect data by **observation**.
  For example, to find how many people use the dodgem ride at the fair, you would have to watch and count the people on the ride.
- You can collect data by a **controlled experiment**.

Example

Simon throws a dice 50 times. He thinks 'lucky' 6 will happen more often than the other numbers. The results are shown.

3	1	4	3	5	2	6	2	1	3
4	2	2	6	3	1	1	2	3	4
5	6	1	2	1	4	3	5	6	2
1	4	6	3	2	2	1	5	5	6
2	4	3	2	5	6	4	4	6	3

Complete the **data collection sheet** to show the dice scores for this controlled experiment.

Score	Tally	Freq.
1		
2		
3		
4		
5		
6		

Score	Tally	Freq.
1	𝍸 III	8
2	𝍸 𝍸 I	11
3	𝍸 IIII	9
4	𝍸 III	8
5	𝍸 I	6
6	𝍸 III	8
		50

Check the frequencies add to 50.

Sometimes it is impossible to collect data from all the population and so a **random sample** is used.

For example, instead of asking every student in your school, you could choose a random sample of 50 students. The 50 students are chosen so that the sample is not **biased**.
The larger the sample, the more accurate the data will be.

Exercise D1.2

1 Decide whether each of these data collections is an observation or a controlled experiment.

a Rolling a dice

b Spinning a coin

c Whether people walk under a ladder

d The colour of vehicles

e Spinning a spinner

f The number of birds in a garden

g The punctuality of trains

h The choices of a school meal

i Choosing chocolates out of a selection box

j The ages of teachers at a school

2 The number of passengers in passing cars are counted. The results are shown.

1 0 2 0 1 0 1 3 1 0 3 2
0 1 0 0 2 1 0 0 0 1 2

a Is this data collection an observation or a controlled experiment?

b Copy and complete the data collection sheet to show this information.

c Calculate the total number of cars that passed.

Number of passengers	Tally	Number of cars
0		
1		
2		
3		

3 A spinner, labelled A to E, is spun and the letter is recorded. The results are shown.

C B E D E A B D A E
B C D A E B C B A E
A E D E B C B A D E
E C B D A B A C D C

a Is this data collection an observation or a controlled experiment?

b Copy and complete the frequency table to show the results.

c How many times was the spinner spun altogether?

d Do you think the spinner is biased? Explain your answer.

e How could you improve the reliability of your answer?

Letter	Tally	Frequency
A		
B		
C		
D		
E		

D1.3 Surveys

This spread will show you how to:

- Collect data using various methods including questionnaires and surveys

Keywords
Data collection sheet
Questionnaire
Survey

Surveys are used to find people's views and opinions.

2 out of 3 boys like chocolate but only

- **You can collect data with a survey using**
  - **a data collection sheet**
  - **a questionnaire.**

You can collect all the data on one **data collection sheet**.

Age group under 16/16+	M/F	Like chocolate?
16+	F	no
under 16	F	yes
16+	M	yes

You will need one **questionnaire** for each person in your survey.

Chocolate questionnaire

Age group

Under 16 ☐ 16 or more ☐

Gender M ☐ F ☐

Do you like chocolate? Y ☐ N ☐

Any other comments?

- You must be careful what questions you ask in a survey.
  - Never ask a personal question. For example, did you brush your teeth yesterday? What is your age?
  - Never ask a leading question. For example, what do you think of this beautiful wood being chopped down to build a noisy road?
  - Never ask a vague question. For example, how often do you eat meat?
  - Never ask a question that will give too many answers. For example, what did you eat yesterday?

Example

One question in a questionnaire is

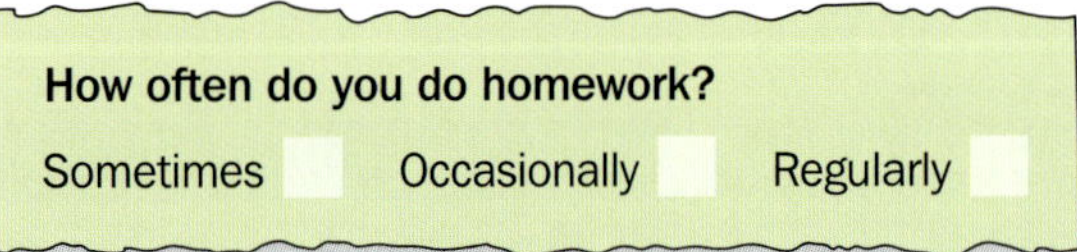

Explain why the suggested answers are not satisfactory.

1 Always/never are not given.
2 Sometimes/occasionally mean the same.
3 Regularly could mean every day, every week or every month.

Exercise D1.3

1 James uses a data collection sheet for a survey to find his class's favourite soup.
He limits the choice to Tomato (T), Vegetable (V), Fish (F) or Other (O). His completed sheet is

T T T V F V V T T V
F O T V O O T V T F
O T V O T V O O O O

a Copy and complete the tally chart to show the data.

b Calculate the number of students in James's class.

c State the most favourite soup.

Type of soup	Tally	Number of students
Tomato (T)		
Vegetable (V)		
Fish (F)		
Other (O)		

2 Andrew uses a questionnaire for a survey about eating habits.

His questionnaire is shown.
He stops people in the street and asks them to answer the questions.

Name:
Age:
What is your favourite meal?

The first response from an elderly lady is shown.
Write three criticisms of Andrew's questionnaire.

Name: Anonymous
Age: Older than you!
What is your favourite meal?
Prawn cocktail, soup, meat and two veg, dumplings, gravy, chocolate gateau with ice cream, cheese and
PTO

3 These questions appeared on a questionnaire.
Write one criticism of each question.

a What is your favourite sweet?

b How often do you use a computer?

c How much pocket money/allowance are you given?

d Did you have a shower this morning?

e These buses are always late. What do you think of the bus service?

f How much do you spend on clothes?

g How tall are you?

h Do you like shopping?

i Where do you live?

j How many DVDs do you own? [loads] [a lot] [many]

D1.4 Grouped data

This spread will show you how to:

- Understand the difference between primary and secondary data
- Understand and use frequency tables to show data
- Identify which primary data you need to collect in what format, including grouped data, considering equal class intervals

Keywords

Class interval
Group
Primary data
Secondary data

- **Primary data** is data you collect yourself, for example, you count the number of heads when spinning a coin.
- **Secondary data** is data someone else has already collected, for example, National Census, information from newspapers or the internet.

Some surveys produce data with too many different values. For example, 36, 44, 18, 27, 23, 6, 73, 25, 19, 31, 80, 46, 51, 55, 65 could be the exam marks for 15 students.

A data collection sheet or a frequency table would have too many categories.

- You can **group** the data into **class intervals** to avoid this.

Example

The number of cars in a car park was recorded every day for one month. The results are shown.

8	34	10	15	24	49	0	13	25	19
23	31	45	0	15	3	21	22	27	47
0	9	24	36	17	19	45	0	18	5

Complete the grouped frequency table.

Number of cars	Tally	Freq.
0 to 9		
10 to 19		
20 to 29		
30 to 39		
40 to 49		

class intervals

Number of cars	Tally	Freq.
0 to 9	𝍸 III	8
10 to 19	𝍸 III	8
20 to 29	𝍸 II	7
30 to 39	III	3
40 to 49	IIII	4
		30

𝍸 = 5

Check that the frequencies add to 30.

Exercise D1.4

1 Decide whether these data collection methods give primary data or secondary data.

a The times of the goals in football matches from a newspaper

b Measuring heights of students in your class

c The rainfall each month in Paris from the internet

d The number of Heads when spinning a coin

e The results of an experiment you do in a science lesson

f The number of telephone calls your class made yesterday evening

g The number of people who went to the theatre in 2005 from information on the internet

h The population of Switzerland from a book

i The reaction times of students in your class by an experiment

j The times of low and high tides from a book.

2 **a** Copy and complete the frequency table using these weights of people, in kilograms.

67	40	56	65	57	42	45
56	66	69	42	51	58	63
65	69	61	44	67	55	43
58	63	68	54	57	49	48
47	42					

Weight (kg)	Tally	Number of people
40 to 44		
45 to 49		
50 to 54		
55 to 59		
60 to 64		
65 to 69		

b Calculate the number of people shown in the frequency table.

3 **a** Copy and complete the frequency table using these exam marks.

45	36	34	56	71	38	55
63	72	80	14	25	44	37
51	58	35	47	22	10	33
37	54	61	77	24	27	29
31	35	27	28	32	36	52
58	59	60	50	35	29	18
66	55	32	35	21	53	67
79						

Exam mark	Tally	Frequency
1 to 10		
11 to 20		
21 to 30		
31 to 40		
41 to 50		
51 to 60		
61 to 70		
71 to 80		

b Calculate the number of people who took the exam.

4 The heights, in centimetres, of students are shown.

148	143	148	152	155	160	171	144
132	133	161	172	133	149	150	164
168	170	153	150	138	139	144	151
163	165	180	180	155	160	165	155
145	133	138	161	168	136	147	145

a Draw and complete a frequency table, using suitable class intervals.

b Calculate the total number of students in the frequency table.

D1.5 Two-way tables

This spread will show you how to:

- Use two-way tables for discrete and grouped data
- Collect data using various methods

Keywords
Column
Frequency table
Row
Total
Two-way table

You can summarise data in a **frequency table**.

Food	Number of people
Pizza	8
Burger	7
Curry	5

7 people prefer a burger.

You can show more detail in a **two-way table**.

A two-way table links two types of information, for example, food and gender.

	Men	Women
Pizza	3	5
Burger	4	3
Curry	3	2

3 women prefer a burger.

You can extend the two-way table by adding
– an extra **row** to give the total of men, women and people
– an extra **column** to give the total of each food.
The extra row and column headings are **totals**.

	Men	Women	Totals
Pizza	3	5	8
Burger	4	3	7
Curry	3	2	5
Totals	10	10	20

$3 + 5 = 8$

$10 + 10 = 20$

$3 + 4 + 3 = 10$

Example

Some children are asked, 'Do you ever eat fruit?'
The results of a survey are shown in the two-way table.

a How many girls never eat fruit?
b How many boys eat fruit?
c How many children answered the survey?

Do you ever eat fruit?

	Yes	No
Boys	6	8
Girls	9	7

a 7 girls
b 6 boys
c $6 + 8 + 9 + 7 = 30$ children

Exercise D1.5

1 The results of an eye colour survey are shown.
Use the two-way table to find the number of

a blue-eyed boys

b brown-eyed girls

c boys

d girls

e blue-eyed children

f brown-eyed children.

	Eye colour		
	Blue	**Brown**	**Other**
Boys	35	24	14
Girls	47	36	4

2 At a local school, students have the opportunity to study French and Spanish.
The table shows the choice for all Year 10 students.
Find the number of students that study

a French and Spanish

b French but not Spanish

c neither French nor Spanish

d Spanish

e French.

	Spanish	**Not Spanish**
French	16	55
Not French	51	14

3 There are two cinemas in a Cinecomplex.
The number of people in each cinema is shown in the two-way table.
Calculate the number of

a adults in the Cinecomplex

b children in the Cinecomplex

c people in Cinema 1

d people in Cinema 2

e people in the whole Cinecomplex.

	Cinema	
	1	**2**
Adult	31	47
Child	12	8

4 In a traffic survey, the colour and speed of 100 cars are recorded.
The results are summarised in the two-way table.

a State the number of cars that are

i red and over the speed limit

ii not speeding and not red.

b Calculate the number of cars that are

i red **ii** over the speed limit.

c Calculate, as a simplified fraction, the number of cars that are

i not speeding **ii** not red.

d Maria claims that drivers of red cars tend to break speed limits.
Use the two-way table to decide whether you agree with Maria.

	Not speeding	**Over the speed limit**
Red	5	55
Not red	10	30

D1 Exam review

Key objectives

- Identify which primary and secondary data you need to collect and in what format, including grouped data, considering appropriate equal class intervals
- Collect data using various methods, including observation, controlled experiment and data logging
- Design and use data collection sheets for discrete and grouped data
- Use two way tables for discrete data

1 Rachel did a survey of her friends' favourite month of the year, shown here.

June	August	September	December	April
August	November	May	June	December
March	February	August	December	November
May	July	August	July	June

a Copy and complete the frequency table to show these results. (3)

b Amongst Rachel's friends, which is the most popular month? (1)

Month	Frequency
January	
February	
March	
April	
May	
June	
July	
August	
September	
October	
November	
December	

2 80 students each study one of three languages. The two-way table shows some information about these students.

	French	German	Spanish	Total
Female	15			39
Male		17		41
Total	31	28		80

Copy and complete the two-way table. (2)

(Edexcel Ltd., 2005)

N3 Fractions, decimals and percentages

This unit will show you how to

- Calculate fractions of quantities and one number as a fraction of another
- Simpify, add and subtract fractions
- Recognise the equivalence of fractions, decimals and percentages
- Convert terminating decimals and percentages into fractions
- Convert fractions into decimals and percentages
- Order decimals, percentages and fractions
- Understand and interpret percentages

Before you start ...

You should be able to answer these questions.

1 What fraction of this shape is shaded?

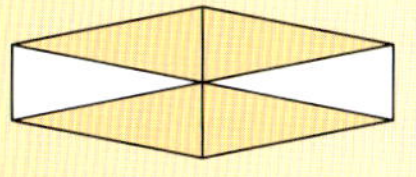

2 Write down two fractions equivalent to $\frac{1}{2}$.

3 Copy and complete

	decimal		fraction
a	0.5	=	______
b	0.25	=	______
c	______	=	$\frac{1}{5}$

4 What percentage of this shape is shaded?

5 Copy and complete

	percentage		fraction		decimal
a	50%	=	______	=	______
b	10%	=	______	=	______
c	______	=	$\frac{45}{100}$	=	______

Review

Key stage 3 (Question 1)

Key stage 3 (Question 2)

Key stage 3 (Question 3)

Key stage 3 (Question 4)

Key stage 3 (Question 5)

N3.1 Fractions

This spread will show you how to:

- Use fraction notation and vocabulary
- Calculate a given fraction of a given quantity
- Express a given number as a fraction of another

Keywords
Denominator
Equal
Fraction
Numerator

In real life you don't just use whole numbers.

When a pizza is divided into 8 equal slices

When you measure someone's height

Each slice is part of the whole pizza. One slice is $\frac{1}{8}$ of the whole pizza.

This person is one whole metre and half of another metre.

When you read the petrol gauge in a car

This car is about $\frac{3}{4}$ full with petrol.

- **You can use a fraction to describe a part of a whole. To use a fraction the whole must be divided into equal sized parts.**

Numerator: the top number shows how many parts you have.

Denominator: the bottom number shows how many equal sized parts the whole has been divided into.

5 out of the 9 equal sections are shaded.

Fraction shaded = $\frac{5}{9}$

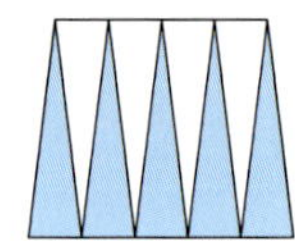

Example

In a class there are 30 students. 19 of the students are girls. What fraction of the class are girls?

19 out of the 30 students are girls.
Fraction of girls = $\frac{19}{30}$

The whole is the entire class of 30.

Example

Here is a fuel gauge from a car. How full is the petrol tank? Give your answer as a fraction.

Empty				Full
■	■	■	□	□

The fuel gauge is divided into five equal sections.
Three of the sections are coloured, and showing fuel.
The car is $\frac{3}{5}$ full.

Exercise N3.1

1 Write the fraction of each of these shapes that is shaded.

a 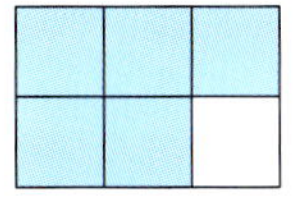
b 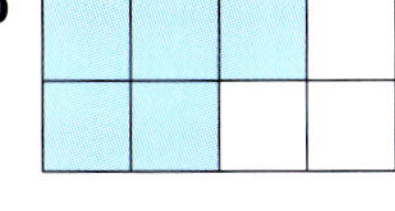
c 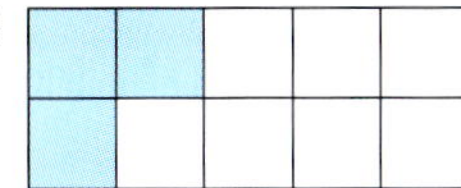
d 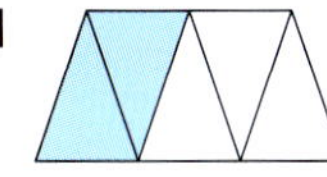
e 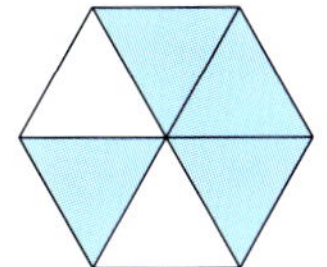
f 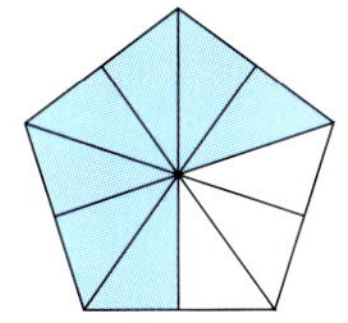

2 **a** There are 28 students in a class. 15 are boys and 13 are girls. What fraction of the class are

i boys **ii** girls?

b Tyrone has eight pairs of brown shoes and seven pairs of black shoes. What fraction of his shoes are

i brown **ii** black?

c Wesley earns £300 a week. He pays £91 of his money each week in tax. He saves £60 each week. What fraction of his weekly wage does Wesley

i pay in tax **ii** save?

d A teacher works for eight hours at school and three hours at home. What fraction of the day does the teacher work

i at school **ii** at home?

e Rory has a collection of 13 CDs, 15 DVDs and nine computer games. What fraction of his collection is

i CDs **ii** DVDs **iii** computer games?

f Irene has 38 hardback books and 77 paperback books. What fraction of her books are

i hardbacks **ii** paperbacks?

3 Write the fraction indicated by each of the pointers.

a 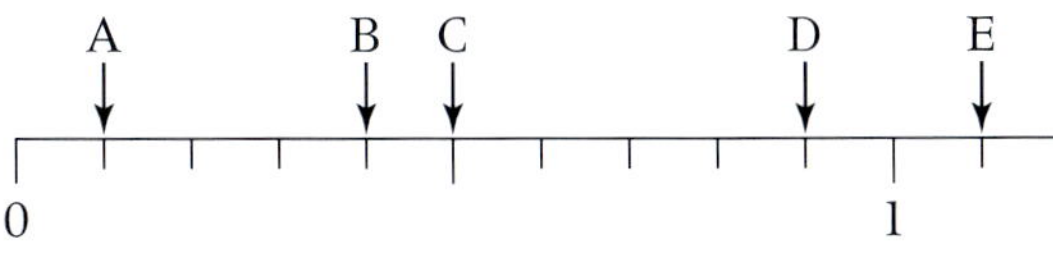

b 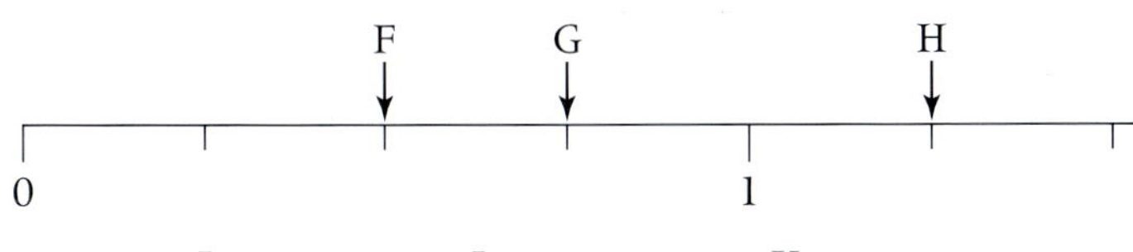

c 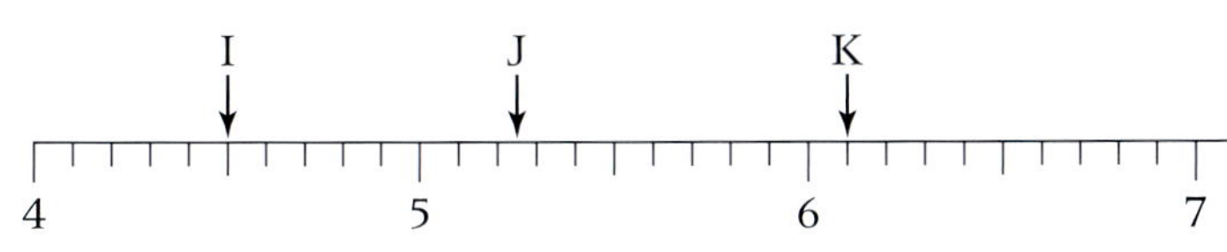

Some of these fractions are greater than one. You write $1\frac{1}{4}$.

N3.2 Equivalent fractions

This spread will show you how to:

- Simplify fractions by cancelling common factors
- Recognise and find equivalent fractions
- Add and subtract simple fractions

Keywords

Cancel
Common factor
Equivalent
Fraction
Simplest form

The same fraction of each rectangle is shaded.

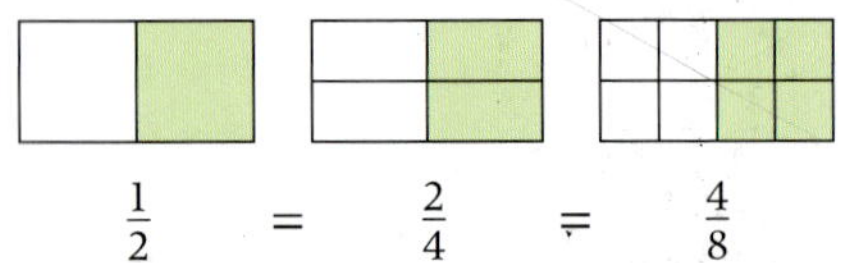

$\frac{1}{2} = \frac{2}{4} = \frac{4}{8}$

These fractions are called **equivalent** fractions.

- **You can find equivalent fractions by multiplying the numerator and denominator by the same number.**

This topic is extended to adding and subtracting fractions on page 364.

Example

Find two equivalent fractions for $\frac{4}{7}$.

$\frac{4}{7} = \frac{12}{21}$ (× 3 numerator and denominator) $\qquad \frac{4}{7} = \frac{20}{35}$ (× 5 numerator and denominator)

So $\frac{12}{21} = \frac{4}{7}$ and $\frac{20}{35} = \frac{4}{7}$

- **You can simplify a fraction by dividing the numerator and denominator by the same number. This process is called cancelling.**

Example

Write each of these fractions in its **simplest form**.

a $\frac{15}{20}$ **b** $\frac{24}{30}$

a $\frac{15}{20} = \frac{3}{4}$ (÷ 5 numerator and denominator)

b

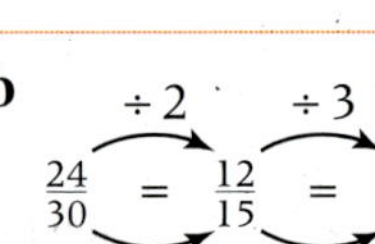

$\frac{24}{30} = \frac{12}{15} = \frac{4}{5}$ (÷ 2, then ÷ 3)

- **You can add or subtract fractions with the same denominator.**

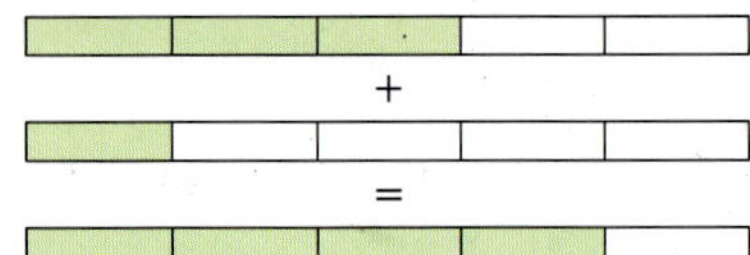

$\frac{3}{5} + \frac{1}{5} = \frac{4}{5}$

When the denominators are the same, you can add or subtract fractions by simply adding or subtracting the numerators.

Exercise N3.2

1 Write the fraction of each of these shapes that is shaded. Give your answer in its simplest form.

a 

b 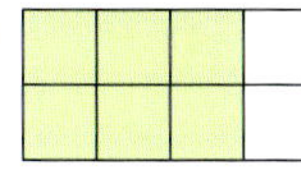

c 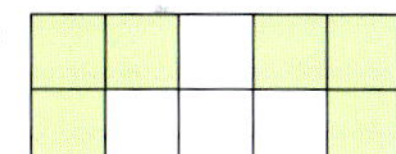

d 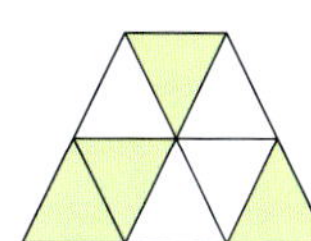

e 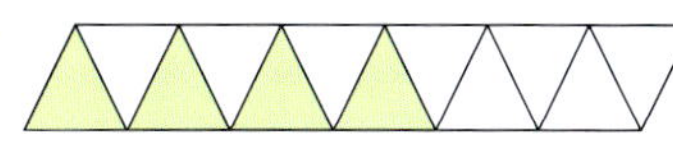

2 Copy and complete each of these equivalent fraction families.

a $\frac{1}{2} = \frac{2}{4} = \frac{?}{6} = \frac{?}{10} = \frac{?}{16}$ **b** $\frac{3}{10} = \frac{6}{20} = \frac{?}{30} = \frac{15}{?} = \frac{30}{?}$

c $\frac{4}{5} = \frac{?}{10} = \frac{12}{?} = \frac{?}{25} = \frac{?}{100}$ **d** $\frac{5}{8} = \frac{10}{?} = \frac{?}{24} = \frac{?}{80} = \frac{75}{?}$

e $\frac{2}{9} = \frac{?}{18} = \frac{6}{?} = \frac{10}{?} = \frac{?}{81}$ **f** $\frac{7}{4} = \frac{?}{8} = \frac{?}{12} = \frac{?}{40} = \frac{77}{?}$

3 Find the missing number in each of these pairs of equivalent fractions.

a $\frac{2}{3} = \frac{?}{9}$ **b** $\frac{4}{5} = \frac{12}{?}$ **c** $\frac{3}{4} = \frac{?}{20}$ **d** $\frac{1}{8} = \frac{?}{40}$

e $\frac{5}{7} = \frac{30}{?}$ **f** $\frac{4}{9} = \frac{?}{63}$ **g** $\frac{7}{8} = \frac{?}{48}$ **h** $\frac{7}{10} = \frac{?}{100}$

i $\frac{12}{15} = \frac{?}{5}$ **j** $\frac{18}{24} = \frac{3}{?}$ **k** $\frac{30}{35} = \frac{?}{7}$ **l** $\frac{14}{35} = \frac{2}{?}$

m $\frac{?}{3} = \frac{16}{24}$ **n** $\frac{4}{?} = \frac{20}{55}$ **o** $\frac{?}{10} = \frac{56}{80}$ **p** $\frac{9}{13} = \frac{?}{65}$

4 Cancel down each of these fractions into their simplest form.

a $\frac{4}{8}$ **b** $\frac{3}{9}$ **c** $\frac{4}{6}$ **d** $\frac{8}{10}$

e $\frac{9}{12}$ **f** $\frac{10}{15}$ **g** $\frac{3}{15}$ **h** $\frac{12}{16}$

i $\frac{14}{16}$ **j** $\frac{13}{16}$ **k** $\frac{12}{18}$ **l** $\frac{8}{20}$

m $\frac{16}{24}$ **n** $\frac{21}{28}$ **o** $\frac{20}{25}$ **p** $\frac{18}{30}$

q $\frac{12}{36}$ **r** $\frac{24}{40}$ **s** $\frac{14}{42}$ **t** $\frac{27}{63}$

5 Find an equivalent fraction for each fraction. Both of your fractions should have the same denominator.

a $\frac{1}{2}$ and $\frac{1}{3}$ **b** $\frac{1}{5}$ and $\frac{1}{3}$ **c** $\frac{1}{2}$ and $\frac{1}{5}$ **d** $\frac{2}{3}$ and $\frac{1}{4}$

e $\frac{3}{10}$ and $\frac{1}{3}$ **f** $\frac{4}{5}$ and $\frac{1}{4}$ **g** $\frac{1}{3}$ and $\frac{3}{7}$ **h** $\frac{5}{6}$ and $\frac{3}{4}$

6 Calculate each of these. Give your answer in its simplest form.

a $\frac{1}{3} + \frac{1}{3}$ **b** $\frac{2}{5} + \frac{1}{5}$ **c** $\frac{3}{10} + \frac{7}{10}$ **d** $\frac{4}{9} + \frac{2}{9}$

e $\frac{7}{8} + \frac{5}{8}$ **f** $\frac{13}{7} - \frac{6}{7}$ **g** $\frac{5}{16} - \frac{1}{16}$ **h** $\frac{13}{18} - \frac{5}{18}$

i $\frac{12}{15} + \frac{8}{15}$ **j** $\frac{19}{16} - \frac{7}{16}$ **k** $\frac{12}{35} + \frac{8}{35}$ **l** $\frac{23}{30} - \frac{7}{30}$

N3.3 Fractions and decimals

This spread will show you how to:

- Convert terminating decimals into fractions
- Convert fractions into decimals
- Order decimals on a number line

Keywords

Decimal
Equivalent
Fraction
Terminating decimal

A **decimal** is another way of writing a **fraction**.

It is often called a decimal fraction.

You should learn some common fractions and their decimal **equivalents**.

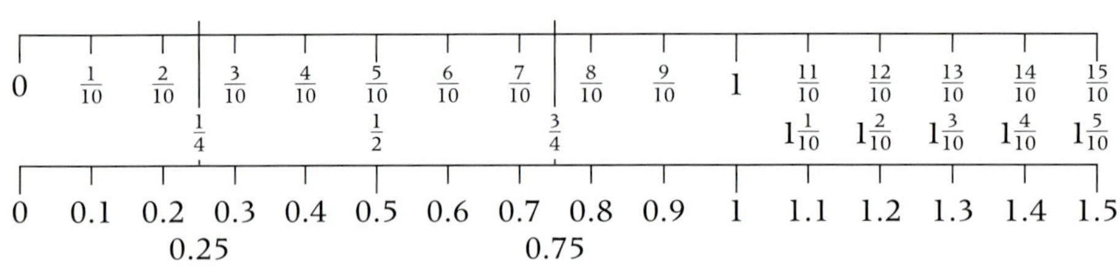

- **You can write a terminating decimal as a fraction by using place value.**

Decimal					Fraction
Units 1	•	tenths $\frac{1}{10}$	hundredths $\frac{1}{100}$	thousandths $\frac{1}{1000}$	
0	•	3			$\frac{3}{10}$
0	•	4	8		$\frac{48}{100}$
3	•	1	5	8	$3\frac{158}{1000}$

$0.48 = \frac{48}{100}$

$\frac{48}{100} \xrightarrow{\div 4} \frac{12}{25}$ ($\div 4$ top and bottom: $\frac{48}{100} = \frac{12}{25}$)

- You can convert a fraction into a decimal.

1 Using equivalent fractions.

$\frac{3}{20} = \frac{15}{100}$ (× 5 numerator and denominator)

Convert the fraction to an equivalent fraction with a denominator of 10, 100, 1000 etc.

$\frac{3}{20} = \frac{15}{100} = 0.15$

Change the equivalent fraction to a decimal.

2 Using division.

$\frac{3}{20} = 3 \div 20 = 0.15$

Divide the numerator by the denominator.

Every number can be represented as a position on a number line.

Example

What number is the arrow pointing to?

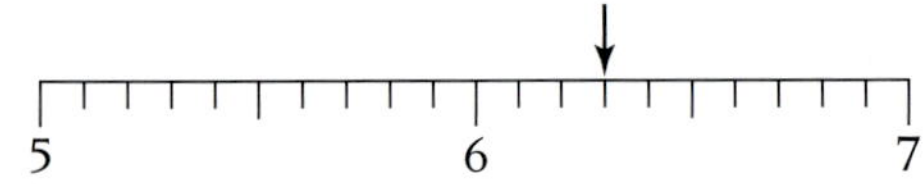

The arrow is pointing between 6 and 7.
There are 10 spaces between 6 and 7. So 10 spaces represent 1 unit.
Each space represents $1 \div 10 = \frac{1}{10} = 0.1$ unit.
The arrow is pointing to the number 6.3.

Exercise N3.3

1 Write these decimals as fractions.

a 0.3 **b** 0.9 **c** 0.23 **d** 0.39

e 0.88 **f** 0.274 **g** 0.814 **h** 0.037

2 Write these decimals as fractions in their simplest form.

a 0.4 **b** 0.8 **c** 0.75 **d** 0.36

e 0.85 **f** 0.08 **g** 0.005 **h** 2.65

3 Change these fractions to decimals without using a calculator.

a $\frac{7}{10}$ **b** $\frac{1}{2}$ **c** $\frac{47}{50}$ **d** $\frac{13}{25}$ **e** $\frac{22}{25}$

f $\frac{11}{10}$ **g** $\frac{31}{25}$ **h** $\frac{145}{500}$ **i** $\frac{2}{8}$ **j** $\frac{32}{40}$

4 Change these fractions into decimals using an appropriate method.
Give your answers to two decimal places where necessary.

a $\frac{19}{50}$ **b** $\frac{1}{3}$ **c** $\frac{7}{20}$ **d** $\frac{3}{50}$

e $\frac{11}{16}$ **f** $\frac{5}{2}$ **g** $\frac{51}{60}$ **h** $\frac{8}{13}$

5 Write the number each of the arrows is pointing to.

a

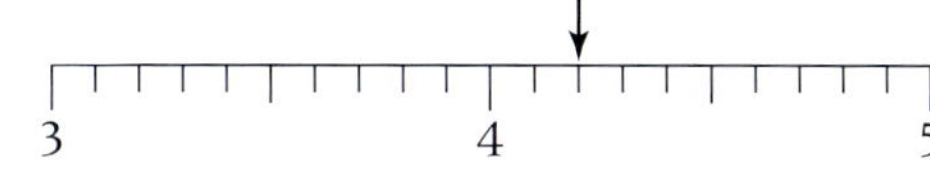

b

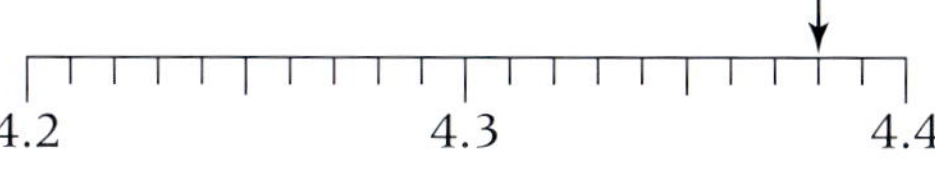

c

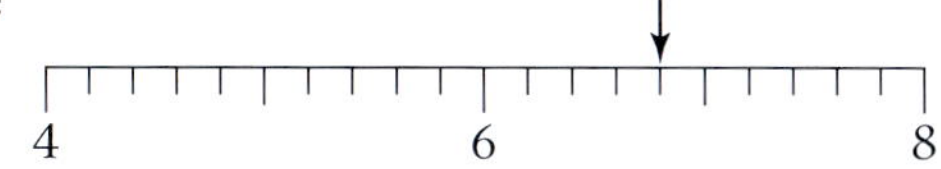

6 **a** Copy this decimal number line.

b Mark on these fractions.

i $\frac{8}{10}$ **ii** $\frac{28}{20}$ **iii** $\frac{7}{8}$

7 Put these lists of numbers in order, starting with the smallest.

a 2.13 2.09 2.2 2.12 2.07

b 0.345 0.35 0.325 0.3 0.309

c 1.32 1.4 1.35 1.387 1.058

d 5.306 5.288 5.308 5.29 5.3

N3.4 Fractions and percentages

This spread will show you how to:

- Understand and interpret percentages
- Convert fractions into percentages
- Convert percentages into fractions

Keywords
Equivalent
Fraction
Percentage

- A **percentage** is a fraction of something. It is written as the number of parts per hundred.

40% is the fraction $\frac{40}{100}$.

40% of this line is shaded. The line is divided into 100 parts and 40 of them are shaded.

- To change a percentage into a **fraction** you write it as a fraction out of a 100 and then simplify.

Some useful equivalents to remember.

$10\% = \frac{10}{100} = \frac{1}{10}$

$20\% = \frac{20}{100} = \frac{1}{5}$

$25\% = \frac{25}{100} = \frac{1}{4}$

$50\% = \frac{50}{100} = \frac{1}{2}$

$75\% = \frac{75}{100} = \frac{3}{4}$

Example

Write these percentages as fractions in their simplest form.

a 60% **b** 45%

a $60\% = \frac{60}{100}$

$\frac{60}{100} \overset{\div 10}{=} \frac{6}{10} \overset{\div 2}{=} \frac{3}{5}$ $60\% = \frac{3}{5}$

b $45\% = \frac{45}{100}$

$\frac{45}{100} \overset{\div 5}{=} \frac{9}{20}$ $45\% = \frac{9}{20}$

To simplify a fraction, divide the numerator and denominator by the same number.

- To change a fraction into a percentage you write it as an equivalent fraction out of 100 and then change it into a percentage.

Example

Change these fractions into percentages.

a $\frac{7}{10}$ **b** $\frac{7}{25}$

a $\frac{7}{10} \overset{\times 10}{=} \frac{70}{100}$ $\frac{70}{100} = 70\%$

b $\frac{7}{25} \overset{\times 4}{=} \frac{28}{100}$ $\frac{28}{100} = 28\%$

You can express something as the percentage of a whole by first finding the fraction of the whole.

Example

What percentage of this shape is shaded?

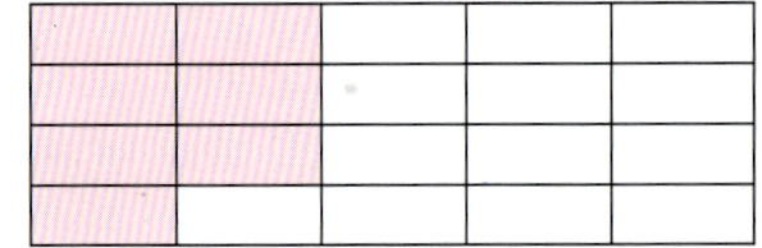

There are 20 equal parts.
7 of the parts are shaded.
The fraction shaded is $\frac{7}{20}$.
Converting to a percentage:
% shaded $= \frac{7}{20} = \frac{35}{100} = 35\%$

Exercise N3.4

1 This rectangle has been divided into 100 parts.

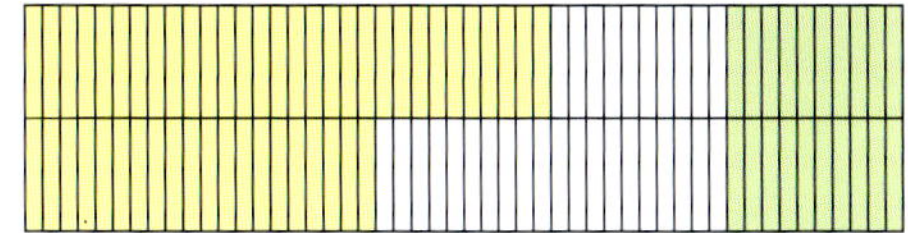

a What percentage of the rectangle is shaded yellow?

b What percentage of the rectangle is shaded green?

c What percentage of the rectangle is not shaded green?

2 Write these percentages as fractions out of 100.

a 35% **b** 10% **c** 67% **d** 43%

e 95% **f** 56% **g** 140% **h** 135%

3 Write the value of each of the letters on these number lines.

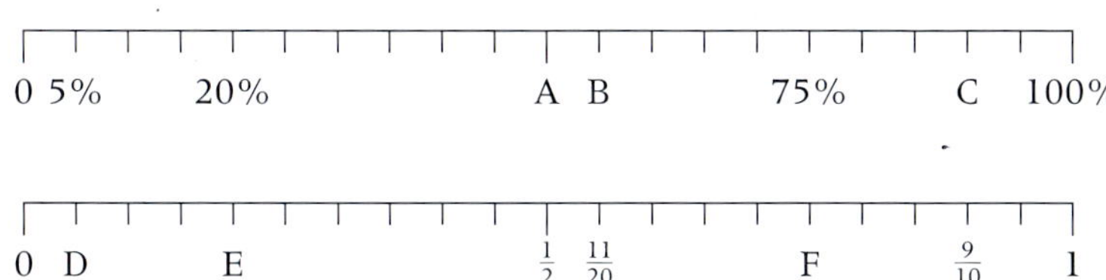

4 Write these percentages as fractions in their simplest form.

a 50% **b** 80% **c** 15% **d** 85%

e 28% **f** 6% **g** 150% **h** 115%

5 Write each of these fractions as a percentage.

a $\frac{27}{100}$ **b** $\frac{1}{2}$ **c** $\frac{7}{10}$ **d** $\frac{11}{25}$ **e** $\frac{3}{4}$

f $\frac{34}{200}$ **g** $\frac{6}{5}$ **h** $\frac{21}{20}$ **i** $\frac{26}{40}$ **j** $\frac{33}{75}$

6 What percentage of each of these shapes is shaded?

a

b

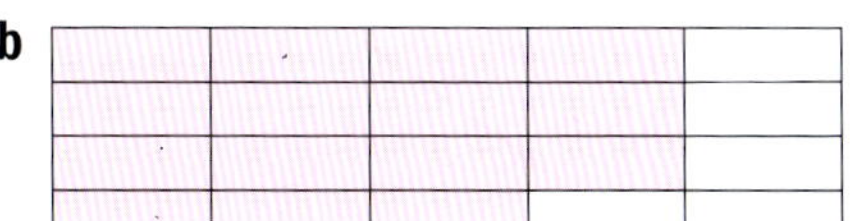

c

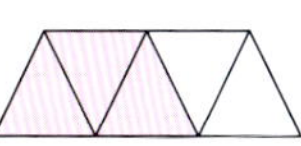

d

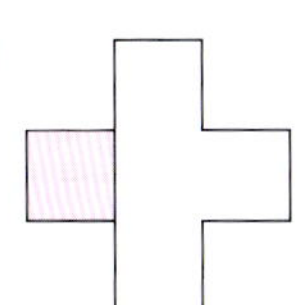

7 a A shirt is 65% polyester and 35% cotton.
Write 35% as a fraction.
Give your answer in its simplest form.

b How would you write $17\frac{1}{2}$% as a fraction in its simplest form?
Show all your working out.

N3.5 Fractions, decimals and percentages

This spread will show you how to:

- Convert between fractions, decimals and percentages
- Order fractions, decimals and percentages

Keywords

Decimal
Equivalent
Fraction
Order
Percentage

You can change between fractions, decimals and percentages.

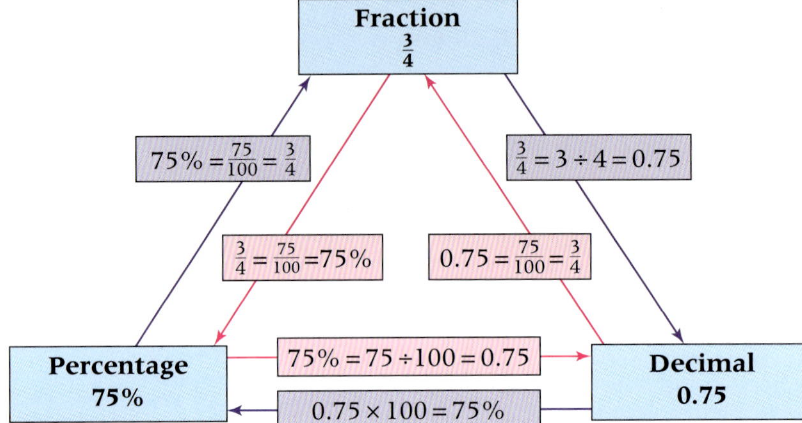

You can compare the size of fractions by placing them on a number line.

Example

Which is bigger, $\frac{3}{5}$ or $\frac{5}{8}$?

Divide a line into five equal pieces and shade three of them. This is $\frac{3}{5}$ of the line.

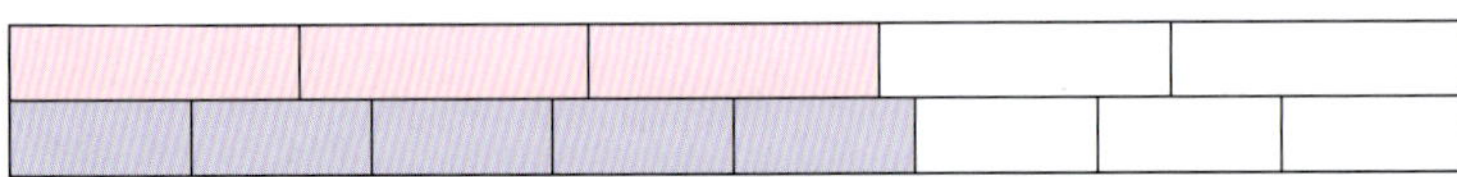

Divide a second line of the same length into eight equal pieces and shade five of them. This is $\frac{5}{8}$ of the line.

$\frac{5}{8}$ is greater than $\frac{3}{5}$.

This topic is extended to comparing fractions on page 364.

- You can **order** fractions, decimals and percentages by converting them into decimals.

Example

Write these numbers in order of size. Start with the smallest number.

0.85 72% $\frac{7}{8}$ $\frac{4}{5}$

$72\% = \frac{72}{100} = 72 \div 100 = 0.72$

$\frac{7}{8} = 7 \div 8 = 0.875$

$\frac{4}{5} = 4 \div 5 = 0.8$

Placing the decimals in order	0.72	0.8	0.85	0.875
So the order is	72%	$\frac{4}{5}$	0.85	$\frac{7}{8}$

Exercise N3.5

1 Write these percentages as decimals.

a 67% **b** 78% **c** 99% **d** 70% **e** 39%

f 88% **g** 150% **h** 125% **i** 99.9% **j** 110%

2 Write these decimals as percentages.

a 0.32 **b** 0.22 **c** 0.85 **d** 0.03 **e** 0.54

f 0.63 **g** 0.38 **h** 0.375 **i** 0.333 **j** 1.25

3 Write these decimals as fractions in their simplest form.

a 0.8 **b** 0.28 **c** 0.325 **d** 0.05 **e** 0.12

4 Change these fractions to decimals. Give your answers to two decimal places as appropriate.

a $\frac{3}{10}$ **b** $\frac{7}{25}$ **c** $\frac{7}{12}$ **d** $\frac{9}{15}$ **e** $\frac{15}{7}$

5 Write these percentages as fractions in their simplest form.

a 25% **b** 40% **c** 65% **d** 15% **e** 145%

6 A shirt is 65% polyester and 35% cotton.

Write 65% as a decimal.

7 Write each of these fractions as percentages.
Give your answers to one decimal place as appropriate.

a $\frac{48}{100}$ **b** $\frac{6}{25}$ **c** $\frac{17}{10}$ **d** $\frac{8}{15}$ **e** $\frac{11}{16}$

Try converting the fraction into a decimal first!

8 Copy and complete this table. Use the most effective method to convert between fractions, decimals and percentages.

Fraction (in its simplest form)		$\frac{5}{9}$		$\frac{13}{5}$		
Decimal (to 3 dp)	0.76				0.125	
Percentage (to 1 dp)			85%			17.5%

dp means decimal places.

9 For each pair of fractions, write which is the larger fraction.

a $\frac{3}{4}$ and $\frac{2}{3}$ (draw a number line 12 cm long)

b $\frac{4}{5}$ and $\frac{3}{4}$ (draw a number line 20 cm long)

c $\frac{4}{5}$ and $\frac{5}{6}$ (draw a number line 30 cm long)

d $\frac{2}{3}$ and $\frac{3}{5}$ (draw a number line 15 cm long)

e $\frac{4}{7}$ and $\frac{2}{5}$ (draw a number line 35 cm long)

Draw a number line and mark each fraction on the number line.

10 Put these lists of numbers in order, starting with the smallest.

a 0.4 43% $\frac{3}{8}$ 0.35

b $\frac{3}{5}$ 0.56 61% $\frac{4}{7}$

c $\frac{3}{4}$ $\frac{2}{3}$ 70% 0.715

N3 Exam review

Key objectives

- Understand equivalent fractions and simplifying a fraction by cancelling all common factors
- Use decimal notation and recognise that each terminating decimal is a fraction
- Order decimals
- Understand that 'percentage' means 'number of parts per 100'. Interpret percentage as the operator 'so many hundredths of'
- Perform short division to convert a simple fraction to a decimal
- Convert simple fractions of a whole to percentages of a whole and vice versa

1 **a** Write down the fraction of the shape that is shaded. (2)
Express your answer in its simplest form.

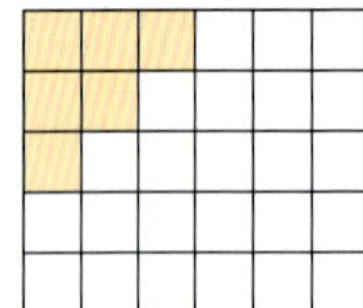

b Express your answer to part **a** as

i a decimal

ii a percentage. (2)

2 Write these numbers in order of size.
Start with the smallest number.

a 75, 56, 37, 9, 59

b 0.56, 0.067, 0.6, 0.65, 0.605

c 5, −6, −10, 2, −4

d $\frac{1}{2}, \frac{2}{3}, \frac{2}{5}, \frac{3}{4}$ (5)

(Edexcel Ltd., 2003)

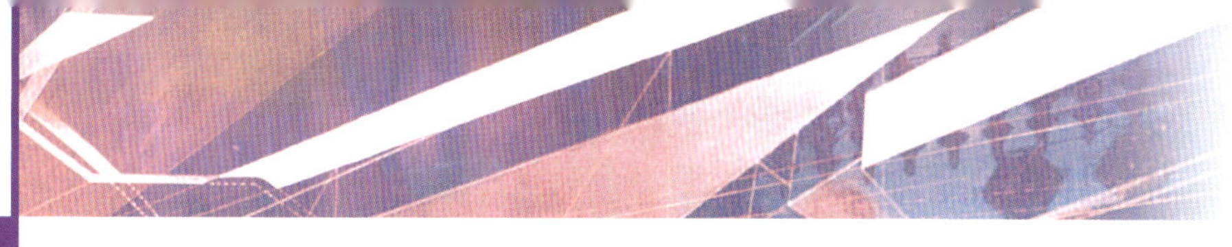

S2 Angles and coordinates

This unit will show you how to

- Identify acute, obtuse, reflex and right angles
- Understand and use properties of parallel and perpendicular lines
- Use angle measure
- Measure and draw angles to the nearest degree
- Estimate the size of an angle in degrees
- Recall and use properties of angles at a point and on a straight line and of opposite angles at a vertex
- Identify right-angled, equilateral, isosceles and scalene triangles
- Use angle properties of equilateral, isosceles and right-angled triangles
- Plot coordinates
- Locate points with given coordinates

Before you start ...

You should be able to answer these questions.

1 Give the coordinates of

a point A

b point B.

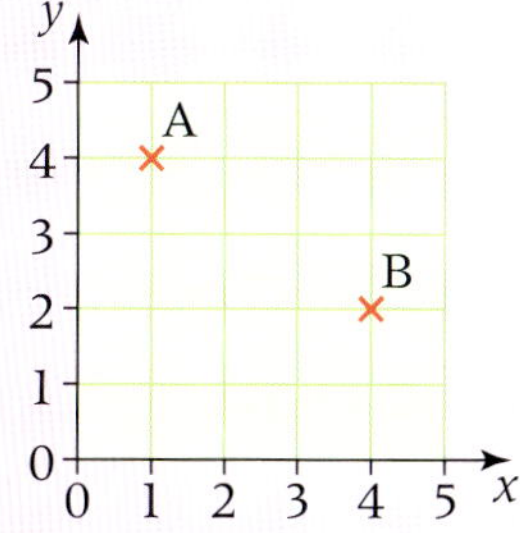

2 Give the reading on the scale.

3 Evaluate each of these.

a 180 – 120 **b** 180 – 45

c 180 – 112 **d** 360 – 270

e 360 – 218

Review

Key stage 3

Unit N1

Unit N2

S2.1 Angles and lines

This spread will show you how to:

- Identify acute, obtuse, reflex and right angles
- Recall and use properties of parallel and perpendicular lines

Keywords
Acute
Angle
Degrees (°)
Intersect
Obtuse
Parallel
Perpendicular
Reflex
Right angle

- An **angle** is a measure of turn. You measure the turn in **degrees**.

Amount of turn	$\frac{1}{4}$ turn	$\frac{1}{2}$ turn	$\frac{3}{4}$ turn	full turn
Angle in degrees	90°	180°	270°	360°

° means degrees.

- **You can describe an angle by its size.**

an **acute** angle is less than 90°

a **right angle** is exactly 90°

an **obtuse** angle is between 90° and 180°

a **reflex** angle is more than 180°

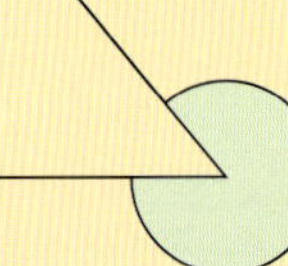

Example

What type of angle is shown by the letter

a x **b** y?

a x is acute. **b** y is reflex.

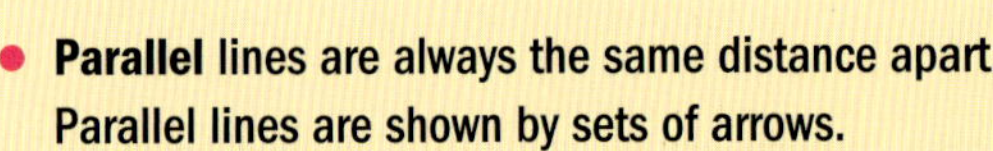

- **Parallel** lines are always the same distance apart.
  Parallel lines are shown by sets of arrows.

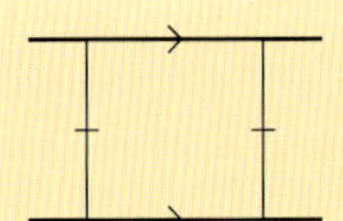

Parallel lines never **intersect** (cross) each other.

- **Perpendicular** lines meet at a right angle.

Example

A line AB is drawn on a grid.

a Draw another line that is parallel to AB.
Label the line CD and mark with arrows (>).

b Draw another line that is perpendicular to AB.
Label the line EF and mark with a square (∟).

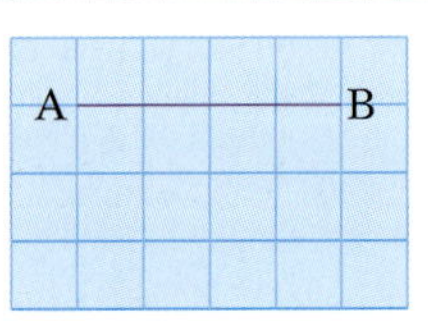

CD is parallel to AB.
EF is perpendicular to AB.

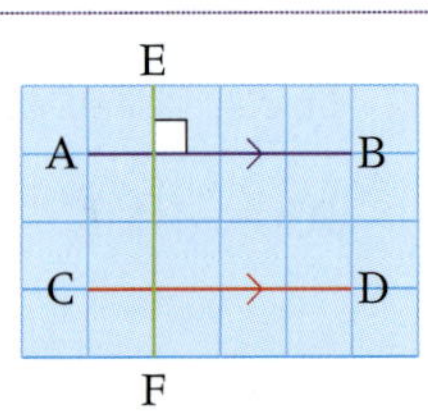

Exercise S2.1

1 How many right angles make up the angle shown in each diagram?

a

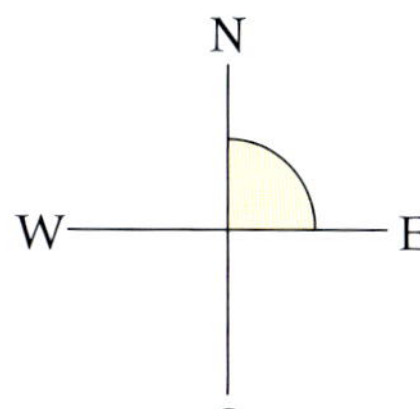

b

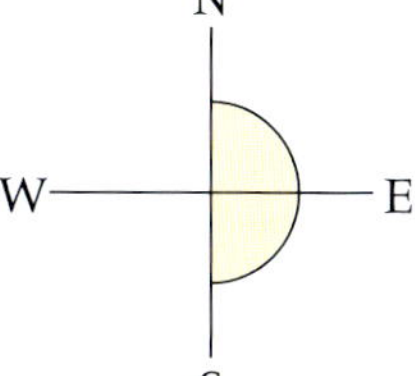

c

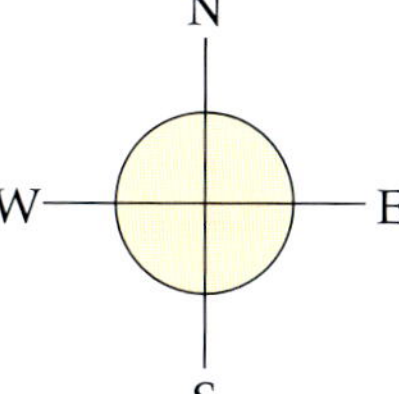

d

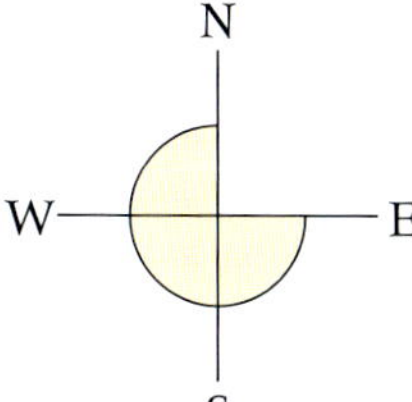

e

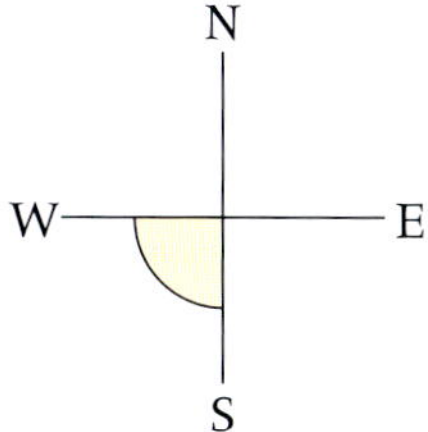

f

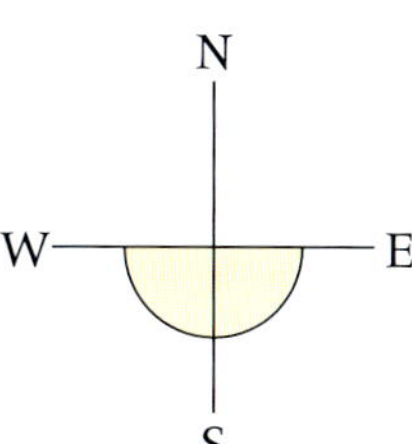

DID YOU KNOW?

Perpendicular and parallel lines are all around you!

2 Choose one of these words to describe each angle.

acute	right angle	obtuse	reflex

a

b

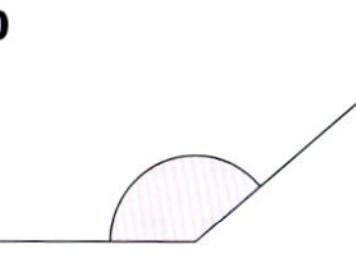

c

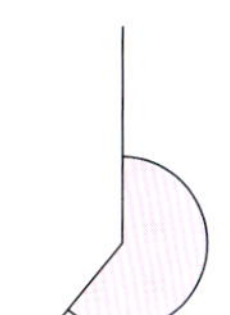

d

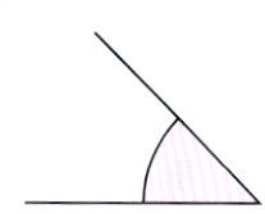

e

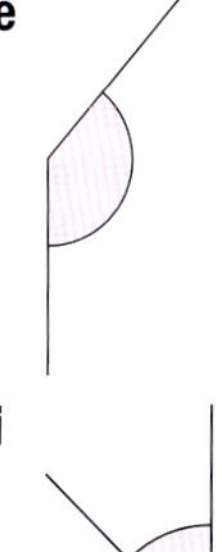

f

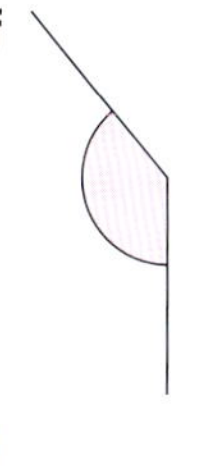

g

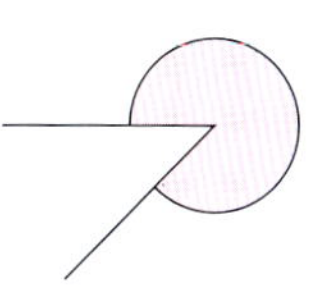

h

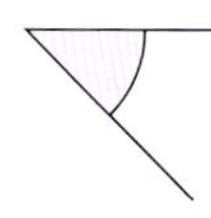

i

j

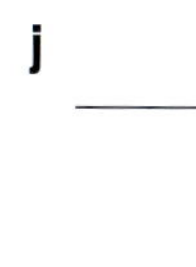

k

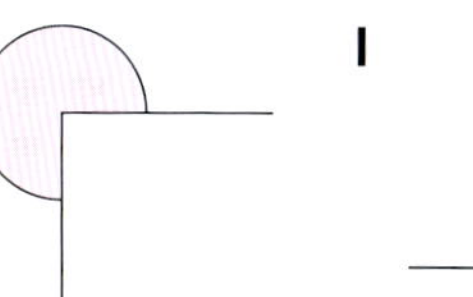

l

3 Choose one of these words to describe each angle.

acute	right angle	obtuse	reflex

a 90° **b** 40° **c** 140° **d** 200° **e** 270° **f** 36°

g 137° **h** 248° **i** 302° **j** 33° **k** 96° **l** 239°

4 **a** Draw two lines that are parallel. Label them with >.

b Draw two lines that are perpendicular. Label them with ∟.

S2.2 Measuring angles

This spread will show you how to:

- Understand angle measure
- Measure and draw angles to the nearest degree
- Estimate the size of an angle in degrees

Keywords
Angle
Degrees (°)
Estimate
Measure
Protractor

You can **measure** and draw an **angle** in **degrees** with a **protractor**.
A protractor measures angles up to 180°.
There are 180° in a half turn.

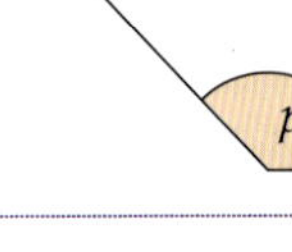

° means degrees.

- **There are 180° on a straight line.**

Example

Measure the size of angle p.

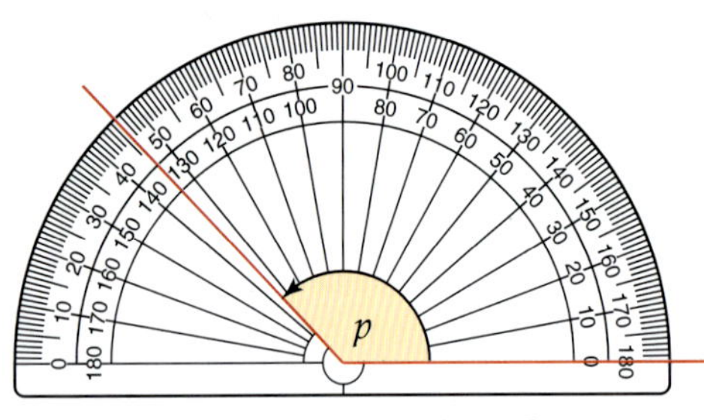

You have to decide which scale to use, either the inner scale or the outer scale.

1. **Estimate** the size of angle p. (Guess 120°, as greater than 90°.)
2. Place the protractor over the angle.
3. The angle point should be at the cross in the protractor.
4. One arm of the angle should be along the zero line.
5. Start counting from this zero line.

   $p = 134°$

For this angle use the right-hand zero scale.

- **You can measure a reflex angle by measuring the associated acute or obtuse angle.**

A full turn is 360°.

Example

Measure the reflex angle $A\hat{B}C$.

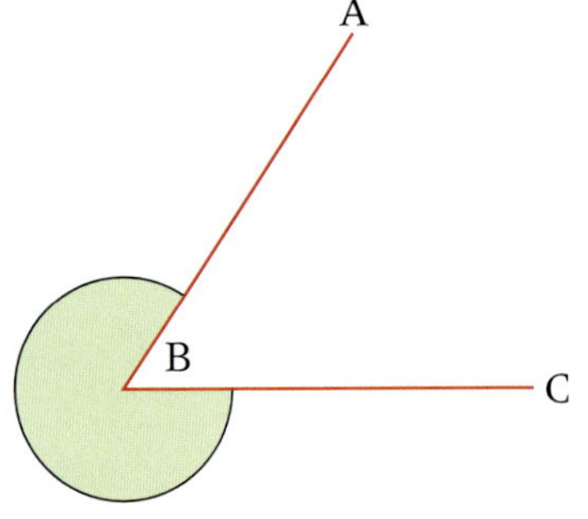

The acute angle $A\hat{B}C = 56°$.
So the reflex angle $A\hat{B}C$ is
$360° - 56° = 304°$

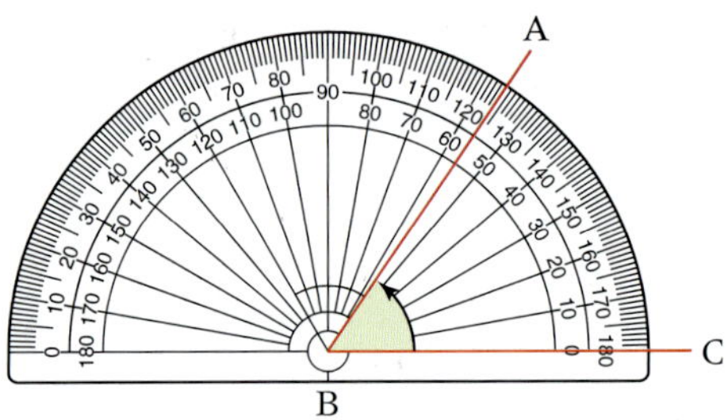

Angles at a point add to 360°.

Exercise S2.2

In questions **1–9**, for each angle state

a the type of angle – acute, right angle, obtuse or reflex
b your estimate in degrees
c the measurement in degrees.

Set out your answers like this

Question	Type of angle	Estimate	Measurement
1	acute	40°	30°
2			

1

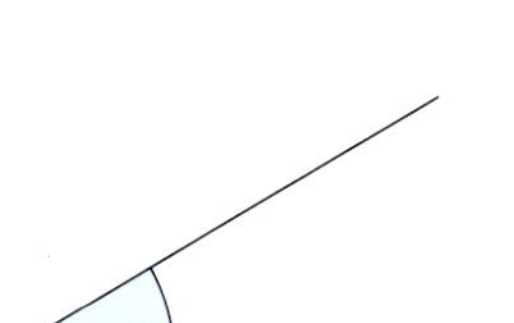

2

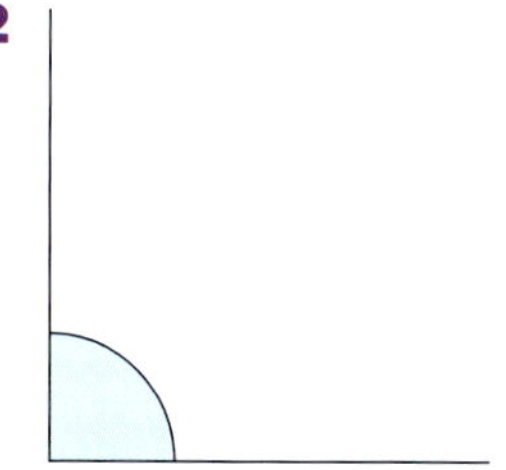

3

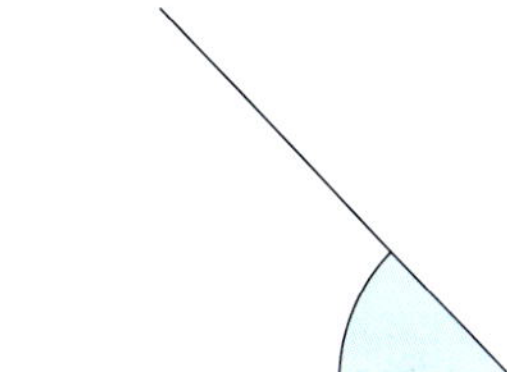

4

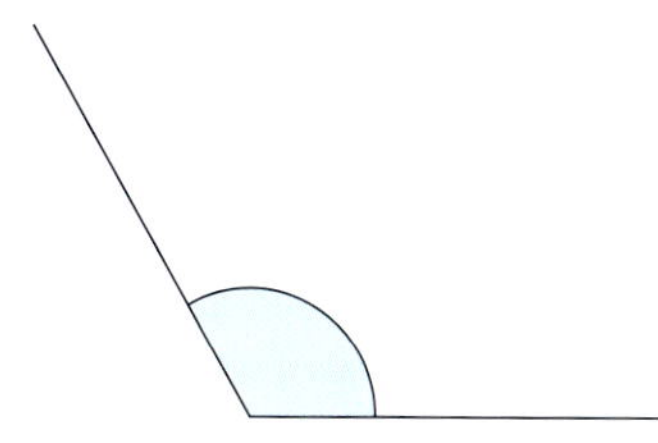

5

6

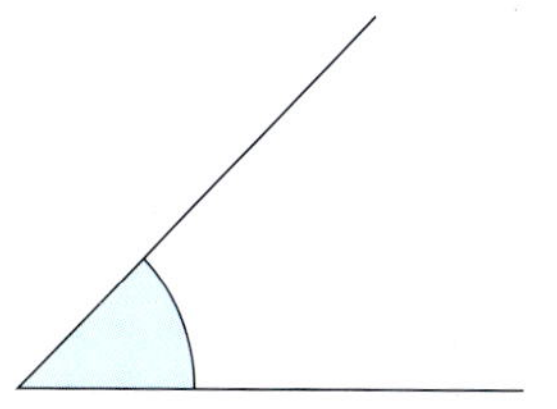

7

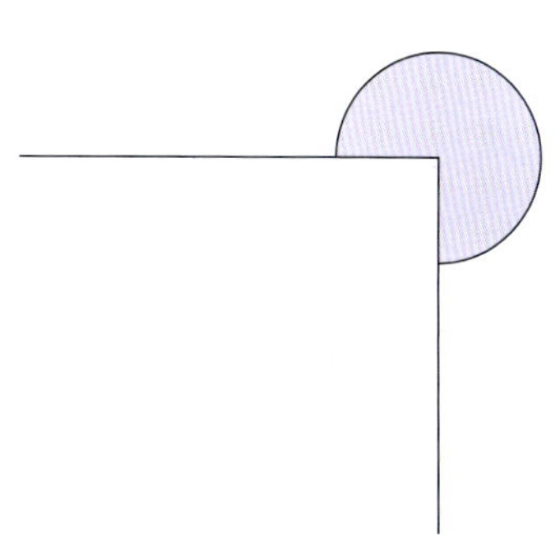

8

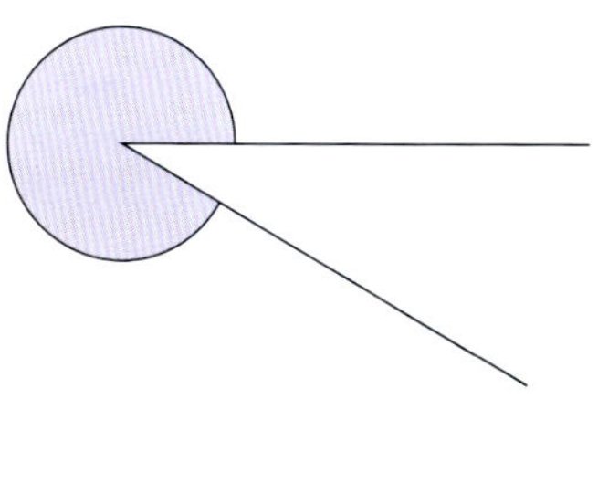

9

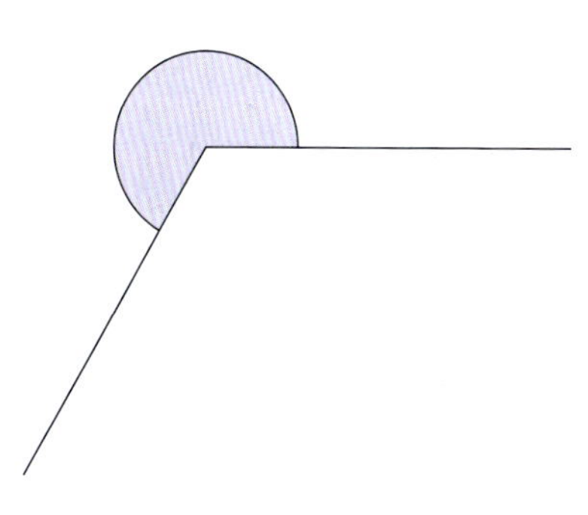

10 Draw and label these angles using a protractor. State whether each angle is acute, obtuse, reflex or a right angle.

a 40° **b** 140° **c** 90° **d** 36°
e 144° **f** 56° **g** 124° **h** 38°
i 142° **j** 85° **k** 300° **l** 200°
m 320° **n** 245° **o** 265°

S2.3 Angle properties

This spread will show you how to:

- Recall and use properties of angles at a point, angles on a straight line, perpendicular lines and opposite angles at a vertex

Keywords
Angle
Degrees (°)
Point
Straight line
Vertically opposite

- **These are 360° in a full turn at a point.**

- **There are 180° on a straight line. This is a half turn at a point.**

- **There are 90° in a quarter turn at a point.**

The two lines are perpendicular.

Example

Calculate the values of p, q and r. Give a reason for each of your answers.

a

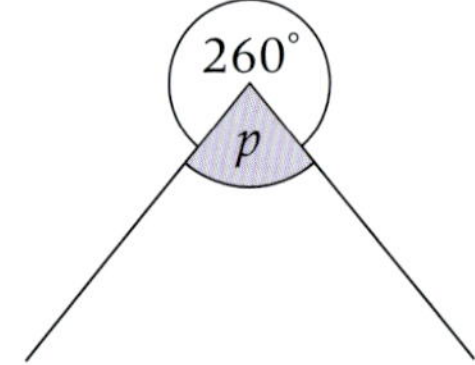

b

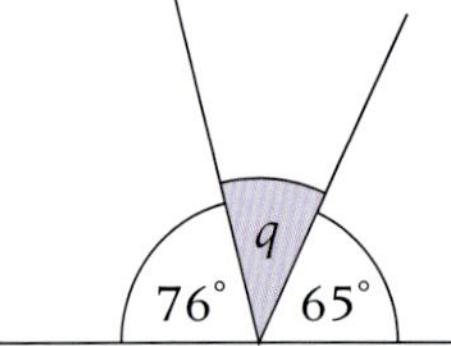

c

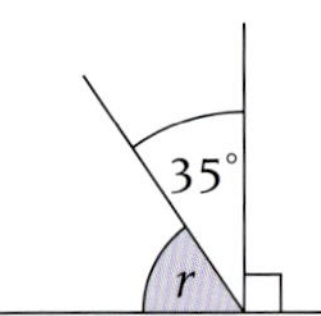

a $360° - 260° = 100°$

$p = 100°$

(angles at a point add to 360°)

b $76° + 65° = 141°$

$180° - 141° = 39°$

$q = 39°$

(angles on a straight line add to 180°)

c $35° + 90° = 125°$

$180° - 125° = 55°$

$r = 55°$

(angles on a straight line add to 180°)

When two lines intersect they make four angles.
The two acute angles are equal.
The two obtuse angles are equal.

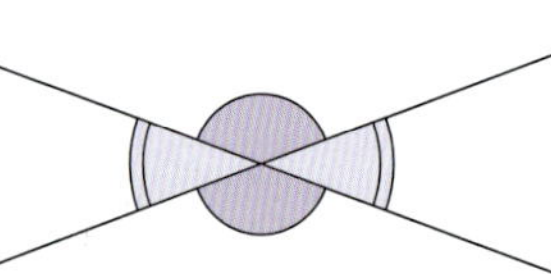

- **Vertically opposite** angles are equal.

This topic is extended to alternate and corresponding angles on page 384.

Example

Calculate the values of x and y.
Give a reason for each of your answers.

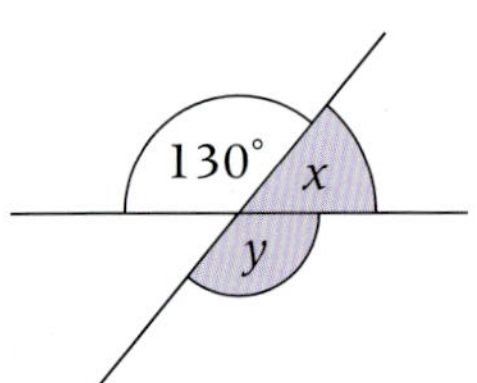

$x = 180° - 130° = 50°$ (angles on a straight line add to 180°)

$y = 130°$ (vertically opposite angles are equal)

Exercise S2.3

1 Give the values in degrees of the coloured angles.

a

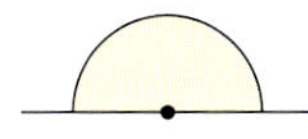

b

2 This diagram is wrong.
Explain why.

3 Calculate the size of the angles marked by letters in each diagram.
Give a reason for each answer. The diagrams are not accurately drawn.

a

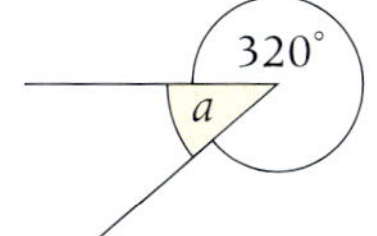

b

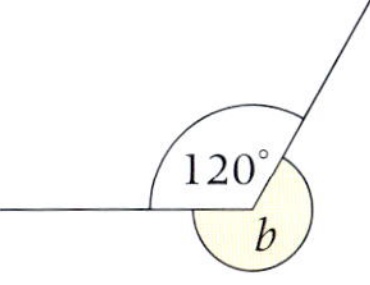

c

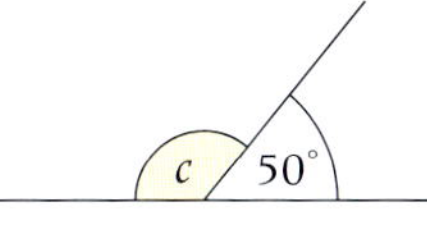

d

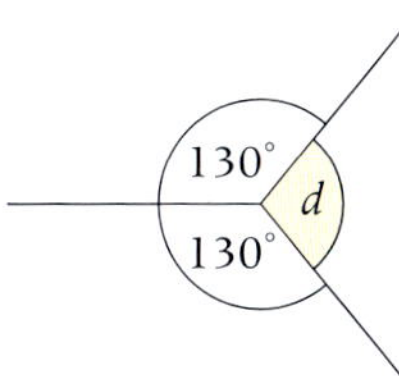

e

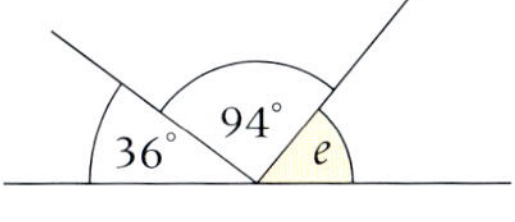

f

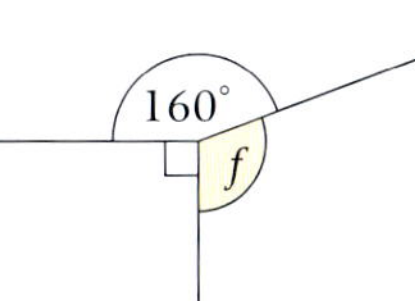

g

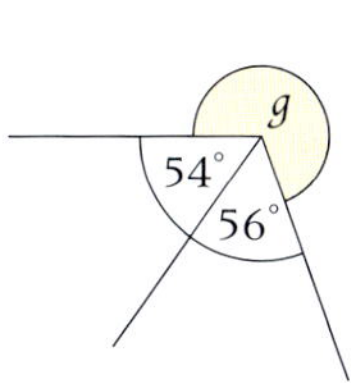

h

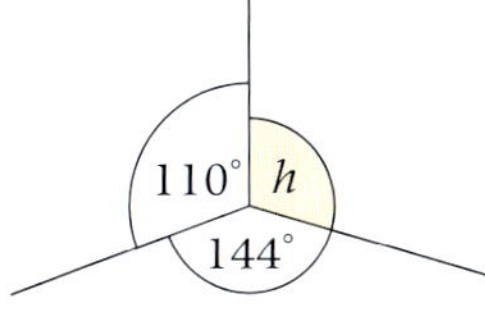

i

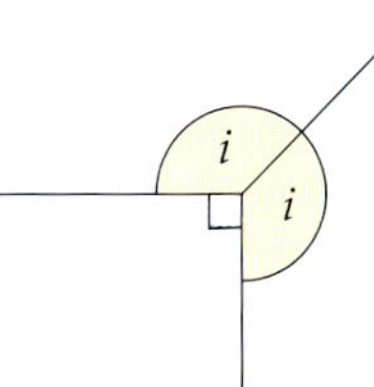

j

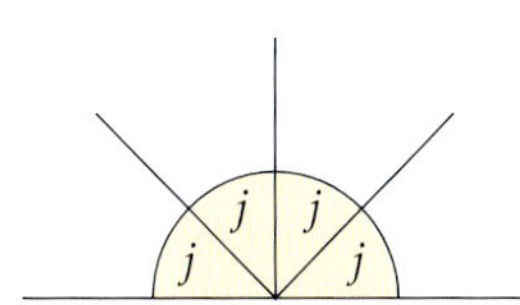

k

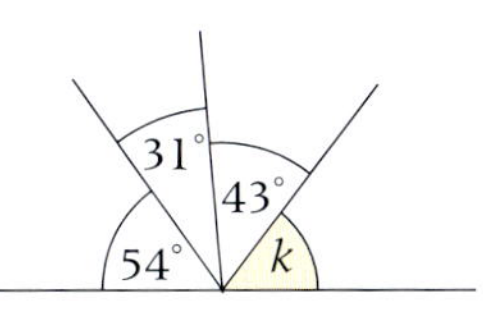

l

m

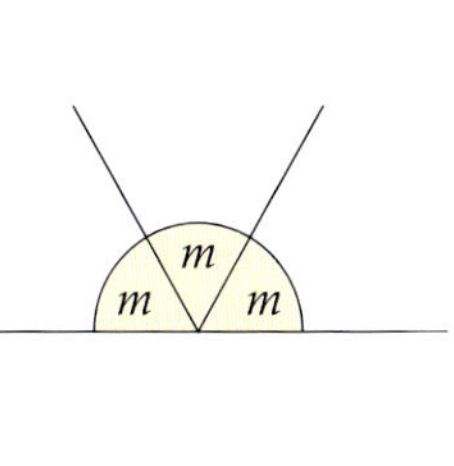

4 Calculate the size of the angles marked by letters in each diagram.
Give a reason for each of your answers.

a

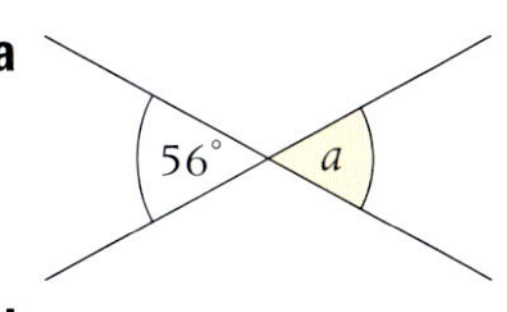

b

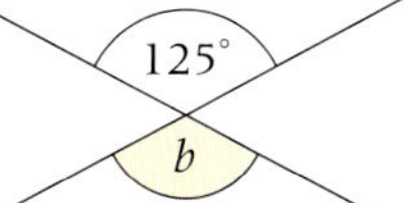

c

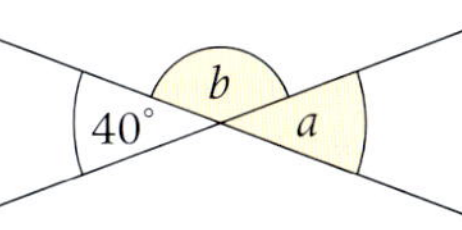

These diagrams are not drawn to scale.

d

e

f

S2.4 Angles in a triangle

This spread will show you how to:

- Use angle properties of equilateral, isosceles and right-angled triangles

Keywords
Angle
Degrees (°)
Equilateral
Isosceles
Right-angled
Scalene
Tessellation
Triangle

- **There are 180° on a straight line.**

You can draw any triangle ... tear off the corners ... and put them together to make a straight line.

- **The angles in a triangle add to 180°.**

Example

Calculate the values of x, y and z. Give a reason for each of your answers.

a

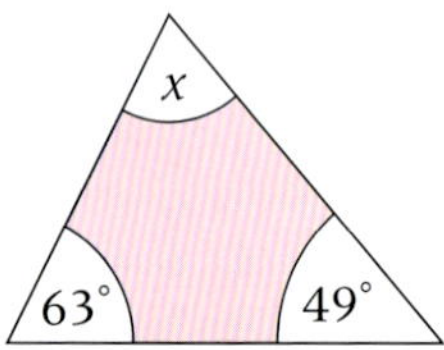

b

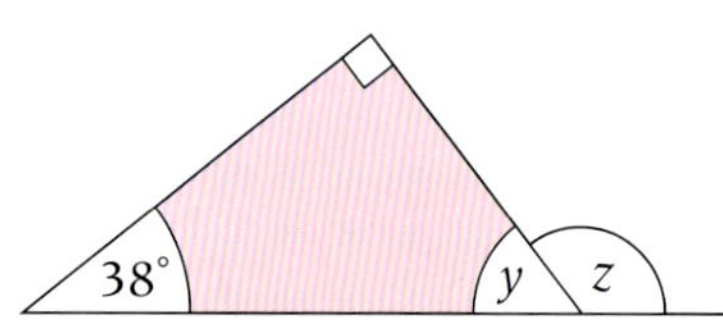

a $63° + 49° = 112°$
$180° - 112° = 68°$
$x = 68°$
(angles in a triangle add to 180°)

b $38° + 90° = 128°$
$180° - 128° = 52°$
$y = 52°$ (angles in a triangle add to 180°)
$180° - 52° = 128°$
$z = 128°$ (angles on a straight line add to 180°)

- **You should know these names for special triangles**

Right-angled	Equilateral	Isosceles	Scalene
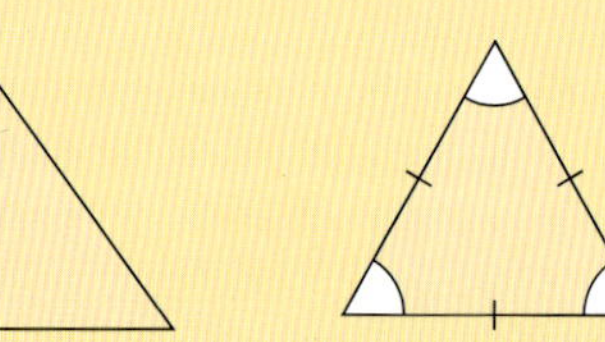	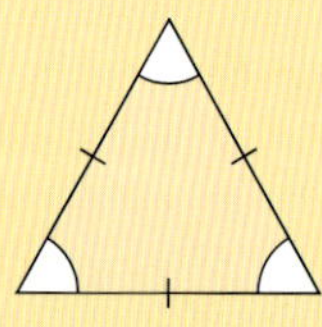	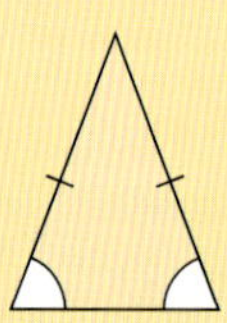 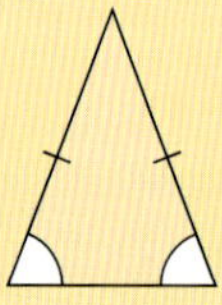	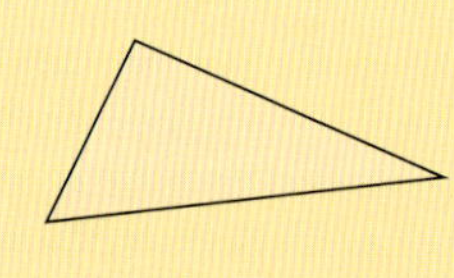
One 90° angle marked ∟	3 equal angles 3 equal sides	2 equal angles 2 equal sides	No equal angles no equal sides

For an equilateral triangle each angle is 60° as $180° \div 3 = 60°$

Lines with the same mark are equal length.

Example

Calculate the values of x and y.
Give reasons for your answers.

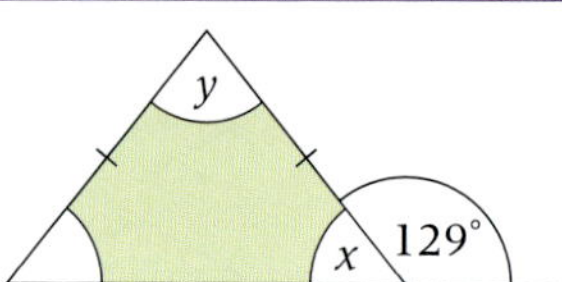

$180° - 129° = 51°$
$x = 51°$ (angles on a straight line add to 180°)
$51° + 51° = 102°$
$180° - 102° = 78°$
$y = 78°$ (angles in a triangle add to 180°)

As the triangle is isosceles, two of the angles are equal.

Exercise S2.4

1 **a** State the total of the three angles in this triangle.

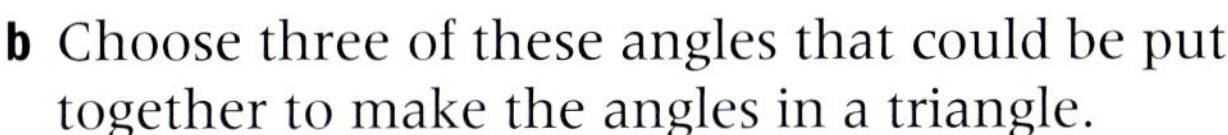

b Choose three of these angles that could be put together to make the angles in a triangle.

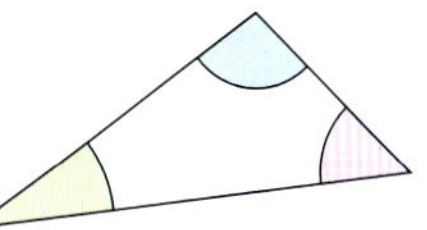

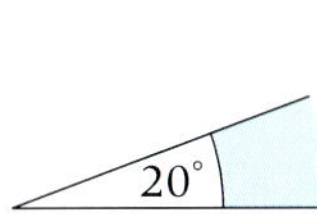

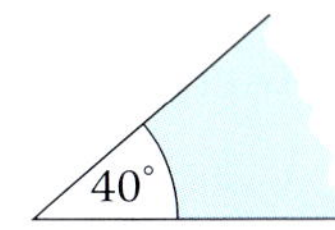

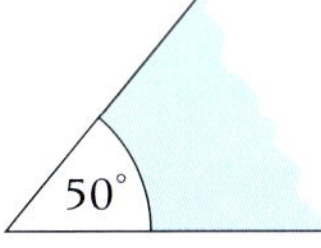

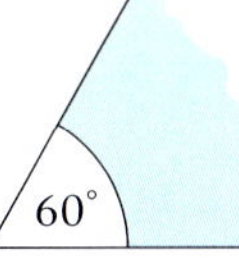

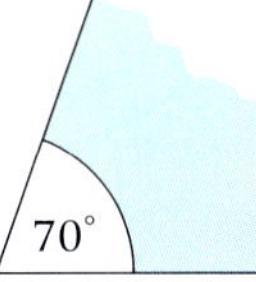

2 Calculate the size of the unknown angles in each diagram. The diagrams are not drawn to scale.

a

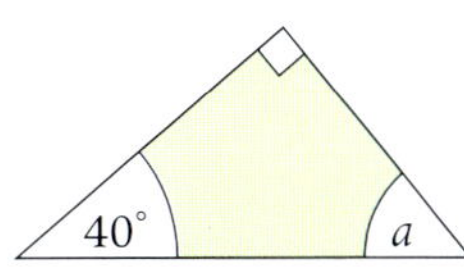

b

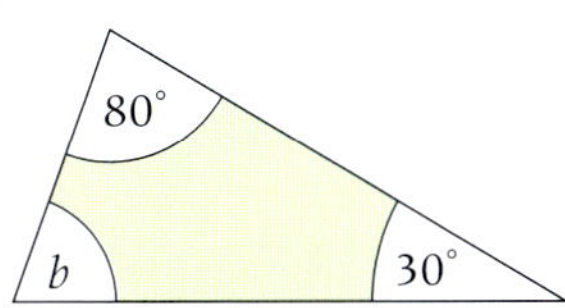

c

d

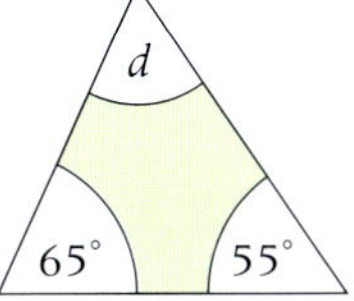

e

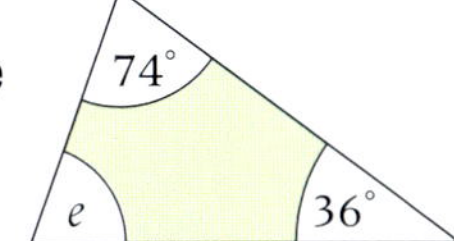

f

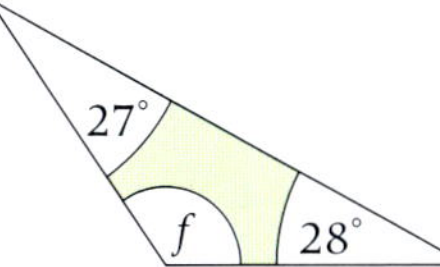

g

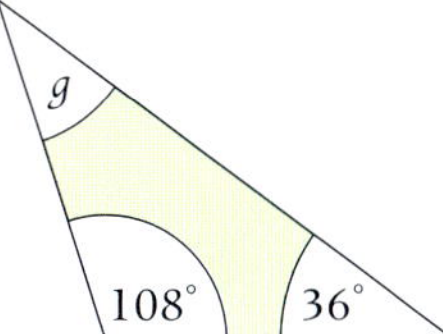

h

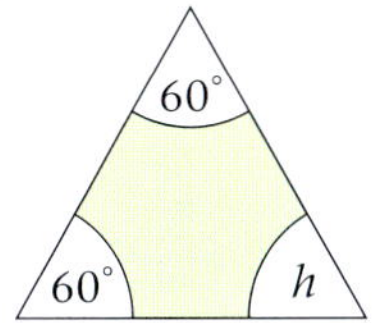

i

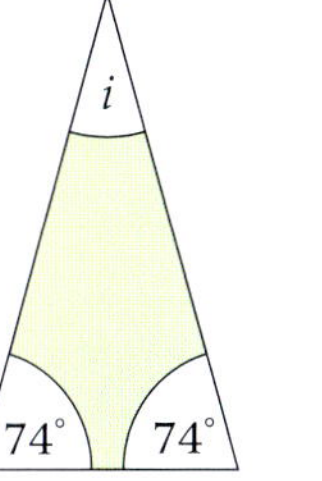

j

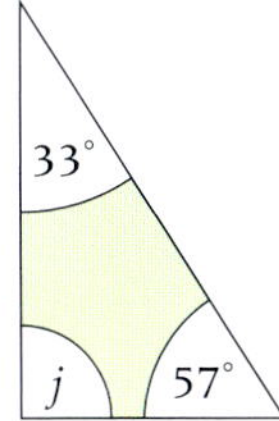

k

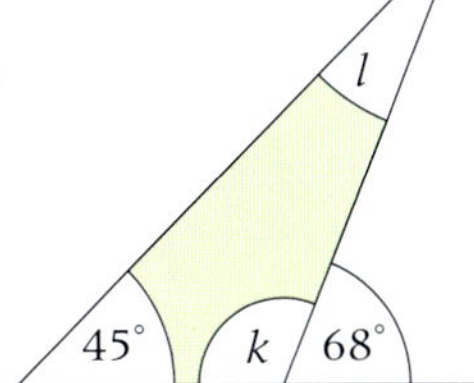

l

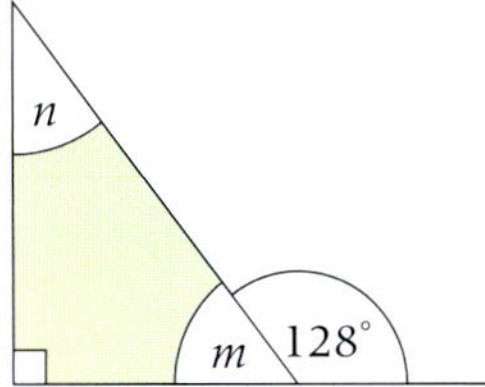

m

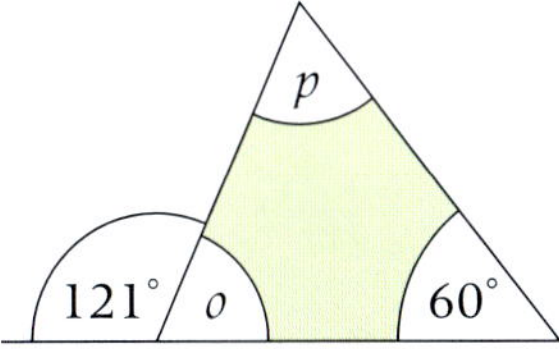

n

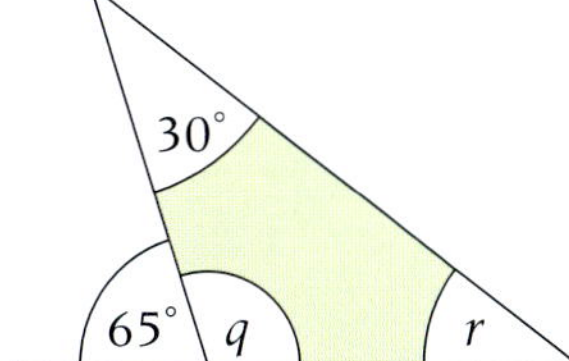

o

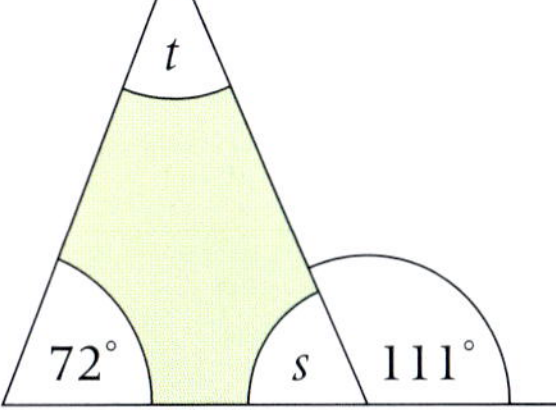

3 List any triangles in question **2** that are

a right-angled **b** isosceles.

S2.5 Coordinates

This spread will show you how to:

- Plot coordinates
- Locate points with given coordinates

Keywords
Coordinates
Negative
Origin
Positive
Quadrant
x-axis
y-axis

- **Coordinates are a pair of numbers (x, y) that fix a point on a grid.**

You can plot coordinates on a grid.

- **A grid has two perpendicular axes: the x-axis and the y-axis.**

You write the coordinates as a pair

(x-coordinate, y-coordinate)

The x-value comes first.

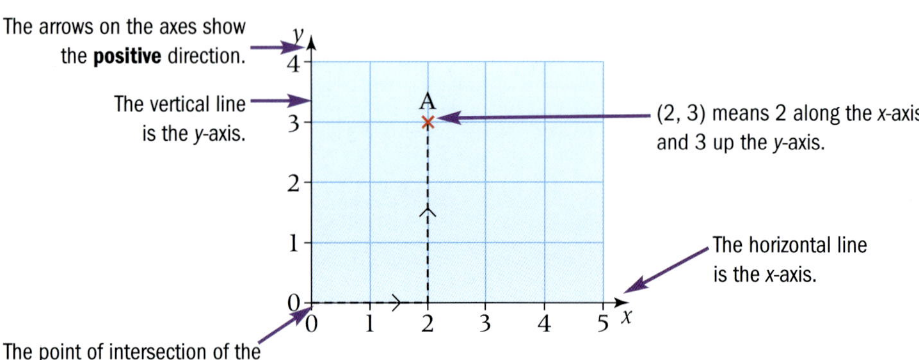

x coordinate is across, y coordinate is up or down.

This topic is extended to the midpoint of a line on page 388.

Example

a Plot and join the points A(1, 3), B(3, 3) and C(3, 1). Name the shape you have drawn.

a

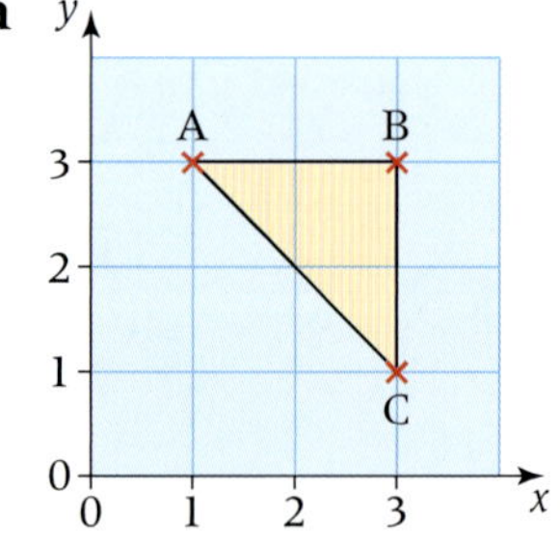

Right-angled triangle.

b Add and join the point D(0, 0). Name the new shape.

b

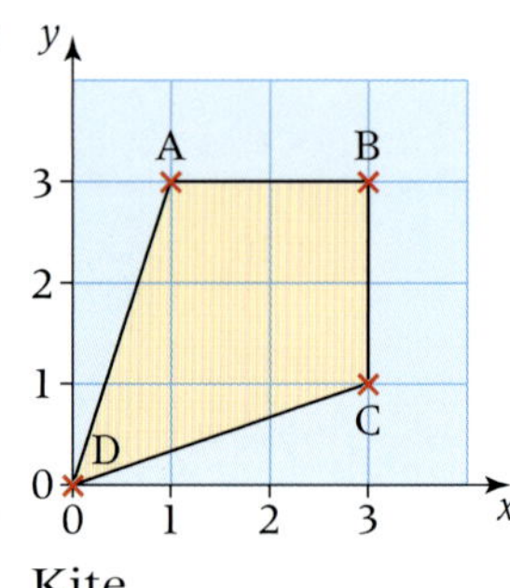

Kite.

Exercise S2.5

1 Give the coordinates of the points A to Z.
The first one is done for you.

A(2, 8)
B(__, __)

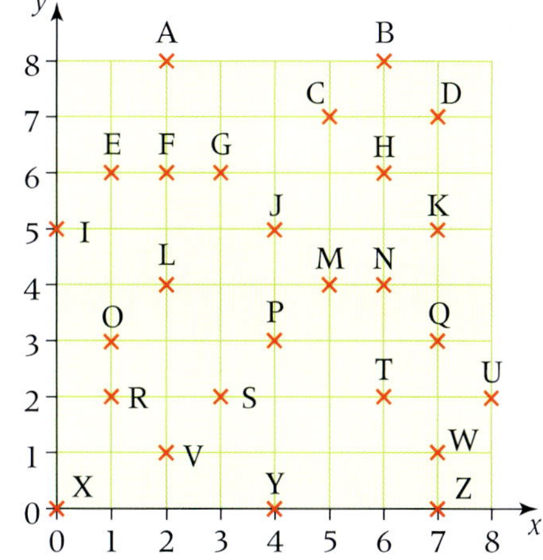

2 **a** Copy this coordinate grid onto square grid paper.

b Plot and label these points.

A(2, 3) B(8, 3) C(8, 9) D(2, 9)

c Join the points in order.

d Write the name of the shape ABCD.

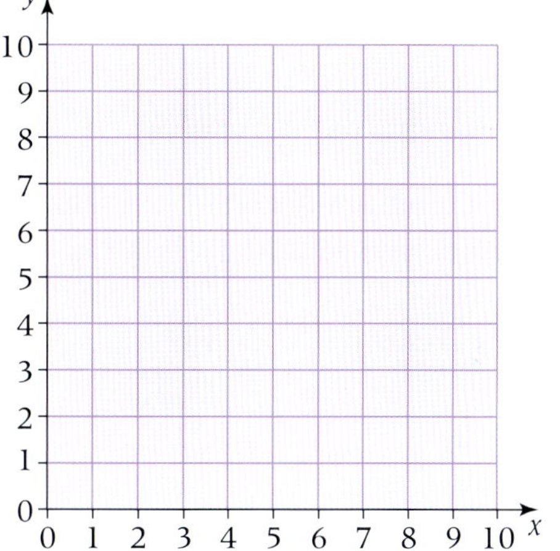

3 **a** Copy the coordinate grid in question **2** onto square grid paper.

b Plot, label and join each of these sets of points, then write the name of the shape you have made.

i A(7, 6), B(7, 9), C(3, 9), D(3, 6)
ii E(5, 1), F(3, 4), G(1, 1)
iii H(8, 3), I(10, 5), J(7, 5), K(5, 3)
iv L(8, 6), M(9, 8), N(8, 9), P(7, 8)

4 Write the coordinates of all the corner points of this shape.
Start at the origin, then point A, ...

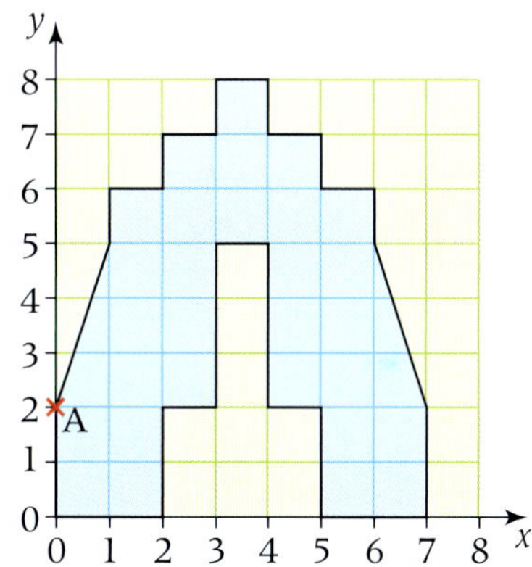

S2 Exam review

Key objectives

- Distinguish between acute, obtuse, reflex and right angles
- Estimate the size of an angle in degrees
- Use angle properties of equilateral, isosceles and right-angled triangles
- Use parallel lines
- Recall and use properties of angles at a point, angles on a straight line, perpendicular lines, and opposite angles at a vertex
- Use axes and coordinates to specify points
- Locate points with given coordinates

1 a The diagram shows a triangle.
What type of triangle is it? (1)

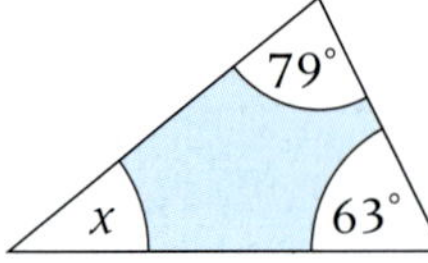

b Work out the value of the missing angle, x. (2)

2 a Write down the special name for this type of angle. (1)

b Write down the special name for this type of angle. (1)

c This diagram is wrong.
Explain why. (1)

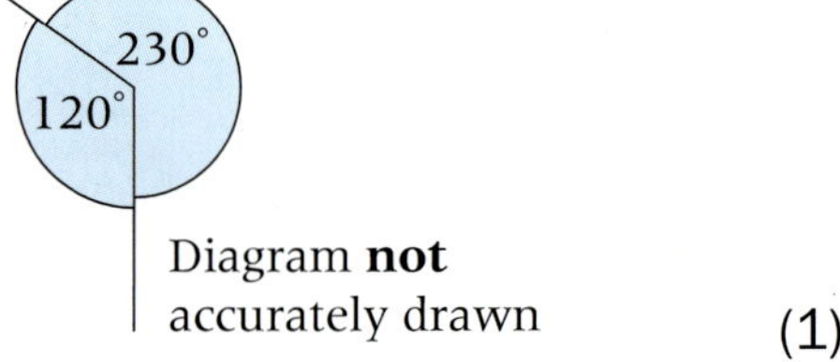

(Edexcel Ltd., 2004)

A3 Sequences

This unit will show you how to

- Understand and use the words multiple, factor, even, odd, prime and square
- Find the factors and multiples of a number and the common factors and common multiples of two numbers
- Understand and use the vocabulary associated with sequences
- Generate terms of a sequence using a term-to-term rule
- Use term-to-term rules to work out missing terms in a sequence
- Generate terms of a sequence using a position-to-term rule
- Describe a sequence by giving its start number and term-to-term rule
- Describe a sequence by comparing it to a sequence of multiples
- Generate and describe sequences derived from patterns

Before you start ...

You should be able to answer these questions.

Review

1 Find the missing numbers in these multiplication calculations. — Unit N2

a $4 \times \square = 12$ **b** $3 \times \square = 27$

c $8 \times \square = 32$ **d** $6 \times \square = 18$

2 Work out the square numbers from 1^2 to 10^2. — Key stage 3

3 Copy and complete. — Unit N2

a $2 + \square = 7$ **b** $16 - \square = 12$

c $11 + \square = 18$ **d** $25 - \square = 19$

4 Copy and complete. — Unit N2

a $3 \times \square = 6$ **b** $12 \div \square = 6$

c $7 \times \square = 21$ **d** $18 \div \square = 6$

5 Substitute $n = 3$ into these expressions. — Unit A1

a $n + 5$ **b** $2n$ **c** $3n + 3$ **d** $4n - 2$

A3.1 Multiples and factors

This spread will show you how to:

- Understand and use the words multiple, factor, even, odd, prime and square
- Find the factors and multiples of a number and the common factors and common multiples of two numbers

Keywords
Common factor
Common multiple
Even
Factor
Multiple
Odd
Prime

Here is the multiplication table up to 6 × 6.

×	1	2	3	4	5	6
1	1	2	3	4	5	6
2	2	4	6	8	10	12
3	3	6	9	12	15	18
4	4	8	12	16	20	24
5	5	10	15	20	25	30
6	6	12	18	24	30	36

- The **multiples** of a number are all the numbers in its times table.

Multiples of 6: 6, 12, 18, 24, 30, 36, ...

Multiples of 4: 4, 8, 12, 16, 20, 24, ...

- The **common multiples** of two numbers are the numbers common to both sets of multiples.

12 and 24 are common multiples of 6 and 4.

6 and 4 have these multiples in common.

- A **factor** of a number is any number that divides into it exactly.

Example

a List the factors of 12. **b** Find the common factors of 8 and 12.

a $12 = 1 \times 12 = 2 \times 6 = 3 \times 4$
The factors of 12 are 1, ②, 3, ④, 6, 12.
b The factors of 8 are 1, ②, ④, 8.
The **common factors** of 8 and 12 are 2 and 4.

Factors come in pairs. The factor pairs of 12 are 1 and 12, 2 and 6, 3 and 4.

- A **prime number** has only two factors, 1 and itself.

1 is a factor of all numbers. You never list it as a common factor.

7 is a prime number. Its factors are 1 and 7.

- A **square** number is the result of multiplying a number by itself.

$2 \times 2 = 4$ $5 \times 5 = 25$ $7 \times 7 = 49$ 4, 25 and 49 are square numbers.

Square numbers only have 3 factors: 1, the number itself, and its square root. For example, 25 has factors 1, 5 and 25.

Its factor pairs are 1 and 25, 5 and 5.

Example

From the list of numbers 3 6 10 12 17 24 write

a the multiples of 6 **b** the factors of 30 **c** the prime numbers.

a 6, 12, 24 **b** 3, 6, 10 **c** 17

Exercise A3.1

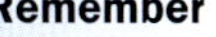

1 Write all the

a odd numbers less than 10

b even numbers between 21 and 29

c odd numbers between 20 and 30

d even numbers less than 35 and greater than 29

e even numbers greater than 10 but less than 20.

> **Remember**
> Even numbers are multiples of 2.

2 Describe each set of numbers in words. The first one is done for you.

a 2, 4, 6, 8 *The even numbers less than 10*

b 11, 13, 15, 17, 19 **c** 16, 18, 20, 22

d 19, 21, 23, 25, 27, 29, 31 **e** 25, 27, 29, 31, 33, 35

3 Here is a list of numbers.

3 4 5 6 7 9 10 12 15 20

From this list write

a the multiples of 3

b the multiples of 5

c the common multiples of 3 and 5.

> Use your answers to parts **a** and **b** to help you.

4 From the numbers in the oval write

a the factors of 18

b the factors of 12

c the common factors of 12 and 18.

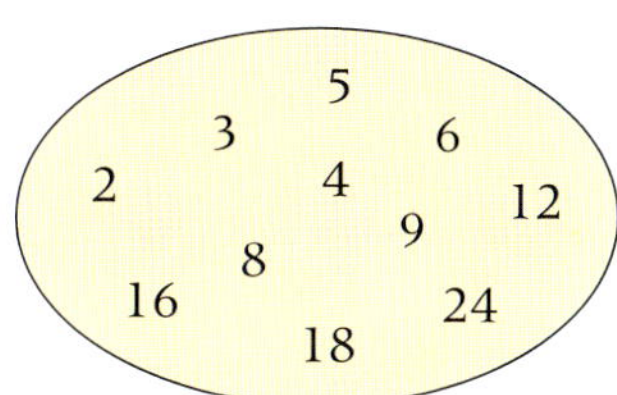

5 **a** Write the odd numbers from 1 to 20.

b From your list, write

i the prime numbers

ii the square numbers

iii the square root of 49.

6 From this set of numbers, write

a the even numbers greater than 10

b the odd numbers less than 15

c all the factors of 24

d all the multiples of 3

e the prime numbers.

4 6 8
2 7 11 12
3 9 16

Three of the numbers in the oval are the square roots of three of the other numbers in the oval.

f Write all six numbers.

A3.2 Number sequences

This spread will show you how to:

- Understand and use the vocabulary associated with sequences
- Generate terms of a sequence using a term-to-term rule

Keyword

Ascending
Consecutive
Descending
Difference
Sequence
Term

- The numbers in a **sequence** follow a pattern.
  Each number in a sequence is called a **term**.

- You can see how a sequence grows by looking at the **differences** between **consecutive** terms.

Consecutive terms are next to each other.

The first five terms of a sequence are

1, 4, 7, 10, 13, ...

Difference: +3 +3 +3 +3

The numbers are getting higher. The sequence is **ascending**.

You can follow the pattern to work out more terms.
13 + 3 = 16, 16 + 3 = 19, etc.

- A linear sequence goes up (or down) in equal sized steps.

Example

The first five terms of a sequence are

14, 12, 10, 8, 6, ...

Work out the next two terms in the sequence.

14, 12, 10, 8, 6, ...

Difference: −2 −2 −2 −2

The next two terms are 6 − 2 = 4 and 4 − 2 = 2.

The numbers are getting lower. The sequence is **descending**.

In a sequence that is not linear, the differences between terms may follow a pattern.

Example

The first five terms of a sequence are

1, 2, 5, 10, 17, ...

Work out the next two terms.

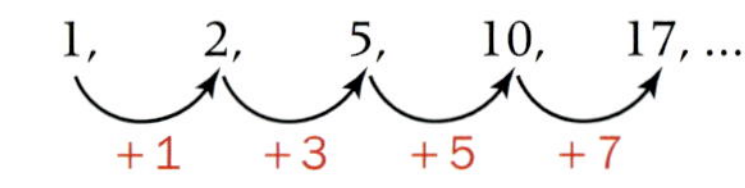

Difference:

The next term is 17 + 9 = 26.
The term after is 26 + 11 = 37.

The difference pattern is the odd numbers. The next odd number is 9. The next odd number is 11.

Exercise A3.2

1 Find the next two terms for each sequence.

a 3, 5, 7, 9, ...   **b** 2, 5, 8, 11, ...

c 1, 5, 9, 13, ...   **d** 4, 6, 8, 10, ...

2 Work out the next two terms for each sequence.

a 26, 21, 16, 11, ...   **b** 32, 28, 24, 20, ...

c 22, 19, 16, 13, ...   **d** 86, 76, 66, 56, ...

3 Find the next two terms for each sequence.

a 6, 9, 12, 15, ...   **b** 15, 13, 11, 9, ...

c 2, 9, 16, 23, ...   **d** 50, 42, 34, 26, ...

4 Write the next two terms of the sequence 15, 17, 19, 21, ...
Explain why 72 is **not** a term in this sequence.

5 For each sequence,

i Work out the differences between consecutive terms.

ii Follow the pattern to work out the next two terms.

a 2, 3, 5, 8, 12, ...   **b** 21, 20, 18, 15, ...

c 4, 5, 8, 13, 20, ...   **d** 20, 15, 11, 8, ...

6 Follow the patterns in these sequences to work out the next two terms.

a 1, 2, 4, ... ($\times 2$, $\times 2$, $\times 2$)   **b** 128, 64, 32, ... ($\div 2$, $\div 2$, $\div 2$)

c 5, 10, 20, ...   **d** 81, 27, 9, ...

7 The first three terms of a sequence are 3, 6, 12, ...
Jim says 'Double a term to get the next one.'
Sophie says 'The differences are +3 then +6. Add 9 to get the next term.'

a Write the first four terms of

i Jim's sequence   **ii** Sophie's sequence.

b Write the first four terms of two different sequences that begin 1, 2, 4, ...

8 Work out the next two terms in each sequence.

a −5, −3, −1, 1, 3, ...   **b** 15, 12, 9, 6, 3, ...

c −4, −3, −1, 2, 6   **d** 21, 19, 16, 12, 7, ...

9 In each part, two sequences have been mixed up.
Write out each sequence in the correct order.

a 2, 4, 6, 6, 8, 10, 10, 12, 14, 18   **b** 1, 4, 5, 7, 9, 10, 13, 13, 17

A3.3 Generating sequences

This spread will show you how to:

- Use term-to-term rules to work out missing terms in a sequence
- Generate terms of a sequence using a position-to-term rule
- Describe a sequence by giving its start number and term-to-term rule

Keywords
*n*th term
position-to-term
term-to-term

- A **term-to-term** rule tells you how to work out the next term in a sequence.

For the sequence

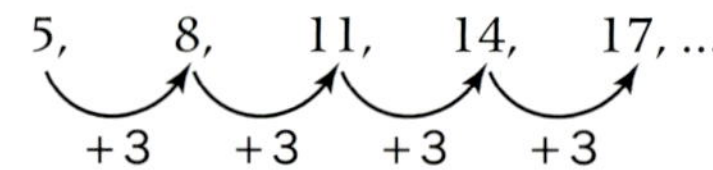

the term-to-term rule is 'add 3'.

- The term-to-term rule links one term to the next term in the sequence.

You can generate a sequence from a start number and a term-to-term rule.

Example

Generate the first five terms of the sequence with start number 7 and term-to-term rule 'add 4'.

7, 11, 15, 19, 23

$7 + 4 = 11$
$11 + 4 = 15$
$15 + 4 = 19$
$19 + 4 = 23$

You can use the term-to-term rule to work out missing terms in a sequence.

Example

Find the missing terms in these sequences.

a 3, 7, 11, ?, 19, ... **b** ?, 8, 11, 14, ...

a
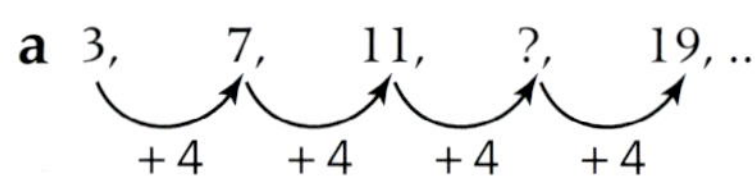

The missing term is $11 + 4 = 15$.

b
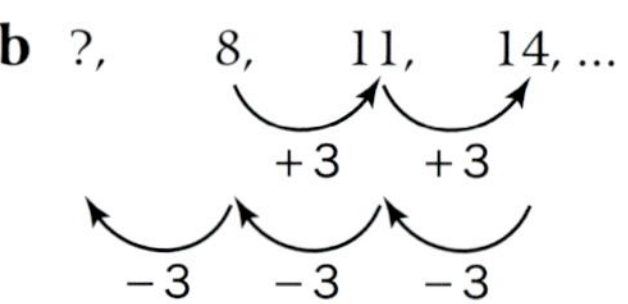

The missing term is $8 - 3 = 5$.

Work backwards using the inverse operation.

- A **position-to-term** rule links a term to its position in the sequence.

The position-to-term rule is often called the ***n*th term**.

$n + 2$, $2n - 1$, $5n$, $4n + 3$ are all examples of *n*th terms.

1st term is in position 1,
2nd term is in position 2,
*n*th term is in position *n*.

Example

The *n*th term of a sequence is $2n + 1$.
Write the first three terms of the sequence.

Substitute the position number into $2n + 1$.

1st term $n = 1$: $2 \times 1 + 1 = 3$
2nd term $n = 2$: $2 \times 2 + 1 = 5$
3rd term $n = 3$: $2 \times 3 + 1 = 7$

This topic is extended to the general term on page 380.

Exercise A3.3

1 Write the first five terms of these sequences.

a start number 4 term-to-term rule 'add 2'

b start number 7 term-to-term rule 'add 3'

c start number 25 term-to-term rule 'subtract 2'

d start number 30 term-to-term rule 'subtract 4'

2 Find the missing terms in these sequences.

a 3, 8, 13, ?, 23

b 18, 15, 12, ?, 6

c ?, 9, 13, 17, 21

d ?, 16, 12, ?, 4

3 Write the first five terms of the sequences with

a start number 5 term-to-term rule 'multiply by 2'

b start number 32 term-to-term rule 'divide by 2'

c start number 3 term-to-term rule 'add consecutive odd numbers'

d start number 40 term-to-term rule 'subtract consecutive even numbers'

4 Generate the first five terms of these sequences.

a start number −5 term-to-term rule 'add 2'

b start number 7 term-to-term rule 'subtract 3'

c start number 3 term-to-term rule 'multiply by −1'

5 Generate the first five terms of the sequences with these nth terms.

a $n + 1$ **b** $n + 3$ **c** $2n - 1$ **d** $2n + 3$

e $3n - 1$ **f** $3n + 2$ **g** $4n + 3$ **h** $5n - 2$

6 Find the missing terms in these sequences.

a −3, 0, 3, ?, 9

b 12, ?, 2, −3, −8

c 2, 3, 5, ?, 12

d ?, 12, 8, ?, −6

7 Generate the first five terms of the sequence with nth term

a $n - 1$ **b** $n - 3$ **c** n^2 **d** $-n + 3$

8 Match each sequence in set A to an nth term in set B.

Set A:

5, 9, 13, 17, ... 16, 14, 12, 10, ... 1, 7, 13, 19, ...

4, 7, 10, 13, ... 21, 17, 13, 9, ... 2, 8, 14, 20, ...

Set B:

$6n - 4$ $4n + 1$ $25 - 4n$ $6n - 5$ $3n + 1$ $18 - 2n$

A3.4 Describing sequences in words

This spread will show you how to:

- Describe a sequence by giving its start number and term-to-term rule
- Describe a sequence by comparing it to a sequence of multiples

Keywords
Multiples
Term-to-term rule

- **You can describe a sequence by giving its start number and term-to-term rule.**

Example

Describe the sequence: 5, 11, 17, 23, 29, ...

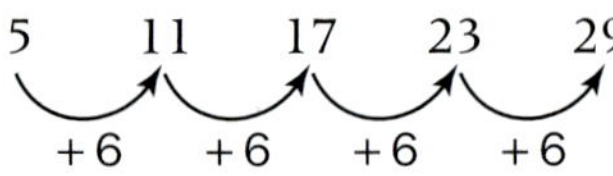

The sequence has start number 5 and term-to-term rule 'add 6'.

Some sequences have special names.

For example:

1, 3, 5, 7, 9, ... are 'the odd numbers'
2, 4, 6, 8, 10, ... are 'the even numbers' or 'the **multiples** of 2'
1, 4, 9, 16, 25, ... are 'the square numbers'.

- **You can describe a sequence by comparing it to a sequence of multiples.**

Example

a Write the first five terms of the sequence 'multiples of 3'.
b Here is a sequence

2, 5, 8, 11, 14, ...

Describe this sequence in words by comparing it to the multiples of 3.

a 3, 6, 9, 12, 15

b Multiples of 3

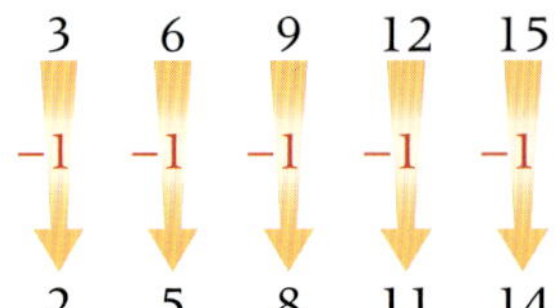

Term of sequence

The sequence is 'one less than the multiples of 3'.

Each term is one less than the corresponding term in the multiples of 3 sequence.

Exercise A3.4

1 Describe each of these sequences in words, by giving the start number and the term-to-term rule.

a 6, 7, 8, 9, 10, ...

b 3, 7, 11, 15, 19, ...

c 9, 15, 21, 27, 33, ...

d 20, 17, 14, 11, 8, ...

e 16, 14, 12, 10, 8, ...

f 33, 28, 23, 18, 13, ...

2 Each set of numbers is a jumbled sequence of multiples.
Write each sequence in the correct order.
Write the name of each sequence in the form 'the multiples of ________'.

a 20, 5, 10, 15, 25, ...

b 21, 35, 7, 14, 28

c 30, 6, 18, 24, 12, ...

d 27, 9, 36, 18, 45

3 Write the first five terms of each of these sequences.

a The multiples of 4

b One more than the multiples of 4

c One less than the multiples of 4

d The multiples of 3

e Two less than the multiples of 3

f Four more than the multiples of 3

4 Describe each of your sequences from question **3** by giving the start number and term-to-term rule.

5 Describe these sequences by giving the start number and the term-to-term rule.

a 10 000, 1000, 100, 10, 1, ...

b 4, 8, 16, 32, 64, ...

c 80, 40, 20, 10, 5, ...

d 3, 9, 27, 81, 243, ...

6 Describe these sequences by giving the start number and the term-to-term rule.

a −10, −4, 2, 8, 14, ...

b 15, 11, 7, 3, −1

c 100, 60, 20, −20, −60, ...

d −5, 0, 10, 25, 45, ...

7 Describe these sequences in words by comparing them to the multiples of 4.

a 6, 10, 14, 18, 22, ...

b 1, 5, 9, 13, 17, ...

8 The nth term of a sequence is $2n$.
Generate the first five terms of this sequence.
Write two names for this sequence.

A3.5 Pattern sequences

This spread will show you how to:

- Generate and describe sequences derived from patterns

Keywords
Pattern

Examiners tip
The techniques in this spread are useful in your coursework.

These **patterns** are made from dots.

Pattern 1 Pattern 2 Pattern 3

To get from one pattern to the next in the sequence, you add two more dots.
The term-to-term rule is 'add 2'.
The numbers of dots in the patterns make a sequence

Pattern number	1	2	3	4
Number of dots	2	4	6	8

Pattern 4 is

The number of dots are the multiples of 2.
The number of dots in each pattern is

pattern number × 2

This is the position-to-term rule for the sequence.

The 10th pattern has 10 × 2 = 20 dots.

- **You can use the term-to-term rule and position-to-term rule to work out the number of dots in other patterns.**

Example

Here is a pattern made of tiles.

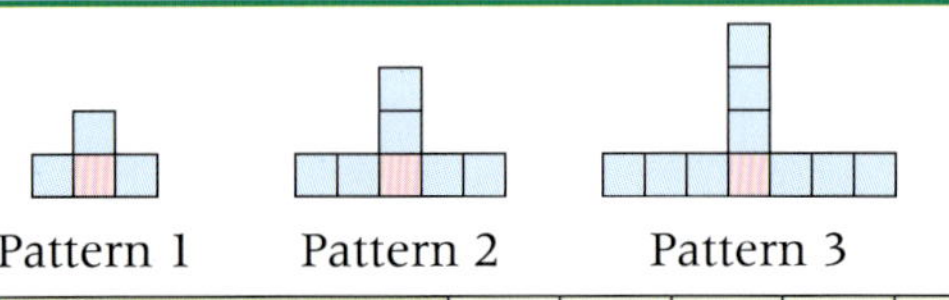

Pattern 1 Pattern 2 Pattern 3

a Copy and complete the table for the pattern sequence.

Pattern number	1	2	3	4	5
Number of tiles					

b Describe how the sequence grows.

c Work out the number of tiles in the 10th pattern.

a

Pattern number	1	2	3	4	5
Number of tiles	4	7	10	13	16

Count the tiles in patterns 1, 2 and 3. Each time you add 3. So pattern 4 has 10 + 3 = 13 tiles. Pattern 5 has 13 + 3 = 16 tiles.

b Add 3 blue tiles each time, one to each arm.

c Pattern 1: 3 = 1 × 3 blue + 1 pink
Pattern 2: 6 = 2 × 3 blue + 1 pink
Pattern 3: 9 = 3 × 3 blue + 1 pink
...
Pattern 10: 10 × 3 blue + 1 pink = 31 tiles

Write out how the pattern grows in 3s. This links the pattern number to the number of tiles.

Exercise A3.5

1 Draw the next pattern in each sequence.

a

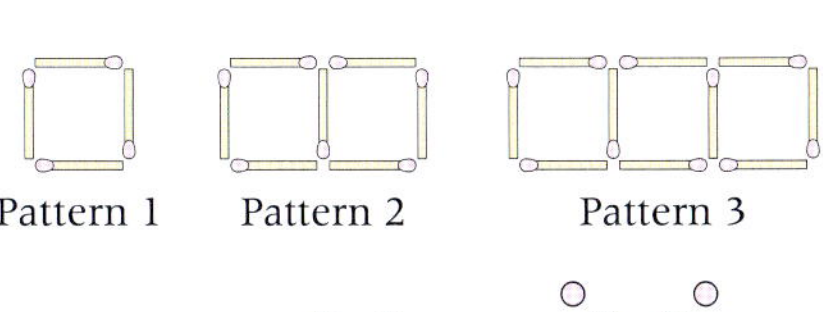

Pattern 1 Pattern 2 Pattern 3

b

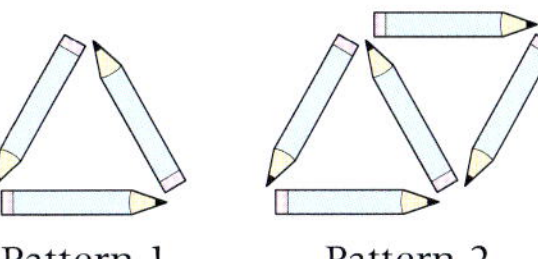

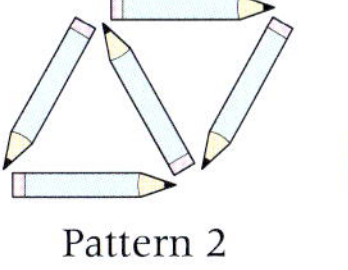

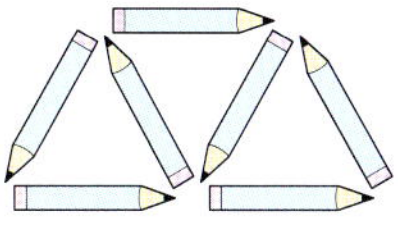

Pattern 1 Pattern 2 Pattern 3

c

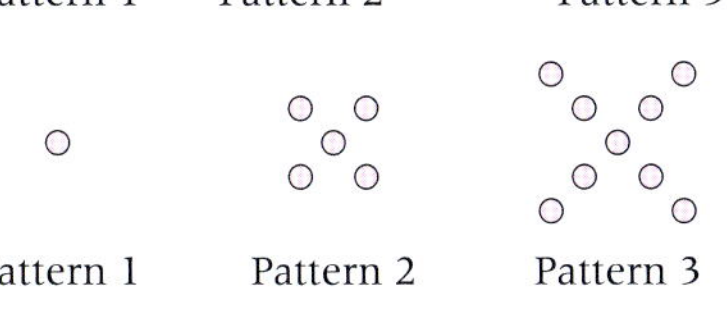

Pattern 1 Pattern 2 Pattern 3

d

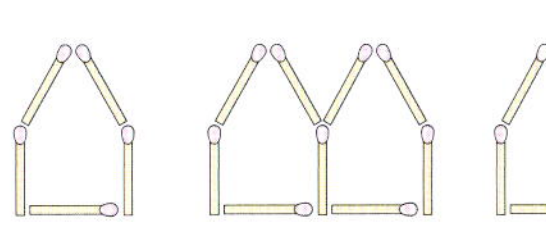

Pattern 1 Pattern 2 Pattern 3

2 **a** Copy and complete this table for pattern **a** in question **1**.

Pattern number	1	2	3	4	5
Number of matches					

b Copy and complete this table for pattern **b** in question **1**.

Pattern number	1	2	3	4	5
Number of pencils					

c Copy and complete this table for pattern **c** in question **1**.

Pattern number	1	2	3	4	5
Number of dots					

d Copy and complete this table for pattern **d** in question **1**.

Pattern number	1	2	3	4	5
Number of matches					

3 For each pattern

i describe how the pattern grows

ii copy and complete these.

Pattern 1 has ________ + ________

Pattern 2 has 2 × ________ + ________

Pattern 3 has 3 × ________ + ________

Pattern 10 has __ × _______ + _______

a

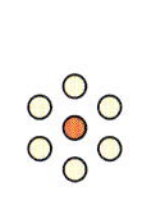

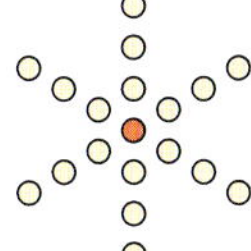

Pattern 1 Pattern 2 Pattern 3

b

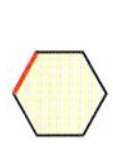

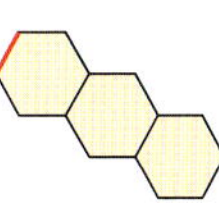

Pattern 1 Pattern 2 Pattern 3

4 Here are some patterns of dots.

a Draw pattern number 4.

b Copy and complete the table.

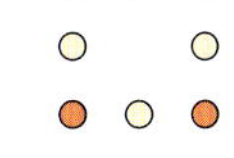

Pattern 1 Pattern 2 Pattern 3

Pattern number	1	2	3	4	5	10
Number of dots						

c Explain how you worked out the number of dots in pattern number 10.

A3 Exam review

Key objectives

- Use the concepts and vocabulary of factor, multiple and common factor
- Generate terms of a sequence using term-to-term and position-to-term definitions of the sequence

1

2	5	7	9	14	19	20	25

From the above list of numbers write

a a prime number (1)

b a square number (1)

c a multiple of 3 (1)

d three numbers that are a factor of 50. (2)

2 Here are the first five terms of a number sequence.

126 122 118 114 110

a Write down the next two terms of the number sequence. (1)

b Explain how you found your answer. (1)

The 20th term of the number sequence is 50.

c Write down the 21st term of the number sequence. (1)

(Edexcel Ltd., 2005)

D2 Displaying data

This unit will show you how to

- Use pictograms, vertical and horizontal bar charts, bar-line charts and pie charts to display data in categories
- Use stem-and-leaf diagrams to display numerical data
- Order data before representing it in a diagram or chart
- Use line graphs for time series data
- Understand the information shown in a time series graph, including individual data values and trends

Before you start ...

You should be able to answer these questions.

Review

1 Calculate each of these. (Review: Unit N2)

a $50 \div 10$ **b** $25 \div 5$

c $20 \div 2$ **d** $40 \div 10$

e $40 \div 5$

2 Find the values of the angle, x. (Review: Unit S2)

a

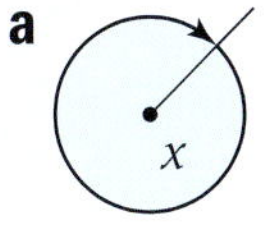

b

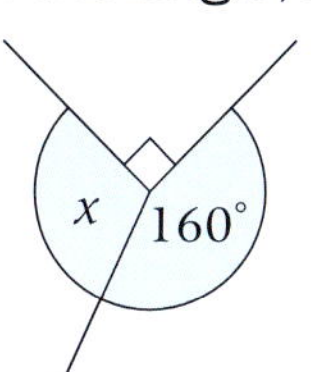

3 Calculate each of these. (Review: Unit N2)

a $360 \div 2$ **b** $360 \div 4$

c $360 \div 10$ **d** $360 \div 5$

e $360 \div 6$ **f** $360 \div 12$

g $360 \div 180$ **h** $360 \div 90$

i $360 \div 20$ **j** $360 \div 18$

4 Put these numbers in order of size, smallest first. (Review: Unit N1, N3)

a 31, 43, 25, 29, 36, 41, 49, 30

b 243, 342, 324, 234, 432, 423

c 21.3, 23.1, 31.2, 13.2, 12.3, 32.1

D2.1 Pictograms

This spread will show you how to:

- Draw and produce pictograms for data in categories

Keywords
Category
Pictogram
Represents

- You can use a **pictogram** to display data.

Pictograms use symbols to give an eye-catching picture of the size of each category.

Number of bottles of milk sold

Monday	
Tuesday	
Wednesday	
Thursday	
Friday	

Key: represents 2 bottles

Always give a key.

- A pictogram shows
  - how each **category** compares with the others
  - all the data, but in categories.

You may have to use part of a symbol to **represent** some quantities.

represents 2 bottles

represents 1 bottle

Example

The number of cars that a car salesman sells is given in the table.

Week	Cars sold
1	4
2	6
3	12
4	10

a Draw a pictogram to illustrate this information.

Use to represent 4 cars.

b In which week did he sell most cars?

a So represents 2 cars.

Week 1	
Week 2	
Week 3	
Week 4	

$4 + 2 = 6$

$4 + 4 + 2 = 10$

Key: represents 4 cars

Don't forget the key.

b In week 3.

Exercise D2.1

1 The number of people who eat different food is shown in the pictogram.
Copy and complete the pictogram to show six people eating chips and five people eating curry.

Key: 🯅 represents one person

Pizza	🯅🯅🯅🯅🯅
Burger	🯅🯅
Chips	
Curry	

2 The colours of flowers in a garden are shown in the pictogram.
Copy and complete the pictogram with this information.

Red	Yellow	Blue	White	Other
5	20	10	15	25

Key: ❀ represents 5 flowers

Red	❀
Yellow	❀❀❀❀
Blue	
White	
Other	

3 The pictogram shows where people use the internet the most.
Copy the pictogram and represent this information on your diagram:

Library 6 School 9 Work 17

Key: 🖥 represents 2 people

Home	🖥🖥🖥🖥
Internet café	🖥🖥 (half)
Library	
School	
Work	

4 This data is from a newspaper survey.

Draw a pictogram using 📰 to represent 10 people.

Daily paper A	40
Daily paper B	50
Daily paper C	35
Daily paper D	25
Other	15

5 The results of a survey about people's favourite hot drink are given in the table.

Letting ☕ represent 20 people, draw a pictogram for this data.

Tea	80
Coffee	60
Hot chocolate	50
Soup	30
Other	20

D2.2 Bar charts

This spread will show you how to:

- Draw and produce bar charts for data in categories
- Draw bar-line charts

Keywords
Bar chart
Bar-line chart
Frequency
Horizontal
Vertical

You can use a **bar chart** to display data.

Bar charts use bars to give a visual picture of the size of each category.

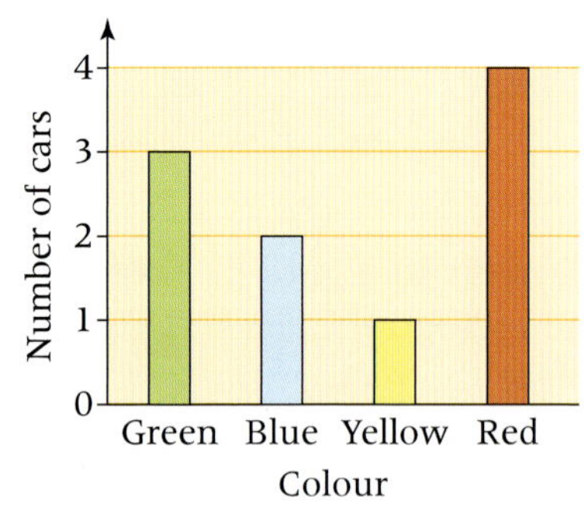

Vertical bars.

Note the equal gaps between the bars.

- A bar chart shows
  - how each category compares with the others
  - all the data, but in categories.

The bars can be horizontal or vertical.

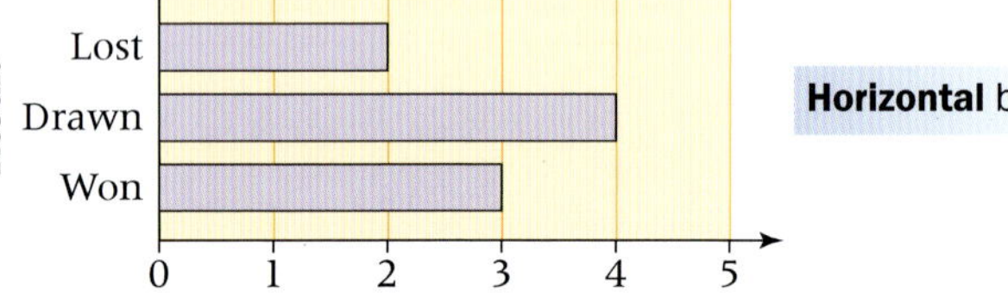

Horizontal bars.

- **Bar-line charts** are a good way to display (discrete) numerical data.

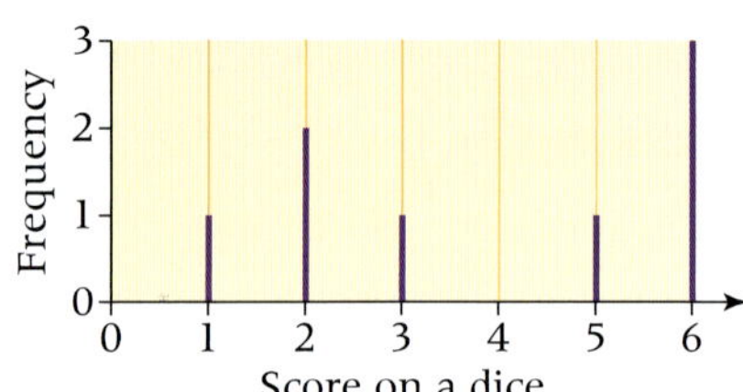

Lines are drawn instead of bars.

Example

A class are asked to name one favourite pet. The results are shown.

Pet	Dog	Cat	Guinea pig	Other
Frequency	16	9	12	3

Draw a bar chart to show this information.

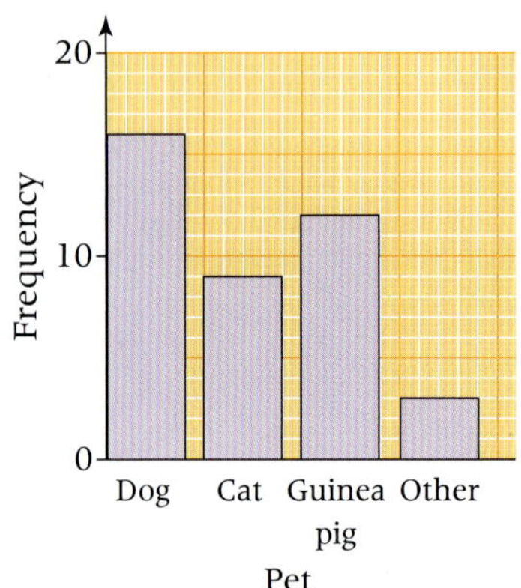

Exercise D2.2

1 The cost of renting a car for a day in different countries is shown.
Copy and complete the bar chart.

Country	Cost
UK	£15
Portugal	£16
Ireland	£21
Germany	£19
France	£17
USA	£16

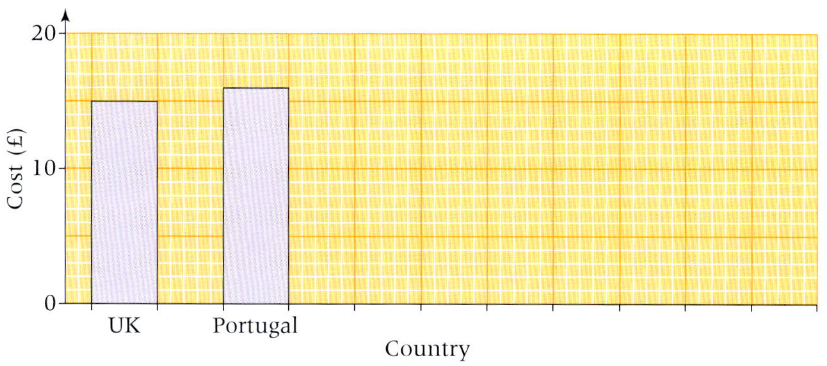

2 The number of concerts held at various venues is given.
Copy and complete the bar chart.

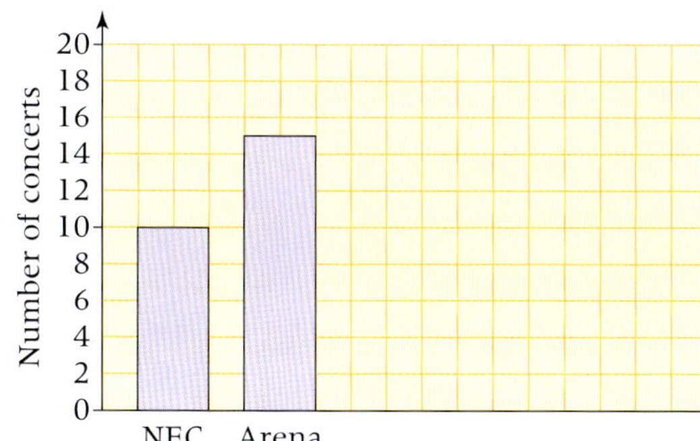

Venue	Number of concerts
NEC	10
Arena	15
MEN	20
NIA	18
Wembley	9

3 The numbers of packets of crisps in a shop are shown in the frequency table.
Draw a vertical bar chart to show this information.

Flavour	Number of packets
Plain	5
Salt'n'Vinegar	3
Cheese & Onion	4
Smokey Bacon	1
Other	8

4 The number of Bank Holidays in different countries is shown.
Draw a bar chart to show this information.

Country	Number of Bank Holidays
UK	8
Italy	16
Iceland	15
Spain	14

5 The cost to fly to certain resorts is given in the frequency table.
Draw a bar chart to show this information.

Resort	Cost (£)
Lisbon	90
Crete	100
Malta	80
Menorca	70
Cyprus	110

D2.3 Pie charts

This spread will show you how to:

- Draw and produce pie charts for data in categories

Keywords
Angle
Category
Pie chart
Proportion
Sector

You can use a **pie chart** to display data.

Pie charts use a circle to give a quick visual picture of all the data. The size of each **angle** shows the size of each **category**.

- A pie chart shows
  - the **proportion** or fraction of each category compared to the whole circle
  - all the data, but in categories.

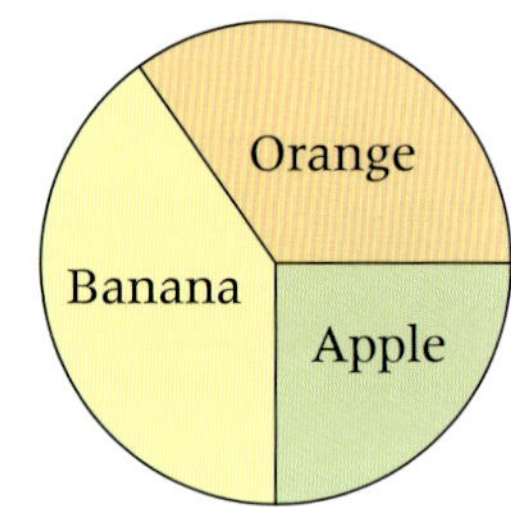

Apple is a quarter of all the data.

Banana is the biggest **sector**.

All the data must be included.

Example

A group of 12 students were asked how they travelled to school that day. The results are shown.

Method of travel	Walked	Bus	Car	Other
Frequency	6	2	3	1

Draw a pie chart to illustrate this information.

Method 1
Calculate the angle for one person:
$360° \div 12 = 30°$
Divide the circle into sectors of 30°.
Colour and label the sectors.

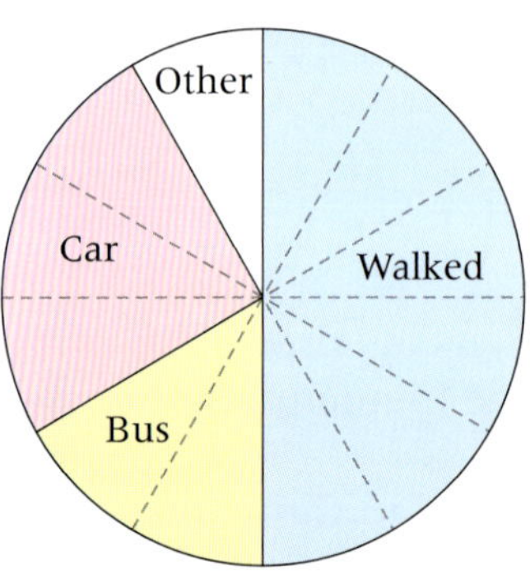

The angles at a point add to 360°.

Method 2
Calculate the angle for one person:
$360° \div 12 = 30°$
Calculate the angles for each category

Walked	$6 \times 30° =$	180°
Bus	$2 \times 30° =$	60°
Car	$3 \times 30° =$	90°
Other	$1 \times 30° =$	30°
	Add to check:	360°

Measure, colour and label the sectors.

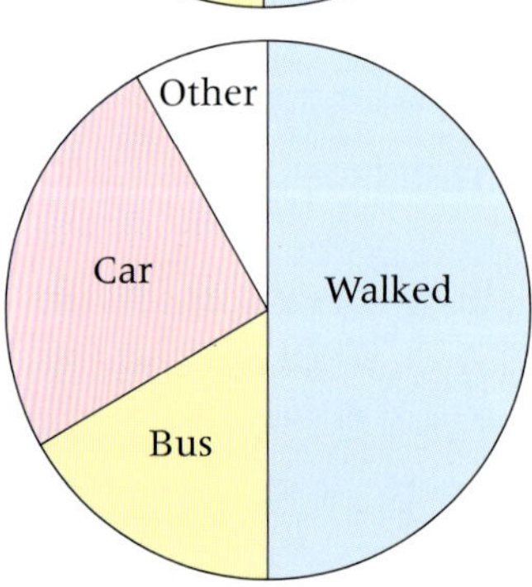

The angles at a point add to 360°.

Check by adding the angles of the sectors.

In the above example, Method 1 is time-consuming and inaccurate. It is best to use Method 2.

Exercise D2.3

1 Alvin makes eight sandwiches

1 tuna
2 cheese and tomato
3 chicken
2 corned beef

Draw a pie chart to show this information.

2 Sophie spends £6 on her hobby of painting.

£1 on paper
£2 on paint brushes
£3 on paint

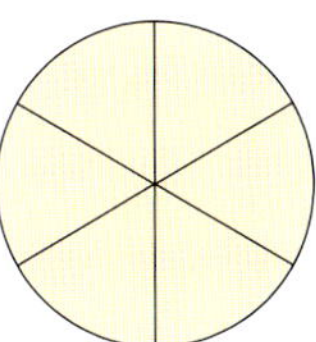

Draw a pie chart to show this information.

3 A survey question asks 'Do you think smoking should be banned in public places?'

The results are:

Yes	5
No	4

a Calculate the number of people who were asked the question.

b Calculate the angle one person represents in a pie chart.

c Calculate the angles to represent Yes and No.

d Draw a pie chart to show the results of the survey.

4 Seven boys and five girls attend an after-school homework club.

a Calculate the total number of students.

b Calculate the angle one student represents in a pie chart.

c Calculate the angles to represent boys and girls.

d Draw a pie chart to show the information.

5 A school fete is open from 10 am to 4 pm.

a Calculate the number of minutes the school fete is open.

A teacher has offered to help. She spends these times on each stall.

b Draw a pie chart to show this information.

Stall	Time
Bat the Rat	30 mins
Hook a Duck	25 mins
Smash a Plate	35 mins
Roll a Coin	80 mins
Tombola	70 mins
Break 1	60 mins
Break 2	60 mins

6 In one week, 180 letters were delivered to a business. The letters were delivered in this pattern.

Mon	Tues	Wed	Thurs	Fri	Sat	Sun
45	30	25	20	50	10	0

Draw a pie chart to show this information.

D2.4 Stem-and-leaf diagrams

This spread will show you how to:

- Draw and produce stem-and-leaf diagrams

Keywords
Ordered
Stem-and-leaf diagram

You can use a **stem-and-leaf diagram** to display numerical data.

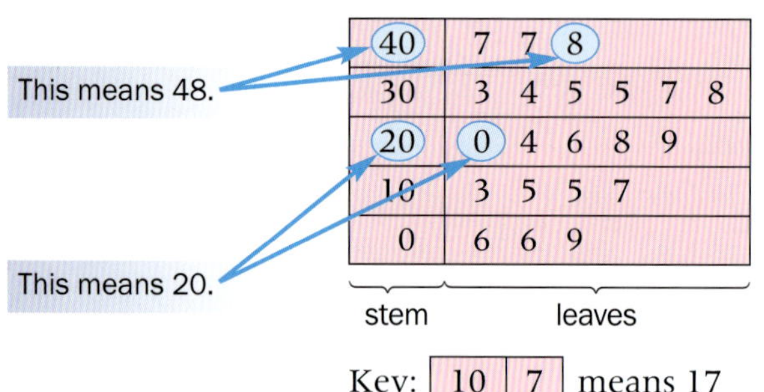

Key: 10 | 7 means 17

Always give the key.

It works by partitioning each number into two parts.

A stem-and-leaf diagram is quick and easy to construct.

- **A stem-and-leaf diagram shows**
  - **the shape of the distribution**
  - **each individual value of the data.**

Gives a 'feel' for the data.

No loss of detail of the data.

This stem-and-leaf diagram is **ordered**, as the data is in numerical order.

A stem-and-leaf diagram is like a bar chart, but with more detail.

stem	leaves
40	7 7 8
30	3 4 5 5 7 8
20	0 4 6 8 9
10	3 5 5 7
0	6 6 9

Key: 10 | 7 means 17

Example

The heights, in centimetres, of 10 students are shown.

168 154 172 167 156
158 163 160 165 154

Show the heights in an ordered stem-and-leaf diagram.

150	
160	
170	

150	4 6 4 8
160	8 7 3 0 5
170	2

Key: 150 | 4 means 154 cm

order →

150	4 4 6 8
160	0 3 5 7 8
170	2

Key: 150 | 4 means 154 cm

You can order the data before you draw the diagram if you want.

Exercise D2.4

1 The numbers of houses on each street of a town are shown.

12	5	25	27	7	35	15	10	22	18
34	30	14	19	20	28	32	4	25	36

a Copy and complete the stem-and-leaf diagram.

0	
10	
20	
30	

Key: | 10 | 8 | means 18 houses

b Redraw the table to give an ordered stem-and-leaf diagram.

2 The weights, in kilograms, of 30 students are shown.

48	47	48	53	61	70	45	56	57	60
42	46	44	55	63	65	49	50	55	65
70	53	64	61	46	47	56	40	41	54

Draw an ordered stem-and-leaf diagram using stems of 40, 50, 60, 70.
Remember to give the key.

3 The exam marks of 40 students are shown.

33	48	18	63	51	66	52	19	55	43
50	35	52	66	43	48	32	18	17	26
5	15	10	25	36	51	61	48	58	68
38	67	11	18	27	30	40	60	46	55

Draw an ordered stem-and-leaf diagram using stems of 0, 10, 20, 30, 40, 50, 60.
Remember to give the key.

4 The attempted heights, in centimetres, during a High Jump event are shown.

214	204	225	230	244
210	207	209	240	232
230	216	242	233	238
206	217	236	216	211
208	230	209	237	241

Draw an ordered stem-and-leaf diagram using stems of 200, 210, 220, 230, 240.

5 The times, in seconds, taken for 24 athletes to run 200 metres are given.

24.5	20.1	20.3	21.4	22.1	22.0	24.8	21.0
25.1	21.5	22.6	23.5	20.9	21.6	23.5	24.0
25.0	21.8	21.8	22.1	21.0	23.5	24.6	21.5

Copy and complete an ordered stem-and-leaf diagram to show this data.

20	
21	
22	
23	
24	
25	

Key: | 20 | 9 | means 20.9 seconds

D2.5 Time series graphs

This spread will show you how to:

- Draw and produce line graphs for time series data

Keywords
Horizontal
Line graph
Time series graph
Trend

You can use a **line graph** to show how data changes as time passes.

The temperature in Belfast is measured every hour from 8 am to 4 pm.

The graph shows the data.

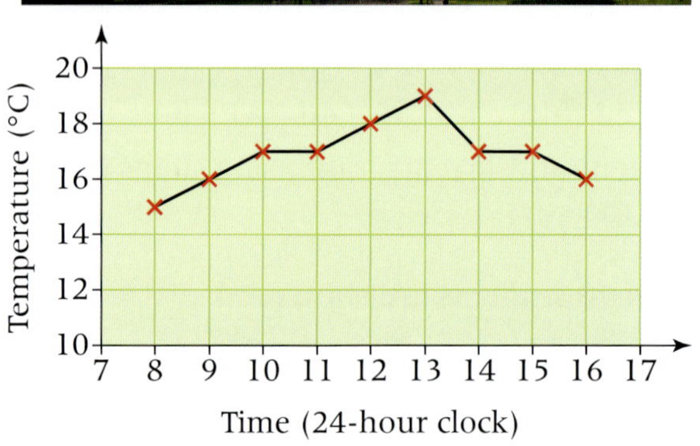

Time is always the **horizontal** axis.

These line graphs are called **time series graphs**.

- A time series graph shows
  - how the data changes, or the **trend**
  - each individual value of the data.

Time could be seconds, minutes, hours, days, weeks, months, years.

Example

The number of Christmas cards Britney received are shown in the table.

Date in December	12	13	14	15	16	17	18	19	20	21	22	23	24
Number of cards	1	2	4	0	3	3	4	3	2	4	7	8	5

Draw a line graph to show this information.

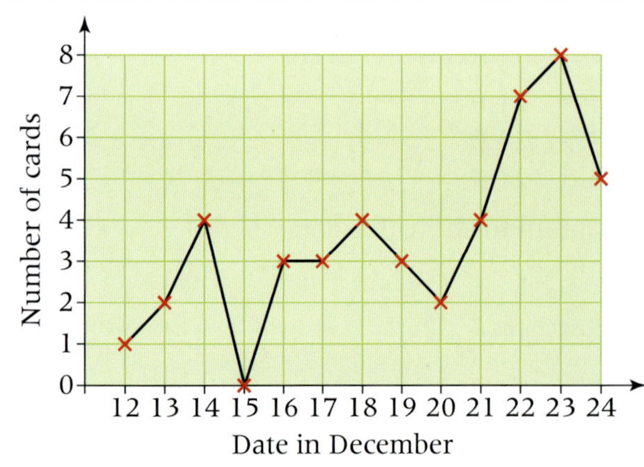

Exercise D2.5

1 The number of photographs taken each day during a 7-day holiday is given.

Sunday	Monday	Tuesday	Wednesday	Thursday	Friday	Saturday
8	12	11	16	19	2	13

Copy and complete the line graph to show this information.

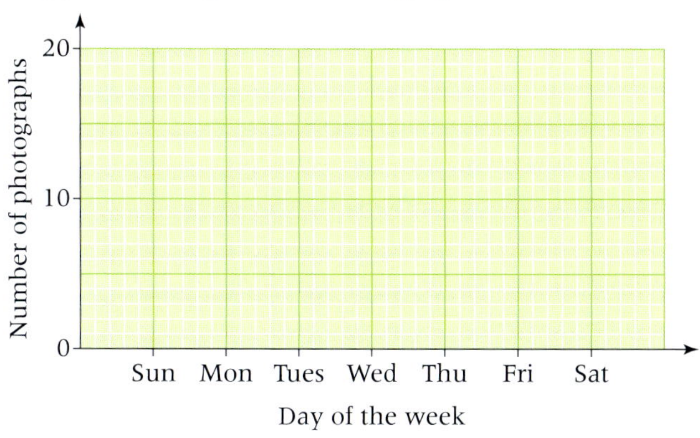

2 The number of sunny days in each month is given.

Jan	Feb	Mar	Apr	May	Jun	Jul	Aug	Sep	Oct	Nov	Dec
9	8	10	15	19	20	21	25	24	18	17	11

Copy and complete the line graph, choosing a suitable vertical scale.

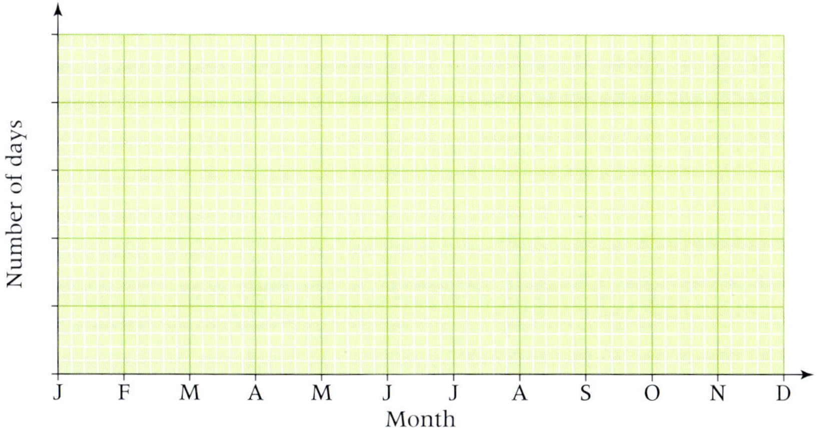

3 The number of points gained by the winners of the Premiership is shown.

Year	1993	1994	1995	1996	1997	1998
Points	84	92	89	82	75	78
Winner	Man Utd	Man Utd	Blackburn	Man Utd	Man Utd	Arsenal

Year	1999	2000	2001	2002	2003	2004	2005
Points	79	91	80	87	83	90	95
Winner	Man Utd	Man Utd	Man Utd	Arsenal	Man Utd	Arsenal	Chelsea

Draw a line graph to show this information.

D2 Exam review

Key objectives

- Draw and produce, using paper and ICT, pie charts for categorical data, and diagrams for data, including line graphs for time series

1 Year 5 from Smalltown School are having a sunflower growing competition.
The final heights of their sunflowers, in centimetres, are shown below

98	80	120	118	92	123
102	98	76	81	99	117
77	93	91	109	125	112
79	84	92	108	122	97

Draw a stem-and-leaf diagram to show this information. (3)

2 The pictogram shows the number of videos borrowed from a shop on Monday and on Tuesday.

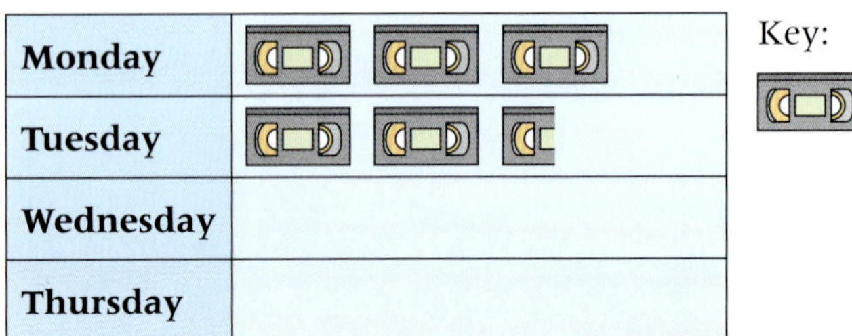

Key: represents 10 videos

a Write down the number of videos borrowed on

i Monday

ii Tuesday. (2)

On Wednesday, 40 videos were borrowed.
On Thursday, 15 videos were borrowed.

b Copy and complete the pictogram, using this information. (2)

(Edexcel Ltd., 2004)

A4 Functions and graphs

This unit will show you how to

- Describe the position of a point on the coordinate grid by giving its coordinates
- Plot and label points in all four quadrants
- Use function machines to solve equations
- Write the input and output values for functions as pairs in a table
- Identify explicit and implicit functions
- Write the linked *x*- and *y*-values as coordinate pairs
- Plot graphs of functions in which *y* is or is not the subject
- Read values from a graph
- Recognise the form of equations of vertical and horizontal lines

Before you start ...

You should be able to answer these questions.

	Review
1 Work out each of these. **a** $2 \times 4 - 1$ **b** $3 \times 5 + 2$ **c** $4 \times 3 + 7$ **d** $6 \times 2 - 11$	Unit N2
2 Work out the value of the letter. **a** $2 + p = 5$ **b** $3 + q = 11$ **c** $5 - r = 3$ **d** $2 - s = -3$	Unit N2, A1
3 Work out the value of each expression when $x = 2$. **a** $x + 6$ **b** $3x - 1$ **c** $2x + 4$ **d** $\frac{x}{4} + 3$	Unit A1
4 Substitute $x = 0$ into these expressions. **a** $2x + 5$ **b** $3x - 4$ **c** $10x + 1$ **d** $4x + 11$	Unit A1
5 Substitute $x = -1$ into these expressions. **a** $2x + 1$ **b** $3x - 2$ **c** $4x + 5$ **d** $2x - 5$	Unit A1

A4.1 Coordinates in all four quadrants

This spread will show you how to:
- Plot points in all four quadrants

Keywords
Axes
Coordinates
Origin
Quadrant
x-coordinate
y-coordinate

A coordinate grid has two **axes**.

The axes meet at the **origin**, O.

- **You can describe the position of a point on the grid by giving its coordinates.**
  The *x*-coordinate is the distance you move along the *x*-axis from O.
  The *y*-coordinate is the distance you move parallel to the *y*-axis from O.

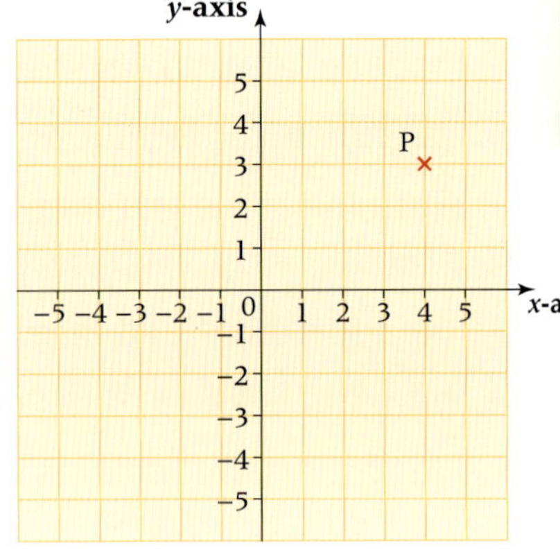

x comes before *y* in the alphabet.

You can extend the *x*- and *y*-axes into negative numbers.
The axes divide the grid into four sections, called **quadrants**.

The coordinates of Q are (−3, 2). From O you move −3 along the *x*-axis and then +2 parallel to the *y*-axis.

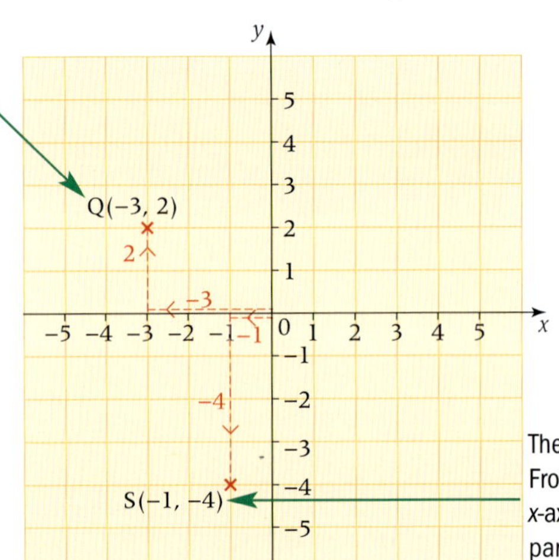

The coordinates of S are (−1, −4). From O you move −1 along the *x*-axis and then −4 (downwards) parallel to the *y*-axis.

- **You can plot and label points in all four quadrants.**

Example

Plot the points
A(1, 4),
B(−2, 3),
C(4, −2) and
D(−2, −3)
on a coordinate grid.

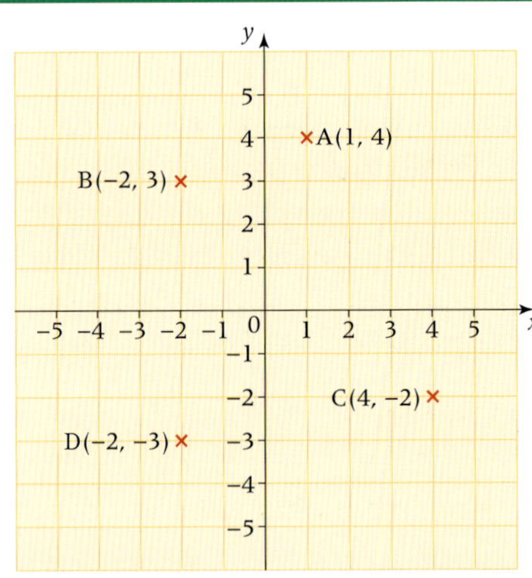

Exercise A4.1

1 Write the coordinates of the points O, P, Q, R, S and T.

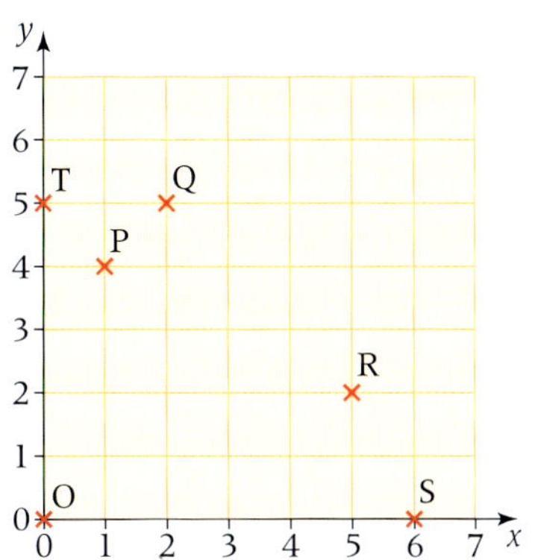

2 Write the coordinates of the points in each diagram.

a

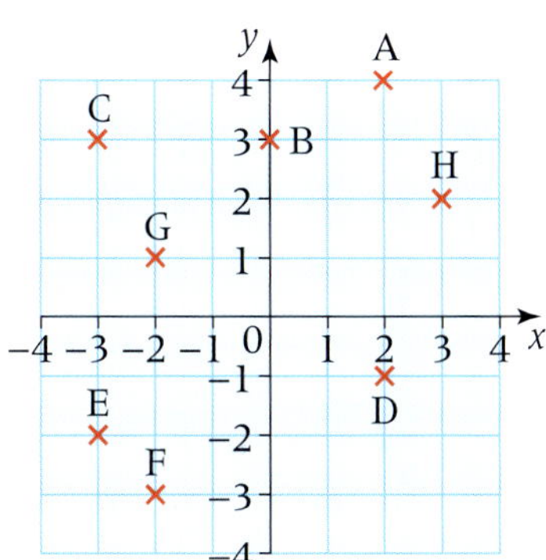

b

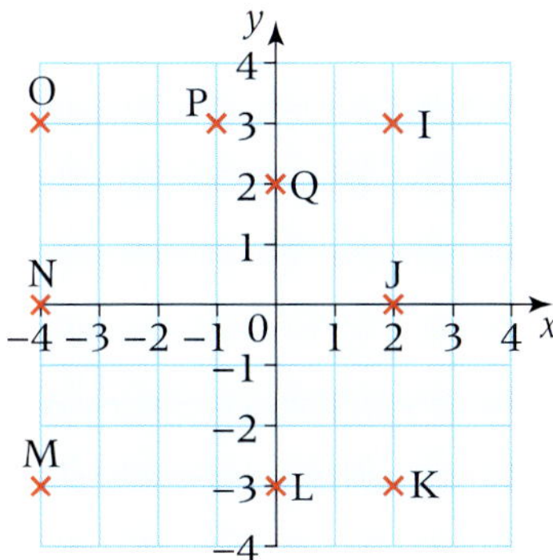

3 **a** Plot these sets of points on a copy of this grid.

i (3, 2) (−1, 2) (−1, −1)
ii (1, 2) (−1, 0) (3, 0)

b Join each set of points in order.

c What is the name of each shape?

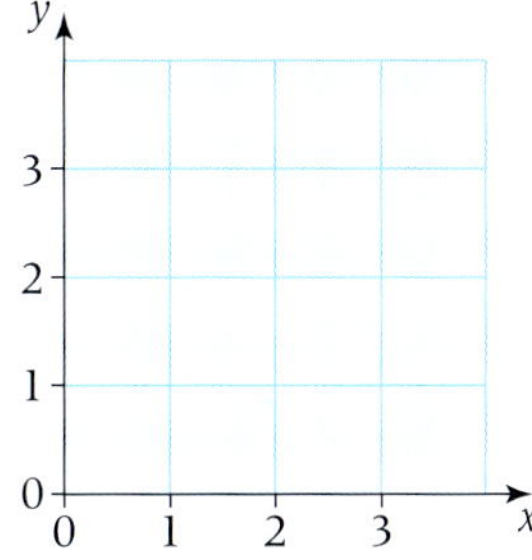

4 **a** Copy a coordinate grid from question **2** onto square grid paper.

b Plot and label these points on your grid.

A(3, 3) B(−2, 3) C(−4, −1)

c A, B and C are three corners of a parallelogram. Write the coordinate of the fourth corner, D.

5 **a** Copy a coordinate grid from question **2** onto square grid paper.

Do **not** copy the points.

b On your grid, draw a triangle with each corner in a different quadrant. Label the corners A, B and C.

c Write the coordinates of A, B and C.

d Repeat for

i a square **ii** a kite, with each corner in a different quadrant.

A4.2 Function machines

This spread will show you how to:

- Use function machines to solve equations
- Write the input and output values for functions as pairs in a table

Keywords
Function
Input
Output

Here is a **function machine**

input		output
0		0
1	→ ×2 →	2
2		4
3		6

When you know the input, you can calculate the output.

In this machine input → output
x → $2x$

You can write this as an equation $y = 2x$.

y is the output.

When you know the value of x you can calculate y.

When $x = 1$, $y = 2 \times 1 = 2$.
When $x = 2$, $y = 2 \times 2 = 4$.

Example

For each function machine, write the output for the inputs given.

a

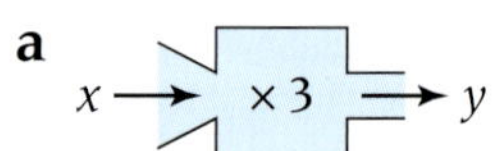

inputs 0, 2, 5, 10

b x → ÷2 → y

inputs 2, 4, 10, 18

a

Input	Output
0	0
2	6
5	15
10	30

b

Input	Output
2	1
4	2
10	5
18	9

- **You can draw a function machine to represent an equation.**

Example

Draw function machines to represent the equations.

a $y = x + 2$ **b** $y = \frac{x}{4}$

a x → +2 → y

b x → ÷4 → y

$\frac{x}{4}$ means $x \div 4$.

Functions can have more than one step.

Example

a Draw a function machine for the equation $y = 3x - 2$.
b Work out the outputs for these inputs: 0, 2, 4.

$3x = 3 \times x$

a x → ×3 → −2 → y

b $0 \rightarrow 0 \times 3 - 2 = 0 - 2 = -2$
$2 \rightarrow 2 \times 3 - 2 = 6 - 2 = 4$
$4 \rightarrow 4 \times 3 - 2 = 12 - 2 = 10$

First multiply x by 3 then subtract 2.

Exercise A4.2

1 Copy and complete the tables for these function machines.

a $+3$

Input	Output
0	3
1	4
4	
7	
9	12

b $\times 3$

Input	Output
0	
3	9
6	
7	
9	27

c

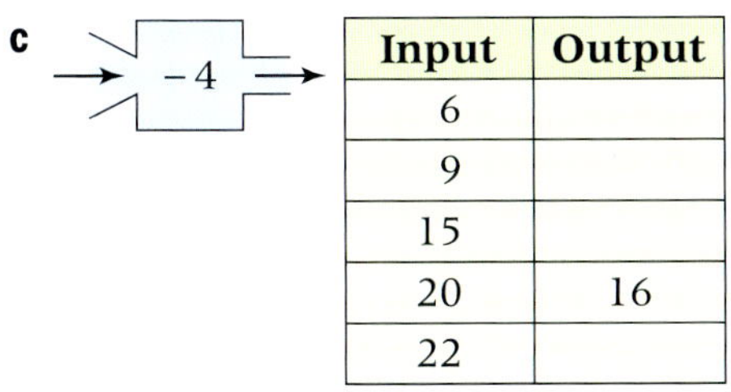

Input	Output
6	
9	
15	
20	16
22	

d $\div 3$

Input	Output
6	
12	
15	
	8
30	

2 Draw function machines for these equations.

a $y = x + 5$ **b** $y = x - 3$ **c** $y = 3x$ **d** $y = x + 1$

e $y = \frac{x}{3}$ **f** $y = x - 2$ **g** $y = 6x$ **h** $y = \frac{x}{2}$

3 Draw a function machine for each equation.
Use your machines to calculate the outputs when

i $x = 1$ **ii** $x = 2$ **iii** $x = 3$

a $y = 2x$ **b** $y = x + 4$ **c** $y = x - 1$ **d** $y = \frac{x}{4}$

4 Copy and complete the tables for these two-step function machines.

a

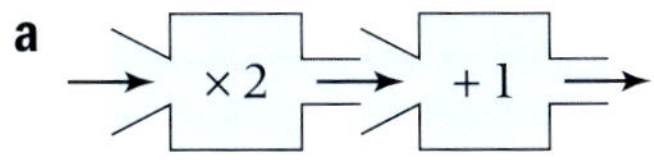

Input	Output
0	1
1	3
2	
3	
4	
10	21

b $\times 3$ -1

Input	Output
2	
4	11
5	
6	
7	20
10	

5 Draw function machines for these equations.
Use your machines to calculate the outputs when

i $x = -2$ **ii** $x = 2$ **iii** $x = 4$

a $y = 2x - 1$ **b** $y = 3x + 2$ **c** $y = \frac{x}{2} - 1$ **d** $y = 4x - 5$

A4.3 Drawing tables of values

This spread will show you how to:

- Write the input and output values for functions as pairs in a table
- Identify explicit and implicit functions

Keyword
Table of values

This function machine represents the equation $y = x - 3$.

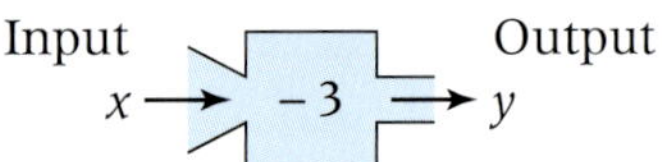

You can work out the value of y for different values of x.

Input		Output
$x = 3$		$y = 0$
$x = 5$		$y = 2$
$x = 7$	→ −3 →	$y = 4$
$x = 12$		$y = 9$
$x = 15$		$y = 12$

You can write the pairs of x and y values in a **table of values**.

x	3	5	7	12	15
y	0	2	4	9	12

Input this x-value and you get this y-value.

Example

Complete the table of values for this function machine.

x → +4 → y

x	0	2	3	6	9
y					

Input each x value in turn.

$x = 0$ → +4 → $y = 4$

$x = 2$ → +4 → $y = 6$

x	0	2	3	6	9
y	4	6	7	10	13

Example

Complete the table of values for the function $x + y = 5$.

x	0	2	4	5
y				

When $x = 0$ the equation is $0 + y = 5$, so $y = 5$
When $x = 2$ $2 + y = 5$, so $y = 3$
When $x = 4$ $4 + y = 5$, so $y = 1$
When $x = 5$ $5 + y + 5$, so $y = 0$

x	0	2	4	5
y	5	3	1	0

$x + y = 5$ is an **implicit** function. x and y appear on the same side of the equals sign.

Exercise A4.3

1 Copy and complete the table of values for each function machine.

a Input $x \rightarrow$ [$+5$] $\rightarrow y$ Output

x	0	2	3	6	9
y					

b Input $x \rightarrow$ [$\times 3$] $\rightarrow y$ Output

x	0	2	3	6	9
y					

c Input $x \rightarrow$ [-6] $\rightarrow y$ Output

x	6	7	8	9	10
y					

d Input $x \rightarrow$ [$\div 4$] $\rightarrow y$ Output

x	4	8	12	16	20
y					

2 Copy and complete the table of values for each function machine.

a Input $x \rightarrow$ [$+3$] $\rightarrow y$ Output

x	−5	−2	3	6	9
y					

b Input $x \rightarrow$ [$\times 6$] $\rightarrow y$ Output

x	−2	0	3	5	7
y					

c Input $x \rightarrow$ [-4] $\rightarrow y$ Output

x	−2	0	3	5	7
y					

d Input $x \rightarrow$ [$\div 3$] $\rightarrow y$ Output

x	−6	−3	3	6	9
y					

3 Match each equation to a table of values.

a $y = 3x$

b $y = x + 4$

c $y = 3x + 4$

d $y = 4x + 3$

i

x	−2	0	2	4	6
y	−2	4	10	16	22

ii

x	−1	2	3	5	7
y	−1	11	15	23	31

iii

x	−2	0	2	4	6
y	−6	0	6	12	18

iv

x	−4	−2	1	2	4
y	0	2	5	6	8

A4.4 Plotting graphs from tables of values

This spread will show you how to:

- Plot graphs of functions in which y is or is not the subject
- Read values from a graph

Keywords
Coordinate pairs
Satisfy

Here is a table of values for the equation

$y = 3x$

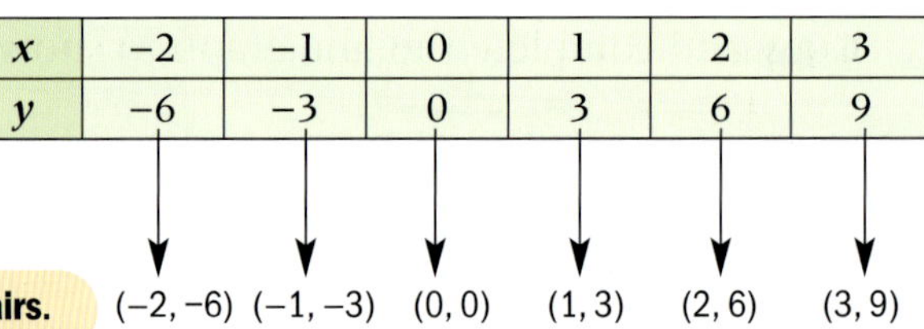

x	−2	−1	0	1	2	3
y	−6	−3	0	3	6	9

In the table, each y-value links to an x-value.

- **You can write the linked x- and y-values as coordinate pairs.**

- **You can plot the coordinate pairs on a grid.**

Plot each point.

Join the points with a straight line.

All the points on the line **satisfy** the equation $y = 3x$. This means that for every point on the line,

$y\text{-value} = x\text{-value} \times 3$

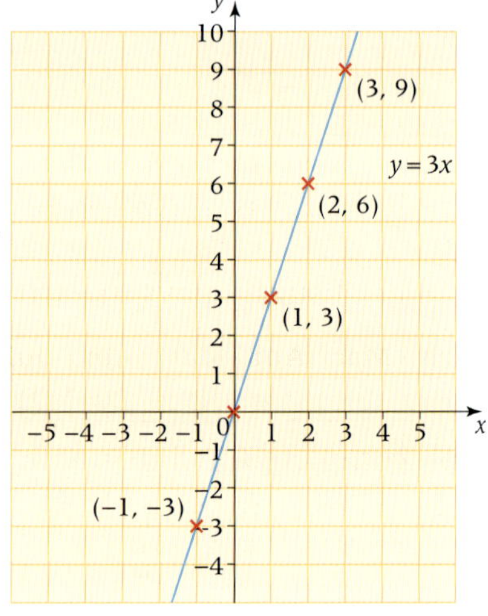

- **You can read values from a graph.**

Example

Here is the graph of $y = 2x + 3$.
Use the graph to find

a the value of x when $y = 7$
b the value of y when $x = 3\frac{1}{2}$.

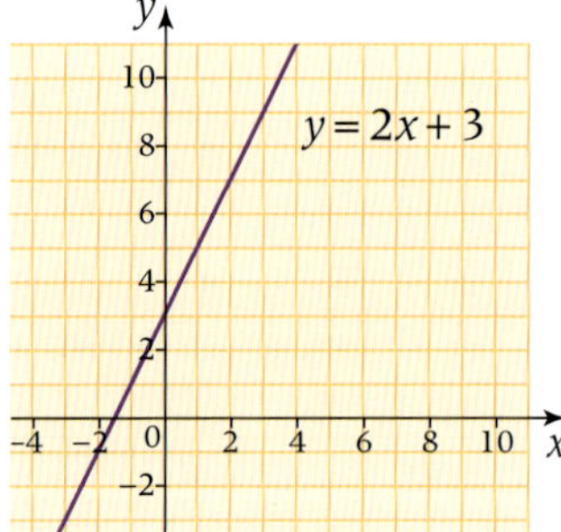

a Find $y = 7$ on the y-axis.
Draw a horizontal line to the graph line.
Draw a vertical line to the x-axis.
Read off the value of x.
$x = 2$

b Find $x = 3\frac{1}{2}$ on the x-axis.
Draw a vertical line to the graph line.
Draw a horizontal line to the y-axis.
Read off the value of y.
$y = 10$

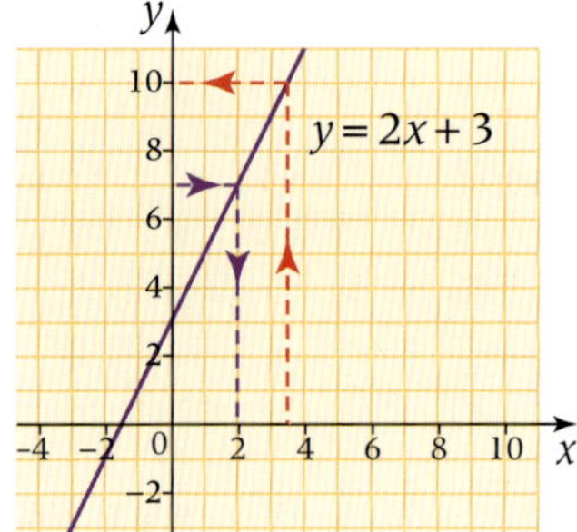

This topic is extended to the intersection of two lines on page 376.

Exercise A4.4

1 Here is the graph of $y = 2x - 1$.

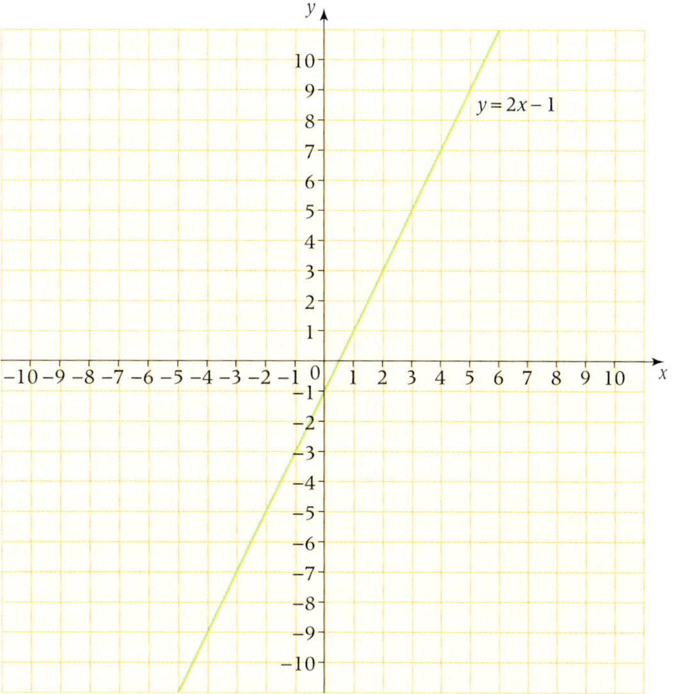

Find the value of x when

a $y = 7$

b $y = 0$

c $y = -9$

d $y = -6$

Find the value of y when

e $x = 4$

f $x = -4$

g $x = 2\frac{1}{2}$

h $x = -1\frac{1}{2}$

2 **a** Copy and complete this table of values for the equation $y = 3x + 1$.

x	−3	−2	0	1	2
y				4	

b Copy and complete this list of coordinate pairs from your table.

(−3, __) (−2, ___) (0, ___) (1, 4) (2, ___)

c Copy the grid in question **1** onto square grid paper.

Do not copy the graph line.

d Plot the points from part **b** on your grid.
Join them with a straight line.

e From your graph, find the value of x when $y = -2$.

3 **a** Copy and complete this table of values for the equation $y = \frac{x}{2}$.

x	−4	−2	0	2	6
y	−2				

b Write a list of coordinate pairs from your table.

c Copy the grid in question **1** onto square grid paper.

Do not copy the graph line.

d Plot the points from part **b** on your grid.
Join them with a straight line.

e From your graph, find the value of y when $x = 9$.

4 Repeat question **3** for the equation $x + y = 7$.

A4.5 Horizontal and vertical lines

This spread will show you how to:

- Recognise the form of equations of vertical and horizontal lines

Keywords
Horizontal
Parallel
Vertical

Horizontal and vertical lines are easy to identify on a graph.
You can also identify them by their equation.

- **The equation of a vertical line is x = a number.**
- **The equation of a horizontal line is y = a number.**

A and B are **vertical** lines.
They are **parallel** to the y-axis.

For line A the points labelled are (1, −4) (1, 2) (1, 7).
Every point on this line has x-coordinate 1.

The equation of line A is $x = 1$.

For line B the points labelled are (−2, −5) (−2, −1) (−2, 6).
Every point on this line has x-coordinate −2.

The equation of line B is $x = -2$.

C and D are **horizontal** lines.
They are parallel to the x-axis.

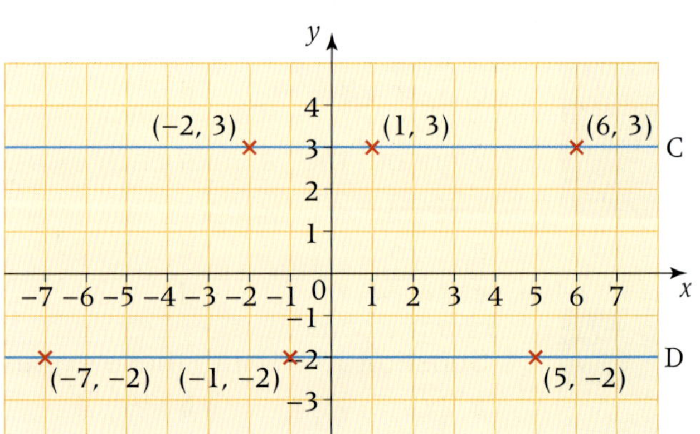

For line C the points labelled are (−2, 3) (1, 3) (6, 3).
Every point on this line has y-coordinate 3.

The equation of line C is $y = 3$.

For line D the points labelled are (−7, −2) (−1, −2) (5, −2).
Every point on this line has y-coordinate −2.

The equation of line D is $y = -2$.

Exercise A4.5

1 Name the lines on the grid.

Copy and complete

The equation of the x-axis is $y =$ _____

The equation of the y-axis is _____

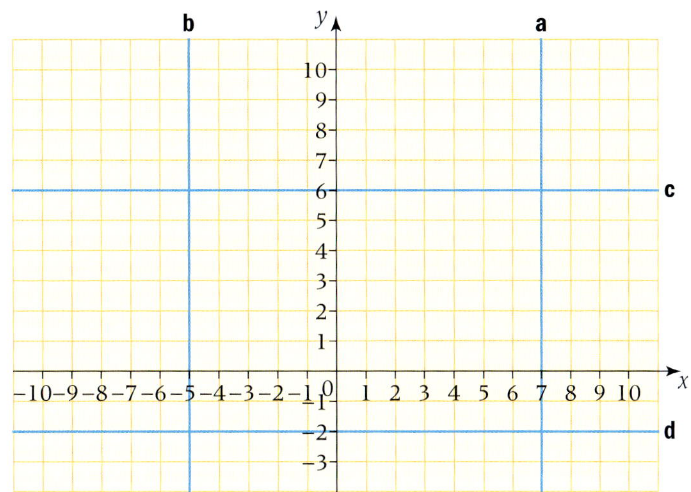

2 Copy the axes from question **1** onto square grid paper.
For each part, plot the points and join them to make a straight line.
Name each line.

Do not copy the graph lines.

a Line 1: (−4, −6) (−4, −3) (−4, 1) (−4, 7)

b Line 2: (−7, 1) (−2, 1) (4, 1) (6, 1)

c Line 3: (−3, −3) (−1, −3) (4, −3) (8, −3)

3 Name these lines without plotting the points.

a Line 1: (−5, 5) (−2, 5) (5, 5) (9, 5)

b Line 2: (−8, −3) (−8, 3) (−8, 6) (−8, 10)

4 Copy the axes from question **1** onto square grid paper.

Do not copy the graph lines.

a Draw these lines on your grid.

$x = -2 \quad y = -4 \quad x = 7 \quad y = 1$

b What shape do your lines enclose?

5 Copy the axes from question **1** onto square grid paper.

Do not copy the graph lines.

a Draw four lines to enclose a square.

b Write the equations of your four lines.

c Write the coordinates of the vertices of your square.

6 Match the points in set B with lines they lie on in set A.

Some points may lie on more than one line.

Set A

$x = 3$	$x = -5$	$x = 6$	$x = -2$
$y = 4$	$y = -2$	$y = 5$	$y = -3$

Set B

(−2, 6)	(3, 3)	(−5, −5)	(6, 5)	(2, 4)
(6, −3)	(−2, −3)	(−5, 4)	(−5, 7)	(3, 6)
(6, 1)	(8, −3)	(3, 4)	(1, −3)	(5, −2)
(6, 3)	(−3, 4)	(3, −3)	(−2, −2)	(−5, 5)

A4 Exam review

Key objectives

- Use the conventions for coordinates in the plane
- Plot points in all four quadrants
- Plot graphs of functions in which y is given explicitly in terms of x or implicitly

1 a Give the coordinates of
i point A **ii** point B. (2)

b Copy the grid and plot point C, with coordinates $(-2, 5)$. (2)

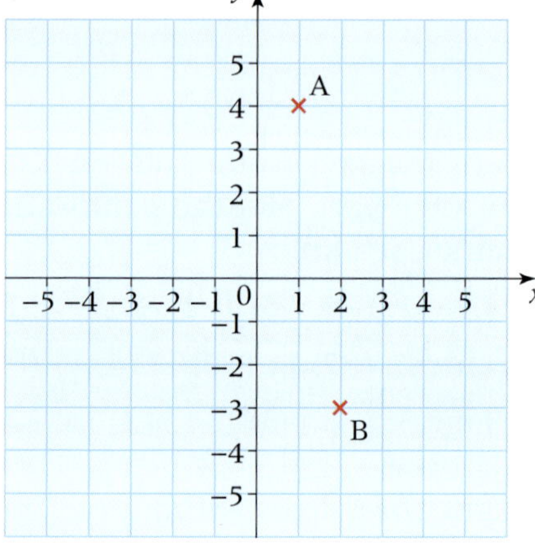

2 a Copy and complete the table of values for $y = 3x + 2$. (2)

x	-2	-1	0	1	2
y		-1		5	

b Copy the grid and draw the graph of $y = 3x + 2$. (2)

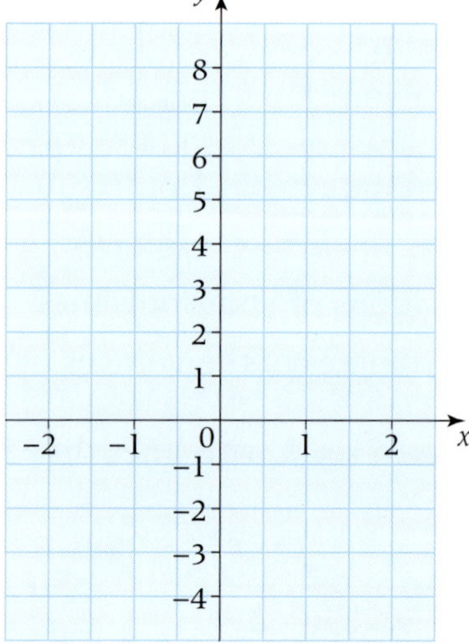

(Edexcel Ltd., 2005)

D3 Probability

This unit will show you how to

- Understand and use the vocabulary of probability
- Understand and use the probability scale
- List all outcomes for events in a systematic way
- Calculate probabilities
- Understand and use estimates or measures of probability
- Identify different mutually exclusive events and know that the sum of the probabilities of all these outcomes is 1

Before you start ...

You should be able to answer these questions.

Review

1 Convert these decimals to fractions. Express your answers in their simplest form. — Unit N3

a 0.1 **b** 0.5 **c** 0.2
d 0.7 **e** 0.6

2 Order these decimals in size, smallest first. — Unit N3

a 0.25 0.2 0.3
b 0.7 0.8 0.75
c 0.85 0.8 1

3 State the shaded part of each diagram as a fraction. — Unit N3

a 
b 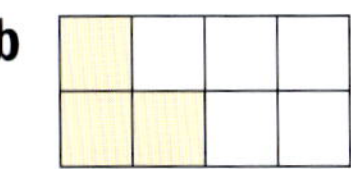

4 Work out each of these. — Unit N3

a $\frac{1}{3} + \frac{2}{3}$ **b** $\frac{3}{10} + \frac{7}{10}$
c $1 - \frac{9}{10}$ **d** $1 - \frac{4}{5}$
e $1 - \frac{3}{4}$

5 Work out each of these. — Unit N3

a $1 - 0.2$ **b** $1 - 0.7$
c $1 - 0.9$

D3.1 Language of probability

This spread will show you how to:

- Understand and use the vocabulary of probability

Keywords
Certain
Chance
Even chance
Event
Impossible
Outcome

- An **event** is an activity, for example, spinning a coin.
- The **outcome** of spinning a coin is either a Head or a Tail.
- The **chance** of an outcome happening can be measured using words, for example, likely, certain, impossible.

The outcome of an event may be

impossible

Choosing a blue ball

possible

Choosing a blue ball

certain

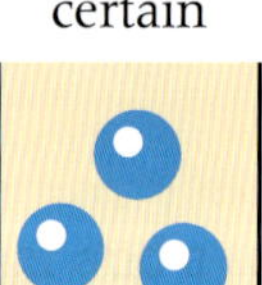

Choosing a blue ball

- **You can use words to describe how likely it is that an outcome will happen.**

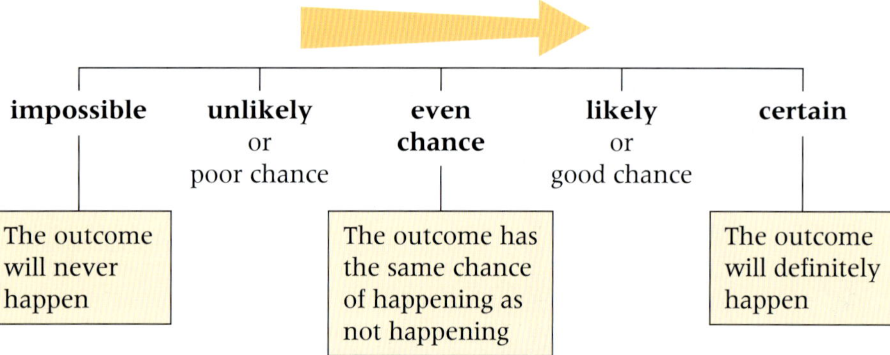

Example

Use impossible, unlikely, even chance, likely or certain to describe these outcomes.

a You get a Head when you spin a coin.
b A baby will be born tomorrow.
c It will snow on the Costa del Sol in Spain next winter. (It did in 2005!)

---

a Even chance
b Likely
c Unlikely

Exercise D3.1

1

impossible	unlikely or poor chance	even chance	likely or good chance	certain

Use impossible, unlikely, even chance, likely or certain to describe these outcomes.

a The day after Christmas Eve is Christmas Day.

b The day after Thursday is Wednesday.

c The sun will rise tomorrow.

d You get a Tail when you spin a coin.

e You are dealt a red card from a shuffled pack of cards.

Hint for **e**:
In a pack of normal playing cards there are 26 red cards and 26 black cards.

f It will rain sometime next year.

g You will climb to the top of Mount Everest tomorrow.

h You will roll a 7 on an ordinary dice.

i You will roll an even number on a dice.

j You will roll a 6 on a dice.

k You will pick a red ball from a bag that contains 8 red balls and 2 blue balls.

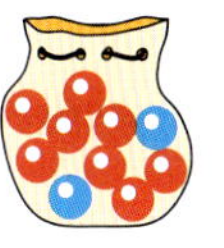

l You will pick a blue ball from a bag that contains 8 red balls and 2 blue balls.

2 Put these events in the order that they might happen, starting with impossible and finishing with certain.

A

Picking a red ball

B

Picking a red ball

C

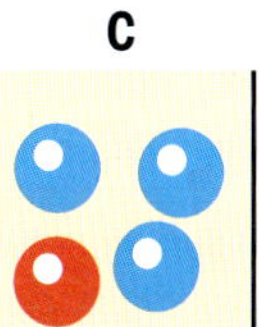

Picking a red ball

D

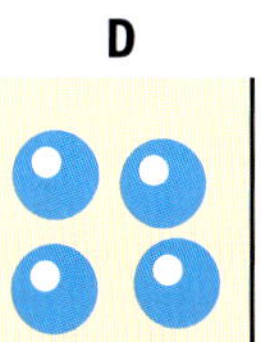

Picking a red ball

E

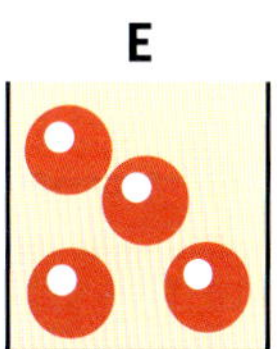

Picking a red ball

3 Five different spinners are shown. For each spinner, state which colour is the most likely. Give a reason for your answers.

a

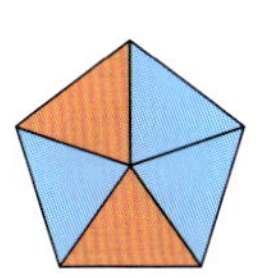

b

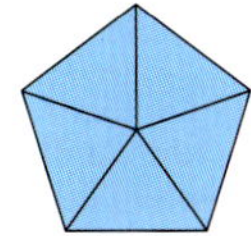

c

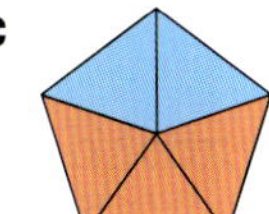

d

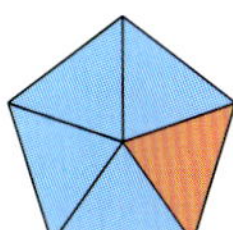

e

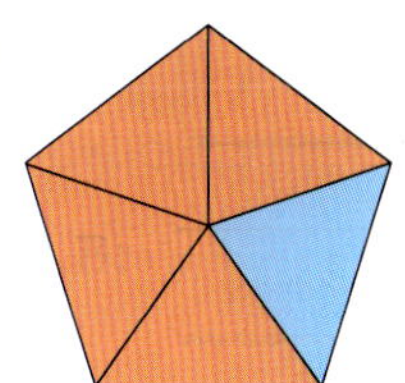

D3.2 Probability scale

This spread will show you how to:

- Understand and use the probability scale

Keywords
Probability
Probability scale

- You can use words to describe how likely it is that an outcome will happen.

more and more likely to happen

impossible — unlikely or poor chance — even chance — likely or good chance — certain

There are gaps in the scale.

- You can use a number scale to be more accurate.

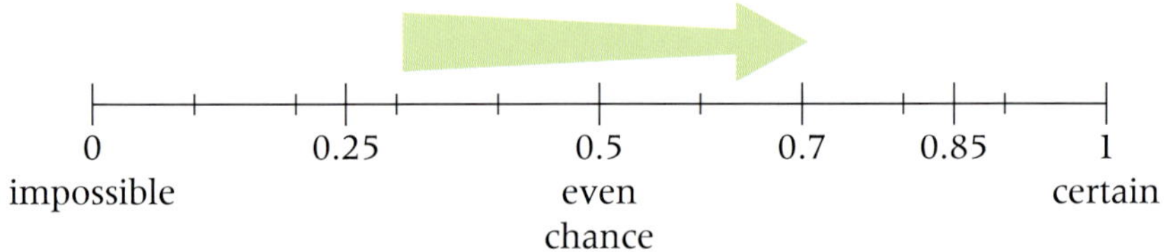

0 means impossible.
1 means certain.

This number is called the **probability**.

The probability measures how likely it is that an outcome will happen.

- **All probabilities have a value between 0 and 1 and can be marked on a probability scale.**

You can use fractions, decimals or percentages on the probability scale.

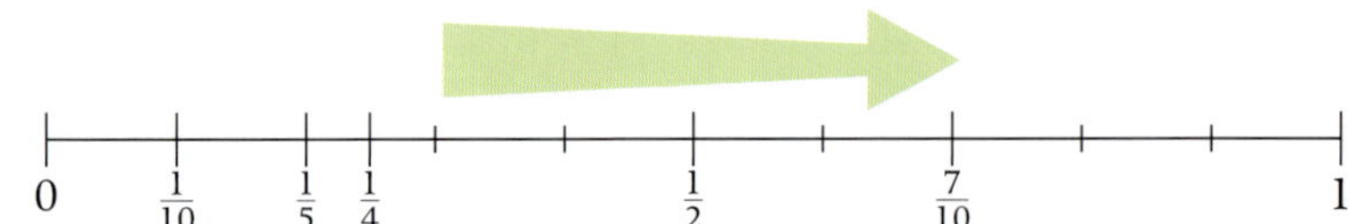

$\frac{1}{2} = 0.5$

$\frac{7}{10} = 0.7$

Example

Mark the position of these outcomes on a probability scale.

a The sun setting today.

b An even number when you roll a dice.

c A glass breaking if dropped onto a stone floor.

0 — 0.5 (**b**) — **c** — 1 (**a**)

c is an approximate answer.

Exercise D3.2

1 Draw a 10 cm line. Put a mark at every centimetre.
Label the marks 0, 0.1, 0.2, ..., 0.8, 0.9, 1 as shown.

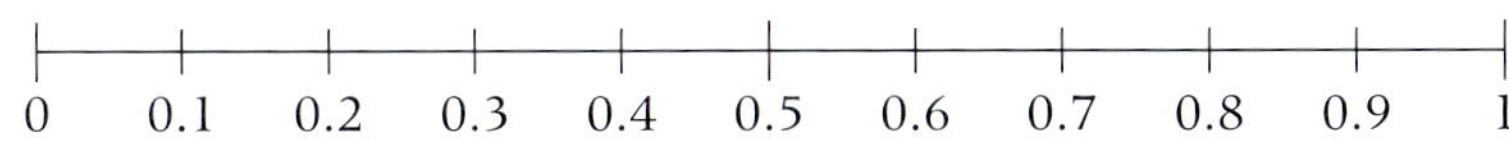

On your probability scale, mark the position of

a an impossible outcome

b a certain outcome.

2 Draw a probability scale. Label it 0, 0.5 and 1.
Mark the points **a**, **b** and **c** on the scale to show the probability of these outcomes.

a You get a Tail when you spin a coin.

b Tomorrow will be Friday.

c You will eat tomorrow.

3 Draw a probability scale as in question **2**.
Mark these probabilities on your scale.

a You roll a 7 on a dice.

b You roll a number 6 or less.

c You roll an odd number.

4 Draw a probability scale as in question **2**.
Mark these probabilities on your scale.

a You will be absent from school tomorrow.

b You will get wet on your way home today.

c You will watch television tonight.

5 Draw a probability scale as in question **2**.
Mark these probabilities on your scale.

a The sun will shine in Spain this summer.

b The bottom card of a shuffled pack of playing cards is Red.

c You roll a 6 on a dice.

6 Wayne says that the probability that he will go to school tomorrow is 1.2. Explain why the number 1.2 must be wrong.

7 A bag contains 1 green and 2 red balls. One ball is picked out. Draw a probability scale. Label it 0, 0.5 and 1.
Mark the points **a**, **b** and **c** on the scale to show the probability of these outcomes.

a a blue ball **b** a green ball **c** a red ball.

D3.3 Equally likely outcomes

This spread will show you how to:

- List all outcomes for single events in a systematic way
- Understand and use estimates or measures of probability including equally likely outcomes

Keywords
Equally likely
Event
Outcome
Probability
Systematically

- An **event** is an activity, for example, rolling a dice.
- The possible **outcomes** are 1, 2, 3, 4, 5 and 6.

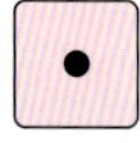 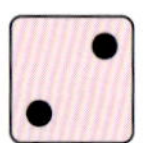    

The six outcomes are drawn in order or **systematically**.

Each outcome is **equally likely** as the faces of the dice are identical in size and shape.

The **probability** is a number that measures how likely it is that an outcome will happen.

You can calculate the probability using this formula.

- **Probability of an outcome happening** $= \dfrac{\text{number of ways the outcome can happen}}{\text{total number of all possible outcomes}}$

All probabilities have a value between 0 and 1.

0 means impossible.
1 means certain.

Example

A counter is taken out of the bucket.

a List the possible outcomes for this event.
b Find the probability that a blue counter is taken out.
c Find the probability that a red counter is taken out.
d Find the probability that a green counter is taken out.

a Blue, Blue, Blue, Red, Red

b The number of ways of taking a blue counter is 3.
The total number of all possible outcomes is 5.
The probability of taking a blue counter is $\frac{3}{5}$.

c The number of ways of taking a red counter is 2.
The total number of all possible outcomes is 5.
Probability of taking a red counter is $\frac{2}{5}$.

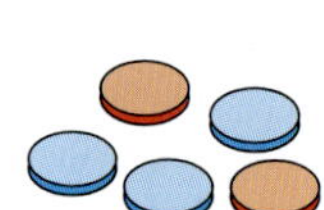

d The number of ways of taking a green counter is 0.
The total number of all possible outcomes is 5.
Probability of taking a green counter is $\frac{0}{5} = 0$.

Exercise D3.3

1 List all the possible outcomes for these events.

a spinning a coin

b spinning this spinner

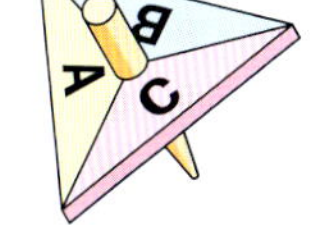

c rolling an ordinary dice

d trying to catch a ball

e picking a letter from Q W E R T Y

2 List the three possible outcomes when a ball is taken from the bag.

3 List the three possible outcomes when a ball is taken from the bag.

4 List the three possible outcomes when a ball is taken from the bag.

5 List the possible outcomes when the spinner is spun.

For the questions **6–10**, the outcomes are equally likely.

6 **a** Which colour is the most likely?

b Calculate the probability of spinning orange.

c Calculate the probability of spinning pink.

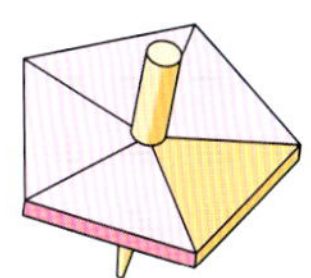

7 Calculate the probability of spinning a Head with a coin.

8 An ordinary dice is rolled. Calculate the probability of rolling

a a 3 **b** an even number **c** a 7.

9 Some children cannot decide whether to go swimming, skating or bowling. Three cards are put into a tin. One card is taken out to decide the activity.
Calculate the probability that the children go bowling.

10 A tetrahedron dice has only four faces.
The four outcomes are 1, 2, 3 or 4.
Calculate the probability that the score is

a a 3 **b** an even number **c** an odd number **d** a 5.

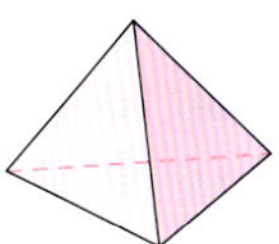

D3.4 Calculating probabilities

This spread will show you how to:

- Calculate probabilities
- Understand and use the probability scale

Keywords
Probability
Probability scale

You can measure how likely it is that an outcome will happen by using a number between 0 and 1 called the **probability**.

0 means impossible.
1 means certain.

The probability can be represented on a **probability scale**.

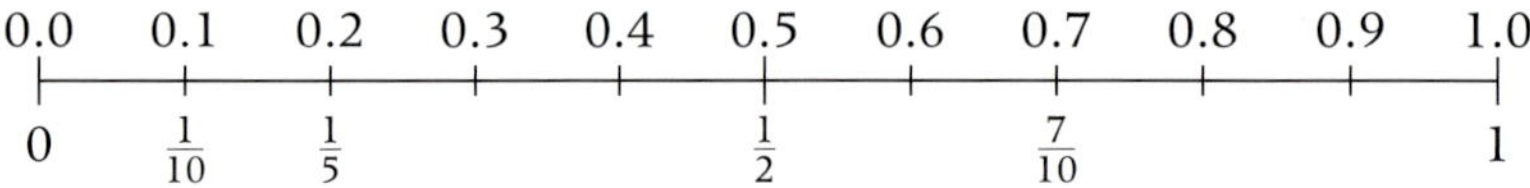

You can use decimals or fractions.

You can calculate the probability using this formula.

- **Probability of an outcome happening** $= \dfrac{\textbf{number of ways the outcome can happen}}{\textbf{total number of all possible outcomes}}$

This formula was introduced on page 138.

The probability of an outcome can be written as P(outcome).

Example

The numbers 1 to 10 are put into a bag.

1 **2** **3** **4** **5** **6** **7** **8** **9** **10**

Karen picks one number out of the bag without looking.
Calculate the probability she picks

a the number 6

b a number greater than 7

c a multiple of 4

d a square number.

---

a There are 10 possible outcomes.
There is one 6.
$P(6) = \frac{1}{10}$

b There are three numbers greater than 7.
$P(\text{greater than } 7) = \frac{3}{10}$

7 is not included in 'greater than 7'

c There are two numbers that are multiples of 4 (4, 8).
$P(\text{multiple of } 4) = \frac{2}{10} = \frac{1}{5}$

d There are three square numbers (1, 4, 9).
$P(\text{square number}) = \frac{3}{10}$

Exercise D3.4

For the following questions, the outcomes are equally likely.

1 A bag contains one yellow ball and four red balls. One ball is taken out of the bag. Calculate the probability that the ball is

a red **b** yellow **c** blue.

2 A bag contains five red balls. One ball is taken out of the bag. Calculate the probability that the ball is

a red **b** blue.

3 A bag contains three red and two yellow balls. One ball is taken out of the bag. Calculate the probability that the ball is

a red **b** yellow **c** blue.

4 This spinner is spun. Calculate the probability that the arrow lands on

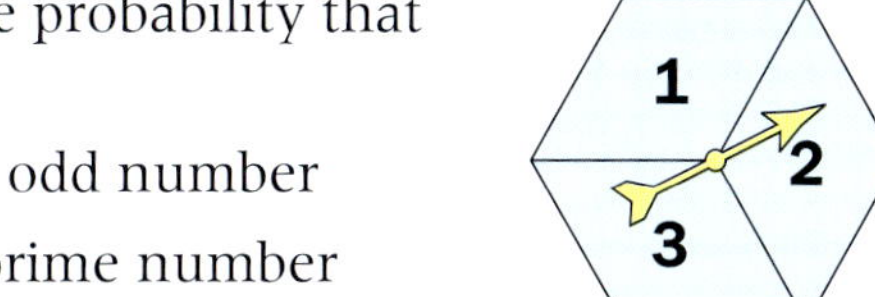

a a 2 **b** an odd number

c a square number **d** a prime number

e a number 3 or less.

5 If one letter is chosen at random from the word ISOSCELES, what is the probability that the letter is

a a C **b** an E **c** an S

d a vowel **e** a consonant?

6 If one letter is chosen at random from the word PARALLELOGRAM, what is the probability that the letter is

a an O **b** an A **c** an L

d a vowel **e** a consonant?

7 These shapes are drawn on six cards.
One card is picked at random.
What is the probability that the shape on the card

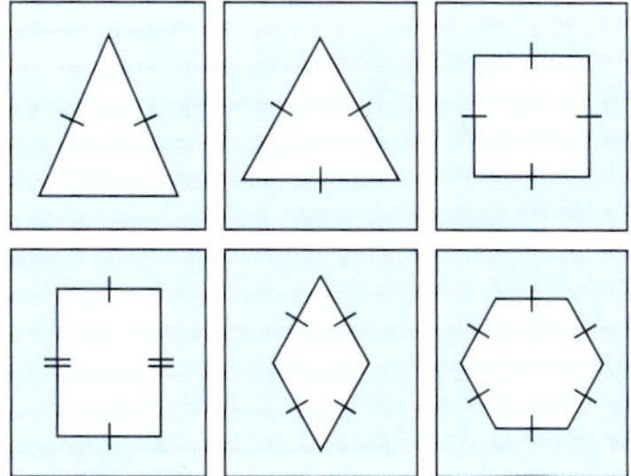

a is an isosceles triangle **b** is a triangle

c is a quadrilateral **d** is a square

e is an octagon **f** has more than four sides

g has all of its sides equal **h** is a regular shape?

8 There are 14 boys and 16 girls in a class. If one student is chosen at random, what is the probability that the student is

a a boy **b** a girl?

D3.5 Probably not

This spread will show you how to:

- Identify different mutually exclusive outcomes and know the sum of the probabilities of all these outcomes is 1
- Understand and use the probability scale

Keywords
Event
Mutually exclusive
Outcome

An **event** is an activity, for example, picking one ball from a box.

The possible **outcomes** are Green, Green and Red.

Probability of picking a green ball = $\frac{2}{3}$

Probability of picking a red ball = $\frac{1}{3}$

The probabilities of the outcomes add up to 1.

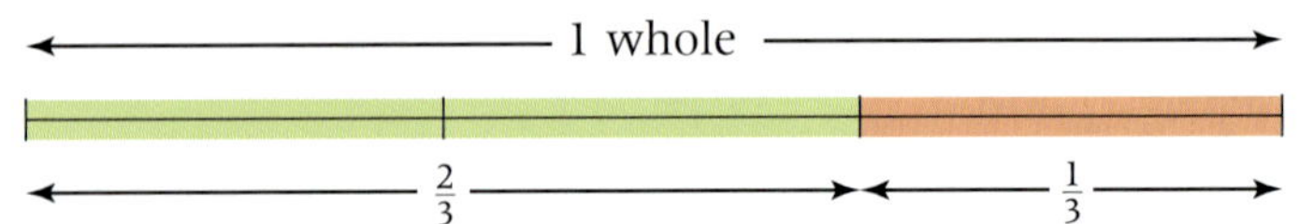

$\frac{2}{3} + \frac{1}{3} = 1$

- **The probabilities of all possible outcomes of an event add up to 1.**

Probability of picking a green ball = $\frac{2}{3}$.

Probability of **not** picking a green ball = $1 - \frac{2}{3} = \frac{1}{3}$.

- **Probability of an outcome not happening = 1 − Probability of the outcome happening**

Example

The probability that Emma does her homework is $\frac{7}{10}$.
Calculate the probability that she does not do her homework.

$1 - \frac{7}{10} = \frac{3}{10}$

Probability she does not do her homework = $\frac{3}{10}$.

These outcomes are **mutually exclusive** because if you get one outcome you cannot get the other one.

- **Mutually exclusive outcomes cannot occur at the same time.**

This topic is extended to calculating mutually exclusive outcomes on page 396.

Example

Which of these outcomes are mutually exclusive?

A	B	C
Spinning a coin: Head, Tail	Pressing a light switch: Light on, light off	Eating breakfast: Cornflakes, muesli

A, **B**.
C is not mutually exclusive as you could choose another type of cereal.

Exercise D3.5

1 A bag contains one blue and four red balls. One ball is taken out.

a Calculate the probability that the ball is blue.

b Calculate the probability that the ball is red.

c Calculate the sum of these answers.

2 A bag contains three blue and seven green balls. One ball is taken out.

a Calculate the probability that the ball is blue.

b Calculate the probability that the ball is green.

c Calculate the sum of these answers.

3 An ordinary dice is rolled.

a Calculate the probability of rolling

i a 1 **ii** a 2 **iii** a 3 **iv** a 4 **v** a 5 **vi** a 6.

b Calculate the sum of the answers.

4 A box contains many red and blue counters.
One counter is taken out. The probability that a red counter is taken out is 0.4.
Calculate the probability that a blue counter is taken out.

5 The probability of winning a raffle is 0.1.
Calculate the probability of not winning the raffle.

M122756Y
016
M122756Y

6 A spinner is made from blue, green, yellow and pink triangles.

a Calculate the probability of spinning blue.

b Calculate the probability of not spinning blue.

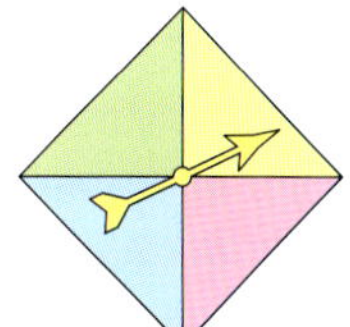

7 A drawing pin is dropped. The probability that the pin lands point up is $\frac{4}{5}$.
What is the probability that the pin does not land point up?

8 The probability of rain at Styhead Tarn in the Lake District is $\frac{7}{10}$.
Calculate the probability of it not raining at Styhead Tarn.

9 The letters of the word EQUILATERAL are put in a bag. One letter is taken out.

a Calculate the probability of choosing an E.

b Calculate the probability of not choosing an E.

D3 Exam review

Key objectives

- Understand and use the probability scale
- Understand and use estimates or measures of probability from theoretical models (including equally likely outcomes)
- List all outcomes for single events, and for two successive events, in a systematic way

1 The diagram shows a fair 5-sided spinner. When the spinner is spun, it lands on one of the numbered sections.

Copy the probability scale below.

a Mark on your scale the probability that the spinner will land

i on the number 3

ii on an even number

iii not on number 5

(label the points i, ii and iii)

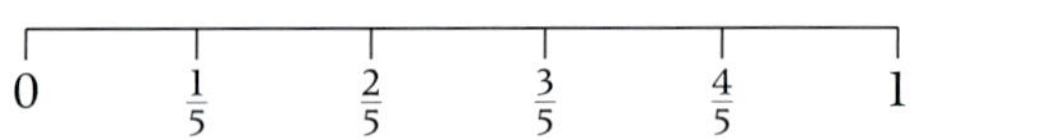

(3)

b Write your probabilities from part **a** as fractions. (3)

2 Michael picks one number from Box A.
He then picks one number from Box B.

Box A **Box B**

7
1 5

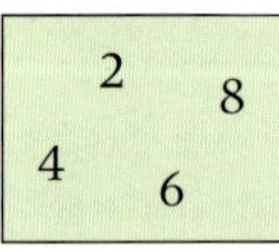

List all the pairs of numbers he could pick. (2)
One pair (1, 2) is shown.

(1, 2)

(Edexcel Ltd., 2003)

Proportion

This unit will show you how to

- Express a part of a whole as a fraction or a percentage
- Convert between fractions, decimals and percentages
- Use fractions to express a number as a proportion of another number
- Express a proportion as a fraction, decimal or percentage
- Compare the size of two objects using a ratio
- Calculate and simplify ratios
- Calculate missing amounts when two quantities are in direct proportion
- Calculate rates and scales, including conversion and exchange rates
- Solve problems involving rates and scales, including conversion and exchange rates

Before you start ...

You should be able to answer these questions.

Review

1 The table shows the favourite types of sandwiches in the school canteen.

Type	Frequency
Cheese	12
Salad	8
Ham	10
Total	30

What proportion of the class surveyed chose ham sandwiches?
Write your answer in its simplest form.

Units D1, N3

2 10 litres of white paint cost £12.
Work out the cost of 20 litres of paint.

Unit N3

3 Bart works for 4 hours. He gets paid £20.
How much does he get paid per hour?

Unit N2

4 The exchange rate for pounds into Australian dollars is £1 = AU\$2.
How many Australian Dollars would you get for £5?

Unit N2

N4.1 Proportion

This spread will show you how to:

- Convert between fractions, decimals and percentages
- Express a quantity as a proportion of another

Keywords
Decimal
Equivalent
Fraction
Percentage
Proportion

- A **proportion** is a part of the whole. It is usually written using a **fraction** or a **percentage**.

You can use fractions to express one number as a proportion of another number.

Example

Steven has £40 in his wallet. He spends £30 on a new shirt.
What proportion of the money in his wallet did Steven spend?
Give your answer as a fraction in its simplest form.

Steven had £40 (the whole).

Steven spent £30.

Fraction spent $= \frac{£30}{£40} = \frac{30}{40} = \frac{3}{4}$.

Cancel by the common factor 10.

So Steven spent $\frac{3}{4}$ of the money in his wallet.

- You can express a proportion as a percentage of a whole in three steps
  1. Write the proportion as a fraction.
  2. Convert the fraction to a decimal by division.
  3. Convert the decimal to a percentage by multiplying by 100.

You can compare proportions by converting them to percentages.

Example

Skye took two tests.
In German she scored 35 out of 50, and in French she scored 60 out of 80.
In which test did she do the best?

German

35 out of 50 $= \frac{35}{50}$
$= 35 \div 50$
$= 0.7$
$= 70\%$

French

60 out of 80 $= \frac{60}{80}$
$= 60 \div 80$
$= 0.75$
$= 75\%$

Skye did better in French.

Exercise N4.1

1 Write the proportion of each of these shapes that is shaded.
Write each of your answers as a fraction in its simplest form.

a 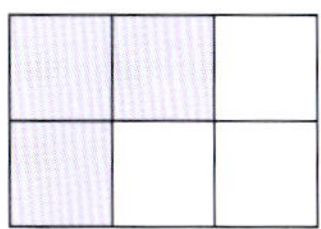

b 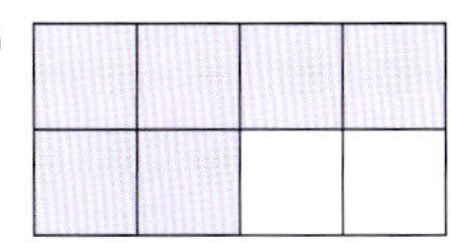

c 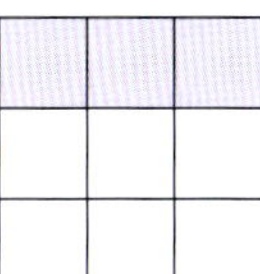

d 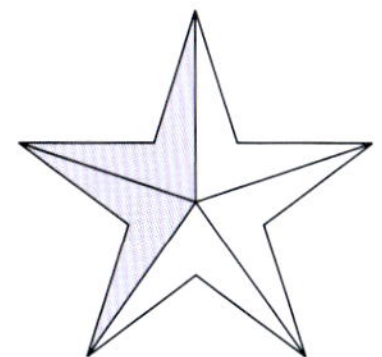

e 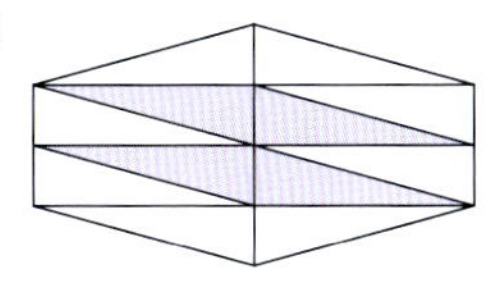

f 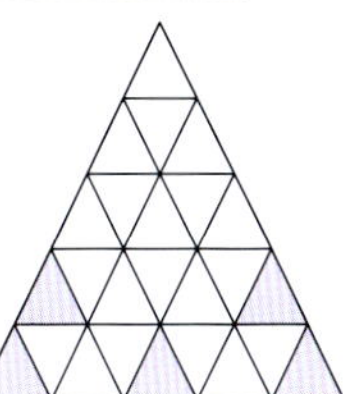

2 Give your answers to each of these questions as fractions in their simplest form.

a There are 20 students in a class. 16 are boys and 4 are girls. What proportion of the class are **i** boys **ii** girls?

b Tina has 12 T-shirts and 8 blouses. What proportion of her clothes are **i** T-shirts **ii** blouses?

c Joachim earns £600 a week. He pays £180 of his money each week in tax. He saves £100 each week. What proportion of his weekly wage does Joachim **i** pay in tax **ii** save?

3 Write each of your answers as a percentage.

a In a class there are 40 students. 25 of the students have brown hair. What proportion of the class have brown hair?

b In a survey of 80 people, 35 said they enjoyed school meals. What proportion of the people said they enjoyed school meals?

c At a rugby club there are 38 members. 15 players have been picked for the first team. What proportion of the members have been picked for the first team?

Give your answer to one decimal place.

4 Put these in order of size starting with the smallest first.

Convert them all to percentages.

a 72% $\frac{3}{4}$ 0.74

b 0.29 31% $\frac{3}{10}$

c 0.8 83% $\frac{7}{8}$ 0.78

d $\frac{2}{5}$ 0.39 41% $\frac{3}{7}$

e $\frac{8}{10}$ $\frac{3}{4}$ 83% 0.84

f 28% $\frac{3}{11}$ 0.3 $\frac{7}{24}$

5 Sunita took three tests. In Maths she scored 48 out of 60, in English she scored 39 out of 50 and in Science she scored 55 out of 70.
In which subject did she do

a the best? **b** the worst?

N4.2 Direct proportion

This spread will show you how to:

- Simplify a ratio
- Calculate missing amounts when two quantities are in direct proportion

Keywords
Direct proportion
Ratio

You can compare the size of two objects using a **ratio**.

Example

Henry the snake is only 30 cm long.
George the snake is 90 cm long.
Express this as a ratio.

The ratio of Henry's length compared to George's length

= Henry's length : George's length
= 30 cm : 90 cm
= 30 : 90

- You can simplify a ratio by dividing both parts of the ratio by the same number.

Example

Express the ratio 30 : 90 in its simplest form.

30 : 90
÷ 10 (both parts)
= 3 : 9
÷ 3 (both parts)
= 1 : 3

The ratio 1 : 3 means that 90 is three times bigger than 30.

When a ratio cannot be simplified any further it is said to be in its simplest form.

- When two quantities are in **direct proportion**, if one of the quantities changes, the other quantity changes by the same proportion.

You can use direct proportion to solve simple problems.

Example

Four pizzas cost £2.60.
Each pizza costs the same.
What is the cost of 12 pizzas?

4 pizzas cost £2.60
× 3 (both)
12 pizzas cost 3 × £2.60
= £7.80

Example

15 boxes of chocolates cost £22.50.
Each box of chocolates costs the same.
What is the cost of three boxes of chocolates?

15 boxes cost £22.50
÷ 5 (both)
3 boxes cost £22.50 ÷ 5
= £4.50

Exercise N4.2

1 Write each of these ratios in its simplest form.

a 2 : 6 **b** 15 : 5 **c** 6 : 18

d 4 : 28 **e** 5 : 50 **f** 30 : 6

g 24 : 8 **h** 2 : 30 **i** 7 : 56

2 Write the number of blue squares to the number of red squares as a ratio in its simplest form for each of these shapes.

a 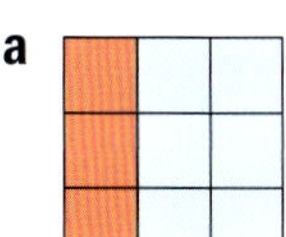

b 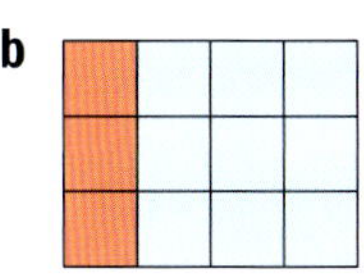

c 

3 How many times bigger than

a 15 is 60? **b** 3 is 21? **c** 12 is 72?

d 20 is 180? **e** 18 is 36? **f** 9 is 45?

g 16 is 80? **h** 6 is 72? **i** 20 is 240?

j 25 is 625? **k** 10 is 15? **l** 36 is 90?

4 Copy and complete this table for working out the cost of buying carpet.

Area of carpet (m^2)	Cost (£)
$6\,m^2$	£39.00
$12\,m^2$	
$18\,m^2$	
$30\,m^2$	
$60\,m^2$	
$3\,m^2$	

5 Copy and complete this table for converting inches and centimetres.

Inches (in)	Centimetres (cm)
6	15
3	
	30
	45
60	
1	

6 **a** 5 pizzas cost £12.50. What is the cost of 10 pizzas?

b 12 boxes of eggs cost £12. What is the cost of 4 boxes of eggs?

c 3 packets of seeds cost £4.50. What is the cost of 9 packets of seeds?

d 2 boxes of cornflakes cost £4.40. What is the cost of 8 boxes of cornflakes?

e 4 tennis balls cost £1.80. What is the cost of 8 tennis balls?

f 5 chocolate bars cost £3.00. What is the cost of 15 chocolate bars?

g 7 bags of wood cost £27.93. What is the cost of 14 bags of wood?

N4.3 Unitary method

This spread will show you how to:

- Calculate missing amounts when two quantities are in direct proportion using the unitary method

Keywords
Direct proportion
Ratio
Unitary method

You can find the value of one unit of a quantity using division.

Example

Three bags of crisps cost 93 pence. What is the price of 1 bag of crisps?

Total cost of crisps = 93 pence
Number of bags of crisps = 3

The price of 1 bag of crisps = $\frac{\text{total cost of crisps}}{\text{number of bags of crisps}}$
= 93 ÷ 3
= 31 pence

This answer is another **ratio**! For every 1 bag of crisps it costs 31 pence. You could say that the cost in pence is 31 × the number of bags.

- **You can use the unitary method to solve direct proportion problems. In this method you find the value of 1 unit of a quantity.**

Example

Here is a recipe for blackcurrant squash for 5 people

Blackcurrant squash
(for 5 people)
400 g of blackcurrants
1200 ml of water
100 g sugar
250 ml blackcurrant juice

Work out the number of grams of blackcurrants needed to make squash for 8 people.

The recipe is for 5 people. First find the number of grams of blackcurrants needed for 1 person. Then multiply this by 8.

Number of people	Grams of blackcurrants
5	400
÷ 5 ↓ 1	80 ↓ ÷ 5
× 8 ↓ 8	640 ↓ × 8

So for 8 people you need 640 g blackcurrants.

Here the number of grams of blackcurrants is in **direct proportion** to the number of people.

Example

15 boxes of chocolates cost £22.50. Each box of chocolates costs the same. What is the cost of 2 boxes of chocolates?

Number of boxes	Cost
15	£22.50
÷ 15 ↓ 1	£1.50 ↓ ÷ 15
× 2 ↓ 2	£3.00 ↓ × 2

So 2 boxes of chocolates will cost £3.

Exercise N4.3

1 **a** 2 pizzas cost £6.00. What is the cost of 1 pizza?

b 4 sweets cost 20p. What is the cost of 1 sweet?

c 10 packets of seeds cost £18. What is the cost of 1 packet of seeds?

d 4 tennis balls cost £2. What is the cost of 1 tennis ball?

e There are 48 biscuits in 3 packets. How many biscuits are there in 1 packet?

f 10 kg of apples cost £9.00. What is the cost of 1 kg of apples?

g There are 320 MB of memory on 5 identical memory sticks. How much memory is there on each stick?

2 **a** There are 24 inches in 2 feet. How many inches are there in 1 foot?

b There are 2000 ml in 2 litres. How many ml are there in 1 litre?

c There are 24 pints in 3 gallons. How many pints are there in 1 gallon?

d There are 120 hours in 5 days. How many hours are there in 1 day?

3 **a** Vince works for 4 hours. He gets paid £24.
How much money is he paid each hour?

b On average Barry fits 36 radiators in 3 days.
How many radiators does he fit each day?

c An athlete runs 240 metres in 30 seconds.
How far does she run in 1 second?

d Rashid drives his car 250 miles and uses 10 gallons of petrol.
On average, how far does the car travel on each gallon of petrol?

4 **a** A recipe for cake uses 400 g of sugar for 5 people. What weight of sugar is needed for

i 1 person **ii** 8 people **iii** 12 people
iv 14 people **v** 30 people?

b A recipe for three bean chilli uses 840 g of beans for 7 people.
What weight of beans is needed for

i 3 people **ii** 6 people **iii** 17 people
iv 24 people **v** 100 people?

c Frank works for 8 hours a day and earns £128.
He is paid the same amount each hour.
How much will he get paid for working

i 40 hours **ii** 35 hours **iii** 168 hours
iv 10 days **v** 3 hours **vi** $\frac{1}{2}$ hour?

N4.4 Rates and tables

This spread will show you how to:

- Calculate with rates and scales

Keywords
Rate
Ratio
Scale

- You can express a **ratio** in the form $1 : n$ using division. This is often called a **scale**.

Example

A photograph is 8 cm tall. An enlargement of the same photograph is 24 cm tall. What is the ratio of the height of the original to the height of the enlargement? Express your answer as a scale in the form $1 : n$.

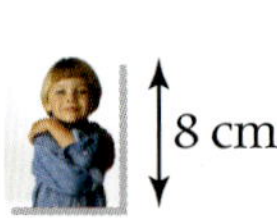

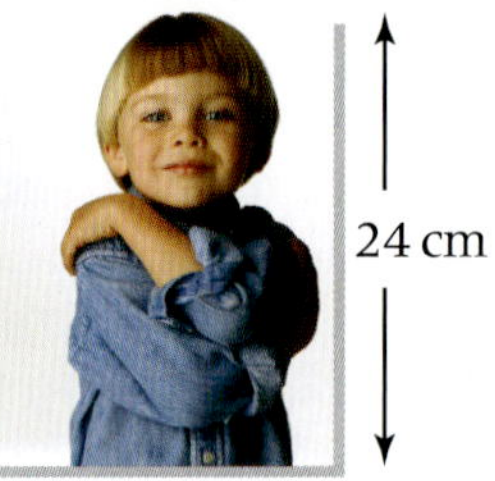

Ratio of height of original : height of enlargement

$= 8 \text{ cm} : 24 \text{ cm}$

$= 8 : 24$

$$\frac{\text{enlargement height}}{\text{original height}} = \frac{24}{8} = 3$$

Scale = ratio of height of original : height of enlargement = 1 : 3

This means that the enlargement is three times taller than the original photograph.

- A **rate** is a way of comparing two quantities.

You can solve problems involving scales or rates by multiplying or dividing by the scale or rate.

Example

a Bill goes for a walk. On his map he travels 40 cm. The scale of his map is 1 : 2000. How far does he really walk?

b Sarah travels in her car. The petrol consumption of her car is 10 miles per litre. She travels 290 miles. How many litres of petrol does she use?

Petrol consumption is a **rate**.

a Map scale

1 = 2000 (× 2000 from left to right; ÷ 2000 from right to left)

$40 \text{ cm} = 40 \times 2000 \text{ cm}$

$= 80\,000 \text{ cm}$

$= 800 \text{ m}$

Bill walks 800 m.

Multiply by the scale.

b Petrol consumption

10 miles = 1 litre (÷ 10 from left to right; × 10 from right to left)

$290 \text{ miles} = 290 \div 10 \text{ litres}$

$= 29 \text{ litres}$

Sarah uses 29 litres of petrol.

Divide by the rate.

Exercise N4.4

1 Express each of these ratios as a ratio in the form 1 : n (a scale).

a 2 : 6 **b** 3 : 12 **c** 10 : 20 **d** 8 : 40

e 3 : 6 **f** 5 : 15 **g** 4 : 20 **h** 12 : 36

i 30 : 60 **j** 9 : 45 **k** 45 : 90 **l** 20 : 120

2 In each of these questions work out the ratio and then express this as a ratio in the form 1 : n (a scale).

a A photograph is 6 cm wide. An enlargement of the same photograph is 30 cm wide. What is the ratio of the width of the original to the width of the enlargement?

b On a model plane the wing span is 2 m. In real life the wing span of the plane is 40 m. What is the ratio of the model wing span to the wing span of the real plane?

3 Work out the hourly rate for each of these people.

a Wilf works for 3 hours. He gets paid £21.
What is his hourly rate of pay?

Work out how many pounds per hour.

b Aaron works for 10 hours. He is paid £55.
What is his hourly rate of pay?

c Gary is a plumber. On average he fits 8 radiator valves every 4 hours. What is his hourly rate of fitting radiator valves?

Work out how many radiators he fits each hour.

4 **a** Kerry has a plan of her house. On her plan she walks 40 cm. The scale of her plan is 1 : 20. How far does she really walk?

b John makes a scale drawing of his kitchen. On his scale drawing the cooker is 6 cm wide. The drawing has a scale of 1 : 10. What is the real width of the cooker?

c Gustav builds a model plane. The scale is 1 : 25. On his model the wing span is 40 cm. What is the real wing span of the plane?

5 Here is the nutritional information for a 500 g serving of pizza.
Copy and complete the nutritional amount for every 100 g of pizza.

Typical values	Amount in a 500 g serving	Amount per 100 g
Energy	1500 kcal	
Protein	40 g	
Carbohydrate	150 g	
Fat	44 g	
Fibre	20 g	

N4.5 Conversion and exchange rates

This spread will show you how to:

- Calculate conversion and exchange rates
- Solve problems using conversion and exchange rates

Keywords
Conversion rate
Exchange rate

- A **conversion rate** is a way of converting between two different units of measurement.

Example

May was driving her car on holiday.
The conversion rate for miles into kilometres is 1 mile = 1.6 kilometres.

a May travelled 300 miles in the UK.
Work out the number of kilometres she travelled in the UK.

b May then travelled 1280 km in France.
Work out the number of miles she travelled in France.

Conversion rate

$$1 \text{ mile} \underset{\div 1.6}{\overset{\times 1.6}{=}} 1.6 \text{ kilometres}$$

a 300 miles = 300 × 1.6 kilometres
= 480 kilometres
May travelled 480 km in the UK.

b 1280 kilometres = 1280 ÷ 1.6 miles
= 800 miles
May travelled 800 miles in France.

Examiner's Tip
You are expected to remember the conversion from miles to km in the exam.

- An **exchange rate** is a way of comparing two currencies. It tells you how many units of one currency there are compared to one unit of another currency.

You can solve problems involving currency by multiplying or dividing by the exchange rate.

Example

Steve went to Austria.

a He changed £500 into euros. The exchange rate was £1 = €1.50.
Work out the number of euros Steve got.

b He had €120 left at the end of the holiday. He changed them back into pounds. How many pounds did he get?

Exchange rate

$$£1 \underset{\div 1.5}{\overset{\times 1.5}{=}} €1.50$$

a £500 = 500 × 1.5
= €750
Steve got €750.

b €120 = 120 ÷ 1.5
= £80
Steve got £80.

Exercise N4.5

1 **a** There are 48 inches in 4 feet. How many inches are there in 1 foot?

b There are 30 cl in 3 litres. How many cl are there in 1 litre?

c There are 40 pints in 5 gallons. How many pints are there in 1 gallon?

d There are 25 cm in 10 inches. How many centimetres are there in 1 inch?

2 **a** There are 1800 Rwandan francs in £2. How many Rwandan francs are there in £1?

b There are 50 Canadian dollars in £25. How many Canadian dollars are there in £1?

c There are 16 000 Belarussian rubles in £4. How many Belarussian rubles are there in £1?

d There are 75 Ethiopian birrs in £5. How many Ethiopian birrs are there in £1?

3 **a** David went to France. He changed £500 into €750. What was the exchange rate for pounds into euros?

b Juan lives in Spain. He changes €400 into AUS$1000. What was the exchange rate for euros into Australian dollars?

c There are approximately 50 squatches in 20 morcks. What is the conversion rate for changing squatches into morcks?

DID YOU KNOW?

The Chinese first used paper money in the 7th century. Paper money was not printed in the UK until the 17th century!

4 Each of these people change amounts of money from pounds into euros. The exchange rate is £1 = €1.60. Work out the number of euros each person receives.

Person	Amount (£)	Exchange rate (£1 = €1.60)	Amount (€)
Basil	£10	£1 = €1.6	
Peter	£200	£1 = €1.6	
Clark	£80	£1 = €1.6	
Kathy	£150	£1 = €1.6	
Harry	£2300	£1 = €1.6	
Rudolph	£265	£1 = €1.6	

5 Use the fact that **1 mile = 1.6 km** to answer each of these questions.

a Convert these distances into kilometres.

i 10 miles **ii** 50 miles **iii** 230 miles **iv** 48 miles

b Convert these distances into miles.

i 32 km **ii** 640 km **iii** 512 km **iv** 1352 km

N4 Exam review

Key objectives

- Calculate a given fraction of a given quantity
- Express a given number as a fraction of another
- Convert simple fractions of a whole to percentages of a whole and vice versa
- Divide a quantity in a given ratio
- Solve problems and word problems, including those involving ratio and proportion

1 Debbie buys a 500 g bar of chocolate.
On the first day she eats $\frac{1}{5}$ of the bar of chocolate.

a Write the amount of chocolate Debbie eats on the first day in grams, g. (1)

On the second day Debbie eats 75 g of the chocolate.

b Write the amount of chocolate Debbie eats on the second day as a proportion of the whole chocolate bar.
Express your answer as

i a fraction

ii a percentage. (2)

Debbie decides to give the remaining chocolate to her friend Martin.

c How much chocolate does Debbie give to Martin? (3)
Give your answer in grams, g.

2 Ken and Susan share £20 in the ratio 1 : 3.
Work out how much money each person gets. (2)

(Edexcel Ltd., 2003)

S3 Transformations

This unit will show you how to

- Recognise and visualise reflections, rotations and translations
- Understand that reflections are specified by a mirror line
- Understand that rotations are specified by a centre of rotation and an angle and direction of turn
- Understand that translations are specified by a distance and direction
- Transform triangles and other shapes by reflection, rotation and translation
- Recognise reflection symmetry of 2-D shapes
- Recognise rotation symmetry of 2-D shapes
- Understand the relationship between the number of lines of symmetry of a regular polygon and its number of sides

Before you start ...

You should be able to answer these questions.

	Review
1 Give the value of these angles.	Unit S2

a

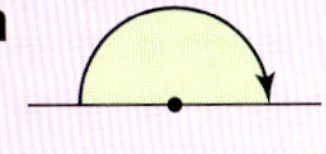

b

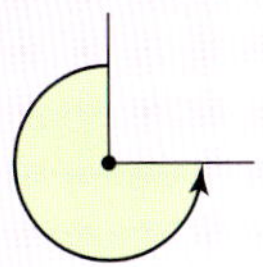

c

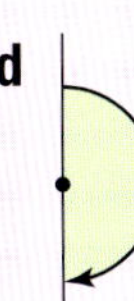

d

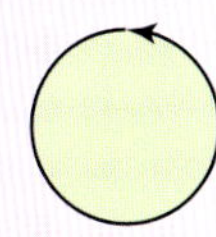

	Review
2 State the direction of the turn.	Unit S2

a **b**

	Review
3 Give the coordinates of	Unit A4

a A **b** B

c C **d** D

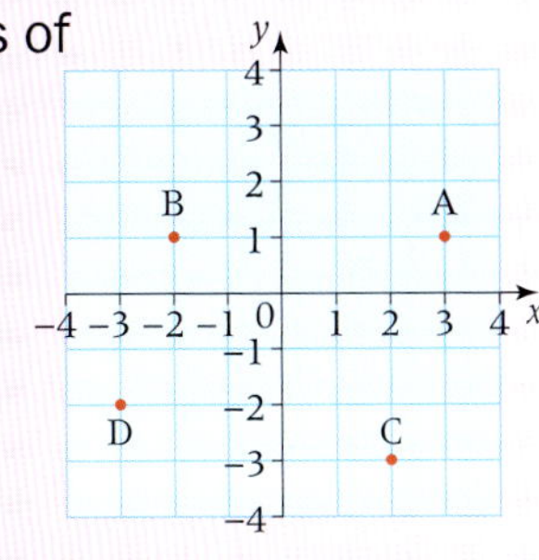

S3.1 Reflections

This spread will show you how to:

- Recognise and visualise reflections
- Understand that reflections are specified by a mirror line
- Transform triangles and other shapes by reflection

Keywords

Mirror line
Reflection
Transformation

A **transformation** can change the size and position of a shape.

- **A reflection flips the shape over.**

When you look at yourself in a mirror, you are seeing a reflection.

You specify a **mirror line** or reflection line to reflect the object.

To find the position of the reflection line you choose corresponding points on the object and the image.

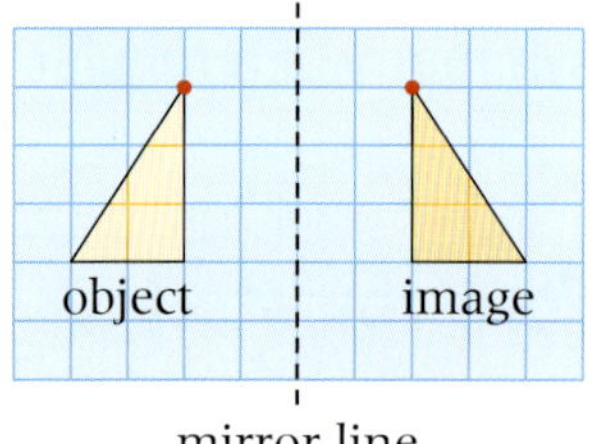

Each dot is 2 units from the mirror line.

- **The image is the same distance from the mirror line as the object.**

Example

Draw the mirror line so that shape B is a reflection of shape A.

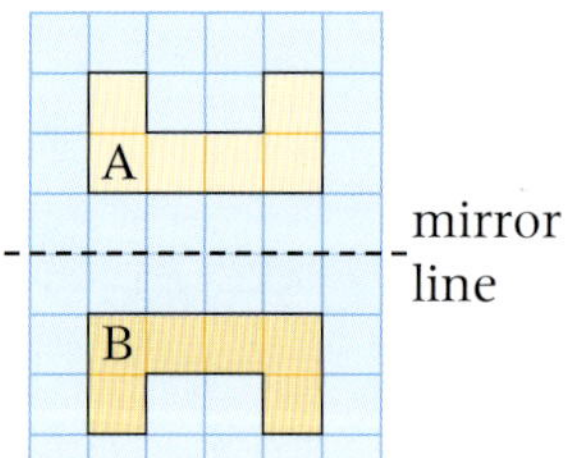

You can rotate the page to make the mirror line vertical.

Example

Draw the reflection of the shape using the mirror line.

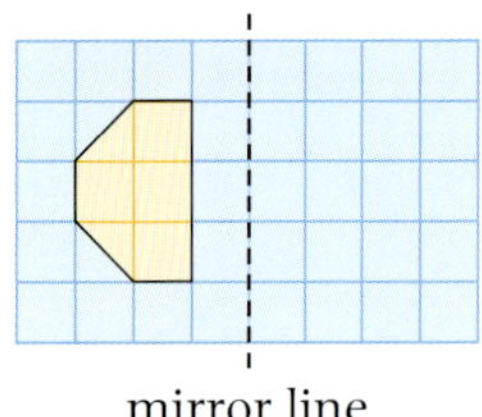

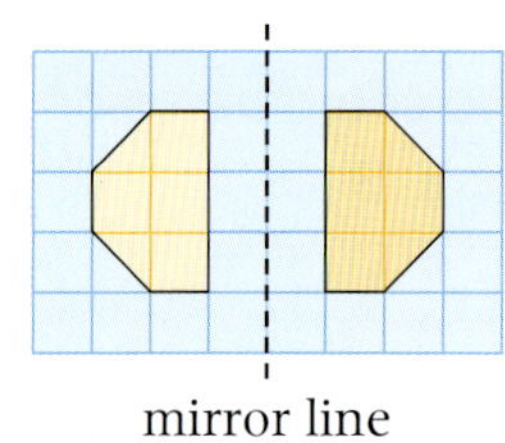

Exercise S3.1

1 Copy the diagrams and draw the mirror lines.

a b c d e

f g h i j

2 Copy and complete each diagram to show the reflection of the shape in the mirror line.

a b c d e

f g h i j

3 Copy the diagrams.
Reflect the shapes in both mirror lines to create a pattern.

a b c d e

4 Give the equation of the mirror line for each reflection.

a

b

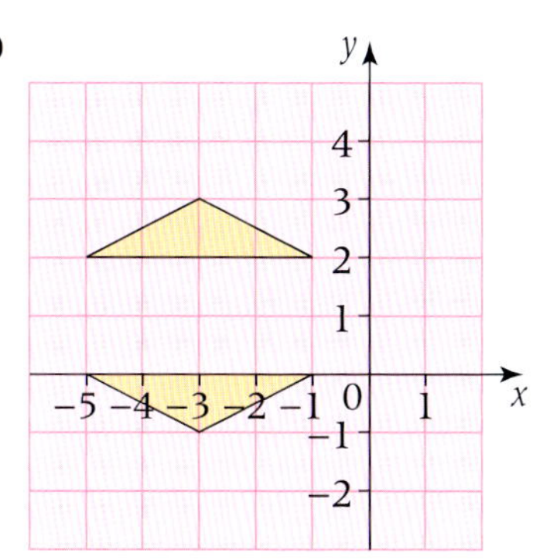

S3.2 Rotations

This spread will show you how to:

- Recognise and visualise rotations
- Understand that rotations are specified by a centre of rotation and an angle and direction of turn

Keywords
Anticlockwise
Centre of rotation
Clockwise
Degrees (°)
Rotation
Turn

- A **rotation** turns a shape.

To describe a rotation you give

- the **centre of rotation** – the point about which it **turns**
- the angle or measure of turn
- the direction of turn – either **clockwise** or **anticlockwise**.

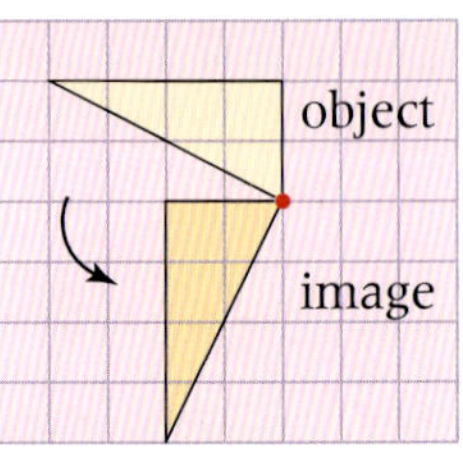

The dot is the centre of rotation.
The turn is 90° or $\frac{1}{4}$ of a turn.
The direction is anticlockwise.

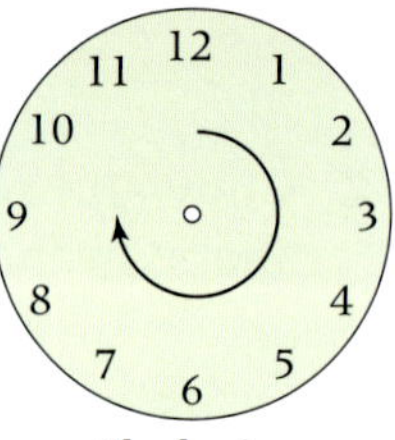

Clockwise

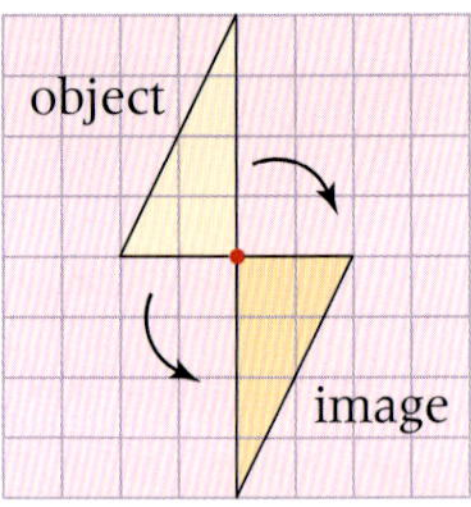

The dot is the centre of rotation.
The turn is 180° or $\frac{1}{2}$ a turn.
The direction is either clockwise or anticlockwise.

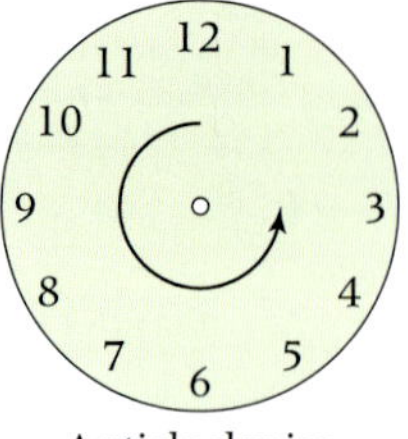

Anticlockwise

Example

a Draw the position of the green triangle after a rotation of 90° clockwise about the dot (•).

b What rotation would return the triangle to its starting position?

---

a

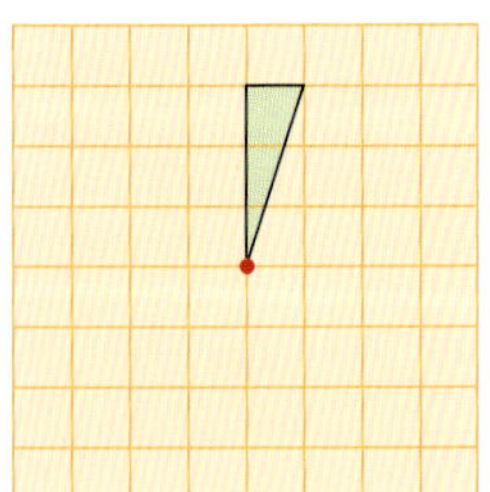

b Rotation of 90° anticlockwise about the dot (•).

Use tracing paper to find the position of the blue triangle.

Exercise S3.2

1 State the angle and direction of turn for each of these rotations (green shape to blue shape).

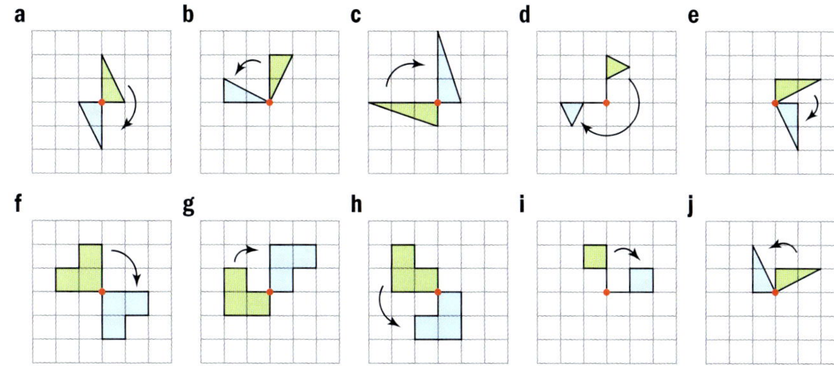

2 Copy these shapes onto square grid paper.
Rotate the shapes through the given angle and direction about the dot (•).

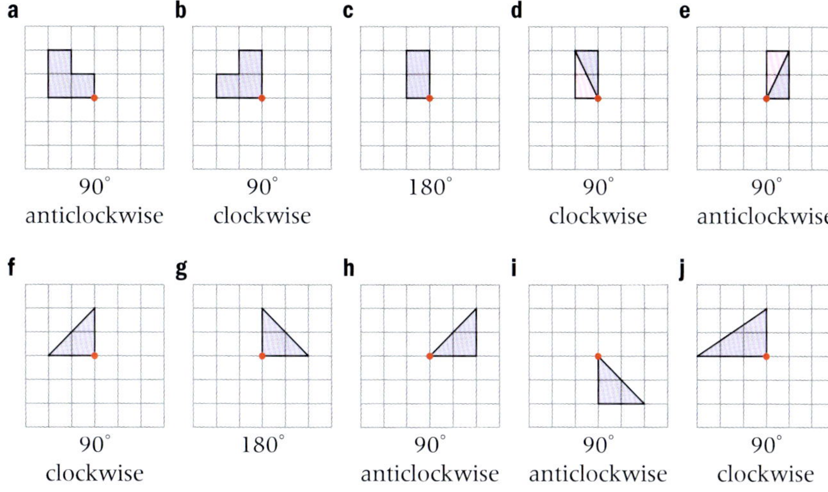

3 Copy the shapes onto square grid paper.

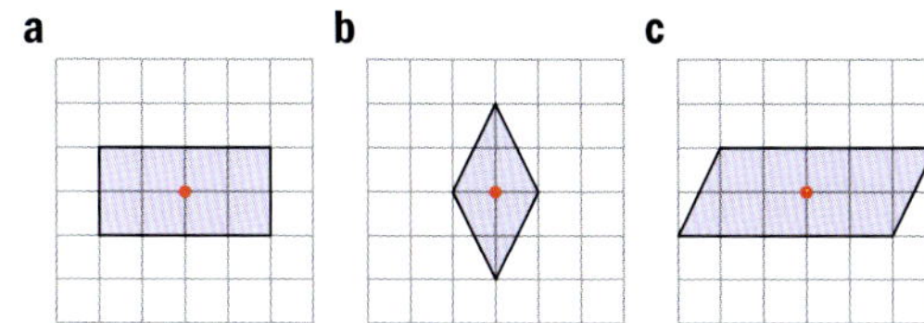

Rotate each shape through 180° about the dot (•).
Describe your results.

4 Copy the triangles onto square grid paper.
Rotate each triangle through 180° about the dot (•).
Give the mathematical name of the new shape you have created.

a

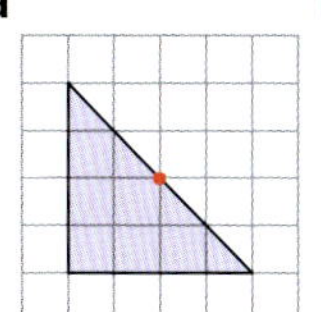

b

S3.3 Translations

This spread will show you how to:

- Recognise and visualise translations
- Understand that translations are specified by a distance and direction

Keywords
Left
Right
Slide
Translation

- **A translation is a sliding movement.**

To describe a translation you give:

- the distance moved **right** or **left**, then
- the distance moved **up** or **down**.

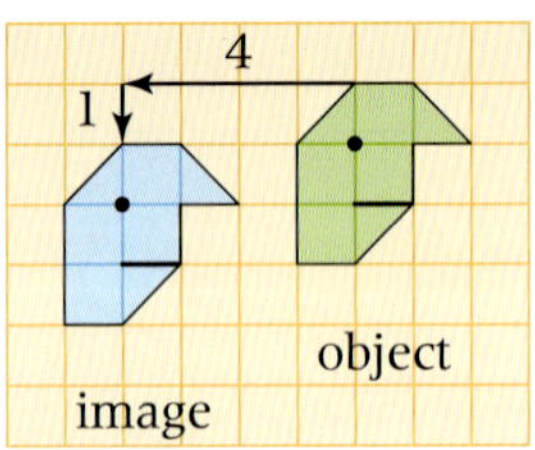

Translate the object 4 units left and 1 unit down.

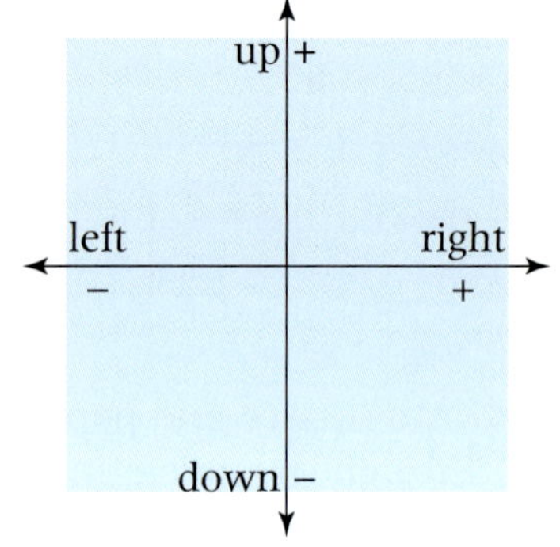

Right and up are positive directions.

You choose corresponding points on the object and image to work out the translation.

Example

Which triangles are translations of the black triangle?

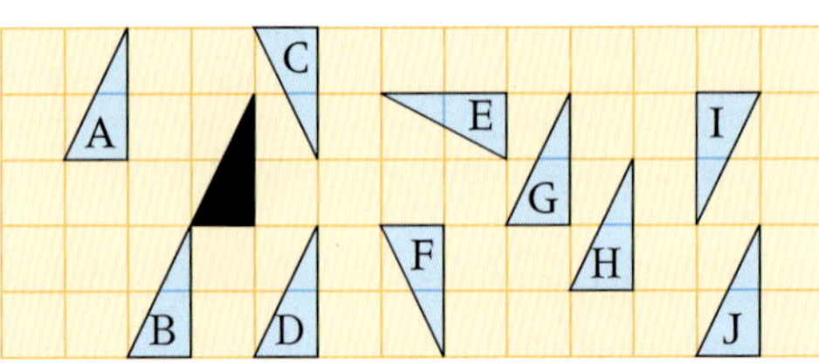

A, B, D, G, H, J because they are all in the same orientation.

Same orientation means the same way up.

Example

a Give the coordinates of the point marked by a dot (•) in triangle A.

b Describe the transformation that moves triangle A to triangle B.

c Give the coordinates of the point in triangle B that corresponds to the dot (•) in triangle A.

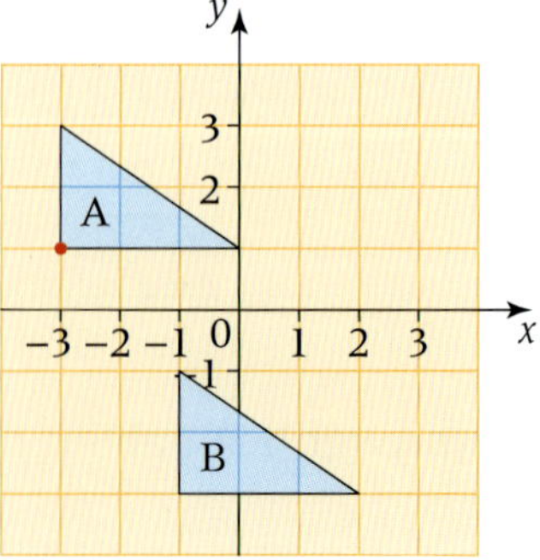

a (−3, 1)
b Translation, 2 units right and 4 units down.
c (−1, −3)

Exercise S3.3

1 Which shapes are translations of the green shape?

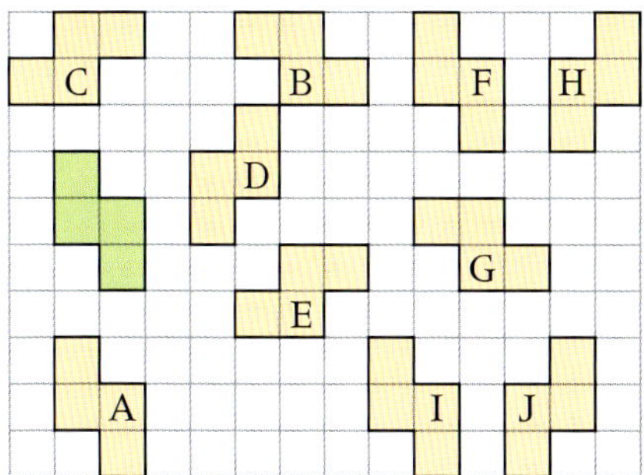

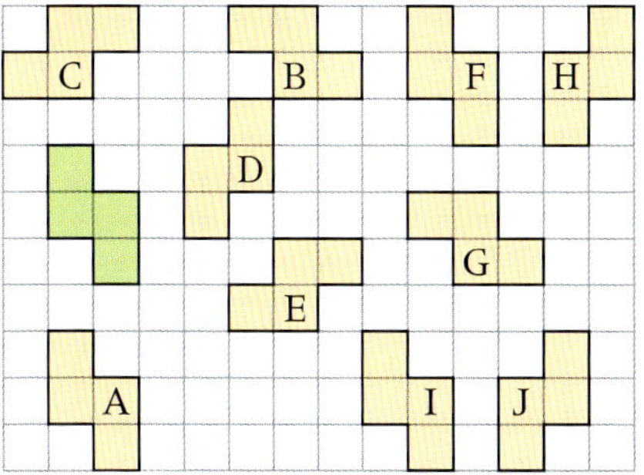

DID YOU KNOW?

The process of turning one language into another is one more definition of the word *translation*.

2 Describe these translations.

a E to D	**b** E to C
c A to B	**d** A to C
e B to D	**f** C to A
g C to E	**h** B to A
i E to A	**j** A to E

3 **a** Give the coordinates of point A.

b Describe the transformation that moves the green shape to the blue shape.

c Give the coordinates of the point in the blue shape that corresponds to point A in the green shape.

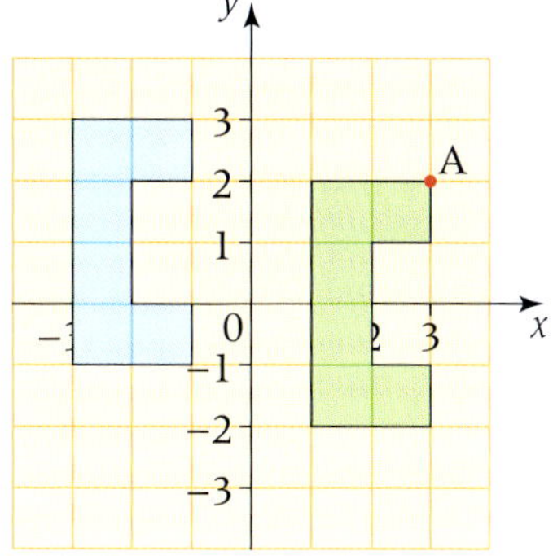

4 Describe these translations.

a D to B	**b** A to B
c A to C	**d** D to E
e B to E	**f** B to C
g E to D	**h** E to A
i B to D	**j** B to A
k E to B	**l** A to D
m C to B	**n** E to C
o C to A	**p** A to E
q C to D	**r** D to C
s C to E	**t** D to A

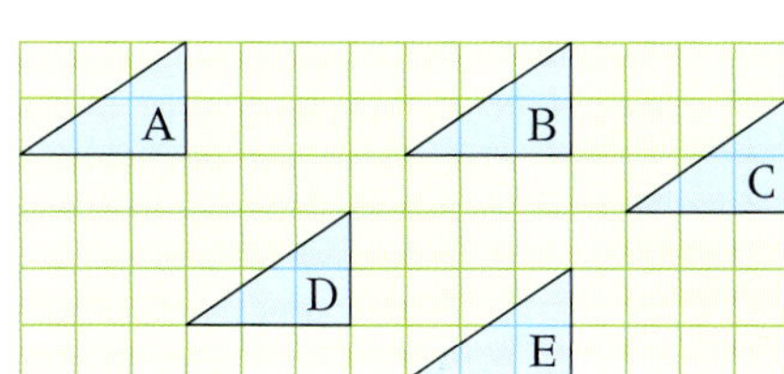

S3.4 Reflection symmetry

This spread will show you how to:

- Recognise reflection symmetry of 2-D shapes

Keywords
Line of symmetry
Reflection symmetry
Regular polygon

A shape has **reflection symmetry** if

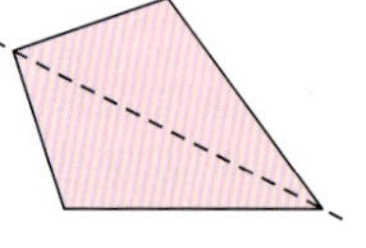

The dotted line is the **line of symmetry**.

- the shape divides into two identical halves
- you can fold the shape so that one half fits exactly on top of the other

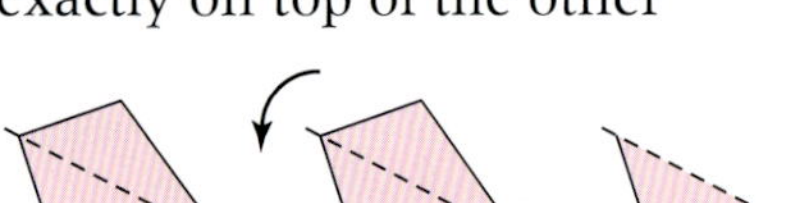

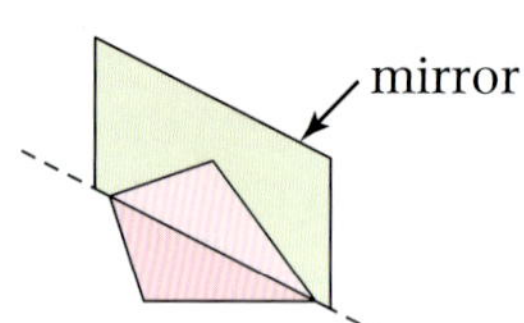

- a mirror reflects half the shape to give the completed shape.

- **A line of symmetry divides the shape into two identical halves.**

You can find symmetry in nature.
This butterfly has one line of symmetry.

Example

a Add one extra square so that the shaded shape has 1 line of symmetry.
b Draw the line of symmetry.

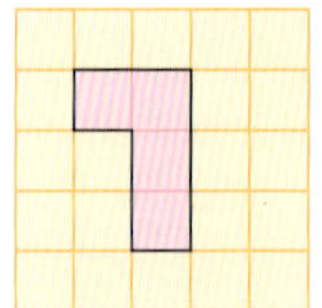

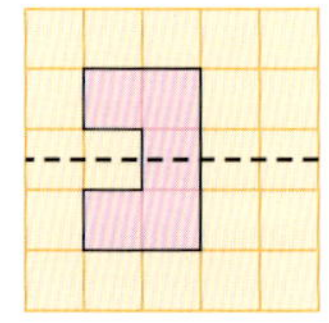 or 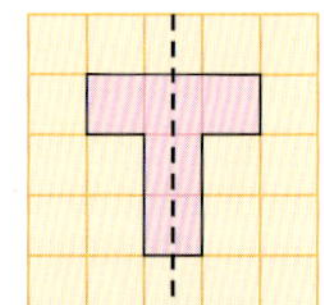 or 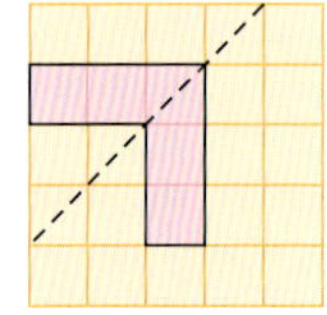

A shape can have more than one line of symmetry.

- **A regular polygon with *n* sides has *n* lines of symmetry.**

Regular polygons have equal sides and equal angles.

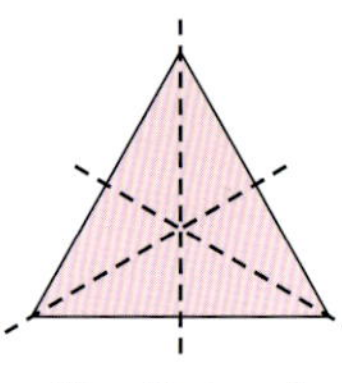

Equilateral triangle
3

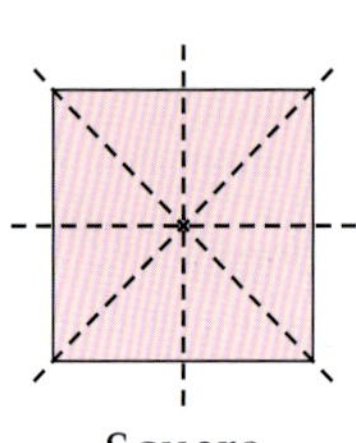

Square
4

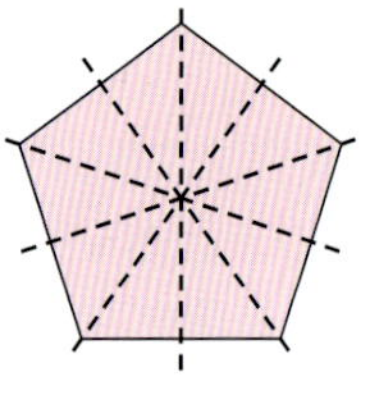

Regular pentagon
5

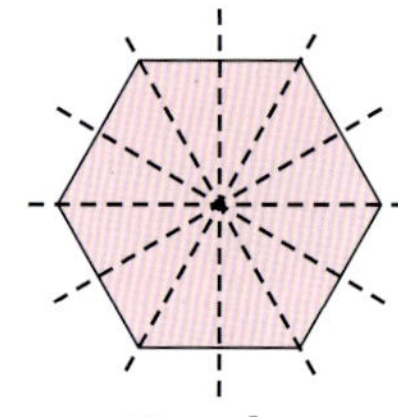

Regular hexagon
6

Exercise S3.4

1 Copy these 2-D shapes and draw in the lines of symmetry.

a **b** **c** **d** **e**

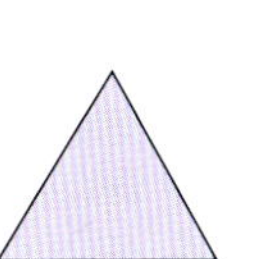
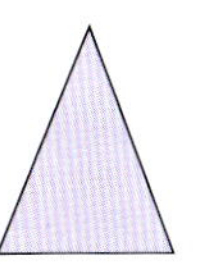
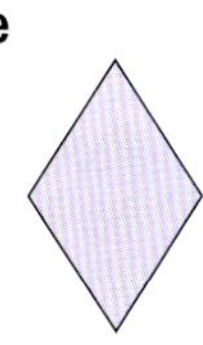

f **g** **h** **i** **j**

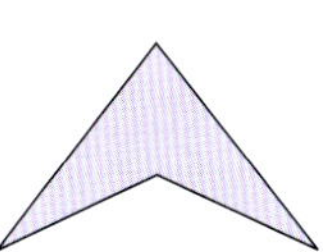
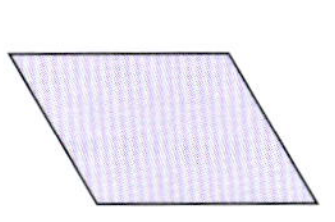
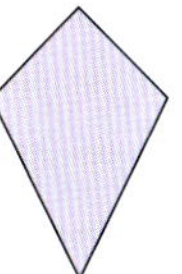
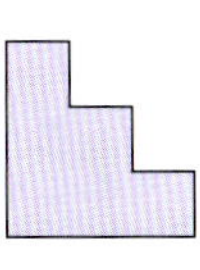

2 State the number of lines of symmetry for each of these regular polygons.

a **b** **c** **d** **e**

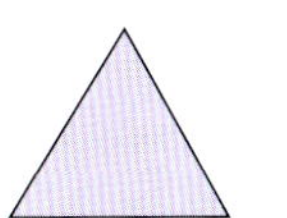

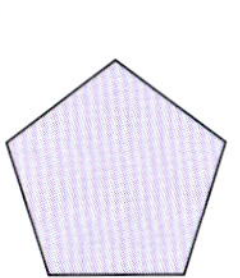
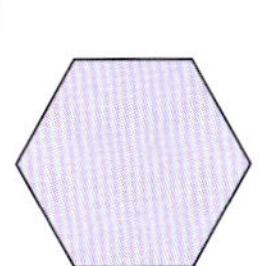
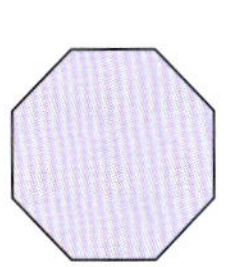

3 Draw these shapes on square grid paper. Add one square to give a shape with the required number of lines of symmetry. Draw the lines of symmetry for the new shape.

a **b** **c** **d** **e**

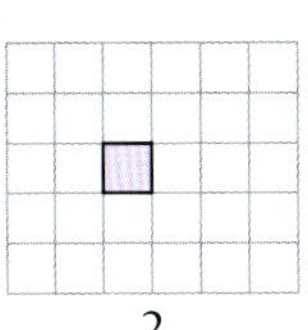
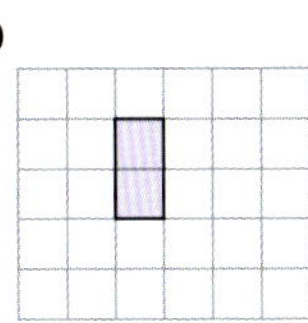
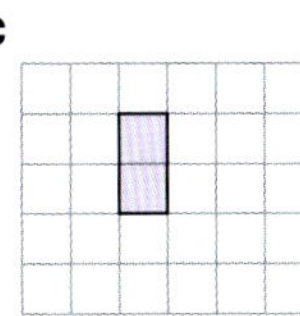
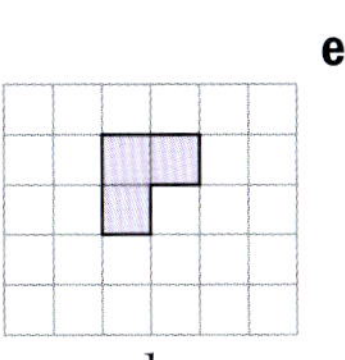
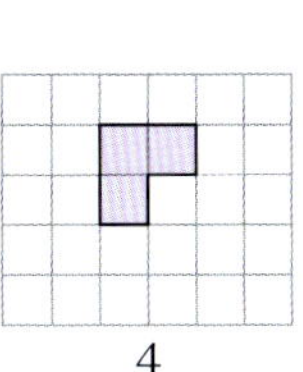

2 1 2 1 4

f **g** **h** **i** **j**

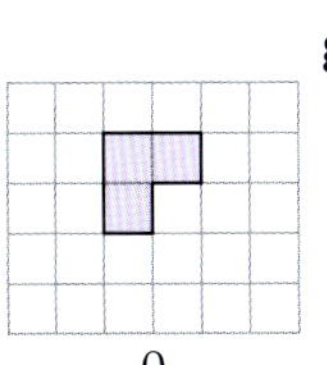
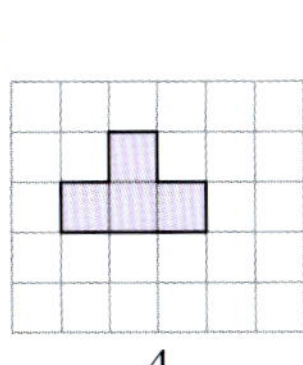
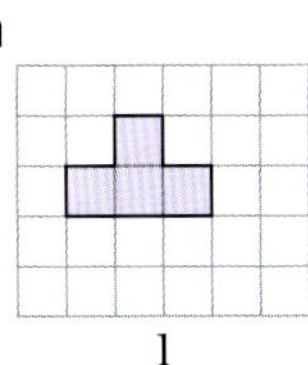
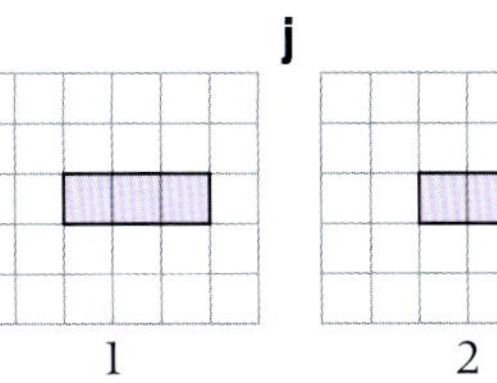

0 4 1 1 2

4 Draw these shapes on isometric paper. Add one triangle to give a shape with the required number of lines of symmetry.
Draw the lines of symmetry for the new shape.

a **b** **c** **d** **e**

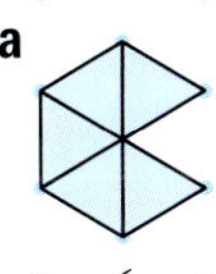
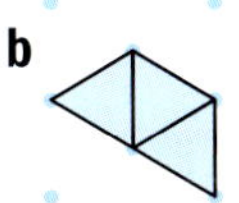
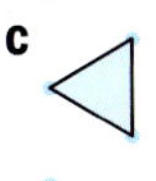
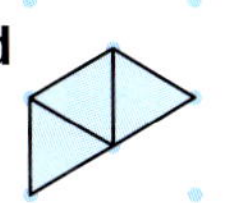
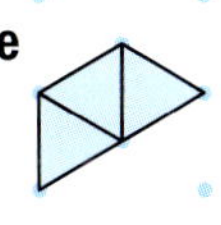

6 3 2 1 0

S3.5 Rotational symmetry

This spread will show you how to:

- Recognise rotational symmetry of 2-D shapes

Keywords
Order of rotational symmetry
Regular polygon
Rotational symmetry

- **A shape has rotational symmetry if the shape looks like itself more than once in a full turn.**

The **order of rotational symmetry** is the number of times a shape looks exactly like itself in a complete turn.

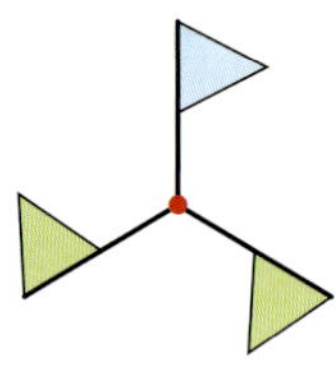

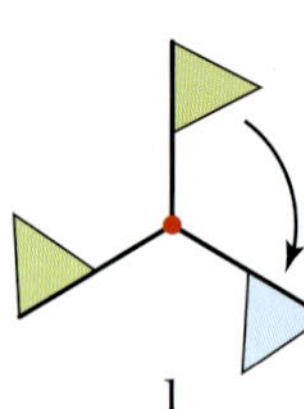

1

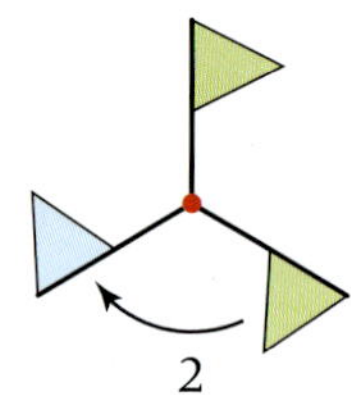

2

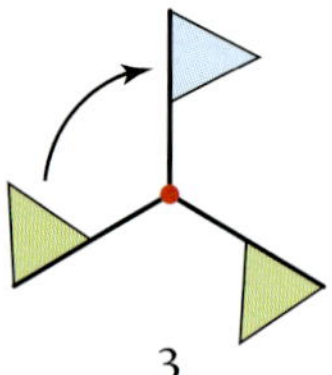

3

The order of rotational symmetry is 3.

Example

Add one extra square so that the shaded shape has rotational symmetry of order 2.

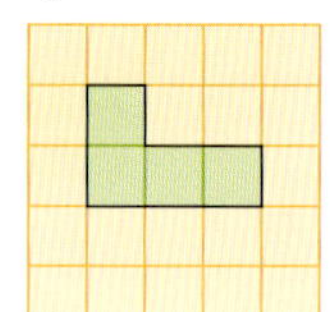

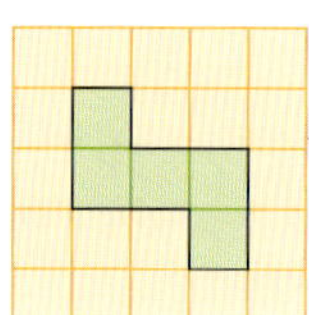

Example

Name these shapes. State the order of rotational symmetry for each one.

a

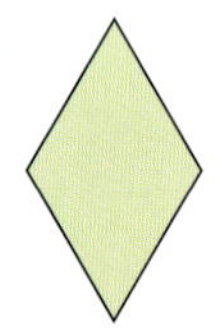

b

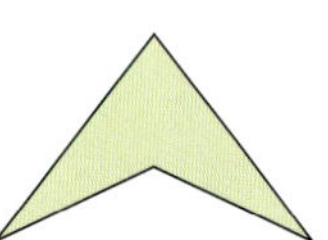

c

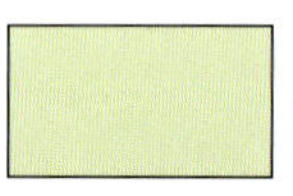

d

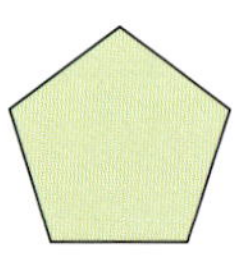

a rhombus
2

b arrowhead
1

c rectangle
2

d regular pentagon
5

- **A regular polygon with *n* sides has rotational symmetry of order *n*.**

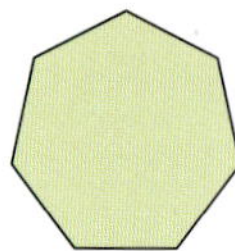

regular heptagon

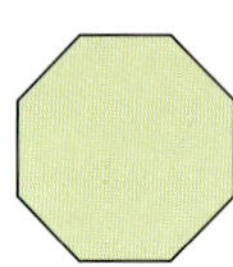

regular octagon

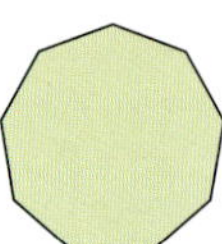

regular nonagon

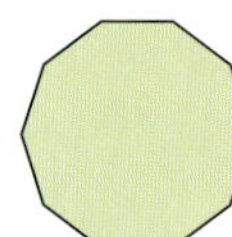

regular decagon

Turn the page around to see the rotational symmetry.

Exercise S3.5

1 Copy these 2-D shapes and state the order of rotational symmetry for each shape.

a 

b 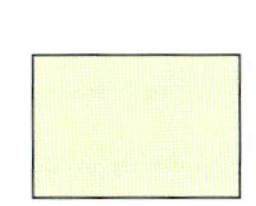

c 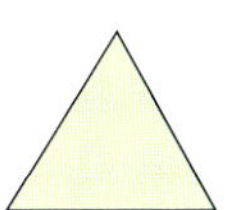

d 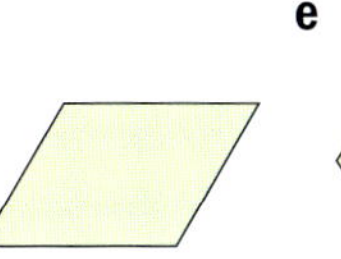

e 

f 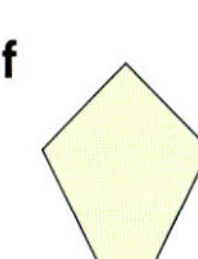

g 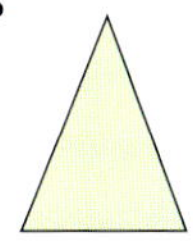

h 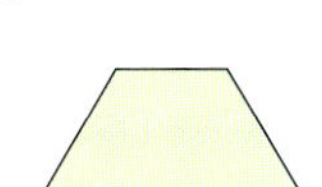

i

j 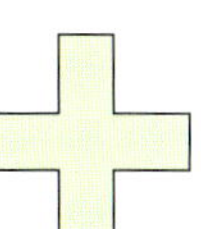

2 State the order of rotational symmetry for these regular polygons.

a 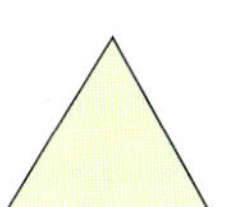

b 

c 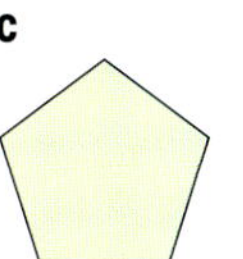

d 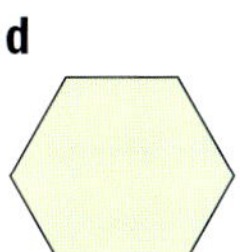

e 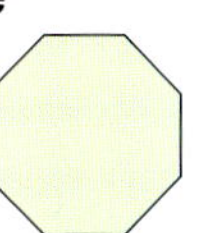

3 Draw these shapes on square grid paper. Add one square to each to give a shape with the given order of rotational symmetry.

a 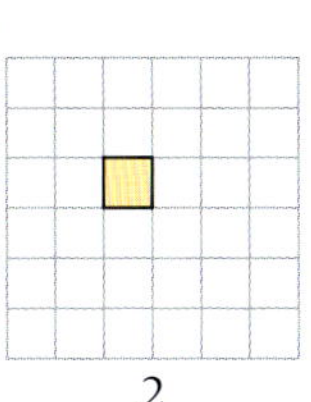
2

b 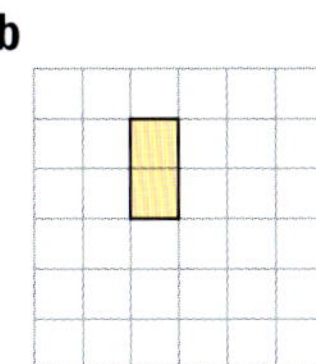
2

c 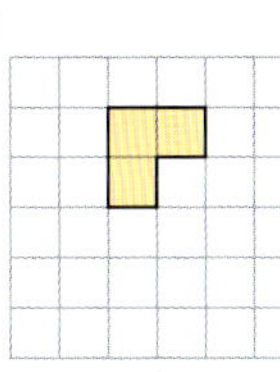
4

d 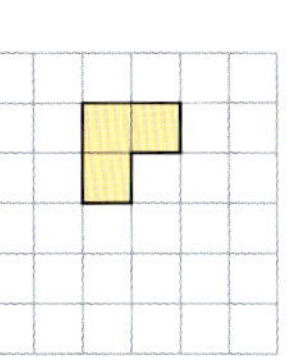
2

e 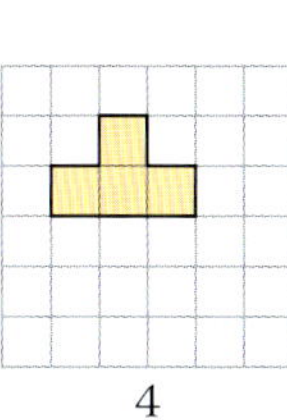
4

f 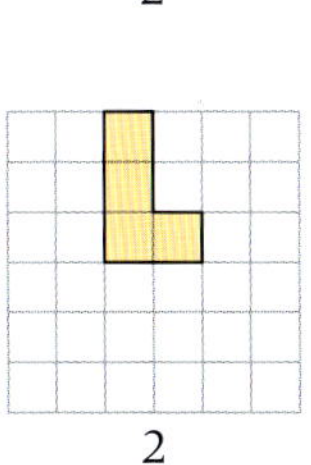
2

g 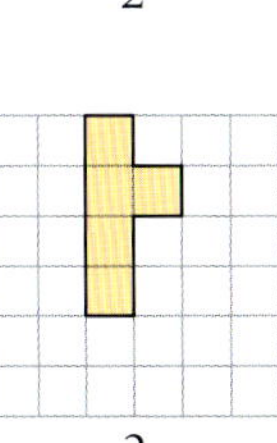
2

h 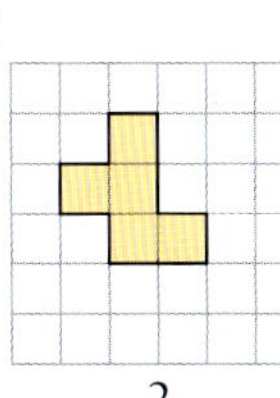
2

i 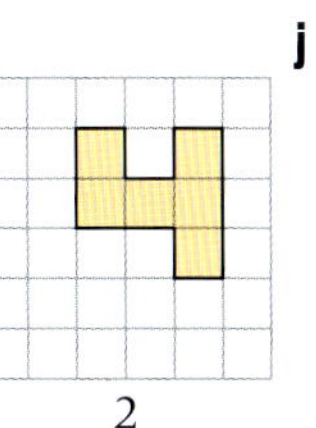
2

j 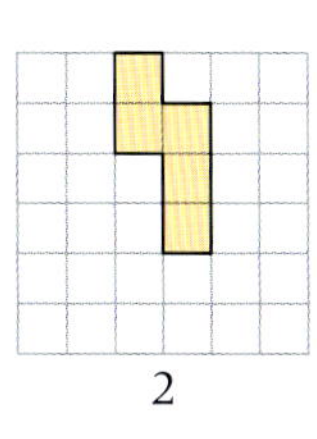
2

4 Draw these shapes on isometric paper. Add one triangle to give a shape with the given order of rotational symmetry.

a 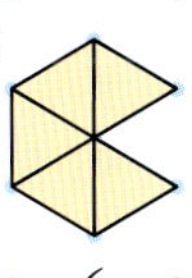
6

b 
2

c 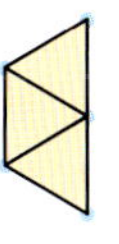
2

d 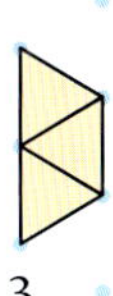
3

e 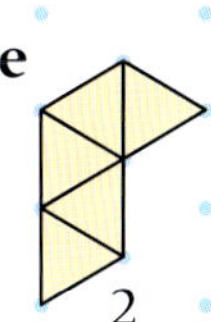
2

Exam review

Key objectives

- Understand that reflections are specified by a mirror line
- Understand that rotations are specified by a centre and an (anticlockwise) angle
- Understand that translations are specified by giving a distance and direction
- Recognise and visualise rotations, translations and reflections, including reflection symmetry of 2-D shapes, and rotation symmetry of 2-D shapes
- Transform triangles and other 2-D shapes by reflection, rotation and translation

1 Copy the grid and rotate triangle T $\frac{1}{2}$ a turn anticlockwise about centre *O*. Label your new triangle S. (2)

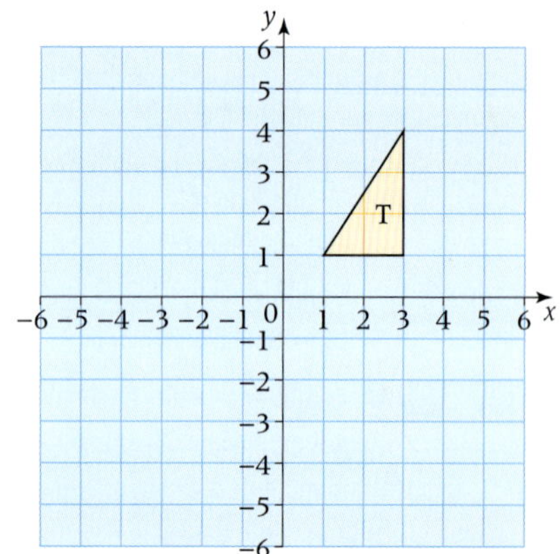

2 A shaded shape is shown on this grid of centimetre squares.

a Work out the perimeter of the shaded shape. (1)

b Work out the area of the shaded shape. (1)

c Copy the grid and reflect the shaded shape in the mirror line. (2)

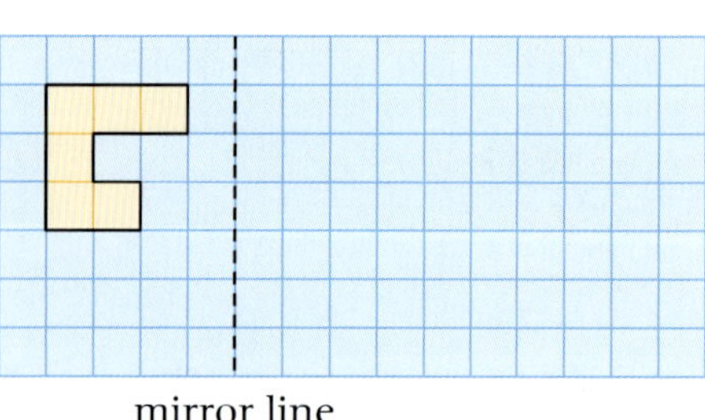

mirror line

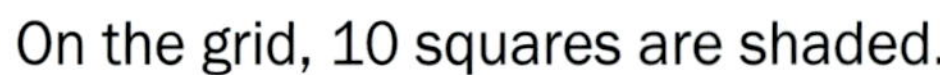

On the grid, 10 squares are shaded.

d Copy the grid and shade one extra square so that the shaded shape has one line of symmetry. (1)

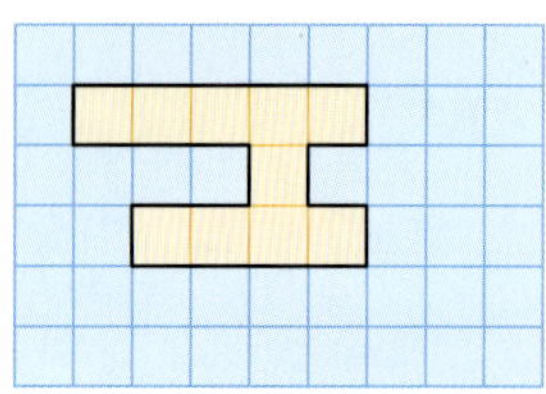

(Edexcel Ltd., 2003)

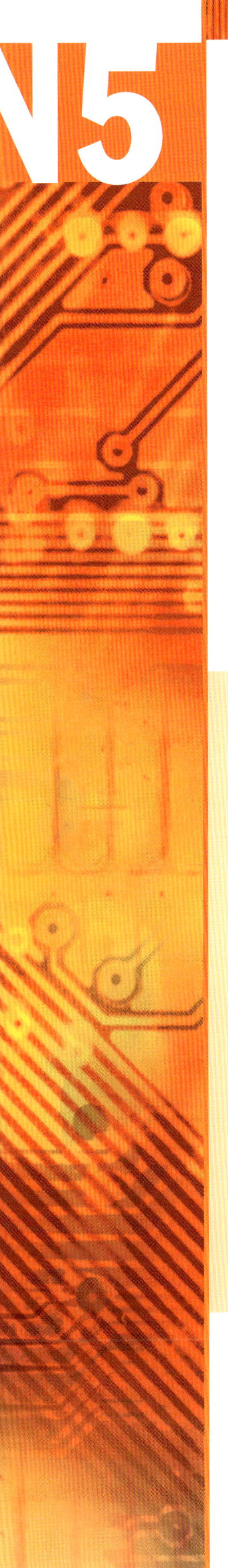

N5 Powers, roots and primes

This unit will show you how to

- Recognise and use square, cube and prime numbers
- Understand and use index notation for small powers
- Use the square, square root, cube and cube root functions of a calculator
- Use square roots and cube roots
- Estimate square roots and cube roots
- Round answers to an appropriate degree of accuracy
- Multiply and divide by powers of 10
- Use simple divisibility tests to check if a number is prime
- Understand and use the terms factor and multiple
- Find all the factors of a number and the highest common factor and least common multiple of two numbers

Before you start ...

You should be able to answer these questions.

Question	Review
1 Calculate each of these. **a** 7×7 **b** $2 \times 2 \times 2$ **c** $10 \times 10 \times 10$ **d** $32 \div 100$	Unit N2
2 Find the missing number in this expression. $\_\_ \times \_\_ = 25$	Unit N2
3 Write all the factors of 12.	Unit N1
4 Does 244 divide by 4? Explain how you know.	Unit N1, N2

N5.1 Squares and cubes

This spread will show you how to:

- Use square and cube numbers
- Use the square and cube functions of a calculator

Keywords
Cube
Index
Power
Square

- **A square number is the result of multiplying a whole number by itself.**

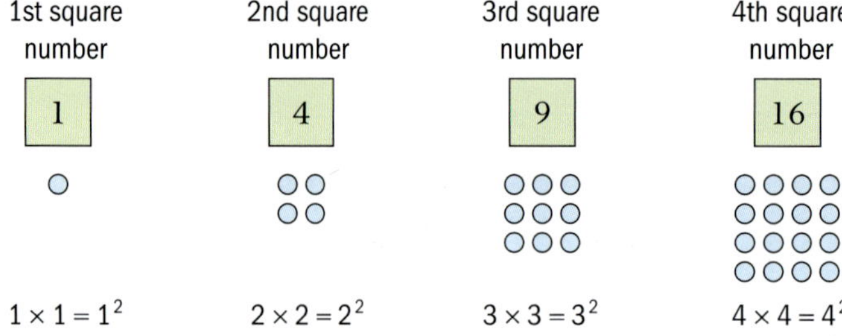

Square numbers can be written using **index** notation.

$$5^2 = 5 \times 5 = 25 \qquad 12^2 = 12 \times 12 = 144$$

Use the x^2 function key on a calculator to find the square of a number.

To find 34^2, type 3   = The display should read 1156.

You can say 5^2 in lots of different ways

5^2 = '5 to the **power** of 2'
= '5 squared'
= 'the square of 5'

- **A cube number is the result of multiplying a whole number by itself and then multiplying by that number again.**

Cube numbers can be represented using 3-D drawings of cubes.

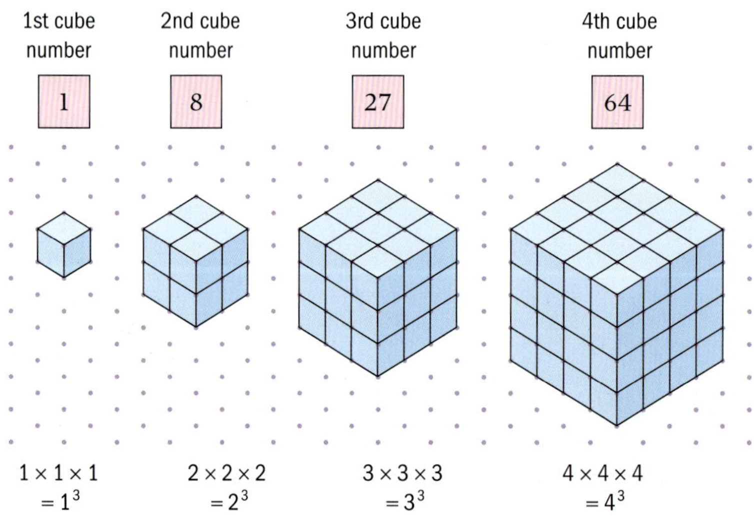

Cube numbers can be written using index notation.

$$4^3 = 4 \times 4 \times 4 = 64 \qquad 14^3 = 14 \times 14 \times 14 = 2744$$

Use the x^3 function key on a calculator to find the cube of a number.

To find 19^3, you would type 1   = The display should read 6859.

You can say 4^3 in lots of different ways:

4^3 = '4 to the power of 3'
= '4 cubed'
= 'the cube of 4'

Exercise N5.1

1 Find these numbers.

a the 4th square number **b** the 8th square number
c the 20th square number **d** the 5th cube number
e the 7th cube number **f** the 10th cube number

2 Use your calculator to work out each of these.

a 6^2 **b** 11^2 **c** 14^2 **d** 23^2
e 31^2 **f** 47^2 **g** 4^3 **h** 6^3
i 8^3 **j** 13^3 **k** 18^3 **l** 21^3

3 Some numbers can be represented as the sum of two square numbers.
For example $1^2 + 2^2 = 1 + 4 = 5$.
Try to find all the numbers less than 50 that can be represented as the sum of two square numbers.

4 Use your calculator to work out each of these.

a $3^2 + 2^2$ **b** $5^2 - 3^2$ **c** $6^2 - 2^3$
d $4^3 + 4^2$ **e** $10^2 - 8^2$ **f** $13^2 + 4^3$
g $14^2 - 5^3$ **h** $6^3 - 13^2$ **i** $12^2 + 13^2 + 14^2$
j $6^3 + 7^3 + 8^3$

5 Use the x^2 and x^3 function keys on your calculator to work out each of these. Give your answer to two decimal places as appropriate.

a 2.5^2 **b** 49^2 **c** 3.2^3 **d** 4.8^2
e 7.3^2 **f** 4.9^3 **g** 1.2^2 **h** 0.5^2
i 9.9^2 **j** 9.9^3 **k** $(5 \text{ cm})^2$ **l** $(4 \text{ m})^3$

6 **a** Work out these.

$3^2 - 2^2 =$ __ $4^2 - 3^2 =$ __ $5^2 - 4^2 =$ __

b Write anything you have noticed about your answers.

c Copy and complete this table and investigate for the next seven pairs of consecutive integers.

Consecutive means 'next to', for example 4 and 5

Square number	Square number		Answer
3^2	2^2	$3^2 - 2^2 = 9 -$ ___	
4^2	3^2	$4^2 - 3^2 = 16 -$ ___	
5^2			
6^2			

N5.2 Square roots and cube roots

This spread will show you how to:

- Estimate and use square roots and cube roots
- Use the square root and cube root functions of a calculator

Keywords
Cube root
Square root

- **A square root is a number that when multiplied by itself is equal to a given number.**

$7 \times 7 = 49$ so you can say that the square root of $49 = 7$.

- Square roots are written using $\sqrt{\ }$ notation. $\sqrt{49} = 7$

You should try to learn the first 10 square numbers, then you will also know their square roots.

1st square number $= 1 \times 1 = 1$ $\sqrt{1} = 1$

2nd square number $= 2 \times 2 = 4$ $\sqrt{4} = 2$

3rd square number $= 3 \times 3 = 9$ $\sqrt{9} = 3$

You can estimate the square root of a square number.

Example

289 is a square number. Work out $\sqrt{289}$ without using a calculator.

$15 \times 15 = 225$ This is too low, so $\sqrt{289}$ is greater than 15

$20 \times 20 = 400$ This is too high, so $\sqrt{289}$ is less than 20

$17 \times 17 = 289$ Correct

So $\sqrt{289} = 17$

Remember to use the grid method if you are stuck!

×	10	7
10	10 × 10 = 100	10 × 7 = 70
7	7 × 10 = 70	7 × 7 = 49

$$\begin{array}{r} 100 \\ 70 \\ 70 \\ +49 \\ \hline 289 \end{array}$$

Use the $\sqrt{x}$ function key on a calculator to find the square root of a number.

To find $\sqrt{289}$, you would type    The display should read 17.

- **A cube root is a number that when multiplied by itself and then multiplied by itself again is equal to a given number.**

You will learn more about the cube root calculator key on page 310.

$2 \times 2 \times 2 = 8$ so the cube root of $8 = 2$ or $\sqrt[3]{8} = 2$

Use the $\sqrt[3]{x}$ function key on a calculator to find the cube root of a number.

To find $\sqrt[3]{1728}$, you would type    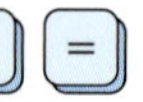 The display should read 12.

Square roots and cube roots are often not whole numbers so you usually round your answer to two decimal places.

$\sqrt{300} = 17.320\,508\,08\ldots$

$= 17.32$ (2 dp)

Exercise N5.2

1 Find these numbers.

a $\sqrt{25}$ **b** $\sqrt{9}$ **c** $\sqrt{16}$ **d** $\sqrt{1}$ **e** $\sqrt{4}$

2 Calculate these using a calculator, giving your answer to 2 dp as appropriate.

a $\sqrt{40}$ **b** $\sqrt{61}$ **c** $\sqrt{180}$ **d** $\sqrt{249}$ **e** $\sqrt{676}$

3 Calculate these using the $\sqrt[3]{\ }$ key on your calculator. Give your answers to 2 dp where appropriate.

a $\sqrt[3]{27}$ **b** $\sqrt[3]{512}$ **c** $\sqrt[3]{3375}$ **d** $\sqrt[3]{100}$ **e** $\sqrt[3]{24\,389}$

4 Harry has mixed up his answers to these questions.

a Without using a calculator, match each of these questions to the correct answer.

Questions	
1	$\sqrt{169}$
2	$\sqrt[3]{343}$
3	$\sqrt{121}$
4	$\sqrt{81}$
5	$\sqrt{64}$
6	$\sqrt[3]{1000}$
7	$\sqrt{196}$
8	$\sqrt[3]{64}$

Estimates	
A	9
B	8
C	7
D	4
E	11
F	13
G	10
H	14

b Check your answers using your calculator.
See how many of the questions and answers you matched correctly.

5 **a** A square has an area of 144 m^2. What is the side length of the square?

b John thinks of a number. He multiplies the number by itself. The answer is 529. What number did John think of?

c Mr Mow designs a paddock. The paddock has to be in the shape of a square. The area of the paddock has to be 4000 m^2. What length should each of the sides of the paddock be? (Give your answer to the nearest metre.)

d Nelly digs a hole in the shape of a cube. The volume of the earth she digs out to make the hole is 64 000 cm^3. How deep is the hole?

6 Do not use a calculator for these questions.

a Between which two numbers does $\sqrt{95}$ lie?

b Between which two numbers does $\sqrt{150}$ lie?

c Between which two numbers does $\sqrt{300}$ lie?

d Between which two numbers does $\sqrt[3]{80}$ lie?

N5.3 Powers of 10

This spread will show you how to:

- Understand and use index notation
- Multiply and divide by powers of 10

Keywords
Index
Index notation
Power
Powers of 10

- **Index notation** is used to represent powers of any number.
  The **index** (or **power**) tells you how many times the number must be multiplied by itself.

  $7^4 = 7 \times 7 \times 7 \times 7 = 2401$
  $6^3 = 6 \times 6 \times 6 \quad = 216$

The small number is the index (or power).

Use your calculator to work out powers of a number.

To work out 13^5, you might type 1 3 y^x  

The display should read 371 293.

Your calculator may work differently. If you are unsure, ask your teacher or check your manual.

- The decimal system is based upon **powers of 10**.

1 ten	= 10	= 10	$= 10^1$
1 hundred	= 100	$= 10 \times 10$	$= 10^2$
1 thousand	= 1000	$= 10 \times 10 \times 10$	$= 10^3$
10 thousand	= 10 000	$= 10 \times 10 \times 10 \times 10$	$= 10^4$
100 thousand	= 100 000	$= 10 \times 10 \times 10 \times 10 \times 10$	$= 10^5$
1 million	= 1 000 000	$= 10 \times 10 \times 10 \times 10 \times 10 \times 10$	$= 10^6$

- It is easy to multiply and divide by **powers of 10**.
  - When you multiply a number by 10^1 all the digits move 1 place to the left of the decimal point.
  - When you multiply a number by 10^2 all the digits move 2 places to the left of the decimal point.

	Thousands	Hundreds	Tens	Units		tenths	hundredths
				3	•	2	
3.2×10^1			3	2	•		
3.2×10^2		3	2	0	•		

The '0' holds the digits in place so that the '3' digit is in the Hundreds column and the '2' digit is in the Tens column.

  - When you divide a number by 10^1 all the digits move 1 place to the right of the decimal point.
  - When you divide a number by 10^2 all the digits move 2 places to the right of the decimal point.

	Thousands	Hundreds	Tens	Units		tenths	hundredths
			4	5	•		
$4.5 \div 10^1$				4	•	5	
$4.5 \div 10^2$				0	•	4	5

Exercise N5.3

1 Find the value of

a 5^2 **b** 2^3 **c** 3^3 **d** 8^2 **e** 12^2

2 Find the value of

a 3^4 **b** 1^5 **c** 2^7 **d** 3^6 **e** 10^6

3 Use the y^x function key on your calculator to work out these.

a 12^3 **b** 6^6 **c** 21^3 **d** 16^5 **e** 13^3

4 Use your calculator to work out these. In each case copy the question and fill in the missing numbers.

a $3^? = 9$ **b** $5^? = 25$ **c** $4^? = 64$ **d** $2^? = 8$ **e** $24^? = 576$

5 Calculate each of these.

a 1.2×10 **b** $655 \div 10$ **c** 3.4×10^3

d $48 \div 100$ **e** 74×1000 **f** $43.3 \div 10^2$

Use a place value diagram.

6 Here are some conversion rates for metric measurements.

1 km = 1000 m	1 tonne = 1000 kg	1 litre = 1000 ml
1 m = 100 cm	1 kg = 1000 g	1 m^3 = 1000 litres
1 cm = 10 mm		

You will need to revise how to change between units to answer these questions.

Change these lengths to the units indicated in brackets.
You will need to multiply and divide by powers of 10.

a 30 mm (centimetres) **b** 7 cm (millimetres)

c 3 km (metres) **d** 2500 m (kilometres)

e 240 cm (metres) **f** 7.2 m (centimetres)

g 2.4 tonnes (kg) **h** 3750 kg (tonnes)

i 450 g (kg) **j** 2.04 kg (grams)

k 330 ml (litres) **l** 4.54 litres (ml)

m 15 m^3 (litres) **n** 2300 litres (m^3)

7 Copy these and fill in the missing numbers.

a $34 \div 10 = \_\_\_$ **b** $2800 \div \_\_\_ = 28$

c $6.1 \times \_\_ = 610$ **d** $5.7 \times 10^3 = \_\_\_$

N5.4 Prime numbers

This spread will show you how to:

- Recognise prime numbers
- Use simple divisibility tests to check if a number is prime

Keywords
Factor
Prime number
Prime factor

Any whole number can be written as the product of two **factors**.

To list all the factors of 12, draw rectangles.

○○○○○○○○○○○○

$1 \times 12 = 12$

○○○○○○
○○○○○○

$2 \times 6 = 12$

○○○○
○○○○
○○○○

$3 \times 4 = 12$

1, 2, 3, 4, 6 and 12 are factors of 12 because all these numbers divide exactly into 12 with no remainder.

The factors of 12 are {1, 2, 3, 4, 6 and 12}.

- **A prime number is a number with only two factors, these are 1 and the number itself.**

29 is a prime number because it has only two factors, the numbers 1 and 29.

The first ten prime numbers are: 2, 3, 5, 7, 11, 13, 17, 19, 23, 29.

- **A prime factor is a prime number that is also a factor of another number.**

Factors of 20 = {1, 2, 4, 5, 10, 20}
Prime factors of 20 = {2, 5}

You can use simple divisibility tests to help you check if a number is a prime number.
Here are the divisibility tests for the first five prime numbers.

This topic is extended to prime factor decomposition on page 366.

÷2	the number ends in 0, 2, 4, 6 or 8
÷3	the sum of the digits is divisible by 3
÷5	the number ends in 0 or 5
÷7	there is no check for divisibility by 7
÷11	the alternate digits add up to the same sum

Example

Which of the numbers in this list are prime numbers: 42, 27, 43, 55?

42 — 2 is a factor (because it ends in a 2)
42 is not a prime number.

27 — 2 is not a factor
3 is a factor (because $2 + 7 = 9$, a multiple of 3)
27 is not a prime number.

43 — 2 is not a factor
3 is not a factor
5 is not a factor
7 is not a factor (because 6×7 is 42)
43 is a prime number.

55 — 2 is not a factor
3 is not a factor
5 is a factor (because it ends in a 5)
55 is not a prime number.

Exercise N5.4

1 Write all the factors of these numbers.

a 8 **b** 12 **c** 11 **d** 14 **e** 28 **f** 30 **g** 40 **h** 50

2 Your task is to find all the prime numbers from 1 to 100.

a Copy this 1–100 number square.

b Follow these instructions.

- 1 is not a prime number so you can cross it out.
- 2 is the lowest prime number. Cross out all the multiples of 2 except for the number 2.
- 3 is the next number not crossed out. It is the next prime number. Cross out all the multiples of 3 except for the number 3.
- 5 is the next number not crossed out. It is the next prime number. Cross out all the multiples of 5 except for the number 5.

1	2	3	4	5	6	7	8	9	10
11	12	13	14	15	16	17	18	19	20
21	22	23	24	25	26	27	28	29	30
31	32	33	34	35	36	37	38	39	40
41	42	43	44	45	46	47	48	49	50
51	52	53	54	55	56	57	58	59	60
61	62	63	64	65	66	67	68	69	70
71	72	73	74	75	76	77	78	79	80
81	82	83	84	85	86	87	88	89	90
91	92	93	94	95	96	97	98	99	100

c Carry on until you have only prime numbers left.

d Make a list of all the prime numbers from 1 to 100.

3 Use the divisibility tests to answer each of these questions. In each case explain your answer.

a Is 5 a factor of 135? **b** Is 5 a factor of 210?

c Is 2 a factor of 321? **d** Is 11 a factor of 231?

e Is 7 a factor of 91?

4 Look at these numbers.

1 2 3 5 6 8 9
10 11 12 13 16 18 20

a Write all the numbers that are factors of 10.

b Write all the numbers that are square numbers.

c Write all the numbers that are prime factors of 44.

d Write all the numbers that are prime numbers.

5 Use the mental method of halving and doubling to calculate each of these. Show the method you have used.

a 4×21 **b** 3×16 **c** 23×4 **d** 16×15

6 Write all the prime factors of each of these numbers.

a 20 **b** 27 **c** 55 **d** 35 **e** 22 **f** 70 **g** 120 **h** 110

Hint for question 6: List all the factors. Find the ones that are prime.

N5.5 Factors and multiples

This spread will show you how to:

- Understand and use the terms factor and multiple
- Understand and use simple divisibility tests
- Understand the terms highest common factor and least common multiple

Keywords
Common factor
Common multiple
Factor
HCF
LCM
Multiple

You can use a range of strategies to find all the **factors** of larger numbers.

Example

Find all the factors of 252.

You can list the factors by factor pairs.
Factors of 252 are

1×252	4×53	9×28
2×126	6×42	12×21
3×84	7×36	14×18

Factors of 252 are
{1, 2, 3, 4, 6, 7, 9, 12, 14, 18, 21, 28, 36, 42, 53, 84, 126, 252}.

You can use doubling and halving to help find factors:
$7 \times 36 = 252$
$14 \times 18 = 252$

You can find the **highest common factor** (**HCF**) of two numbers by listing all the factors of both numbers.

This topic is extended to HCF and LCM on page 366.

Example

Find the HCF of 24 and 60.

The factors of 24 are: 1 2 3 4 6 8 12 24
The factors of 60 are: 1 2 3 4 5 6 10 12 15 20 30 60

1, 2, 3, 4, 6 and 12 are **common factors** of 24 and 60.
12 is the **highest common factor** of 24 and 60.

- **The multiples of a number are those numbers that divide by it exactly, leaving no remainder.**

The multiples of 18 are 18, 36, 54, 72, ...

$1 \times 18 = 18$ $3 \times 18 = 54$
$2 \times 18 = 36$ $4 \times 18 = 72$

You can think of multiples as being the numbers in the 'times tables'.

You can find the **least common multiple** (**LCM**) of two numbers by listing the first few multiples of each number.

Example

Find the least common multiple of 24 and 60.

The first six multiples of 24 are: 24 48 72 96 120 144
The first six multiples of 60 are: 60 120 180 240 300 360

120, 240, 360 are **common multiples** of 24 and 60.
120 is the **least common multiple** of 24 and 60.

Exercise N5.5

1 Write all the factor pairs of each of these numbers.

a 18 **b** 14 **c** 30 **d** 48

2 Look at these numbers.

2	3	4	5	6	8	10
12	15	16	17	18	19	20

a Write all the numbers that are factors of 40.

b Write all the numbers that are factors of 90.

c Write all the numbers that are multiples of 2.

d Write all the numbers that are prime numbers.

3 Write the first three multiples of each of these numbers.

a 6 **b** 11 **c** 19 **d** 25 **e** 65 **f** 105 **g** 187 **h** 308

4 Use divisibility tests to answer each of these questions.
In each case explain your answer.

a Is 5 a factor of 95?

b Is 10 a factor of 710?

c Is 9 a factor of 321?

d Is 11 a factor of 451?

e Is 6 a factor of 98?

5 Find the highest common factor of

a 6 and 4 **b** 12 and 18 **c** 14 and 16

d 28 and 35 **e** 30 and 54 **f** 56 and 64

6 Find the least common multiple of

a 6 and 4 **b** 6 and 8 **c** 12 and 18

d 15 and 25 **e** 21 and 28 **f** 26 and 39

7 **a** Copy and complete this table.

Numbers	Product	HCF	LCM
6 and 4	$6 \times 4 = 24$	2	12
8 and 10	$8 \times 10 =$ ___		
12 and 18			
6 and 9			
15 and 20			
15 and 25			

b Write anything you notice about the numbers in your table.

c Write a quick way to find the LCM if you know the HCF.

N5 Exam review

Key objectives

- Use the terms square, positive square root, cube
- Use index notation for squares, cubes and powers of 10
- Multiply or divide any number by powers of 10
- Use calculators effectively; know how to enter complex calculations and use function keys for squares and square roots
- Use the concepts and vocabulary of factor (divisor), multiple, highest common factor and least common multiple of two numbers

1 **a** Write all the factors of

i 26

ii 34 (4)

b Find the highest common factor of 26 and 34. (1)

2 **a** Work out the value of

i 4^2

ii 2^3 (2)

b Write as a power of 10

$10 \times 10 \times 10 \times 10 \times 10$ (1)

(Edexcel Ltd., 2004)

Formulae

This unit will show you how to

- Use formulae from mathematics and other subjects expressed initially in words
- Write formulae to represent real-life situations
- Use letter symbols to represent quantities in formulae
- Use the rules of algebra to simplify a formula written in symbols
- Simplify formulae by collecting like terms
- Substitute numbers into formulae
- Solve simple equations

Before you start ...

You should be able to answer these questions.

Review

1 Work out each of these. (Unit N6)

a $2 \times 5 + 1$ **b** $3 \times 4 + 5 \times 2$

c $\frac{6}{2} + 10$ **d** $7 \times 8 - 4 \times 3$

2 Simplify each of these. (Unit A1)

a $x + x + x + x$ **b** $m \times m$

c $4 \times p$ **d** $x \times 6$

3 Collect like terms for each. (Unit A1)

a $2p + q + 3p$ **b** $4x - 2y + x + 3y$

c $m + 2n - 3p + 3n - 2m - p$

4 Solve each of these. (Unit A2)

a $5t = 25$ **b** $14 + x = 21$

c $\frac{m}{6} = 2$

A5.1 Formulae in words

This spread will show you how to:

- Use formulae from mathematics and other subjects expressed initially in words
- Substitute numbers into formulae

Keywords
Formula
Formulae
Substitute

Sarah takes a job as a shop assistant.

In her first week she works 10 hours.
In her second week she works 14 hours.

Sarah uses this **formula** to work out her pay

Pay = number of hours worked × hourly rate

First week: Pay = 10 × £4.20 = £42
Second week: Pay = 14 × £4.20 = £58.80

- **You can substitute values into a formula given in words.**

Replace the words with numbers.

Sarah's formula has a one-step calculation.
Formulae can have two or more steps.

The plural of formula is **formulae**.

Example

Jeremy works in telesales.

His hourly rate of pay is £4.80.
He gets a bonus for every sale he makes.

He uses this formula to work out his pay

Pay = number of hours worked × hourly rate + bonus

a In week 1 Jeremy works 18 hours and gets £24 bonus.
Work out his pay for the week.

b In week 2 Jeremy earns £120 and gets a £24 bonus.
How many hours does he work?

---

a Pay = 18 × £4.80 + £24
= £86.40 + £24
= £110.40

BIDMAS
Multiplication before addition.

b Subtract the bonus: 120 − 24 = 96
£96 at £4.80 per hour: 96 ÷ 4.80 = 20

Jeremy works 20 hours.

Exercise A5.1

1 Here is part of a hockey league table:

	Number of matches played	Goals for	Goals against
Woodford	15	32	13
Chorton	15	27	17
Digley	14	24	18

To find a team's goal difference you can use the formula:

Goal difference = goals for − goals against

Work out the goal difference for

a Woodford **b** Chorton **c** Digley

2 A music shop has a one-day sale.
In the sale, every CD costs £9.99.
The total price for a number of CDs is worked out using the formula:

Total price = number of CDs × £9.99

Work out the total price for

a 10 CDs **b** 4 CDs **c** 7 CDs

3 A group of volunteers are painting the village hall.
They use this formula to work out how much paint they need.

Number of cans of paint = area to be painted (square metres) ÷ 15

The area to be painted is 285 square metres.
How many cans of paint do they need?

4 Mick works in an electrical shop and is paid £5.50 per hour.
He calculates his pay using the formula:

Pay = number of hours worked × hourly rate + commission on goods sold

One week Mick works 16 hours and earns £15 commission.
Work out his pay for the week.

5 Super Snacks catering calculate their prices using this formula.

Price = cost per head × number of guests

For a buffet, their cost per head is £4.50.

a Work out the cost of a Super Snacks buffet for 75 guests.

Tasty Treats catering calculate their prices using this formula.

Price = basic charge £45 + cost per head + number of guests

For a buffet, their basic charge is £45 and their cost per head is £3.

b Work out the cost of a Tasty Treats buffet for 75 guests.

c Stephen is organising a buffet for 75 people.
Which company should he choose?

A5.2 Writing formulae in words

This spread will show you how to:

- Use formulae from mathematics and other subjects expressed in words
- Write formulae to represent real-life situations

Keywords
Formula

Katie makes and sells bead necklaces and bracelets.

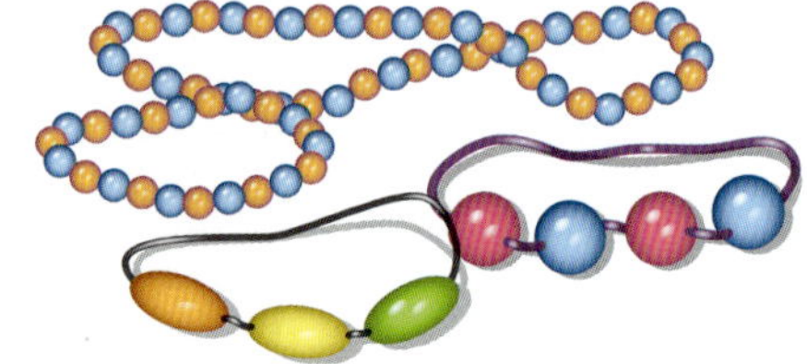

For glass beads, the price is 22p per centimetre.
For a 30 cm glass bead necklace, the price is
$30 \times 22\text{p} = 660\text{p} = £6.60$.

For plastic beads, the price is 18p per centimetre.
For a 20 cm plastic bead bracelet, the price is
$20 \times 18\text{p} = 360\text{p} = £3.60$.

Katie writes a formula to work out the price of any necklace or bracelet.

Price = cost per cm × length in cm

To write a formula, you can try a few examples with numbers first, to see a pattern in the calculations.

Example

Here is a sequence of patterns made from sticks.

Pattern 1

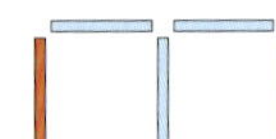
Pattern 2

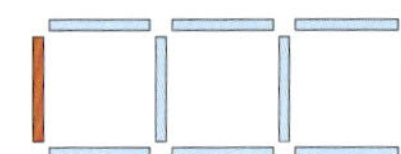
Pattern 3

a Write the number of blue sticks in
i 1 square **ii** 2 squares **iii** 3 squares.

b Write a formula connecting the number of blue sticks with the number of squares.

c Write the number of red sticks in
i 1 square **ii** 2 squares **iii** 3 squares.

d Use your answers to **b** and **d** to write a formula connecting the total number of sticks to the number of squares.
Use your formula to find the number of sticks in 15 squares.

---

a i 1 square: 3 blue sticks
ii 2 squares: 6 blue sticks
iii 3 squares: 9 blue sticks

b Number of blue sticks = 3 × number of squares

c i 1 square: 1 red stick
ii 2 squares: 1 red stick
iii 3 squares: 1 red stick

d Total number of sticks = number of blue sticks + number of red sticks
= 3 × number of squares + 1

e In 15 squares
Total number of sticks = $3 \times 15 + 1$
$= 45 + 1 = 46$

Exercise A5.2

1 Plain ribbon costs 30p per metre.
Tartan ribbon costs 42p per metre.

a How much does 2 m of plain ribbon cost?

b How much does 3 m of tartan ribbon cost?

c Write a formula connecting

cost of ribbon | length of ribbon | price per metre

2 Sareeta is making a row of coins in the High Street for charity.
She collects coins from the public and arranges them like this on the pavement

a What is the value of each vertical column?

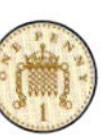

b A 10 cm length includes four 10p pieces.
What is the total value of a 10 cm length?

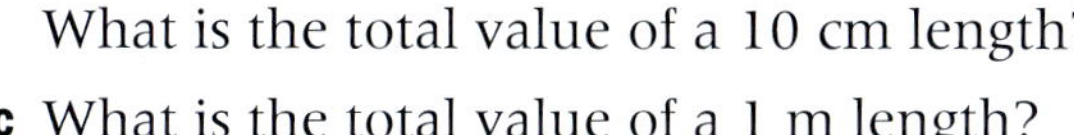

10 cm

c What is the total value of a 1 m length?
Give your answer in pounds (£).

d Copy and complete:
Total value of row = _____ × length of row in metres

3 Employees in a factory are paid by the hour.
Under 18s and adults are paid different hourly rates.
Write a formula for any employee, connecting

number of hours worked | hourly rate | pay

4 To fix a car, Tony charges £20 per hour for labour. He also charges the cost of any parts used.
Write a formula connecting

cost of the repair | cost of parts | number of hours worked | hourly rate for labour

A5.3 Formulae using letters

This spread will show you how to:

- Use letter symbols to represent quantities in formulae
- Use the rules of algebra to simplify a formula written in symbols

Keywords
Symbol
Variable

- **You can use letter symbols to represent quantities in a formula.**

The formula to work out the area of a rectangle is

Area = length × width

length

width

You can write this using letter symbols as

$$A = l \times w$$

$$A = lw$$

where A is the area, l is the length and w is the width.

In algebra you do not write the × sign.

You can use this formula to calculate the area of any rectangle.

When you write a formula, you need to use a different letter symbol for each variable.

l, w and A are called **variables**. l and w can take **any** values. The value of A is determined by the values of l and w.

Example

Write these formulae using letter symbols.
Explain what each letter symbol represents.

a goal difference = goals for − goals against
b taxi fare = basic charge + distance in miles × 0.2
c speed = distance ÷ time

a $D = F - A$
D = goal difference, F = goals for, A = goals against
b $F = b + d \times 0.2$
$F = b + 0.2d$
F = taxi fare, b = basic charge, d = distance in miles
c $s = d \div t$
$s = \frac{d}{t}$
Where s = speed, d = distance and t = time.

Write the number first and leave out the × sign.
$d \times 0.2 = 0.2d$

In algebra, write division $a \div b$ as $\frac{a}{b}$

Example

Recordable CDs come in packs of five.

a Write a formula in words for working out the cost of one CD.
b Write your formula using letter symbols.

a Cost of one CD = cost of pack of CDs ÷ 5
b $C = \frac{p}{5}$
Where C = cost of one CD, p = cost of pack of five CDs.

Exercise A5.3

1 Write these formulae using letter symbols.
Explain what each letter symbol represents.

a Repair cost = labour cost + parts cost

b Cost of electric cable = price per metre × length in metres

c Monthly cost = annual cost ÷ 12

d Cost of apples = price per kg × number of kg

e Distance in metres = distance in km × 1000

f Length in metres = length in centimetres ÷ 100

2 A mobile phone bill is calculated by adding the cost of calls made and the cost of texts sent.

a Write a formula to calculate a mobile phone bill in words.

b Write your formula from part **a** using algebra.
Explain what each letter symbol represents.

3 In a school hall, chairs are arranged in rows.
There are 15 chairs in each row.

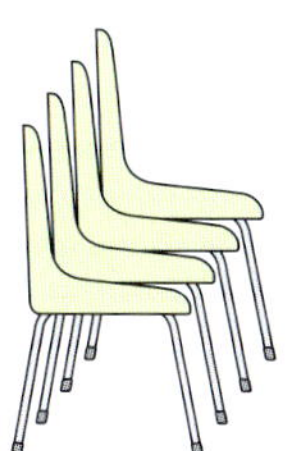

a Write a formula to work out the number of chairs in the hall.

b Write you formula in algebra.
Explain what each letter symbol represents.

c Use your formula to work out the number of chairs in the hall when there are 13 rows of chairs.

4 Eggs are packed in boxes of six.

a How many boxes are needed for 18 eggs?

b Write a formula in words to work out the number of boxes needed for any number of eggs.

c Write your formula using letter symbols.

d Use your formula to work out the number of boxes needed for

i 132 eggs **ii** 75 eggs.

5 In a café, each table has four chairs.
Write a formula connecting the number of tables and the number of chairs. Use t for the number of tables and c for the number of chairs.

6 In a library, each shelf holds the same number of books.
Write a formula, using letter symbols, connecting the number of books and the number of shelves.
Explain what each letter symbol represents.

7 Duvets cost £30 each and pillows cost £4.50 each.
Write a formula for the cost of m duvets and n pillows.

A5.4 More formulae

This spread will show you how to:

- Use the rules of algebra to simplify a formula written in symbols
- Simplify formulae by collecting like terms

Keywords
Like terms
Simplify
Square

You can use the rules of algebra to simplify a formula written in letter symbols.

$a + a = 2a$
$3 \times b = 3b$
$m \times m = m^2$

Example

a Write a formula in words for the area of a square.
b Write your formula using letter symbols. Give your formula in its simplest form.

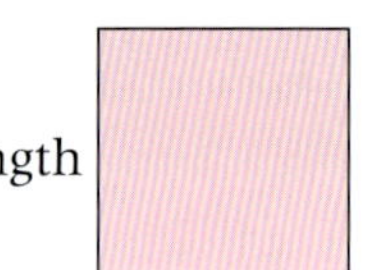

a Area = length × length
b $A = l \times l$
$A = l^2$

This topic is extended to expanding and factorising on page 372.

- **You can simplify a formula by collecting like terms.**

Like terms have exactly the same letter.

Example

Write a formula for the perimeter of this rectangle.
Simplify your formula as much as possible.

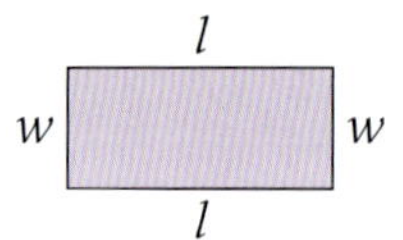

Perimeter $P = l + w + l + w$
$= l + l + w + w$
$P = 2l + 2w$

A formula may connect quantities in different units.
You need to explain which units to use for each variable.

Example

A plasterer charges a basic fee, plus £2 for each square metre of plaster.
Write a formula to work out the plasterer's charge in pounds.

Charge in pounds =

basic fee in pounds + £2 × area of plaster in square metres

$C = F + 2A$ where C = charge in £
F = basic fee in £
A = area in m^2

The rate per square metre (£2) is in pounds – the charge needs to be in pounds so the basic fee needs to be in pounds.

Exercise A5.4

1 Write a formula for the perimeter of this equilateral triangle.
Simplify your formula as much as possible.

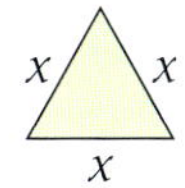

2 Write a formula for the perimeter of this square.
Write your formula in its simplest form.

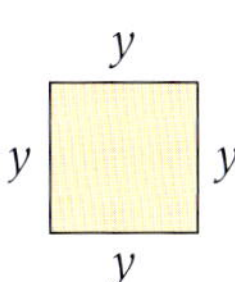

3 A taxi charges a basic fee and an extra £1 per mile.
Write a formula for the cost of a taxi ride in pounds.
Explain what each letter represents, and the units to use for each.

4 To hire a carpet cleaner you pay a fixed amount per day and the cost of the shampoo.
Write a formula for the cost of hiring the carpet cleaner in pounds.
Explain what each letter represents, and the units to use for each.

5 Write a formula for the perimeter of a polygon with n sides, where each side has length z.

Look back at your answers to questions **1** and **2**.

6 To make a cover for a square cushion, you need two squares of fabric.
Write a formula for the area of fabric (in square metres) needed for a square cushion of side k metres.

7 The formula to work out average speed is

$$s = \frac{d}{t}$$

where s = speed in miles per hour, d = distance in miles, t = time in hours.
Ben uses the formula to calculate his average speed when he travels 120 km in 2 hours.

$d = 120$ km
$t = 2$ hours
$s = \frac{120}{2} = 60$

What are the units for Ben's average speed?

8 p magazines are packed in a bundle.

a Write an expression for the number of magazines in m bundles.

q bundles are packed into one box.

b Write a formula to work out the number of magazines in a lorry load of n boxes.

A5.5 Substituting into formulae

This spread will show you how to:

- Substitute numbers into formulae
- Solve simple equations

Keywords
Substitute

- **In a formula, the letters represent quantities.**

For example, in the formula for the perimeter of a rectangle

$P = 2l + 2w$

l represents the length and w represents the width of the rectangle.

If you know the values of l and w, you can **substitute** the values of l and w into the formula.
Then you work out the calculation.

This means you write the formula replacing l and w with their number values.

Example

Use the formula

$P = 2l + 2w$

to work out

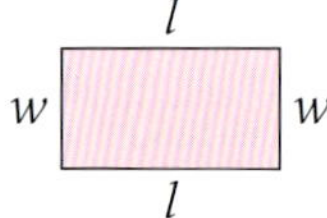

a the perimeter of a rectangular field with length 20 m and width 8 m
b the perimeter of a table mat with length 25 cm and width 30 cm.

a $l = 20$ m and $w = 8$ m
$P = 2l + 2w$
$P = 2 \times 20 + 2 \times 8$
$= 40 + 16$
$P = 56$ m

b $l = 25$ cm and $w = 30$ cm
$P = 2l + 2w$
$P = 2 \times 25 + 2 \times 30$
$= 50 + 60$
$= 110$ cm

$2l$ means $2 \times l$ multiplication before addition.
In part **a**, l and w are in metres, so P is also in metres.

In part **b**, l and w are in centimetres, so P is also in centimetres.

You can calculate a quantity from a formula by substituting quantities you know.

- **Sometimes when you substitute values into a formula you end up with an equation to solve.**

Example

The formula

$v = u + at$

is used in science.
Find the value of t when $v = 16$, $u = 0$ and $a = 4$.

$v = u + at$
$16 = 0 + 4t$ Substitute the values given.
$16 = 4t$
$16 \div 4 = 4t \div 4$ Solve by dividing both sides by 4.
$4 = t$
The value of t is 4.

Exercise A5.5

1 Here is a formula

$p = 5m$

Work out the value of p when

a $m = 4$ **b** $m = 3$ **c** $m = 10$ **d** $m = 2.5$

2 Here is a formula

$y = ax$

Work out the value of y when

a $x = 3$ and $a = 2$ **b** $x = 7$ and $a = 4$ **c** $x = 4.5$ and $a = 3$

Work out the value of x when

d $y = 10$ and $a = 2$ **e** $y = 25$ and $a = 10$

3 Use the formula

$P = 2l + 2w$

where P is perimeter, l is length and w is width, to work out the perimeter of each rectangle.

a 5 cm, 2 cm

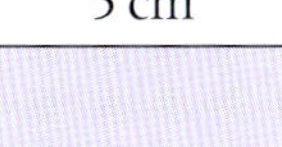

b 40 cm, 25 cm

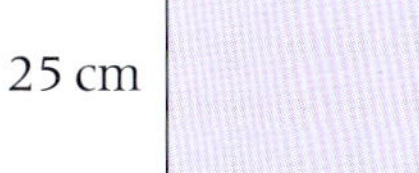

c 4 m, 3 m

d 1.5 m, 0.75 m

Be careful with the units.

4 For each formula, work out the value of y when $x = 4$ and $c = 6$.

a $y = 3x + c$ **b** $y = 4x - c$ **c** $y = \frac{4c}{x}$ **d** $y = cx + 10$

e $y = x^2$ **f** $y = c^2$ **g** $y = x^3$ **h** $y = 2x^2 + c$

5 Use the formula

$s = \frac{d}{t}$

where s = average speed, d = distance and t = time to work out the average speed when

a $d = 150$ miles, $t = 3$ hours **b** $d = 190$ km, $t = 2$ hours

c $d = 500$ metres, $t = 10$ seconds **d** $d = 60$ km, $t = 0.5$ hours

Be careful with the units.

6 Jan uses the formula

$s = ut + \frac{1}{2}at^2$

to calculate s when $u = 0$, $t = 3$ and $a = 2$
Here is her working:

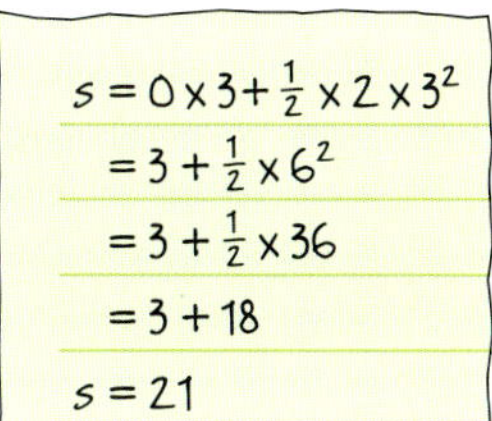

Jan's answer is **wrong**.
Work out the correct value of s.
Explain which **two** mistakes Jan has made in her working.

A5 Exam review

Key objectives

- Use formulae from mathematics and other subjects expressed initially in words and then using letters and symbols
- Distinguish the different roles played by letter symbols in algebra, knowing that letter symbols represent definite unknown numbers in equations and defined quantities or variables in formulae
- Generate a formula
- Understand that the transformation of algebraic expressions obeys and generalises the rules of arithmetic
- Manipulate algebraic expressions by collecting like terms
- Substitute numbers into a formula
- Solve simple equations

1 Simplify

a $n + n + n$ (1)

b $t - 2t$ (1)

c $2x + y - x$ (2)

2 You can use this rule to work out the total number of points a football team got last season.

Multiply the number of wins by 3 and then add the number of draws

Last season Rovers had 10 wins and 0 draws.

a Use the rule to work out the total number of points Rovers got last season. (1)

Last season United had 20 wins and 5 draws.

b Use the rule to work out the total number of points United got last season. (2)

(Edexcel Ltd., 2004)

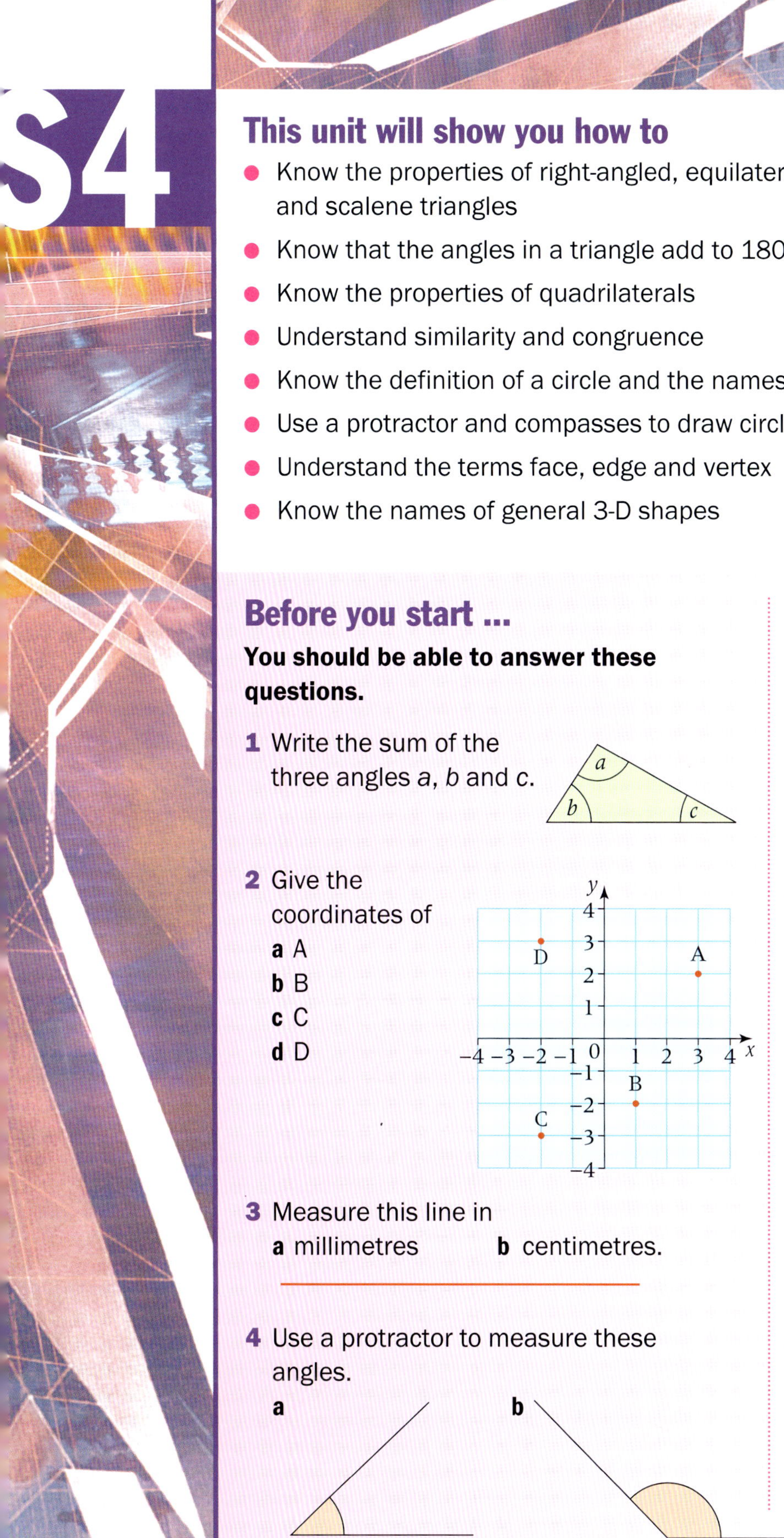

S4 2-D and 3-D shapes

This unit will show you how to

- Know the properties of right-angled, equilateral, isosceles and scalene triangles
- Know that the angles in a triangle add to 180°
- Know the properties of quadrilaterals
- Understand similarity and congruence
- Know the definition of a circle and the names of its parts
- Use a protractor and compasses to draw circles and sectors
- Understand the terms face, edge and vertex
- Know the names of general 3-D shapes

Before you start ...

You should be able to answer these questions.

Review

1 Write the sum of the three angles a, b and c. — Unit S2

2 Give the coordinates of — Unit A4

a A
b B
c C
d D

3 Measure this line in — Unit S1

a millimetres **b** centimetres.

4 Use a protractor to measure these angles. — Unit S2

a

b

S4.1 Properties of triangles

This spread will show you how to:

- Know the properties of right-angled, equilateral, isosceles and scalene triangles
- Know that the angles in a triangle add to 180°
- Understand similarity and congruence

Keywords
Congruent
Equilateral
Isosceles
Right-angled
Scalene
Similar
Triangle

A **triangle** is a 2-D shape with three sides and three angles.

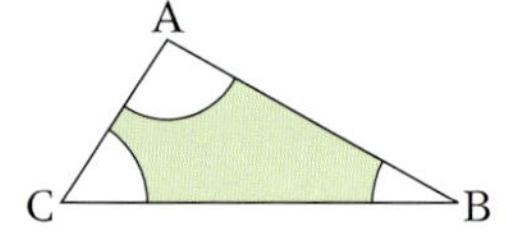

- **The angles in a triangle add to 180°.**

$\angle A + \angle B + \angle C = 180^\circ$

You need to know the properties of these triangles.

Triangle	Properties	Reflection symmetry	Rotation symmetry
Right-angled	One 90° angle marked ∟	No lines of symmetry	Order 1
Equilateral	3 equal angles 3 equal sides	3 lines of symmetry	Order 3
Isosceles	2 equal angles 2 equal sides	1 line of symmetry	Order 1
Scalene	No equal angles No equal sides	No lines of symmetry	Order 1

The marks show equal sides.

- **Similar** shapes are the same in shape but differ in size.
- **Congruent** shapes are the same size and shape.

Example

Here are two similar triangles.

a State the type of triangle.
b How many small congruent triangles will fit inside the large triangle?
c Draw the arrangement.

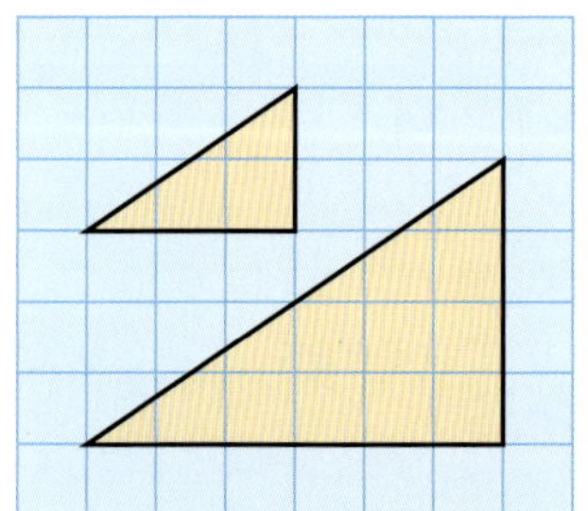

a Right-angled scalene triangle **b** 4 **c**

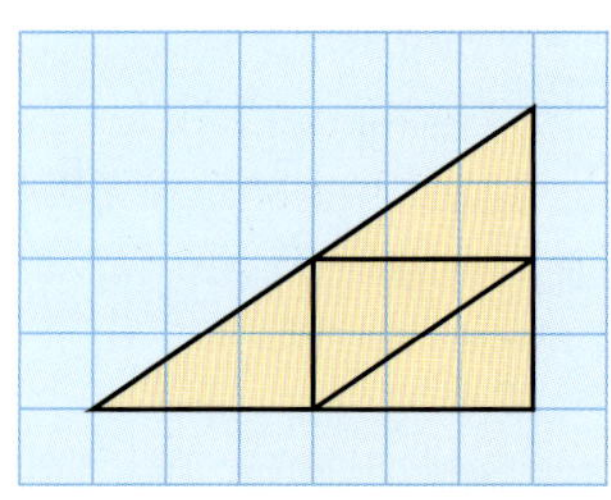

Exercise S4.1

1 State the type of each triangle.

a 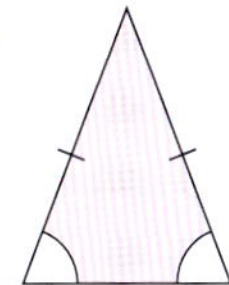 **b** 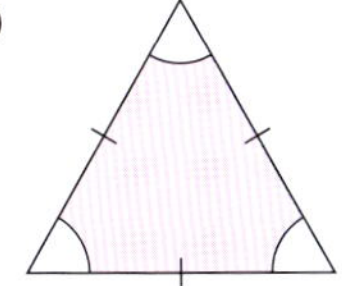 **c** 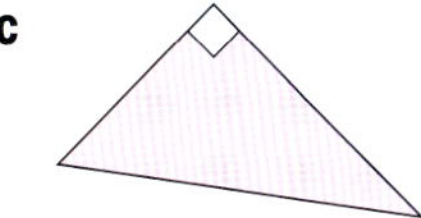 **d** 

2 State the type of each triangle, if the sides of the triangle are

a 8 cm, 8 cm, 8 cm

b 6 cm, 7 cm, 8 cm

c 3 cm, 5 cm, 5 cm

3 Calculate the third angle of the triangle and state the type of each of these triangles.

a 30°, 60°

b 70°, 40°

c 60°, 60°

d 35°, 65°

e 45°, 45°

4 How many equilateral triangles are in this pattern?

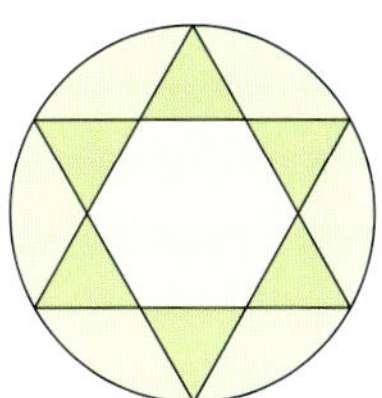

5 Plot and join up each set of points on a separate copy of this grid.

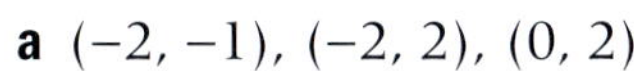

a (−2, −1), (−2, 2), (0, 2)

b (−2, −3), (−1, 0), (0, −3)

c (1, −2), (3, 2), (1, 3)

d (1, −3), (3, −3), (3, −1)

In each case, state the type of triangle.

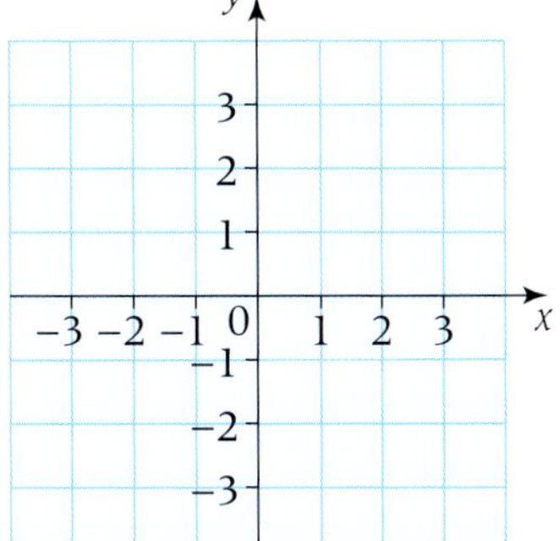

6 Using three more congruent (identical) equilateral triangles, draw a larger similar equilateral triangle on isometric paper.

S4.2 Properties of quadrilaterals

This spread will show you how to:

- Know the properties of quadrilaterals

Keywords
Diagonal
Parallel
Quadrilateral
Rotational symmetry

A **quadrilateral** is a 2-D shape with four sides and four angles.

You need to know the properties of these quadrilaterals.

The equal angles are colour coded.

Square	Rhombus	Parallelogram	Rectangle
4 equal angles 4 equal sides 2 sets parallel sides 4 lines of symmetry Rotational symmetry of order 4	2 pairs equal angles 4 equal sides 2 sets parallel sides 2 lines of symmetry Rotational symmetry of order 2	2 pairs equal angles 2 sets equal sides 2 sets parallel sides 0 lines of symmetry Rotational symmetry of order 2	4 equal angles 2 sets equal sides 2 sets parallel sides 2 lines of symmetry Rotational symmetry of order 2

Trapezium	Isosceles trapezium	Kite	Arrowhead
Usually: No equal angles No equal sides Always has: 1 set of parallel sides 0 lines of symmetry Rotational symmetry of order 1	2 pairs equal angles 1 set equal sides 1 set parallel sides 1 line of symmetry Rotational symmetry of order 1	1 pair equal angles 2 sets equal sides No parallel sides 1 line of symmetry Rotational symmetry of order 1	1 pair equal angles 2 sets equal sides No parallel sides 1 reflex angle 1 line of symmetry Rotational symmetry of order 1

Example

The diagonals on this rhombus are drawn.
Say whether each of these statements is true or false.

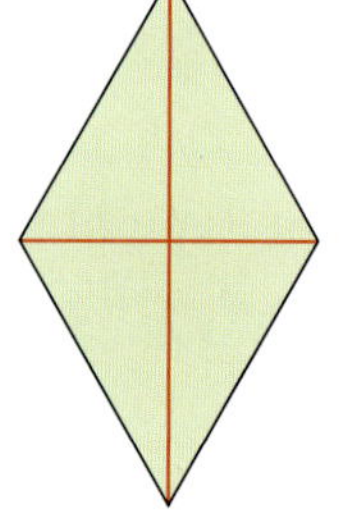

a The diagonals are of equal length.
b The diagonals are perpendicular.
c The diagonals are the lines of symmetry.
d The diagonals bisect each other.
e The diagonals bisect the angles.

Perpendicular lines cross at 90°.

a false **b** true **c** true **d** true **e** true

Exercise S4.2

1 Cut out four congruent (identical) right-angled triangles. Arrange all four triangles to make

4cm

2cm

a a square

b an isosceles trapezium

c a rectangle

d a parallelogram

e a rhombus.

Draw a sketch of each arrangement.

2 Give the mathematical name of each coloured quadrilateral in the regular octagon.

a

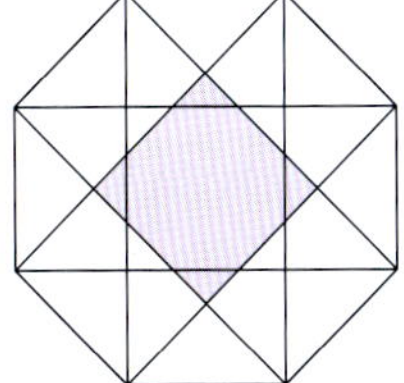

b

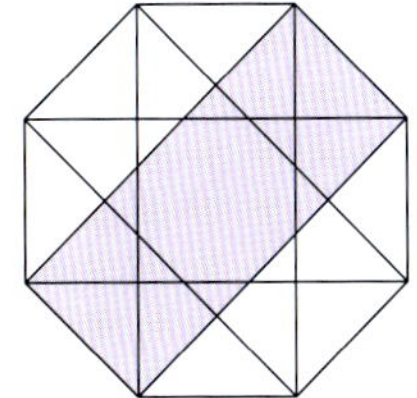

c

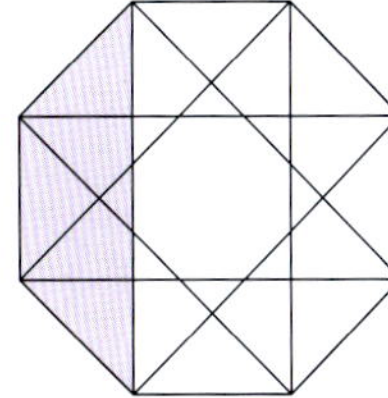

d

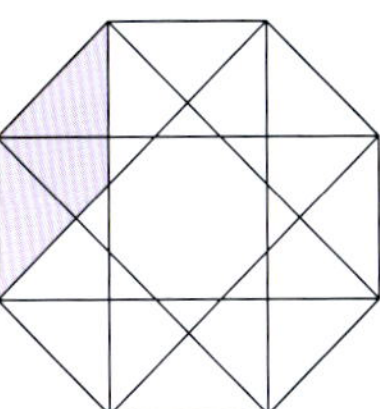

e

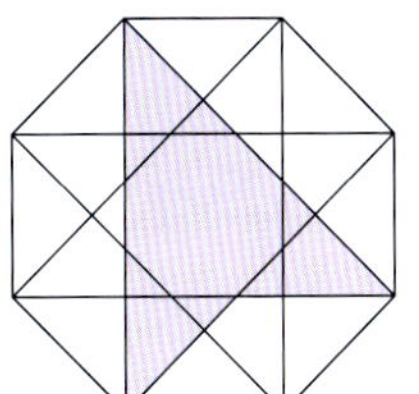

f

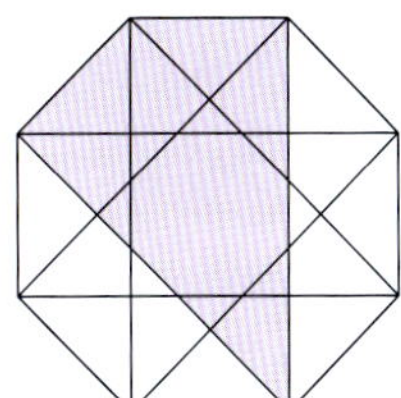

3 On separate copies of this grid, plot and join up each set of points.

a (2, 3) (1, 0) (2, −3) (3, 0) (2, 3)

b (2, −3) (2, 3) (0, 1) (0, −1) (2, −3)

c (3, 2) (−2, 2) (−3, −1) (2, −1) (3, 2)

d (0, −1) (2, 2) (0, 3) (−2, 2) (0, −1)

e (2, 2) (−3, 3) (−1, 2) (−3, 1) (2, 2)

f (1, 1) (−1, 3) (−3, 1) (−1, −1) (1, 1)

g (−2, 3) (−3, 2) (1, −1) (2, 0) (−2, 3)

h (3, −3) (0, −1) (−2, −1) (−3, −3) (3, −3)

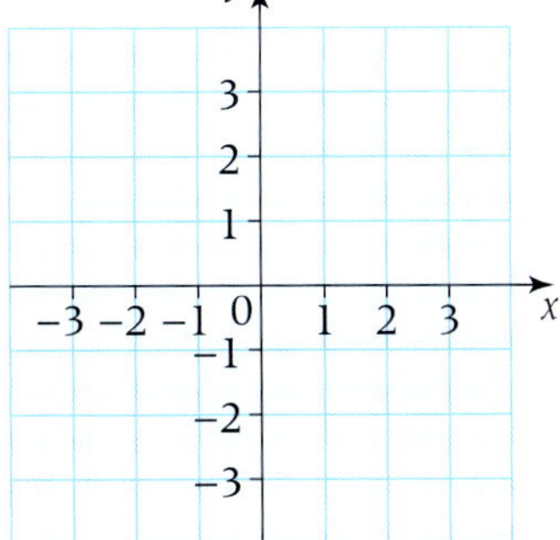

In each case, state the type of quadrilateral.

4 **a** Copy this shape and reflect it in the dotted line.

b Give the mathematical name of the new shape.

S4.3 Congruent shapes

This spread will show you how to:

- Understand congruence

Keywords

Congruent
Corresponding angles
Corresponding sides

- **Congruent shapes are exactly the same size and same shape.**

Example

Which of these shapes are congruent?

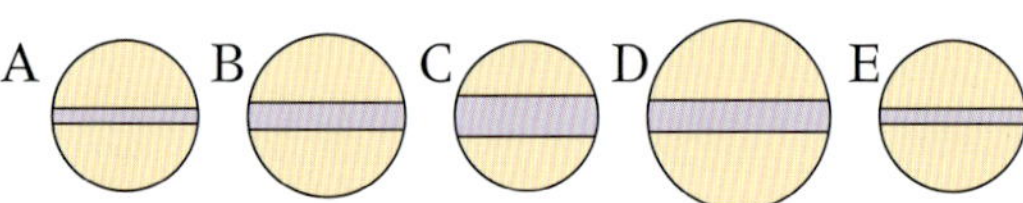

A and E are congruent.

Congruent shapes can be rotated or reflected if necessary.

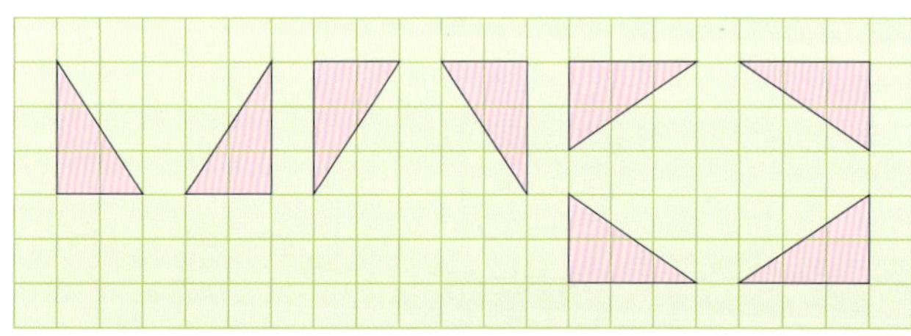

All these triangles are congruent – they fit exactly on top of each other.

- **In congruent shapes**
  - **corresponding angles** are equal
  - **corresponding sides** are equal.

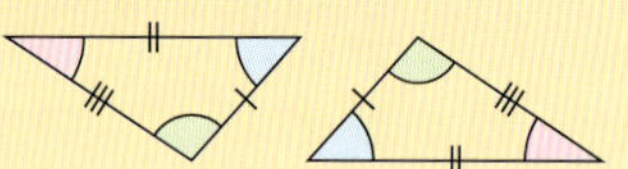

Example

This is triangle A.

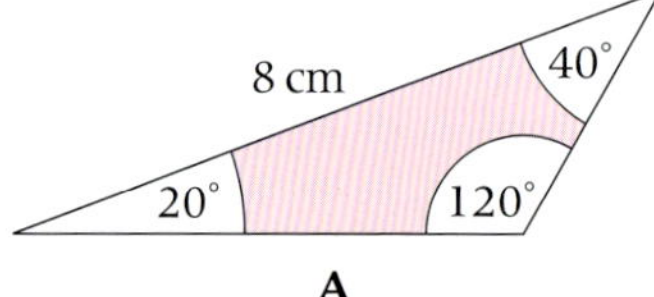

Which triangle is congruent to triangle A?

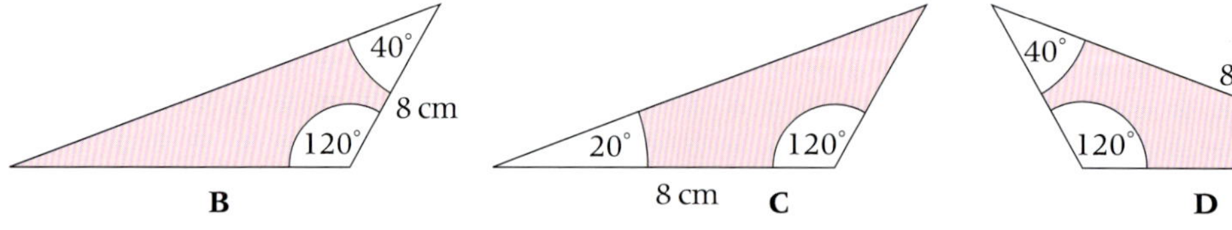

Fill in the missing information.

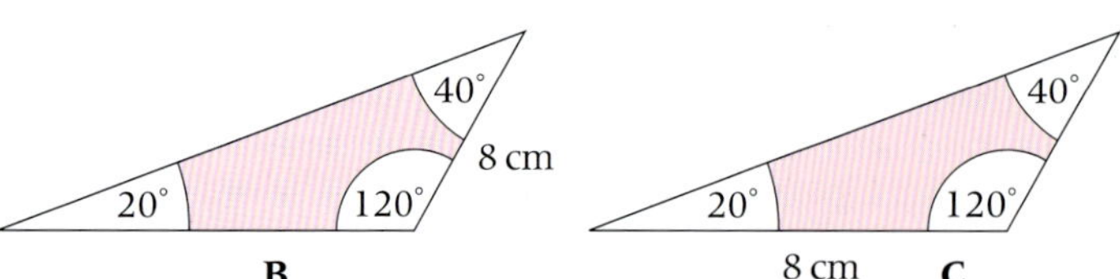

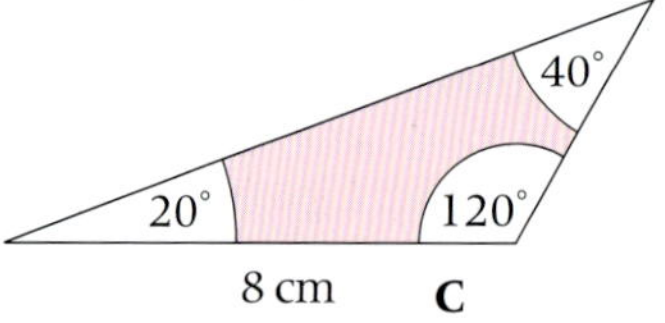

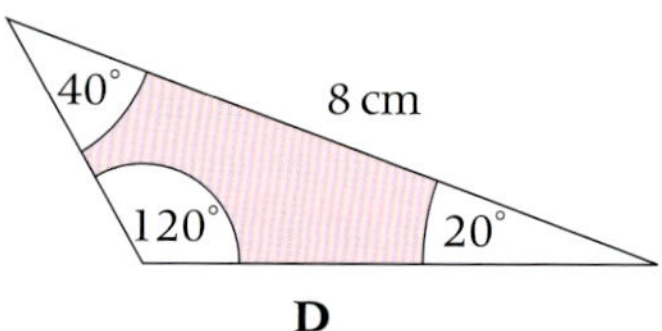

No, 8 cm in the wrong place.

No, 8 cm in the wrong place.

Yes, congruent.

Exercise S4.3

1 State which are the congruent pairs.

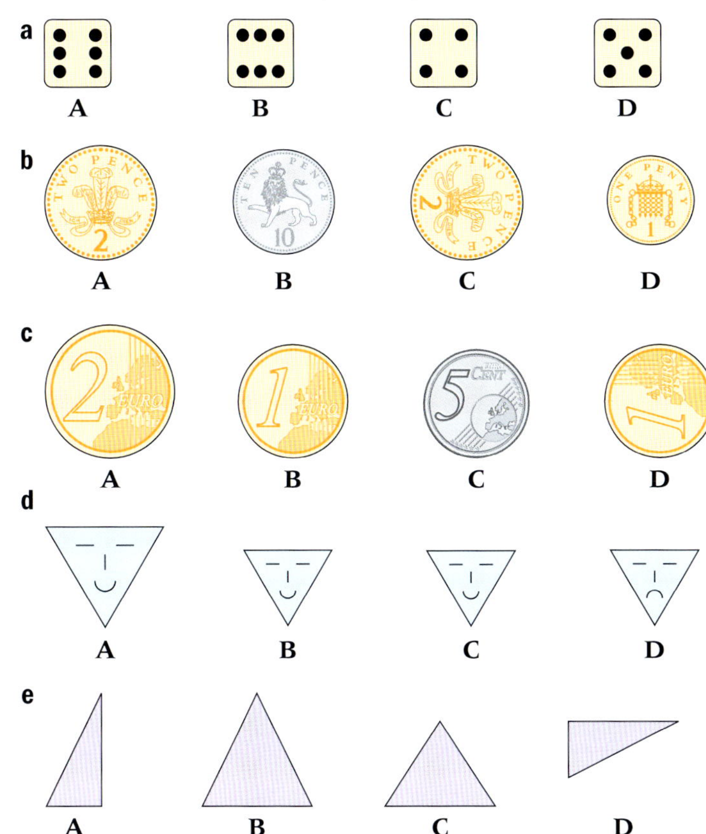

2 **a** Find **all** the pairs of quadrilaterals that are congruent.

b Give the name of the shape for each pair.

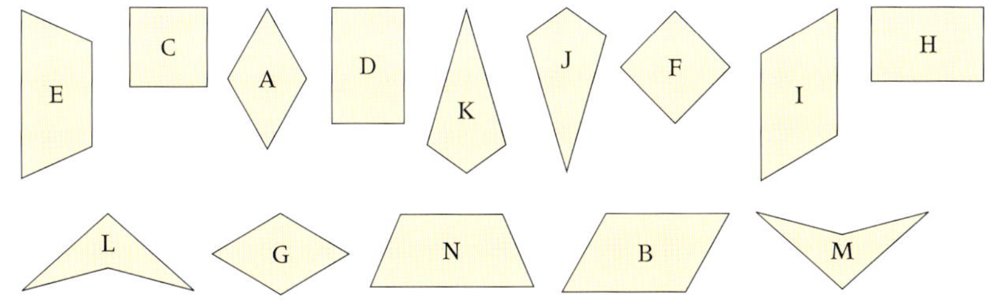

3 Here is triangle A.
Which of these triangles are congruent to triangle A?

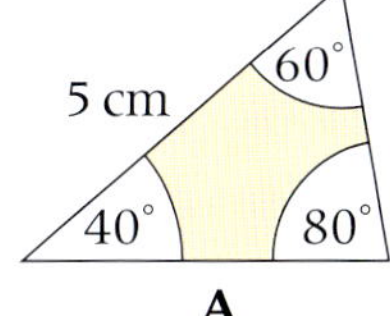

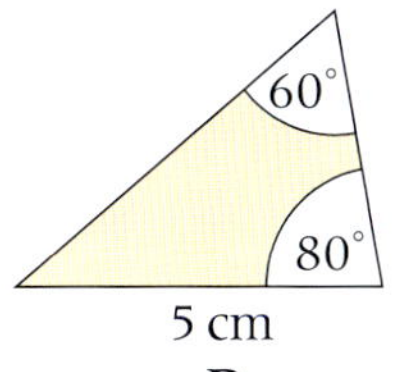

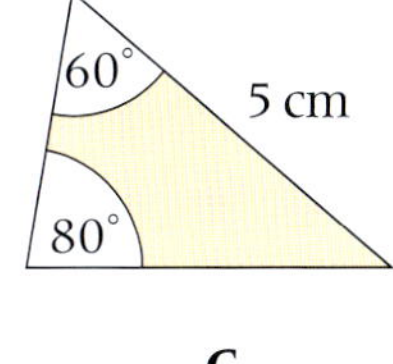

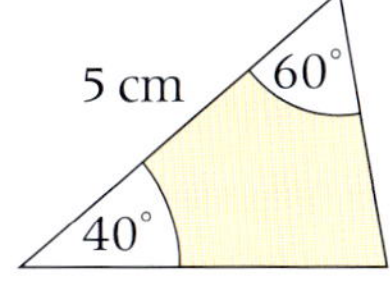

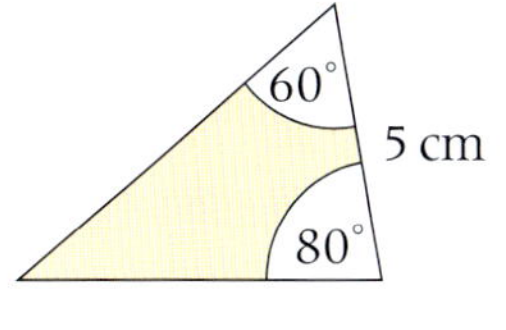

S4.4 Circles

This spread will show you how to:

- Know the definition of a circle and the names of its parts
- Use a protractor and compasses to draw circles and sectors

Keywords
Arc
Centre
Chord
Circle
Circumference
Compasses
Diameter
Equidistant
Radius
Sector
Segment
Semicircle
Tangent

- A **circle** is a set of points **equidistant** from its **centre**.

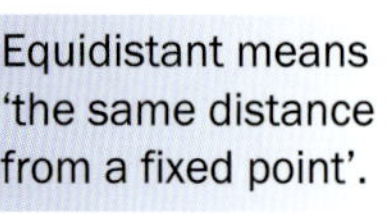
Equidistant means 'the same distance from a fixed point'.

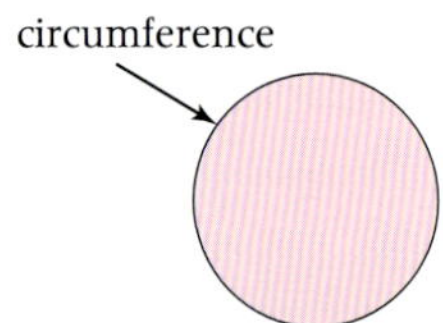

The **circumference** (C) is the distance around the circle.

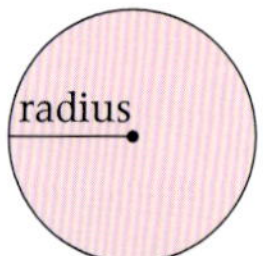

The **radius** (r) is the distance from the centre to the circumference.

Radii is the plural of radius.

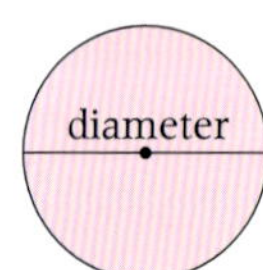

The **diameter** (d) is the distance across the centre of the circle.

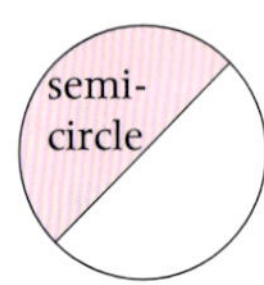

The diameter divides the circle into two **semicircles**.

Example

Draw a circle so that the line AB is the radius. A——B

Put the point of the **compasses** at A or B and open the compasses to the length of AB. Draw the circle.

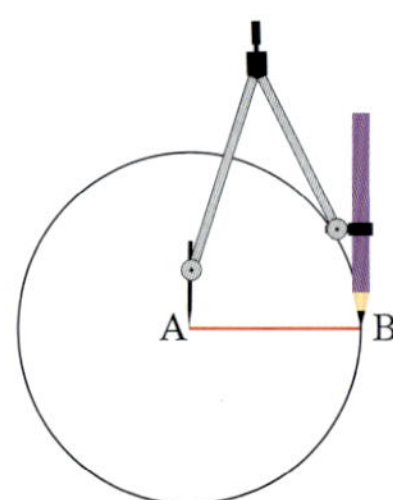

You also need to know these parts of a circle.

An **arc** is part of the circumference.

A **tangent** is a line that touches the circle at a single point.

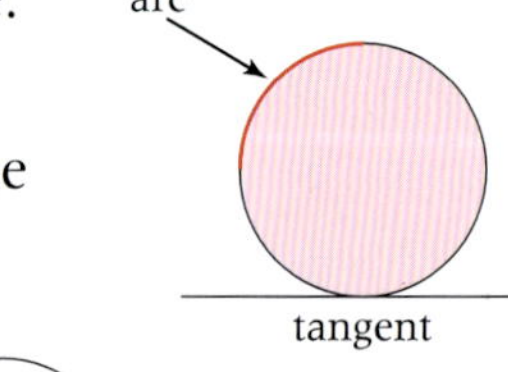

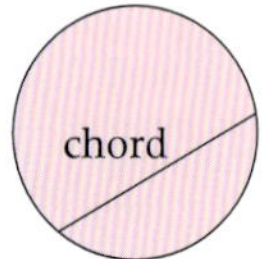

A **chord** is a line joining two points on the circumference.

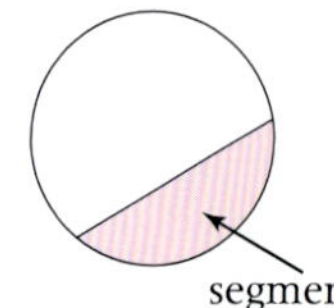

A **segment** is the region enclosed by a chord and an arc.

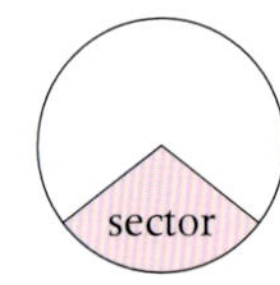

A **sector** is the region enclosed by an arc and two radii.

Exercise S4.4

1 **a** Draw two circles that intersect at A and B.

b Draw the diameters AC and AD.

c Is CBD a straight line?

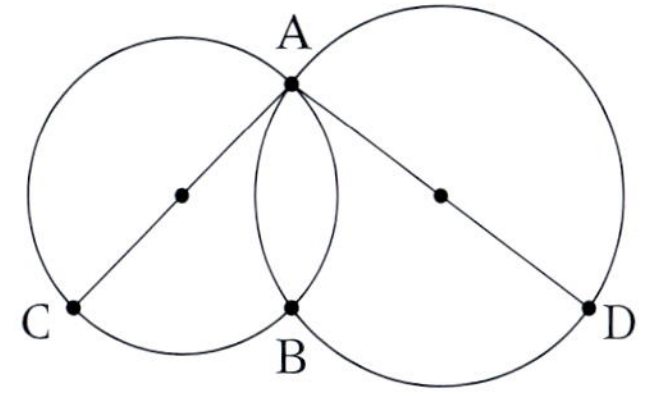

Intersect means cross.

2 Measure

a the diameter of the circle

b the radius of the circle.

3 **a** Draw a circle with a radius of 4 cm.

b Draw a chord of length 5 cm inside the circle.

4 Draw a circle with a diameter of 10 cm.

5 **a** Draw a 4 cm line AB.

b Draw a circle so that AB is the diameter.

c Find the radius of the circle.

A ———————— B
4 cm

6 Use a protractor and compasses to construct these sectors.

a

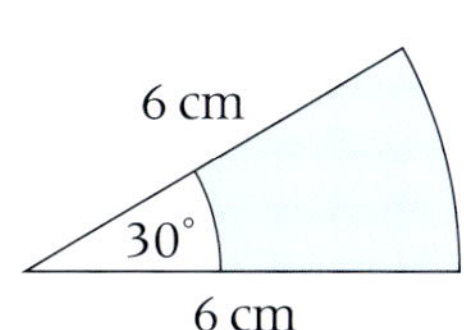

b

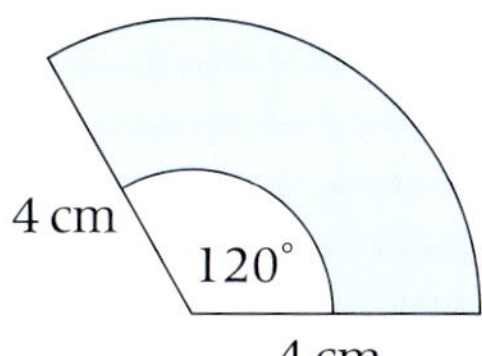

7 Two circles have the same centre.
One has a radius of 3.5 cm and the other has a radius of 2.5 cm.
Construct and colour this diagram for the two circles.

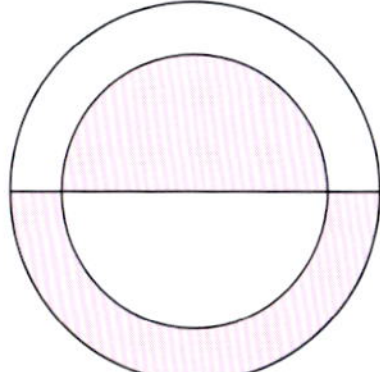

8 Explain why these circles are similar.

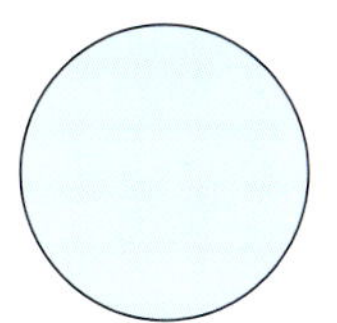

S4.5 3-D shapes

This spread will show you how to:

- Understand the terms face, edge and vertex
- Know the names of general 3-D shapes

Keywords
Cube
Cuboid
Edge
Face
Prism
Pyramid
Solid
Three-dimensional (3-D)
Vertex

A **solid** is a **three-dimensional (3-D)** shape.

The three dimensions are length, width and height for a cuboid.

- **In a 3-D shape**
  - **a face is a flat surface of the solid**
  - **an edge is the line where two faces meet**
  - **a vertex is a point at which two or more edges meet. The vertex is a corner of the shape.**

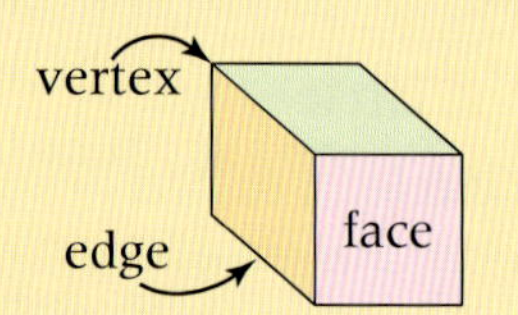

The plural of vertex is vertices.

A **cube** has 6 faces that are squares
12 edges
8 vertices.

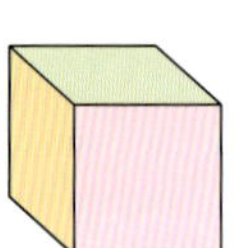

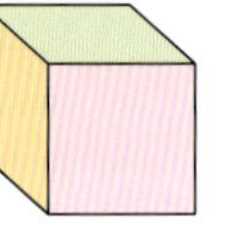

A **cuboid** has 6 faces that are rectangles
12 edges
8 vertices.

A **prism** has a constant cross-section.

You name a prism by the shape of the cross-section.

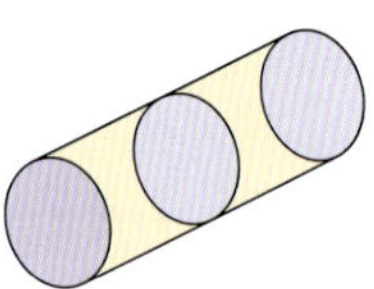

cylinder

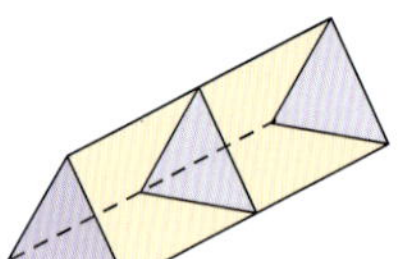

triangular prism

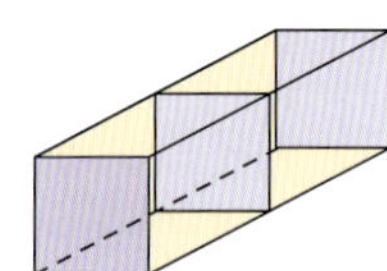

square-based prism

A **pyramid** has faces that meet at a common point.

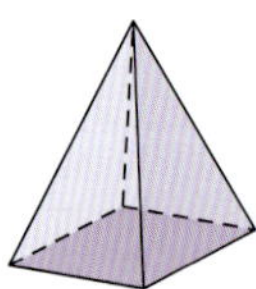

square-based pyramid

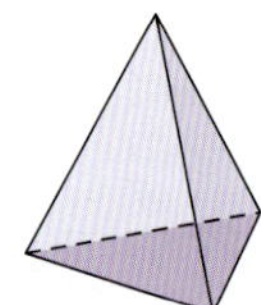

triangular-based pyramid

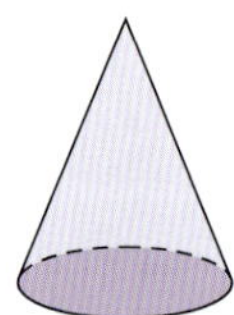

cone

A tetrahedron is a pyramid made from four equilateral triangles.

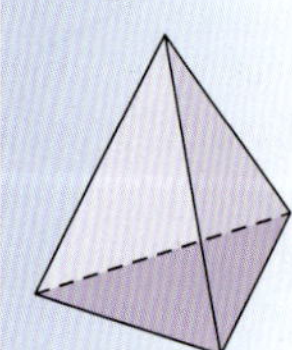

Example

For this solid, write

a its mathematical name
b the number of faces
c the number of edges
d the number of vertices.

---

a Pentagonal prism
b 7 faces
c 15 edges
d 10 vertices (corners)

Exercise S4.5

1 Decide whether these shapes are prisms, pyramids or neither of these.

a 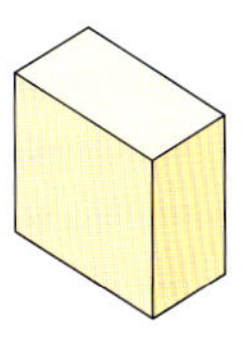

b 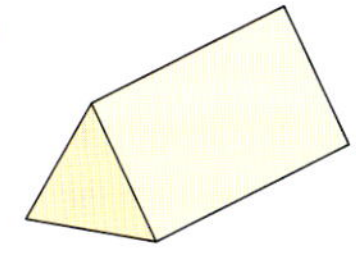

c 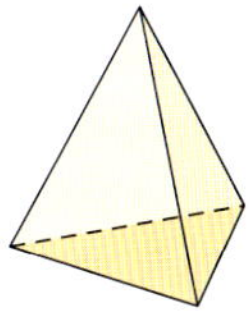

d 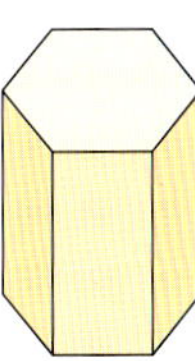

e 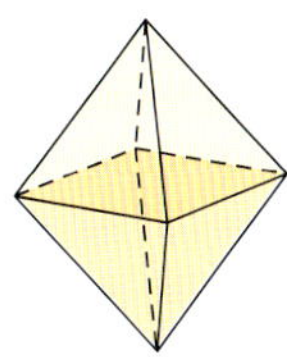

f 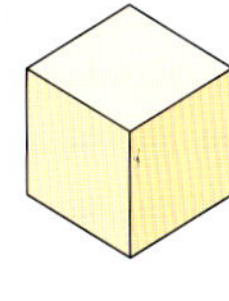

g 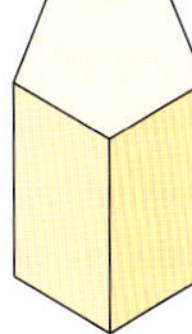

h 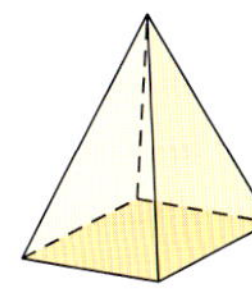

i 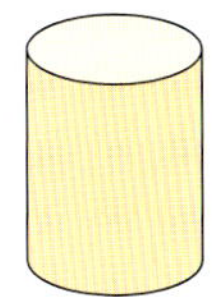

j 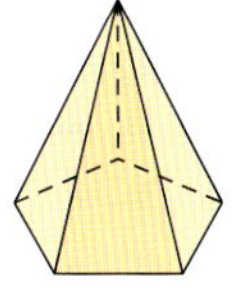

k 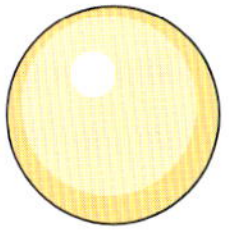

l 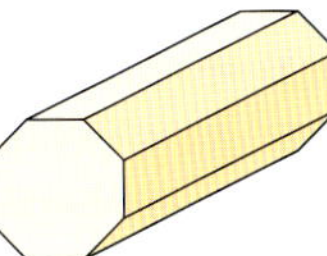

m 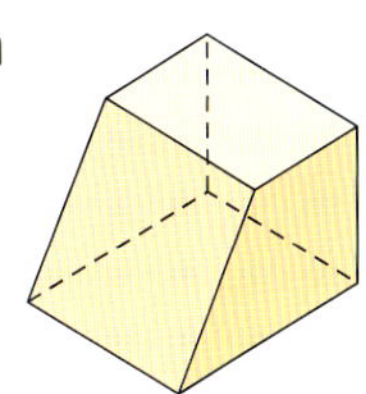

n 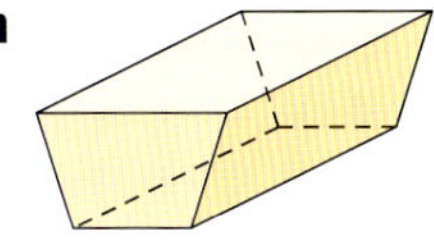

o 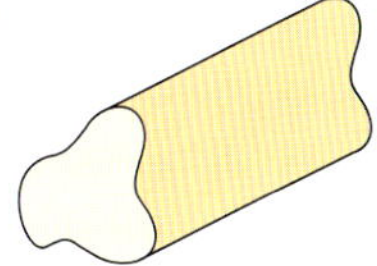

2 This solid is made from eight isosceles triangles and one octagon. For this solid, write

a the mathematical name

b the number of faces

c the number of edges

d the number of vertices.

3 **a** Copy and complete this table.

Name of solid	Number of faces (f)	Number of edges (e)	Number of vertices (v)
pentagonal prism	7	15	10
cuboid			
pentagonal pyramid			
square-based pyramid			
cube			
hexagonal prism			
tetrahedron			
triangular prism			
hexagonal pyramid			
octagonal-based prism			

b Write a relationship between f, e and v.

S4 Exam review

Key objectives

- Use angle properties of equilateral, isosceles and right-angled triangles
- Understand congruence
- Recall the essential properties of special types of quadrilateral, including square, rectangle, parallelogram, trapezium and rhombus
- Recall the definition of a circle and the meaning of related terms, including centre, radius, diameter, circumference
- Explore the geometry of cuboids (including cubes), and shapes made from cuboids

1 **a** Draw a line of length 5 cm. (2)

b Draw a circle, using your line from part **a** as the diameter of the circle. (2)

2 Here are 7 quadrilaterals.

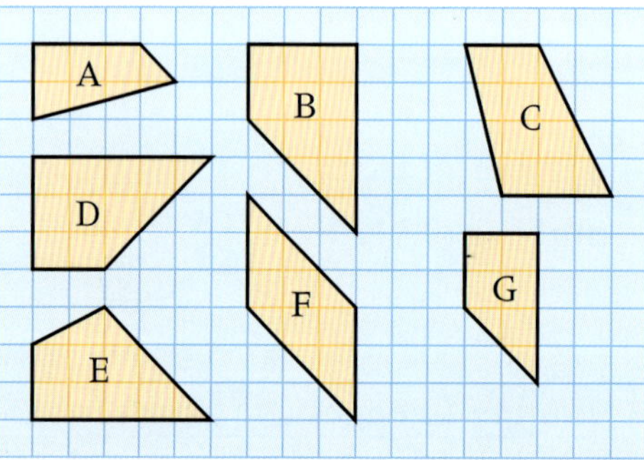

a Write down the letter of a parallelogram. (1)

Two of the quadrilaterals are congruent.

b Write down the letters of these quadrilaterals. (1)

(Edexcel Ltd., 2004)

N6 Decimal calculations

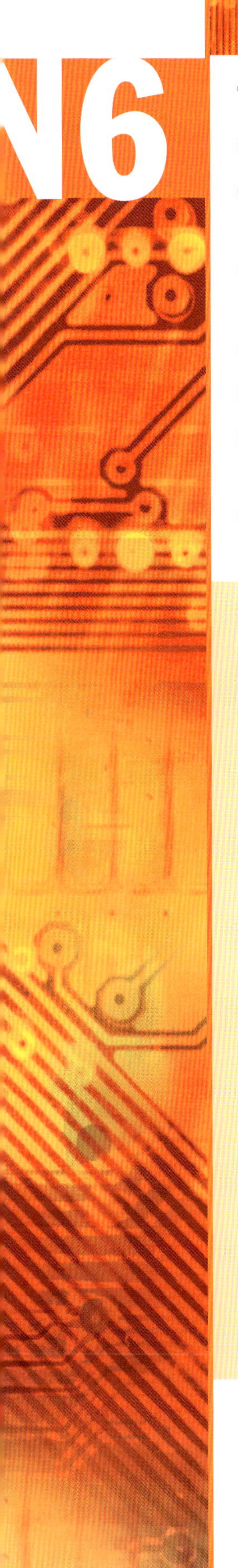

This unit will show you how to

- Know and use the order of operations, including brackets
- Round numbers to any given power of 10, the nearest integer and to one significant figure
- Use a range of written methods for addition, subtraction, multiplication and division of integers and decimals
- Estimate answers to calculations
- Choose an appropriate method for calculations
- Carry out more complex calculations involving decimals using the functions of a calculator
- Give answers to a given degree of accuracy

Before you start ...

You should be able to answer these questions.

	Review
1 Calculate $5 + 3 \times 4$.	Unit N2
2 Give an approximate answer to $6.8 \times (2.3 + 1.89)$	Unit N3
3 Calculate each of these. **a** $3.4 + 4.25$ **b** $12.6 - 3.8$	Unit N3
4 Calculate these using a mental or written method. **a** 28×5 **b** 2.5×12 **c** $107 \div 8$ **d** 23×48 **e** $23 \div 100$	Unit N2, N3
5 Use your calculator to work out these. **a** $(3.2)^2$ **b** $\frac{4.7 - 2.95}{0.51}$	Unit N3, N5

N6.1 Order of operations

This spread will show you how to:

- Know and use the order of operations, including brackets

Keywords
Brackets
Index
Order of operations
Power

Eve and Sarah are working out this calculation.

$2 + 3 \times 4$

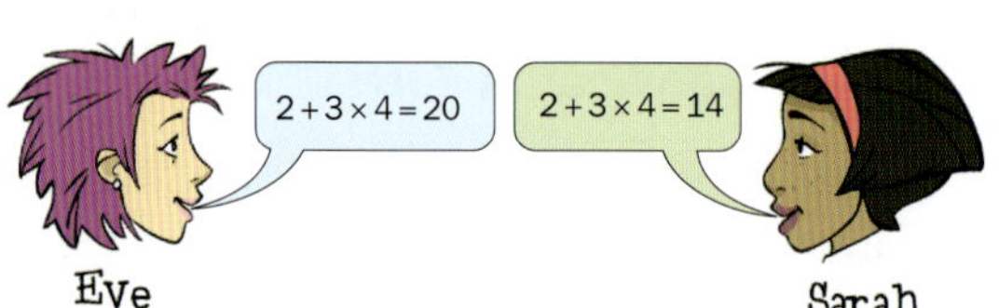

To find out who is correct, use the **order of operations**.

Brackets
First work out the contents of any brackets.

Powers or indices
Then work out any powers or roots.

Multiplication and division
Then work out any multiplications and divisions.

Addition and subtraction
Finally work out any additions and subtractions.

$2 + 3 \times 4$
$= 2 + 12$
$= 14$

Sarah is correct because she has followed the order of operations.

Example

Put brackets into this expression to make the answer correct.

$2 \times 3 + 4 - 5 = 9$

By inserting a pair of brackets

$2 \times (3 + 4) - 5 = 2 \times 7 - 5$ (work out the multiplication)
$= 14 - 5$ (work out the subtraction)
$= 9$ ✓ the correct answer

For lots of additions and subtractions (or multiplications and divisions), work them out from left to right.

Calculations involving brackets need to be interpreted carefully.

Example

Calculate each of these.

a $\dfrac{160}{5 + 11}$

b $21 + 2(8 + 7)$

a $\dfrac{160}{5 + 11} = \dfrac{160}{(5 + 11)}$
$= 160 \div (5 + 11)$
$= 160 \div 16$
$= 10$

b $21 + 2(8 + 7) = 21 + 2 \times 5$
$= 21 + 30$
$= 51$

Rewrite division caclulations using brackets, like part **a**.

Exercise N6.1

1 Match each of these calculations with one that gives the same answer.

A $7 + 3$ **1** $7 - 4$
B $2 + 3 \times 4$ **2** $33 - 9$
C $18 \div 6$ **3** $6 \times 3 - 4$
D 6×4 **4** $15 - 5$

2 Calculate these using the order of operations.

a $2 + 6 \times 3$ **b** $2 \times 12 - 7$ **c** $8 + 20 \div 4$ **d** $2 \times 3 + 4 \times 2$
e $14 \div 2 + 3$ **f** $2 + 7 \times 5$ **g** $2 + 3 \times 3 - 4$ **h** $4 \times 3 + 2 \times 6$

3 Calculate these using the order of operations.

a $(2 + 6) \times 3$ **b** $2 \times (12 - 7)$ **c** $(8 + 20) \div 4$ **d** $2 \times (3 + 4) \times 2$
e $15 \div (2 + 3)$ **f** $(2 + 7) \times 5$ **g** $(2 + 3) \times 3 - 4$ **h** $4 \times (3 + 2) \times 6$

4 Calculate these using the order of operations.

a $2 + 3^2 \times 4$ **b** $2^2 \times 3 - 7$ **c** $4^2 + 20 \div 4$
d $2 \times 3^2 + 4 \times 2$ **e** $16 \div 2^2 + 3$ **f** $2 + 7 \times 5^2$

5 Calculate these using the order of operations.

a $(2 + 6) \times 3^2$ **b** $2^2 \times (12 - 7)$ **c** $(8^2 - 20) \div 4$
d $2 \times (3^2 + 4)$ **e** $33 \div (2 + 3^2)$ **f** $(2^2 + 7) \times 5$

6 Copy these calculations. Insert brackets where necessary to make each of them correct.

a $3 \times 2 + 1 = 9$ **b** $5 \times 3 - 1 = 10$ **c** $3 + 5 \div 2 = 4$ **d** $2 + 3^2 \times 4 = 44$
e $4^2 \div 5 + 3 = 2$ **f** $4 \times 5 + 5 \times 6 = 240$

7 Calculate each of these.

a $\frac{3 + 7}{5}$ **b** $\frac{14 - 8}{2 + 1}$ **c** $\frac{5 \times 4}{2 \times 5}$
d $\frac{2 \times 3 + 4}{5}$ **e** $\frac{24}{2 + 3 \times 2}$

8 Calculate these.

a $3(2 + 5)$ **b** $4(2 + 6)$ **c** $5 + 4(7 - 3)$ **d** $3^2 + 2(7 - 4)$

9 Use the numbers 3, 4, 5, 7 and 8 to make as many of the numbers as you can from 10 to 40.
You may use each digit only once in each calculation.
Use the correct order of operations for each calculation.
Copy and complete this table to show all your results

Number	Calculation
10	$= 3 + 7$
11	$= 4 + 7$
12	$=$
13	$= 8 - 7 + 3 \times 4$

Number	Calculation
26	$=$
27	$= (4 + 5) \times 3$
28	$=$
29	$= 3 \times 7 + 8$

N6.2 Estimation and rounding

This spread will show you how to:

- Round numbers to any given power of 10
- Round to the nearest integer and to one significant figure
- Estimate answers to problems involving decimals

Keywords
Decimal places (dp)
Estimate
Rounding
Significant figure

You can **round** a number to the nearest 10, 100, 1000, and so on.

You can also round a number to the nearest whole number or to a given number of **decimal places**.

Example

Round 16.473

a to the nearest whole number **b** to 1 decimal place
c to 2 decimal places **d** to the nearest 10.

a 16.473 ≈ 16. Look at the tenths digit.
b 16.473 ≈ 16.5. Look at the hundredths digit.
c 16.473 ≈ 16.47. Look at the thousandths digit.
d 16.473 ≈ 20. Look at the units digit.

Look at the next digit. If it is 5 or more then the number is rounded up, otherwise it is rounded down.

- **The first digit that is not zero in a number is called the first significant figure. It has the highest value in the number.**

This topic is extended to more significant figures on page 362.

You can round a number to one significant figure.

Example

Round the numbers to one significant figure.

a 7560 **b** 52.3 **c** 1.5

	Thousands	Hundreds	Tens	Units	•	tenths
a	7	5	6	0	•	
b			5	2	•	3
c				1	•	5

a First significant figure is 7, so round up to 8000.
b First significant figure is 5, so round down to 50.
c First significant figure is 1, so round up to 2.

You can **estimate** the answer to a calculation by first rounding the numbers in the calculation.

Example

Estimate the answer to this calculation. $\frac{4.23 \times 5.89}{9.7}$

You can round each of these numbers to the nearest whole number.

$$\frac{4.23 \times 5.89}{9.7} \approx \frac{4 \times 6}{10} = \frac{24}{10} = 2.4$$

You have to decide how much to round the numbers to make a good estimate for the calculation.

Exercise N6.2

1 Write the value of the red digit in each of these numbers.

a 1324 **b** 21 894 **c** 234 897 **d** 54 327

2 Round each of these numbers to the nearest

i 10 **ii** 100 **iii** 1000.

a 2568 **b** 4297 **c** 7853 **d** 1432
e 12 473 **f** 18 258

3 Here is a table showing the populations of five cities.

Town	Population
London	7 172 091
Paris	2 142 800
New York	8 085 742
Mumbai	16 368 084
Beijing	7 441 000

Round the population of each city to the nearest 10 000 and then place the cities in order of size from smallest to largest.

4 Round each of these numbers to the nearest

i 10 **ii** 100.

a 458.2 **b** 1329.5 **c** 342.52 **d** 354.82

5 Round each of these numbers to the nearest whole number.

a 3.7 **b** 8.7 **c** 18.63 **d** 69.49
e 109.9 **f** 6.899

6 Round each of these numbers to 1 decimal place.

a 0.27 **b** 2.89 **c** 3.82 **d** 12.48

7 Round each of these numbers to

i 2 decimal places **ii** one significant figure.

a 0.327 **b** 2.869 **c** 3.802 **d** 14.458

8 **a** Irwin estimates the value of

$$\frac{47.3 \times 18.9}{8.72}$$ to be 100.

Write three numbers Irwin could use to get his estimate.

b Sarah estimates the value of

$$\frac{21.4 \times 4.87}{49.8}$$ to be 2.

Write three numbers Sarah could use to get her estimate.

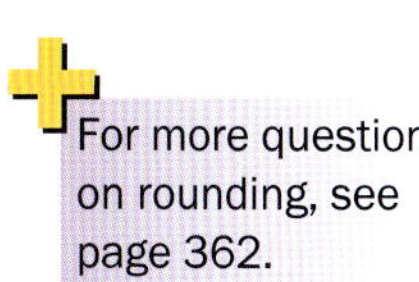
For more question on rounding, see page 362.

N6.3 Adding and subtracting decimals

This spread will show you how to:

- Use written methods for addition and subtraction of whole numbers and decimals
- Add and subtract mentally numbers with up to two decimal places

Keywords
Addition
Decimals
Subtraction
Written method

There is a standard written method for **addition** of **decimals**.

Example

Liam measures the lengths of three vehicles.

Car 2.4 m Lorry 5 m Motorbike 1.68 m

What is the total length of the three vehicles?

Estimate the answer first.

$2.4 + 5 + 1.68 \approx 2 + 5 + 2$
$= 9$ m

Set out the calculation in columns, lining up the decimal points.

	Units	•	tenths	hundredths
	2	•	4	
	5	•		
+	1_1	•	6	8
=	9	•	0	8

So the combined length = 9.08 m.

Start with the smallest place value (in this case the hundredths) and add hundredths to hundredths, tenths to tenths, units to units and so on.

There is a standard written method for **subtraction** of decimals.

Example

Calculate 27.8 litres – 14.45 litres.

Estimate the answer first.

$27.8 - 14.45 \approx 28 - 14$
$= 14$ litres

Set out the calculation in columns, lining up the decimal points.
You could write the numbers in a place value table.

```
    7 1
  27.80
 -14.45
  13.35
```

It is a good idea to add a zero here so that both of the numbers have the same number of decimal places.

Start with the smallest place value (in this case hundredths), and subtract hundredths from hundredths, tenths from tenths, and so on.

27.8 litres – 14.45 litres = 13.35 litres

You could also use 'shopkeepers subtraction': count up from the smallest to the largest number and add up how much you have counted.

Exercise N6.3

1 Use an appropriate method for each of these calculations.

a 83 + 57 **b** 93 + 32 **c** 275 + 958

d 843 + 872 **e** 838 + 697 **f** 834 + 787

2 Use an appropriate method for each of these calculations.

a 62 − 47 **b** 83 − 68 **c** 487 − 356

d 852 − 728 **e** 548 − 387 **f** 589 − 387

3 Use an appropriate method for each of these calculations.

a 25 + 38 + 68 **b** 123 + 76 − 58

c 173 − 27 + 56 **d** 327 + 176 − 255

4 Use a mental or written method to work out these calculations.

a 33.4 + 15.2 **b** 34.6 + 13.7 **c** 19.8 + 8.8 **d** 18.7 + 26.5

5 Use a mental or written method to work out these calculations.

a 8.7 − 2.5 **b** 15.8 − 8.4 **c** 26.3 − 7.9 **d** 53.6 − 27.8

6 Jodie buys

3 pens at 14p each
2 writing pads at £1.89 each
1 calculator at £2.79

She pays with a £10 note.
Work out how much change Jodie should get from £10.

7 Use a written method to work out these additions.

a 3.52 + 4.6 **b** 13.62 + 2.9 **c** 8.5 + 14.81 **d** 75.8 + 28.39

8 Use a written method to work out these subtractions.

a 17.3 − 4.22 **b** 16.6 − 3.47 **c** 37.7 − 18.86 **d** 57.28 − 38.4

9 Use a written method for each of these calculations.

a 16.4 + 9.87 **b** 49.2 + 7.72 **c** 9.42 − 5.9

d 26.9 + 9.82 **e** 36.57 − 8.59 **f** 36.28 − 17.4

10 Use a mental or written method to solve each of these problems.

a Sean sells olives at the weekend. On Saturday he sells 8.9 kg; on Sunday he sells 3.38 kg. What weight of olives has Sean sold altogether during the weekend?

b Naomi is building a wooden cold frame for growing vegetables. She needs three lengths of wood to finish the frame. The lengths are 1.82 m, 1.3 m and 1.79 m. What is the total length of wood she will need to buy?

N6.4 Multiplying and dividing decimals

This spread will show you how to:

- Use written methods for multiplication and division of whole numbers and decimals
- Use approximation to estimate the answers to problems

Keywords
Divisor
Estimate
Grid method
Standard method
Whole number

You can multiply decimals by replacing them with an equivalent **whole number** calculation that is easier to work out.

Example

Bruce buys 28 lengths of wood. Each length is 4.8 m long.
What is the total length of wood he buys?

Estimate the answer first.

$28 \times 4.8 \approx 30 \times 5$
$= 150$ m

Rewrite as a whole number calculation $28 \times 4.8 = 28 \times 48$

Work out 28×48 using the standard method.

```
             48
            ×28
20 × 48     960
 8 × 48    +384
28 × 48 =  1344
```

Bruce buys

$28 \times 4.8 = 28 \times 48 \div 10$
$= 1344 \div 10$
$= 134.4$ m of wood

You could also use the grid method:

×	40	8
20	800	160
8	320	64

$28 \times 48 =$
$800 + 320 + 160 + 64$
$= 1344$

Adjust your answer for the decimals.

You can divide any number including decimals by a whole number using the 'chunking method'.

Example

Andrew packs tins into boxes. Each box will hold 12 tins.
Andrew has 159 tins. How many boxes will he fill?

Estimate the answer first.

$159 \div 12 \approx 150 \div 10$
$= 15$ boxes

```
12)159
   −120     12 × 10
     39
    −36     12 × 3
      3     159 ÷ 12 = 13 remainder 3
```

Andrew will fill 13 boxes with 3 tins left over.

Exercise N6.4

1 Calculate these using an appropriate written method.
Remember to do a mental approximation first.

a 13×16 **b** 18×13 **c** 13×13
d 16×19 **e** 17×19

2 Use an appropriate method of calculation to work out each of these.

a 23×8 **b** 9×64 **c** 8×147
d 17×21 **e** 16×28 **f** 24×23
g 12×114 **h** 25×135 **i** 28×154

3 Use an appropriate method of calculation to work out each of these.
Where appropriate leave your answer in remainder form.

a $56 \div 6$ **b** $84 \div 8$ **c** $95 \div 5$
d $116 \div 7$ **e** $123 \div 9$ **f** $144 \div 8$
g $155 \div 12$ **h** $185 \div 14$ **i** $284 \div 12$

4 **a** Beatrice plants 8 seeds in a pot. She has 114 seeds. How many pots will she fill with seeds? How many seeds will there be left?

b Melinda packs 7 boxes into every crate. She has 103 boxes.
How many crates will she be able to fill?

5 Use an appropriate method of calculation to work out each of these.

a 13×2.1 **b** 3.7×13 **c** 5.4×13
d 23×3.3 **e** 28×3.2 **f** 6.3×29

6 Use an appropriate method of calculation to work out each of these.

a $25.9 \div 7$ **b** $34.8 \div 6$ **c** $66.4 \div 8$
d $73.8 \div 6$ **e** $117.9 \div 9$ **f** $122.4 \div 8$

7 Use an appropriate method of calculation to work out each of these.

a 12×1.64 **b** 13×1.42 **c** 13×1.35
d 15×1.32 **e** 18×1.43 **f** 28×1.52

8 **a** Wally buys 13 bags of potatoes. Each bag costs £1.23.
How much does this cost in total?

b Ashley buys six CDs. Each CD costs £8.79. How much is this in total?

c Christina buys 12 ready meals. Each meal costs £1.79.
What is the cost of the 12 ready meals?

9 Use an appropriate method of calculation to work out each of these.

a $54.16 \div 4$ **b** $85.32 \div 6$ **c** $130.72 \div 8$
d $62.23 \div 7$ **e** $188.7 \div 6$ **f** $219.51 \div 9$

N6.5 Calculator methods

This spread will show you how to:

- Carry out more complex calculations using the functions of a calculator
- Give an answer to a given degree of accuracy

Keywords
Approximation
Brackets
Order of operations

You can use a scientific calculator to carry out more complex calculations that involve decimals.

Example

Use you calculator to work out this calculation

$3.46 + 2.9 \times 4.8$

Give your answer to one decimal place.

Estimate

$3.46 + 2.9 \times 4.8 \approx 3 + 3 \times 5$ (using the order of operations)

$= 3 + 15 = 18$

You type [3] [·] [4] [6] [+] [2] [·] [9] [×] [4] [·] [8] [=]

The calculator should display 3.46+2.9×4.8 17.38

So the answer is $17.38 = 17.4$ (1 decimal place)

A scientific calculator has algebraic logic, which means that it understands the **order of operations.**

In this case the calculator automatically works out the multiplication before the addition.

A scientific calculator has **bracket** keys.

Example

a Use your calculator to work out $(3.9 + 2.2)^2 \times 2.17$.
Write all the figures on your calculator display.

b Put brackets in this expression so that its value is 16.26.

$1.4 + 3.9 \times 2.2 + 4.6$

---

a Estimate

$(3.9 + 2.2)^2 \times 2.17 \approx (4 + 2)^2 \times 2$ (using the order of operations)

$= 6^2 \times 2$

$= 36 \times 2$

$= 72$

You type [(] [3] [·] [9] [+] [2] [·] [2] [)] [x^2] [×] [2] [·] [1] [7] [=]

The calculator should display (3.9+2.2)²×2.17 80.7457 $= 80.7457$

b $(1.4 + 3.9) \times 2.2 + 4.6$

The calculator should display 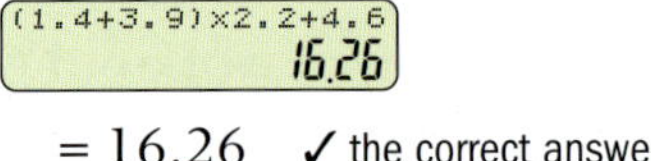

$= 16.26$ ✓ the correct answer

Exercise N6.5

1 Use your calculator to work out these.
In each case, first write an estimate for your answer.
Write the answer from your calculator to one decimal place.

a $3.4 + 6.2 \times 2.7$

Estimate $3.4 + 6.2 \times 2.7 \approx 3 + 6 \times 3$
$= 3 + 18$
$=$ ____

Using the calculator $3.4 + 6.2 \times 2.7 =$ ________

b $1.98 \times 11.7 - 4.6$

c $7.8 + 19.3 \div 4.12$

d $2.09 \times 2.87 + 3.25 \times 1.17$

e $13.67 \div 1.75 + 3.24$

f $1.2 + 3.7 \times 0.5$

2 Use your calculator to work out each of these calculations.
Write all the figures on your calculator display.

a $(2.3 + 5.6) \times 3^2$

b $2.3^2 \times (12.3 - 6.7)$

c $(2.8^2 - 2.04) \div 2.79$

d $7.2 \times (4.3^2 + 7.4)$

e $11.33 \div (6.2 + 8.3^2)$

f $(2.5^2 + 1.37) \times 2.5$

3 Calculate each of these, giving your answer to one decimal place.

Hint for part **a**:
Type the calculation as $(5.4 + 3.8) \div (4.5 - 2.9) =$

a $\dfrac{5.4 + 3.8}{4.5 - 2.9}$

b $\dfrac{3.8 - 1.67}{4.3 - 2.68}$

c $\dfrac{12.4 + 5.8}{14.5 - 3.9}$

d $\dfrac{13.08 - 2.67}{2.13 + 2.68}$

4 Put brackets into each of these expressions to make them correct.

a $3.4 \times 2.3 + 1.6 = 13.26$

b $3.5 \times 2.3 - 1.04 = 4.41$

c $2.6 + 6.5 \div 1.3 = 7$

d $1.4^2 - 1.2 \times 2.3 = 1.748$

e $2.4^2 \div 1.8 \times 3.2 + 1.6 = 15.36$

f $3.2 + 5.3 \times 2.4 - 1.2 = 10.2$

5 Use your calculator to work out each of these. Write all the figures on your calculator display.

a $\dfrac{462.3 \times 30.4}{(0.7 + 4.8)^2}$

b $\dfrac{13.58 \times (18.4 - 9.73)}{(37.2 + 24.6) \times 4.2}$

6 Here is a mathematical calculation.

$$\frac{50.1 \times 29.8}{50.1 - 29.8}$$

a Write **approximate** values for 50.1 and 29.8 that you could use to estimate the value of the mathematical expression.

b Work out an estimate for the mathematical calculation.

c Use your calculator to work out the value of the calculation.
Write all the figures on your calculator display.

N6 Exam review

Key objectives

- Add, subtract, multiply and divide integers and then any number
- Use the hierarchy of operations
- Estimate answers to problems involving decimals, round to a given number of significant figures
- Use calculators effectively and efficiently, knowing how to enter complex calculations
- Understand the calculator display, interpreting it correctly and knowing not to round during the intermediate steps of a calculation

1 **a** Use your calculator to work out

$$(7.6 - 2.4)^2 \div 4.2$$

Write down all the figures on your calculator display. (2)

b Give your answer from part **a** rounded to one significant figure. (1)

2 Nick takes 26 boxes out of his van.
The weight of each box is 32.9 kg.

a Work out the total weight of the 26 boxes. (3)

Then Nick fills the van with large wooden crates.
The weight of each crate is 69 kg.
The greatest weight the van can hold is 990 kg.

b Work out the greatest number of crates that the van can hold. (4)

(Edexcel Ltd., 2004)

Linear equations

This unit will show you how to

- Set up and solve linear equations where the unknown appears on either side using the balance method and inverse operations
- Check a solution to an equation by substituting it back into the equation
- Write equations to represent real-life problems

Before you start ...

You should be able to answer these questions.

Review

1 State the inverse operation for each of these. *(Unit A2)*

a $+3$ **b** -6

c $\times 5$ **d** $\div 4$

2 Simplify each of these. *(Unit A1)*

a $x + x + y + y$ **b** $5 \times p + 2$

c $\frac{m}{3 + 5}$

3 Work out each of these. *(Key stage 3)*

a 6×9 **b** 4×8

c $54 \div 6$ **d** $63 \div 7$

4 Calculate the value of each expression when $x = 4$ and $y = 3$. *(Unit A1)*

a xy **b** $2x + y$

c $\frac{3x}{y}$ **d** $x + 3y$

5 Solve each of these equations. *(Unit A2)*

a $4x + 2 = 10$ **b** $3x - 6 = 9$

c $2y \times 4 = 16$ **d** $\frac{y}{7} = 1$

A6.1 Solving simple equations

This spread will show you how to:

- Solve linear equations where the unknown appears on either side

Keywords
Balance method
Inverse operations
Solve

- You can **solve** equations using the **balance method**.

An equation remains balanced if you do the same to both sides.

You can

- add the same number to both sides

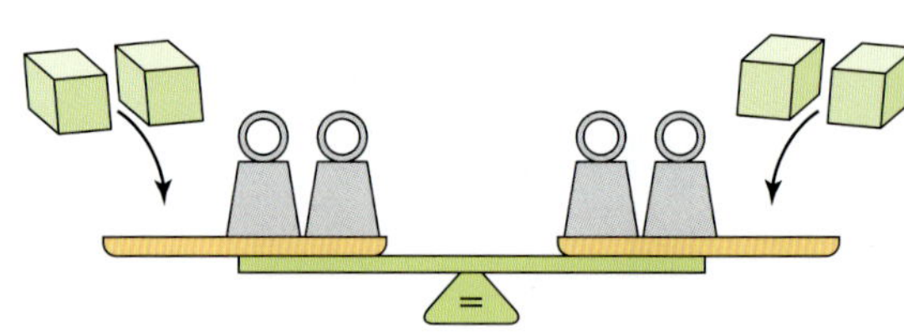

- subtract the same number from both sides

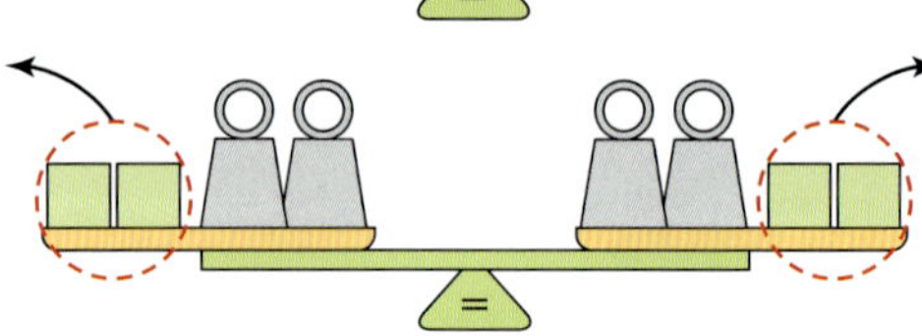

- multiply both sides by the same number

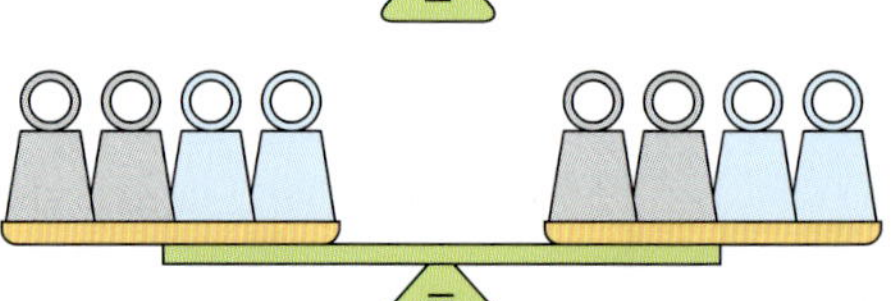

- divide both sides by the same number

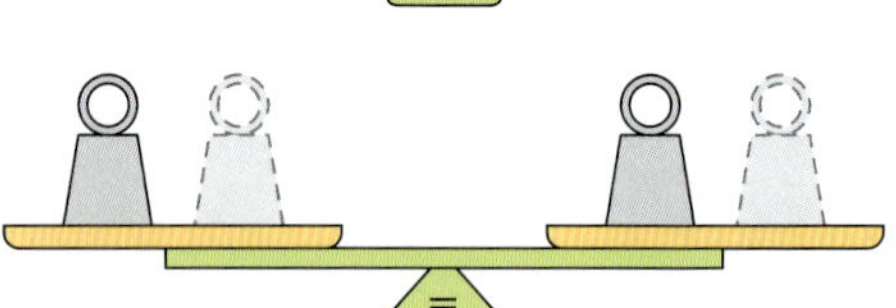

- You use **inverse operations** to get the unknown on its own on one side of the equation.

Example

Solve **a** $x + 3 = 17$ **b** $y - 5 = 27$

c $3s = 24$ **d** $\frac{t}{5} = 4$

a

$x + 3 = 17$
$x + 3 - 3 = 17 - 3$
$x = 14$

b

$y - 5 = 27$
$y - 5 + 5 = 27 + 5$
$y = 32$

c

$3s = 24$
$3s \div 3 = 24 \div 3$
$s = 8$

d

$\frac{t}{5} = 4$
$5 \times \frac{t}{5} = 4 \times 5$
$t = 20$

Example

Solve $15 - x = 18$.

$15 - x = 18$ subtract 15 from both sides
$15 - 15 - x = 18 - 15$
$-x = 3$
$x = -3$

Exercise A6.1

1 Work out the weight of one box on these balances.

a

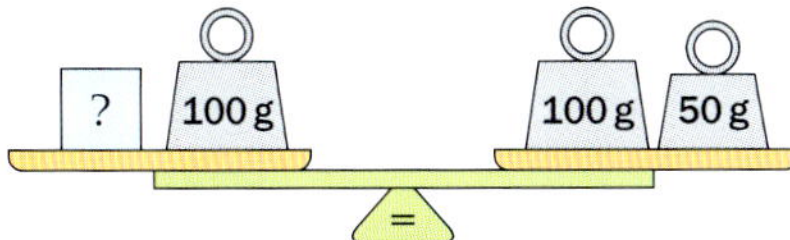

b

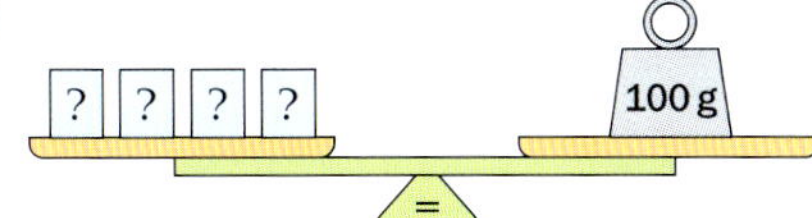

c

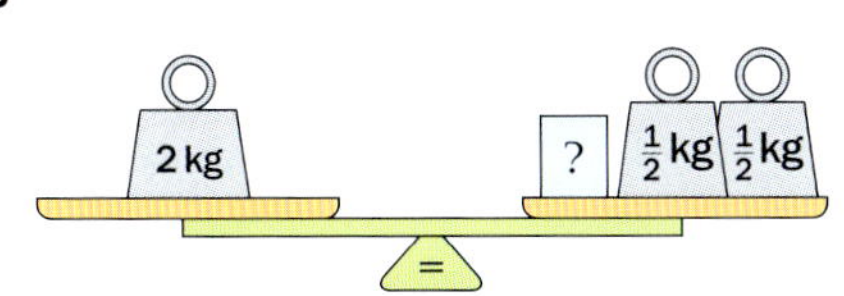

d

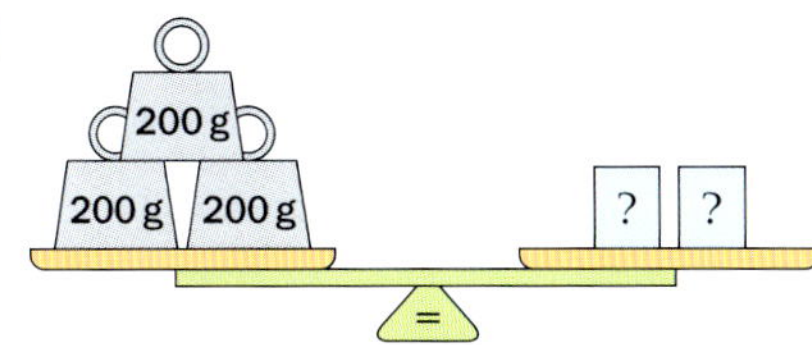

2 Solve these equations using the balance method.

a $x + 4 = 12$ **b** $y - 6 = 15$ **c** $18 = z + 9$

d $r - 89 = 27$ **e** $75 = 40 + s$ **f** $11 - t = 5$

3 Solve these equations.

a $8m = 40$ **b** $7n = 70$ **c** $25 = 5p$ **d** $42 = 6q$

e $9r = 72$ **f** $2t = 3$ **g** $15 = 2u$ **h** $4m = 48$

4 Solve these equations.

a $\frac{m}{5} = 11$ **b** $\frac{n}{6} = 8$ **c** $15 = \frac{p}{2}$ **d** $9 = \frac{q}{7}$

5 Solve these equations.

a $9 + r = 15$ **b** $11 - s = 14$ **c** $-15 = 3t$ **d** $\frac{u}{4} = -10$

e $v - 8 = -6$ **f** $6w = 8$ **g** $4 = \frac{x}{9}$ **h** $17 = 12 - y$

6 a Use these numbers and symbols.

Write all the correct equations you can make.

b Use these numbers and symbols.

Write the different equations you can make.
Solve each equation to find the value of x.

c Use these numbers and symbols.

x 5 2 = × ÷

Write the different equations you can make.
Solve each equation to find the value of x.

A6.2 Two-step function machines

This spread will show you how to:

- Set up and solve equations with two steps

Keywords
Function machine
One-step
Order of operations
Two-step

- You can draw a **function machine** for an equation.

$x + 7 = 20$

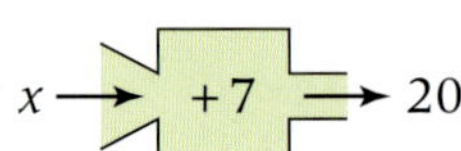

- Some functions have two operations or steps.

In number calculations with two steps, you follow the order of operations.

In the calculation $\qquad 3 \times 2 + 4 = 10$

You multiply first, then you add $\qquad 6 + 4 = 10$

BIDMAS helps you remember the order.

In algebra you follow the same order of operations.

The equation $\qquad 3n + 4 = 10$

means $\qquad 3 \times n + 4 = 10$

You multiply first, then you add.

So the function machine for $3n + 4 = 10$ is

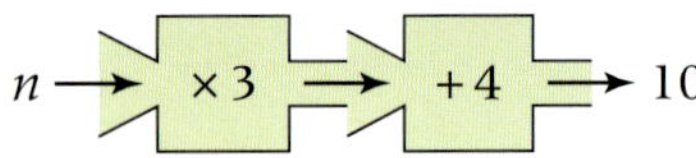

Draw a function box for each operation. Two operations need a two-step function machine.

- To draw a function machine for a two-step function, follow the order of operations.

Example

For each function machine, work out the outputs for the inputs given.

a input $\qquad$ output

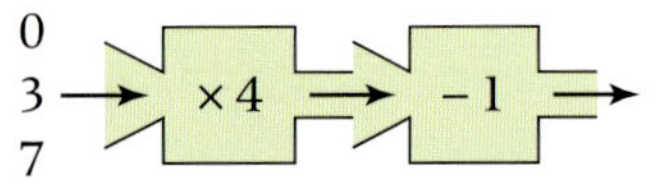

b input $\qquad$ output

1, 5, 9 → ×3 → +2 →

a $0 \times 4 - 1 = 0 - 1 \;\; = -1$
$3 \times 4 - 1 = 12 - 1 = 11$
$7 \times 4 - 1 = 28 - 1 = 27$

b $1 \times 3 + 2 = 3 + 2 \;\; = 5$
$5 \times 3 + 2 = 15 + 2 = 17$
$9 \times 3 + 2 = 27 + 2 = 29$

Example

Draw function machines for these equations.

a $4y + 2 = 26$ $\qquad$ **b** $2x - 5 = 7$

a $4y + 2 = 26$

y → ×4 → +2 → 26

Multiplication then addition.

b $2x - 5 = 7$

x → ×2 → −5 → 7

Multiplication then subtraction.

Exercise A6.2

1 For these number machines, work out the output for each input.

a input output

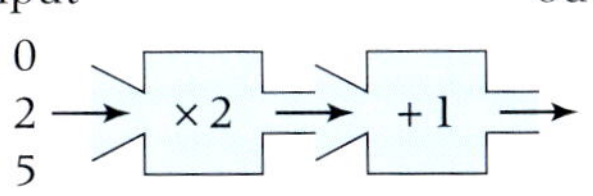

b input output

1, 4, 7 → ×3 → −2 →

c input output

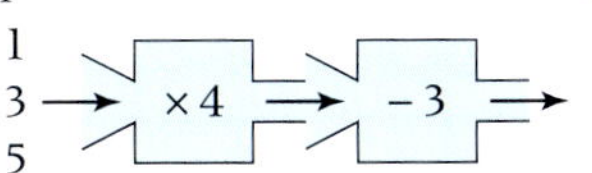

d input output

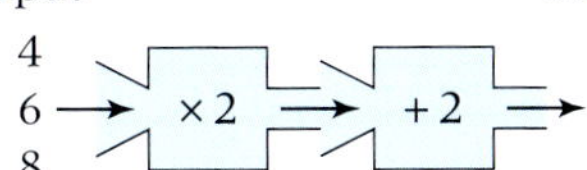

2 Copy and complete the table of input and output values for each function machine.

a

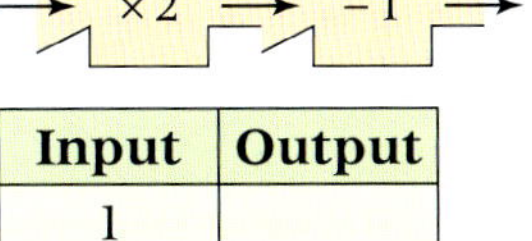

Input	Output
1	
3	
5	
7	
9	

b

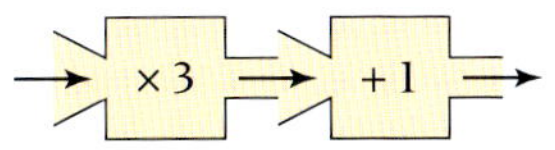

Input	Output
0	
2	
5	
7	
10	

c

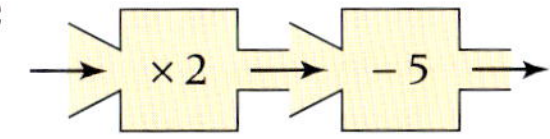

Input	Output
20	
15	
	15
4	
1	

3 Draw function machines for these equations.

a $3x - 1$ **b** $4n + 2$ **c** $5m - 3$ **d** $6p + 1$

e $3y + 5$ **f** $7z - 2$ **g** $4s - 9$ **h** $8t + 3$

4 A two-step function machine has input 4.

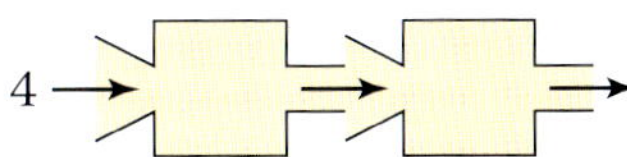

Use any **two** of these function machines

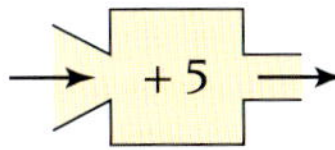

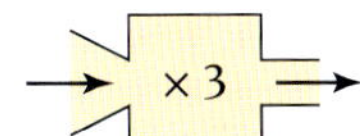

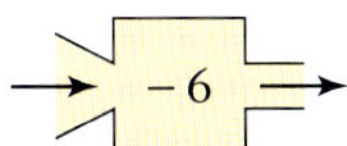

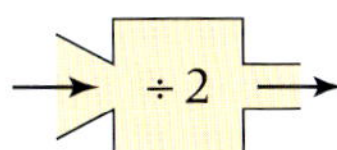

to make a two-step machine that gives the output

a 3 **b** 7 **c** 17

Make as many different outputs as you can. Draw the function machines you use each time.

5 Match each function machine to the equation it represents.

A $3x - 2$ **B** $7x + 6$ **C** $2x + 4$ **D** $2x - 1$

a → ×2 → +4 →

b → ×2 → −1 →

c → ×7 → +6 →

d → ×3 → −2 →

A6.3 Inverse function machines

This spread will show you how to:

- Set up and solve equations using inverse operations

Keywords
Inverse operation

You can draw a function machine for a 'think of a number' problem.

You can work out the unknown number using the **inverse** function machine.

Example

Solve this 'think of a number' problem using a function machine.

I think of a number.
Half my number is 8.
What is my number?

Draw a function machine $n \rightarrow \boxed{\div 2} \rightarrow 8$

Work backwards, using **inverse operations**. $16 \leftarrow \boxed{\times 2} \leftarrow 8$
The number is 16.

Use n for the unknown number.

This is the inverse function machine.

- **To draw an inverse function machine, work backwards through the function machine, using inverse operations.**

Example

I think of a number. I double my number and add 3.
The answer is 17. What is my number?

Function machine

Inverse function machine

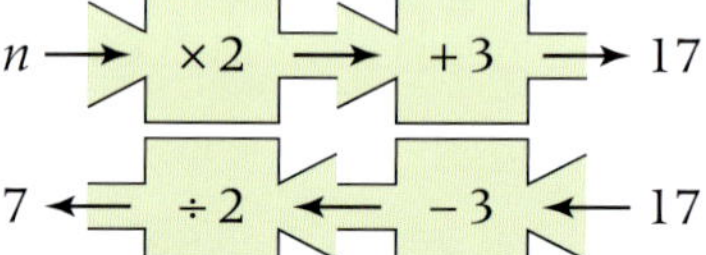

The number is 7.

$17 - 3 = 14$
$14 \div 2 = 7$

You can use inverse function machines to solve equations.

Example

Solve these equations.

a $4x + 2 = 22$ **b** $3x - 5 = 13$

a $x \rightarrow \boxed{\times 4} \rightarrow \boxed{+2} \rightarrow 22$

$5 \leftarrow \boxed{\div 4} \leftarrow \boxed{-2} \leftarrow 22$

$x = 5$

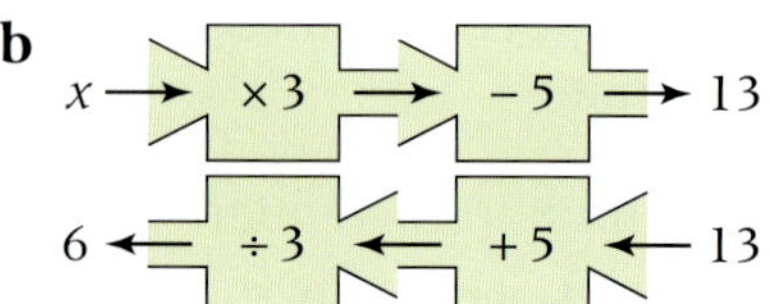

b $x \rightarrow \boxed{\times 3} \rightarrow \boxed{-5} \rightarrow 13$

$6 \leftarrow \boxed{\div 3} \leftarrow \boxed{+5} \leftarrow 13$

$x = 6$

Draw the function machine.
Draw the inverse machine.
$22 - 2 = 20$
$20 \div 4 = 5$

$13 + 5 = 18$
$18 \div 3 = 6$

Exercise A6.3

1 Draw the inverse function machine for each of these function machines.

a → ×4 → +3 →

b → ×3 → −2 →

c → ×6 → −1 →

d → ×7 → +1 →

2 For each of these 'think of a number' problems

- Draw a function machine.
- Draw the inverse function machine.
- Use your inverse machine to work out the number.

a I think of a number. I double it and add 3. The answer is 13. What is my number?

b I think of a number. I double it and subtract 6. The answer is 0. What is my number?

c I think of a number. I multiply it by 3 and add 3. The answer is 15. What is my number?

d I think of a number. I multiply it by 4 and subtract 2. The answer is 26. What is my number?

3 Draw a function machine for each equation.
Use the inverse function machine to solve the equation.

a $3x + 4 = 16$ **b** $5x + 2 = 12$ **c** $4x + 3 = 19$ **d** $3x + 7 = 13$

e $4x + 9 = 17$ **f** $8x + 5 = 21$ **g** $6x + 10 = 22$ **h** $9x + 3 = 30$

4 Use function machines to solve these equations.

a $4x - 3 = 9$ **b** $6x - 4 = 20$ **c** $3x - 5 = 25$ **d** $5x - 7 = 8$

e $2x - 10 = 12$ **f** $7x - 2 = 19$ **g** $4x - 8 = 16$ **h** $5x - 7 = 43$

5 Solve these equations.

a $2x + 7 = 5$ **b** $3x + 8 = 2$ **c** $4x + 15 = 3$ **d** $2x - 5 = -9$

e $4x - 7 = -19$ **f** $2x + 3 = 8$ **g** $4x + 2 = 0$ **h** $2x - 2 = 1$

6 Match each equation in box A with its solution in box B.

There are four 'spare' solutions. Make up an equation with each of these spare solutions.

Box A	Box B	
$2x + 6 = 9$	$x = 1$	$x = -1$
$3x + 1 = 7$	$x = 2$	$x = -2$
$3x + 15 = 6$	$x = 3$	$x = \frac{1}{2}$
$4x - 3 = -5$	$x = 4$	$x = \frac{3}{2}$
$4x + 11 = 3$	$x = 5$	$x = -\frac{1}{2}$
$4x + 7 = 19$	$x = 6$	$x = -3$
$5x - 12 = 13$		
$7x + 8 = 15$		

A6.4 The balance method for two-step equations

This spread will show you how to:

- Solve equations using the balance method
- Check a solution to an equation by substituting it back into the equation

Keywords
Balance method
Substitution

You can work out the unknown weight on a balance by doing the same to both sides.

Take 3 weights off both sides

Divide both sides by 2

The unknown weight equals 4.

This topic is extended to equations with brackets and fractions on page 378.

You can work out the value of x in the equation $2x + 3 = 7$ by doing the same to both sides.

$2x + 3 \quad = \quad 7$

Subtract 3 from both sides.

Divide both sides by 2.

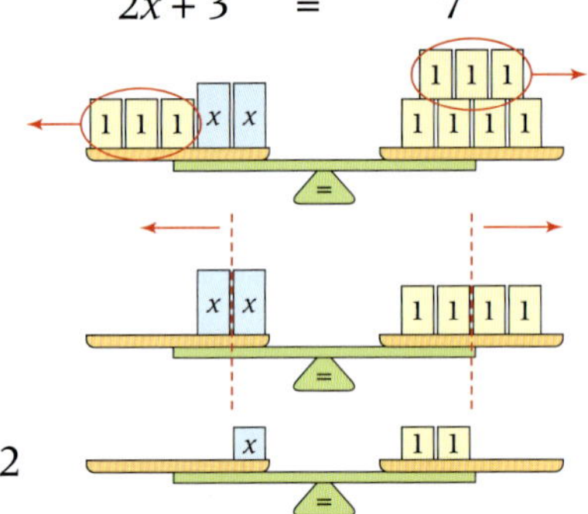

$x = 2$

First get the x term on its own on one side.
This gives the value of $2x$.

Then divide to find the value of x.

- **You can solve two-step equations using the balance method.**

Example

Solve these equations.

a $4y + 2 = 18$ **b** $4x - 3 = 17$

a

$4y + 2 = 18$

$4y + 2 - 2 = 18 - 2$ Subtract 2 from both sides.

$4y = 16$

$4y \div 4 = 16 \div 4$ Divide both sides by 4.

$y = 4$ This gives the value of y.

b

$4x - 3 = 17$

$4x - 3 + 3 = 17 + 3$ Add 3 to both sides.

$4x = 20$

$4x \div 4 = 20 \div 4$ Divide both sides by 4.

$x = 5$

You can check using **substitution**.
In part **b**
$4x - 3 =$
$4 \times 5 - 3 =$
$20 - 3 = 17$

- **You can check a solution by substituting the value back into the equation.**

Exercise A6.4

1 Solve these equations using the balance method.

a $2x + 3 = 13$ **b** $2x + 5 = 9$ **c** $2x + 8 = 20$ **d** $2x + 1 = 21$

e $2x + 3 = 15$ **f** $2x + 7 = 9$ **g** $2x + 12 = 44$ **h** $2x + 5 = 23$

i $2x + 13 = 21$ **j** $2x + 11 = 51$ **k** $2x + 12 = 28$ **l** $2x + 36 = 40$

2 Solve these equations using the balance method.
Check your answers by substitution.

a $2m - 4 = 16$ **b** $3n - 5 = 10$ **c** $5 = 2p + 9$ **d** $5 + 3r = 26$

e $13 = 2s - 1$ **f** $4t + 1 = 1$ **g** $25 - 3x = 10$ **h** $19 - 2y = 7$

3 Solve these equations.

a $4m + 3 = 5$ **b** $4q - 2 = 2$ **c** $8 - 6p = 5$

d $4r + 2 = 12$ **e** $12 = 8q + 20$ **f** $6s - 6 = -3$

Check your answers.

4 Tom and Sacha both solve the equation

$$5x - 7 = 43$$

Tom's solution is

x = 8

Sacha's solution is

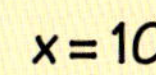

Check their solutions by substitution.
Who is correct?

5 The formula for the perimeter of a rectangle is

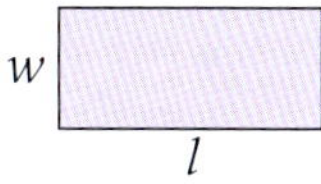

$$P = 2l + 2w$$

where l = length and w = width.

a Work out the perimeter of a rectangle where $l = 10$ cm and $w = 4$ cm.

b Work out the length of a rectangle where $P = 40$ cm and $w = 7$ cm.

c Work out the width of a rectangle where $P = 84$ cm and $l = 30$ cm.

6 Amos uses the formula

$$4.5s - 11.2t = r$$

a Work out the value of r when $s = 13$ and $t = 4$.

b Work out the value of s when $t = 5$ and $r = 48$.

c Work out the value of t when $s = 10$ and $r = 4.68$.

Check your answers by substitution.

A6.5 Writing and solving equations

This spread will show you how to:

- Check a solution to an equation by substituting it back into the equation
- Write equations to represent real-life problems

Keywords
Equation
Solution

Jenny has n books in her school bag.
Her maths teacher gives her 4 more books for revision.
She now has 7 books in her school bag.
How many books did she have to start with?

You can write an equation for this problem

Jenny starts with	$\rightarrow$	adds 4	$\rightarrow$	total 7
n		$+4$	$=$	7

You can solve the equation $n + 4 = 7$ to find n.

$$n + 4 = 7$$
$$n + 4 - 4 = 7 - 4$$
$$n = 3$$

Subtract 4 from both sides.

Jenny had 3 books to start with.

- **You can write an equation to represent a problem.**
- **You can solve the equation to find the solution to the problem.**

Example

Marcus spends £r out shopping.
Pritesh spends twice as much as Marcus.
Tariq spends £6 more than Pritesh.

a Write an expression for the amount each person spends.

Tariq spends £50 in total.

b Write an equation for the amount Tariq spends.
c Solve your equation to find r.
d Work out how much Marcus and Pritesh spend.

a Marcus r
Pritesh $2 \times r = 2r$
Tariq $2r + 6$

b $2r + 6 = 50$

c
$$2r + 6 = 50$$
$$2r + 6 - 6 = 50 - 6$$
$$2r = 44$$
$$r = 22$$

Subtract 6 from both sides.
Divide both sides by 2.

d Marcus £22
Pritesh $2 \times £22 = £44$

Substitute $r = £22$ into the original expressions.

Exercise A6.5

1 A pack of sausages costs £2.

a Write an expression for the cost of x packs of sausages.

For a barbeque, Charlotte spends £14 on sausages.

b Write an equation for the sausages Charlotte buys.

c Solve your equation to find the number of packs of sausages she bought.

2 Sarah, Josh and Millie bring sandwiches to a picnic.
Sarah brings y sandwiches.
Josh brings twice as many sandwiches as Sarah.
Millie brings 4 less than Josh.

a Write an expression for the number of sandwiches each person brings.

Millie brings 12 sandwiches.

b Write an equation for the sandwiches Millie brings.

c Solve your equation to find y.

d Work out the number of sandwiches each person brings to the picnic.

3 For each of these 'think of a number' problems

- write an equation
- solve your equation to find the missing number.

a I think of a number. I double it. Then I add 5. My answer is 17.

b I think of a number. I multiply it by 4. Then I subtract 8. My answer is 8.

c I think of a number. I multiply it by 7. Then I add 1. My answer is 50.

d I think of a number. I multiply it by 3. Then I subtract 17. My answer is 13.

4 Four people go out for a meal.
The meal costs £x each.
The drinks cost £15.

a Write an expression for the total bill for the meal.

The total bill comes to £65.

b Write an equation for the total bill.

c Solve your equation to find x, the cost of one meal.

5 Rory buys four CDs at £y each and a DVD for £16.
His total bill comes to £60.
Write an equation for Rory's total bill.
Solve your equation to find the cost of one CD.

A6 Exam review

Key objectives

- Solve linear equations, with integer coefficients, in which the unknown appears on either side
- Set up simple equations
- Substitute numbers into a formula

1 Solve

a $3x = 21$ (1)

b $2y + 5 = 13$ (2)

2 Andrew, Brenda and Callum each collect football stickers.
Andrew has x stickers.
Brenda has three times as many stickers as Andrew.

a Write an expression for the number of stickers that Brenda has. (1)

Callum has 9 stickers less than Andrew.

b Write an expression for the number of stickers that Callum has. (1)

(Edexcel Ltd., 2005)

Exercise D4.2

1 **a** Write these numbers in order, smallest first.

i 7, 8, 8, 5, 4, 3, 3 **ii** 11, 12, 10, 9, 9

iii 38, 35, 30, 37, 34 **iv** 101, 98, 103, 97, 99, 97, 95

v 3, 2, 0, 0, 1, 2, 1, 2, 3

b Use your answers to find the median of each set of numbers.

2 The weights of five boys are given.

a Arrange the weights in order, smallest first.

b Find the median weight.

c Which boy has the median weight?

Tony	71 kg
Tom	64 kg
Tim	61 kg
Tariq	70 kg
Thomas	66 kg

3 Eleven students were asked how many hours of television they had watched yesterday. The results were

$4, 3\frac{1}{2}, 0, 1, \frac{1}{2}, 5, 2, 2\frac{1}{2}, 10, 1, 1\frac{1}{2}$

Find the median number of hours of television watched.

4 The numbers of tomatoes on each of seven plants in a greenhouse are

12, 6, 9, 15, 10, 9, 7

Find the median number of tomatoes on the plants.

5 The ages, in years, of eight children in a creche are:

2, 4, 2, 3, 5, 1, 3, 4

Find the median age for the children.

6 Ten students were asked to estimate a minute.
Their attempts, in seconds, were

65, 71, 57, 41, 56, 68, 85, 42, 58, 60

Calculate the median time in seconds.

7 The numbers of passengers in 20 cars are recorded. The numbers are

1, 0, 0, 1, 2, 3, 5, 0, 1, 2, 1, 1, 2, 0, 0, 0, 4, 2, 1, 1

Calculate the median number of passengers.

8 Explain why these statements are wrong.

a The median of 3, 2, 1, 5, 6 is 1.

b The median of 1, 1, 2, 3 is 1 or 2.

c You cannot find the median of 8, 8, 9, 10 because there is no middle number.

D4.3 Mode and range

This spread will show you how to:

- Calculate the mode and range of small sets of data

Keywords
Average
Modal
Mode
Range
Representative value
Spread

You can use an **average** to represent a set of data.
One **representative value** is called the mode.

- The **mode** is the value that occurs most often.

Example

In the last ten football matches, Matlock Town scored 2, 5, 2, 3, 4, 1, 3, 1, 1, 1 goals.
Calculate the modal number of goals.

1 goal occurred most often.
Mode = 1 goal

You can represent the goals on a diagram.

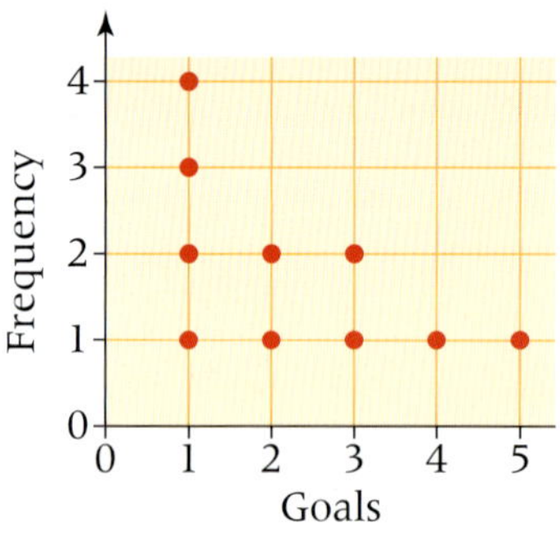

The mode is also called the **modal** value.

The modal number of goals scored is 1.

- You can measure the **spread** of the numbers by calculating the range.
- The **range** is the highest value minus the lowest value.

The range measures the spread of the data.

Example

Find

a the range
b the mode of these numbers of goals.

1, 1, 1, 1, 2, 2, 3, 3, 4, 5, 3, 3

a Range = highest value – lowest value
= 5 – 1
= 4 goals

b There are 2 modes: 1 and 3

Exercise D4.3

1 The shop Tops4U sells 25 tops in one day.
The sizes sold are

M = Medium
L = Large
XL = Extra large

L	M	L	M	XL	L	M	M	XL	M
M	XL	XL	L	M	M	L	L	M	L
L	L	XL	M	M					

a Copy and complete the frequency table for the tops.

Size	Tally	Frequency
Medium (M)		
Large (L)		
Extra large (XL)		

b State the modal size.

2 Calculate the mode and range of each set of numbers.

a 0, 0, 1, 1, 1, 1, 2, 2, 2, 3, 3, 4, 4

b 5, 5, 6, 6, 6, 7, 7, 7, 7, 8, 8, 8, 8, 8

c 10, 11, 11, 11, 12, 12, 13, 14

d 21, 22, 23, 24, 24, 25, 25, 25, 26

e 8, 8, 8, 9, 9, 10, 11

f 4, 3, 5, 5, 6, 6, 4, 3, 4, 5, 6, 5, 3

3 Calculate the mode and range of these sets of numbers.

a

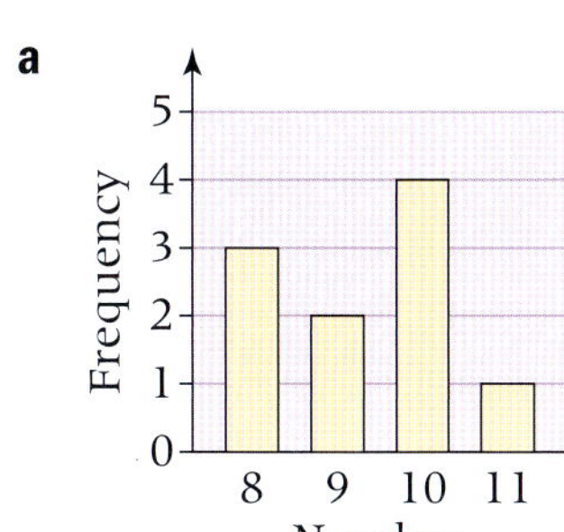

b

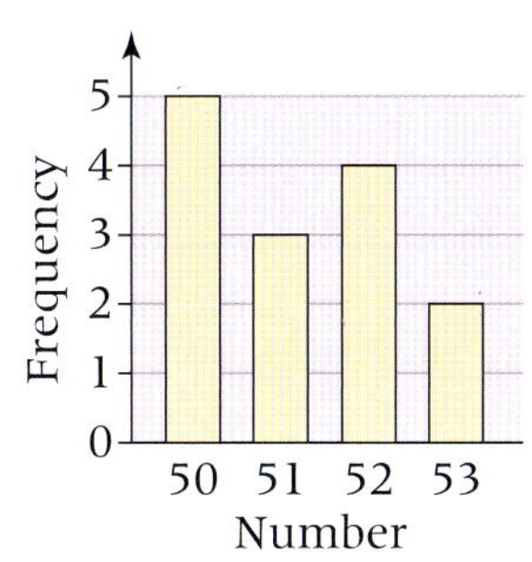

c

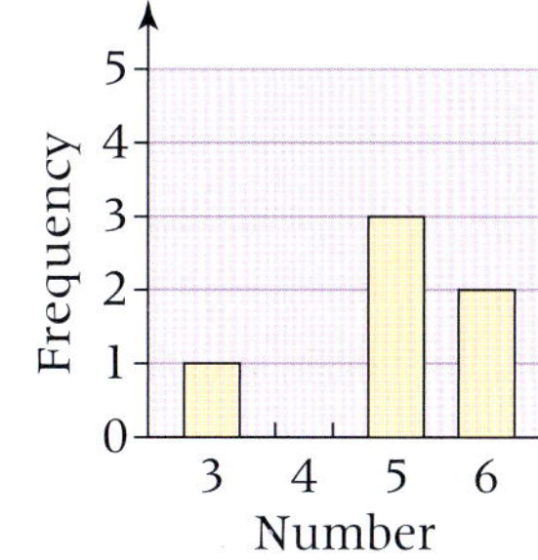

4 The mean monthly temperature, in °F, is shown for Leeds.

Average temperature

	Jan	Feb	Mar	Apr	May	Jun	Jul	Aug	Sep	Oct	Nov	Dec
°F	41	41	44	46	54	58	62	61	56	52	45	44

Source: www.weatherbase.com

Calculate the range of the temperatures.

5 The range of five weights is 18 kg.
The five weights in order are 35 kg, 42 kg, 43 kg, 51 kg, ? kg.
Calculate the value of the unknown weight.

6 The mode of these five numbers is 4.
Find the missing number and calculate the range.

D4.4 Charts and tables

This spread will show you how to:

- Calculate the mean, mode, median and range of sets of data

Keywords
Frequency table
Mean
Median
Mode
Range
Spread

- You can calculate
  - the **mean**
  - the **mode**
  - the **median**
  - the **range**

  from a **frequency table**.

Example

A dice was thrown ten times. The scores were

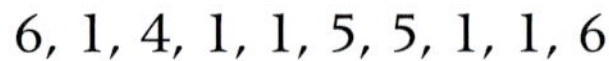

6, 1, 4, 1, 1, 5, 5, 1, 1, 6

Complete the frequency table and calculate the mean, mode, median and range of the ten scores.

Score	Tally	Frequency
1		
2		

This topic is extended to finding the mean from a frequency table on page 392.

Score	Tally	Frequency
1	~~||||~~	5
2		0
3		0
4	|	1
5	||	2
6	||	2
		10

1 was thrown on five occasions.

6 was thrown on two occasions.

The results can be written in numerical order

1, 1, 1, 1, 1, 4, 5, 5, 6, 6

Mean = total of scores ÷ number of throws

$= \frac{31}{10}$

$= 3.1$

Mode = 1 as 1 occurs the most often.

Median = $(1 + 4) \div 2$

= 2.5 as there are two middle scores.

Range = highest value − lowest value

$= 6 - 1$

$= 5$

The mean, mode and median must be between 1 and 6 inclusive.

The range measures the **spread** of the results.

Exercise D4.4

1 The numbers of days that 20 students were absent from school in a week were

1 2 1 0 5 0 5 3 3 1
2 1 1 5 1 2 4 0 0 1

Number of days	Tally	Frequency
0		
1		
2		
3		
4		
5		

a Copy and complete the frequency table.

b Calculate the mean, mode, median and range of the 20 numbers.

2 The numbers of shots taken for each hole for a round of golf were

3 4 3 2 3 4 3 4 3
4 3 2 4 6 3 4 4 4

a Copy and complete the frequency table.

Number of shots	Tally	Frequency
2		
3		
4		
5		
6		

b Calculate the mean, mode, median and range of the 18 scores.

3 Twenty people are asked 'How many televisions do you have in your home?' The results are

1 2 0 2 3 2 1 1 1 2
1 3 3 1 4 2 1 2 1 1

Number of TVs	Tally	Frequency
0		
1		
2		
3		
4		

a Copy and complete the frequency table.

b Calculate the mean, mode, median and range of the 20 numbers.

4 Ten people are asked to choose a number from

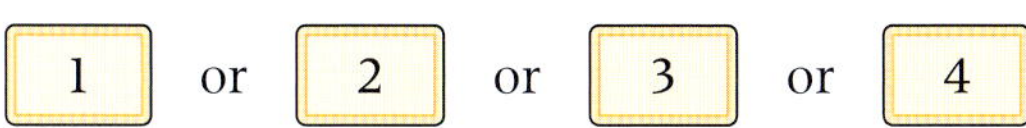

The results are shown in the bar chart.

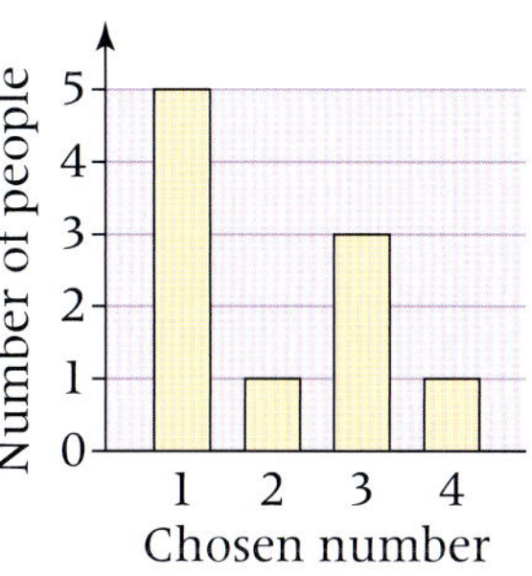

a Write the ten numbers in order of size, smallest first.

b Calculate the mean, mode, median and range of the ten numbers.

D4.5 Comparing data

This spread will show you how to:

- Compare two sets of data using mean, mode, median and range
- Draw and interpret diagrams to represent data

Keywords
Compare
Mean
Median
Mode
Range
Spread

The **mean**, **mode** and **median** each give one typical value to represent the data.
The **range** is a number that measures the **spread** of the data.

- You can **compare** two sets of data using mean, mode, median and range.

Examiner's tip
The techniques in this spread are useful for your Data coursework.

Example

The results of a test are shown in the diagrams.

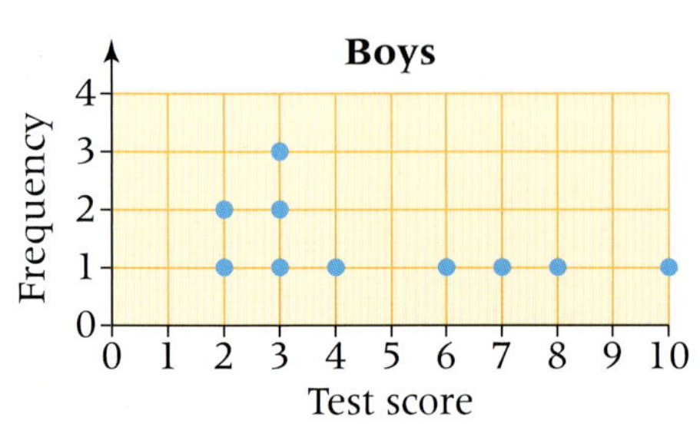

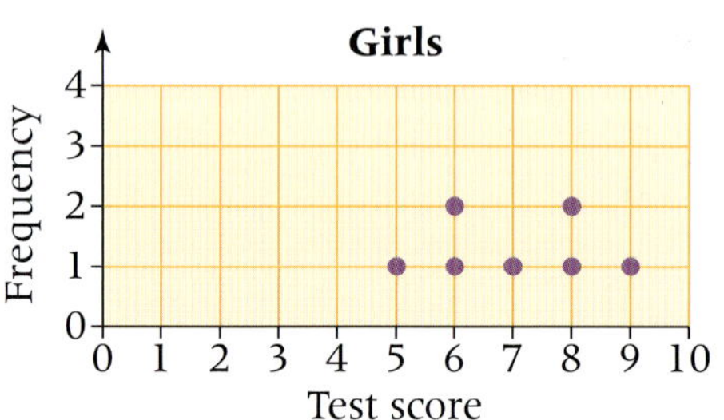

a Calculate the mean and median for each set of data.
b Compare the sets of data using the mean and median.
c Calculate the range for each set of data.
d Compare the sets of data using the range.

a Boys' mean $= (2 + 2 + 3 + 3 + 3 + 4 + 6 + 7 + 8 + 10) \div 10$
$= 48 \div 10 = 4.8$

Boys' median $= (3 + 4) \div 2 = 3.5$ (the mean of the 5th and 6th test scores)

Girls' mean $= (5 + 6 + 6 + 7 + 8 + 8 + 9) \div 7$
$= 49 \div 7 = 7$

Girls' median $= 7$ (the 4th test score)

b The mean and median show on average the girls did better in the test than the boys.

Use the words 'on average'.

c Boys' range $= 10 - 2 = 8$
Girls' range $= 9 - 5 = 4$

d The range shows that the boys' results are more spread out than the girls' results. The girls' results are more consistent.

- **When you compare data, you should use**
  - **a measure of average**
  - **a measure of spread.**

This topic is extended to scatter diagrams on page 394.

Exercise D4.5

1 The numbers of people who live in each house in Windermere Street are

1, 1, 1, 1, 2, 2, 2, 4, 4

a Calculate the mean, mode and median of these numbers.

The numbers of people who live in each house in Coniston Road are

3, 3, 3, 3, 4, 6, 6

b Calculate the mean, mode and median of these numbers.

c Use your answers to compare the average number of people in each house for the two streets.

2 John throws three darts. He scores 40, 20, 20.
Jim throws three darts. He scores 1, 5, 60.

a Calculate the range of the scores for John and Jim.

b Who is more consistent?

3 The ages of ten teachers are 25, 22, 51, 34, 28, 45, 37, 28, 33, 50.

a Calculate the range of these ages.

b All the Year 11 students are either 15 or 16 years old.
Calculate the range of the Year 11 ages.

c Explain the meaning of these two answers.

4 The highest recorded monthly temperatures for St Tropez in France and for Wick in Scotland are shown.

Highest recorded temperature (°C)

	Jan	Feb	Mar	Apr	May	Jun	Jul	Aug	Sep	Oct	Nov	Dec
St Tropez	17	22	22	24	27	30	34	33	29	26	20	19
Wick	13	13	16	19	22	24	23	23	21	19	15	13

a List the temperatures for St Tropez and for Wick in order, smallest first.

b Calculate the median temperature for St Tropez and for Wick.

c Using the answers for the median, compare the two sets of data.

d Calculate the range of the temperatures for St Tropez and for Wick.

e Using the answers for the range, compare the two sets of data.

D4 Exam review

Key objectives

- Calculate mean, range and median of small data sets with discrete data
- Compare discrete distributions and make inferences, using the shapes of distributions and measures of average and range

1 The test results of 7 boys and 5 girls are given.

Boys: 7, 9, 4, 8, 7, 6, 7
Girls: 8, 7, 9, 5, 7

Work out:

a the mode of the boys' results (1)
b the median of the girls' results (2)
c the range of all 12 results (1)
d the mean of all 12 results. (2)

2 Sarah watched a water ride at a theme park.
She counted the number of people in each of 20 boats.
These numbers are shown below:

2 3 1 2 2 3 4 5 4 1
1 2 2 3 2 4 5 4 2 4

a Copy and complete the frequency table. (2)

Number of people in a boat	Tally	Frequency
1		
2		
3		
4		
5		

b Write down the mode of the number of people in a boat. (1)

Emily asked 5 people the number of rides each of them had been on.
The numbers are shown below:

6 8 7 6 10

c Work out the mean number of rides per person. (3)

(Edexcel Ltd., 2004)

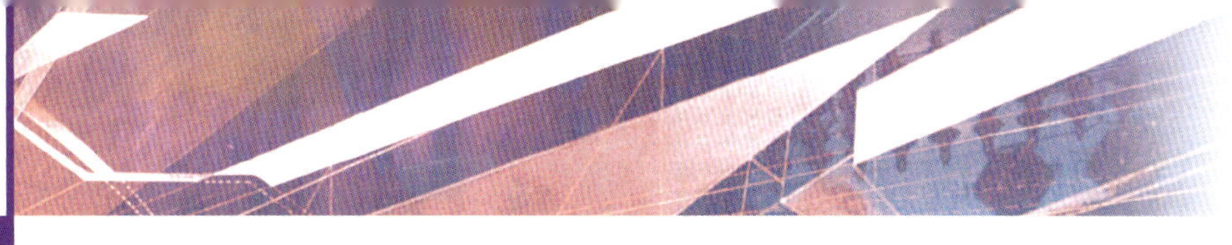

S5 Measuring and constructing

This unit will show you how to

- Measure and draw lines to the nearest millimetre
- Draw triangles and other 2-D shapes using a ruler and protractor, given information about side lengths and angles
- Use and interpret scale drawings
- Understand angle measure and the associated language
- Give bearings accurately
- Interpret scales on a range of measuring instruments
- Recognise the possible inaccuracy of measurements

Before you start ...

You should be able to answer these questions.

Review

1 Convert these measurements to the units stated. — Unit S1

a 10 mm =	cm	**b** 40 mm =	cm
c 45 mm =	cm	**d** 100 cm =	m
e 400 cm =	m	**f** 150 cm =	m
g 1000 m =	km	**h** 6000 m =	km
i 500 m =	km	**j** 1500 m =	km

2 Estimate these angles. — Unit S2

a **b** **c**

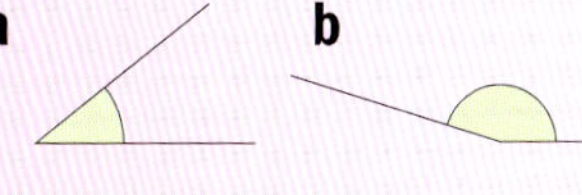

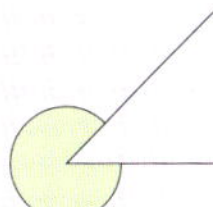

3 State the value of the marked angle. — Unit S2

4 Evaluate each of these. — Unit N3

a 1 ÷ 2 **b** 1 ÷ 5 **c** 1 ÷ 10

5 **a** Calculate the area of the rectangle. — Unit S2

b State the units of your answer.

3 cm

8 cm

S5.1 Lines and shapes

This spread will show you how to:

- Measure and draw lines to the nearest millimetre

Keywords
Centimetre
Diagonal
Length
Measure
Millimetre
Perimeter

A ruler **measures length** in **millimetres** (mm) or **centimetres** (cm).

0 10 20 30 40
mm

0 1 2 3 4
cm

10 mm = 1 cm

This line measures 2.5 cm or 25 mm.

0 1 2 3 4
cm

2.5 cm = $2\frac{1}{2}$ cm

This line measures 2.7 cm or 27 mm.

0 10 20 30 40
mm

H	T	U	•	$\frac{1}{10}$	$\frac{1}{100}$
		2	•	7	

2.7 cm = $2\frac{7}{10}$ cm

To measure a line, line up the ruler so that the zero mark is at the start of the line.

You can construct shapes using a ruler and a protractor.

Example

Using a ruler and a protractor, construct a square of length 5 cm.
Draw and measure the diagonal in **a** centimetres
b millimetres.

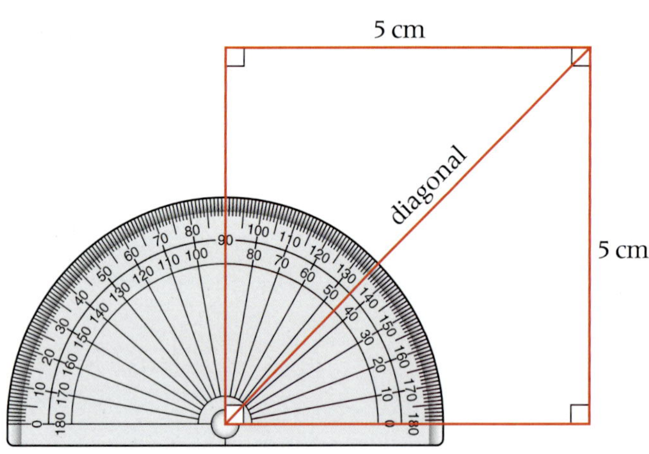

Use the ruler to draw the 5 cm lines.

Use the protractor to construct the 90° angles.

a 7.1 cm
b 10 mm = 1 cm
$10 \times 7.1 = 71$ mm

Exercise S5.1

1 Measure the lengths of these lines in

a centimetres

b millimetres.

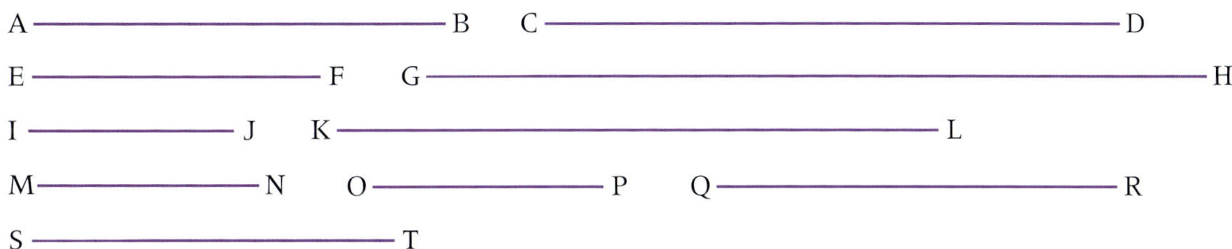

2 **a** Draw a line AB, so that AB = 9 cm.

b Find the mid-point of AB and mark it with a cross.

3 **a** Draw a line CD, so that CD = 11.6 cm.

b Find the mid-point of CD and mark it M.

c Measure CM, stating the units of your answer.

4 Measure and state the diameter of each circle in centimetres.

a

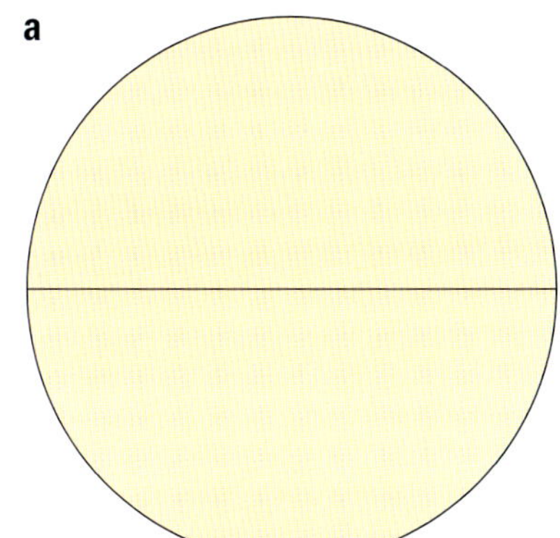

b

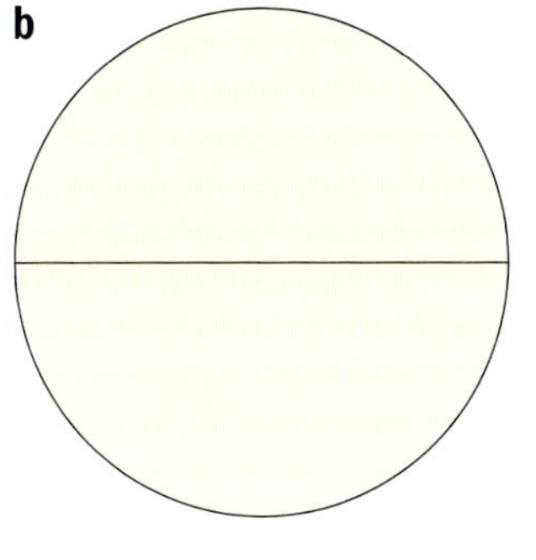

c

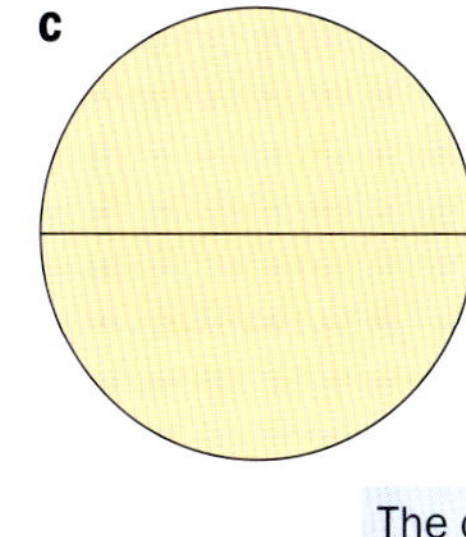

The diameter is the distance across a circle through the centre.

5 Measure and calculate the perimeter of this rectangle in

a centimetres

b millimetres.

The **perimeter** is the **distance** round the edge of a shape.

6 Using a ruler and protractor, construct a rectangle of length 7 cm and width 5 cm. Draw and measure the diagonal in

a centimetres

b millimetres.

S5.2 Constructing triangles

This spread will show you how to:

- Draw triangles and other 2-D shapes using a ruler and protractor, given information about side lengths and angles

Keywords
Base
Construct
Protractor
Straight edge

Sunita and Emma were asked to **construct** a triangle with angles 50°, 40° and 90°. Here are their answers.

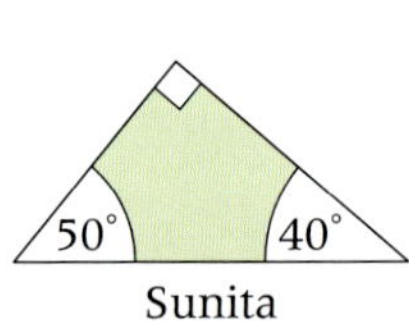

Sunita

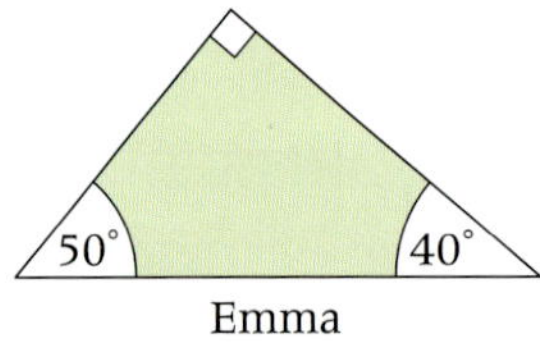

Emma

They are both correct!

They constructed different triangles because not enough information was given.

You will always construct the same triangle if you know

Two sides and the angle between them (SAS) or Two angles and a side (ASA)

A = Angle
S = Side

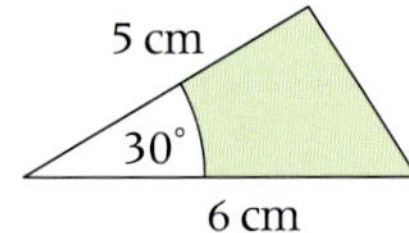

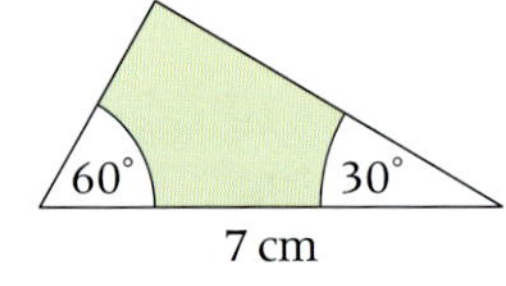

You will need a ruler and a **protractor** to construct SAS and ASA triangles.

A ruler is sometimes called a **straight edge**.

Example

Construct triangle ABC with AB = 6 cm, BC = 4 cm and $\hat{B}$ = 50°.
What is the length of AC?

Sketch the shape. Draw the **base** AB. Construct angle B using a protractor. Mark point C. Join it to A.

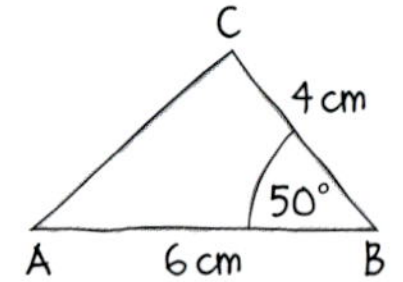

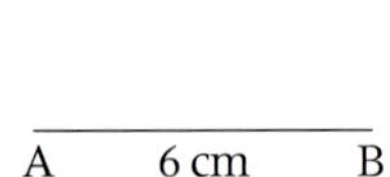

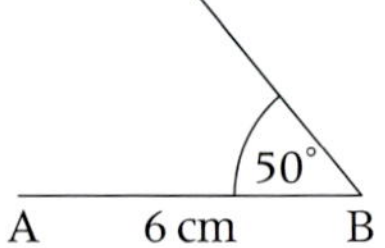

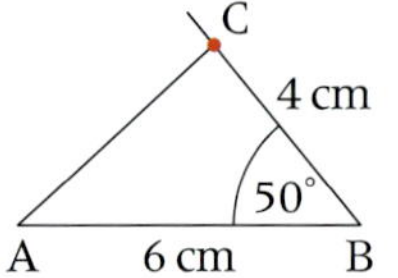

Measuring gives AC = 4.6 cm.

Example

Construct triangle PQR with angle P = 40°, angle Q = 40° and length PQ = 8 cm.
What sort of triangle is PQR?

Sketch the shape. Draw the base PQ. Measure angle P. Measure angle Q.

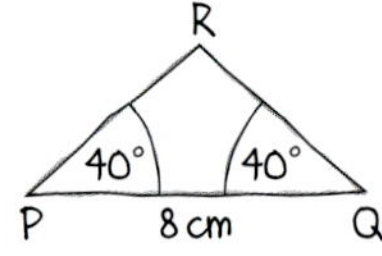

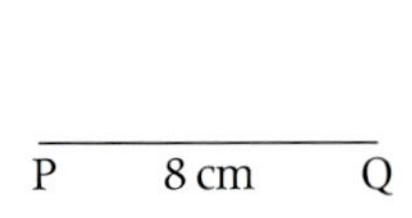

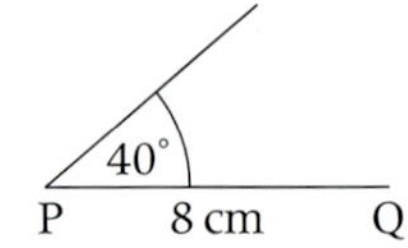

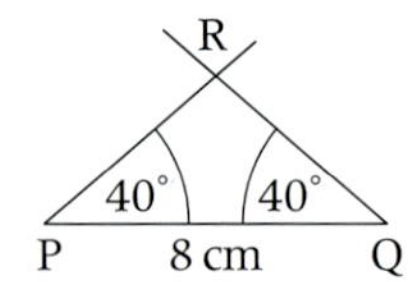

Triangle PQR is isosceles.

Exercise S5.2

1 Make accurate drawings of these triangles (SAS).
Measure the unknown length in each triangle.

a 4 cm, 3 cm

b 3.5 cm, 5.5 cm

c 3.5 cm, 45°, 6 cm

d 4 cm, 120°, 4 cm

e 5 cm, 60°, 5 cm

2 Make accurate drawings of these triangles (ASA).
Measure the two unknown lengths in each triangle.
State the units of your answers.

a 55°, 35°, 5 cm

b 40°, 65°, 6.5 cm

c 70°, 70°, 4 cm

d 60°, 60°, 6 cm

e 100°, 30°, 6.5 cm

3 Make accurate drawings of these triangles.
Measure the unknown lengths in each triangle.

a 53°, 3 cm

b 5 cm, 120°, 3 cm

c 5 cm, 4 cm

4 Make an accurate drawing of this parallelogram.

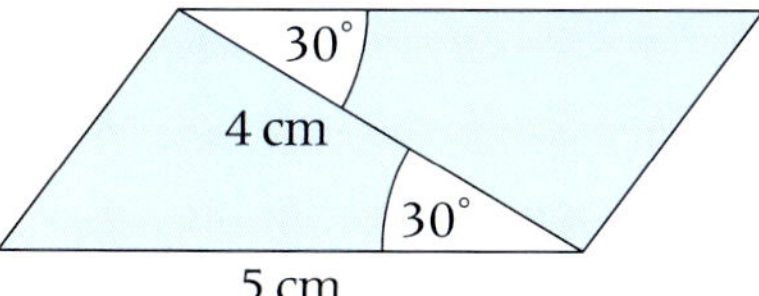

S5.3 Scale

This spread will show you how to:

- Use and interpret scale drawings
- Understand the implications of enlargement for scale drawings

Keywords
Proportion
Reduced
Scale
Scale drawing

Real-life lengths are **reduced** in **proportion** to give a **scale drawing**.

Real-life

Scale: 1 cm represents 1 m

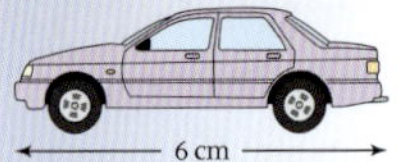

Length in scale drawing = 6 cm
Length in real life = 6 m

Scale drawing

Scale: 1 cm represents 2 cm

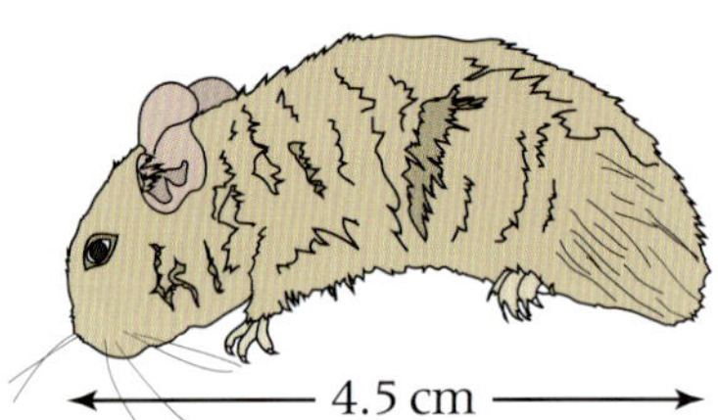

The scale of a drawing gives the relationship between the drawing and the real object.

Example

The scale for this drawing of a chessboard is 1 cm represents 10 cm.

a Calculate the dimensions of the real chessboard.

b Calculate the perimeter of the real chessboard.

c How many squares are on the real chessboard?

---

a 1 cm represents 10 cm.
4 cm represents 4×10 cm = 40 cm.
Dimensions are 40 cm by 40 cm.

b Perimeter $40 + 40 + 40 + 40 = 160$ cm.

c Number of squares $= 8 \times 8 = 64$.

There are exactly the same number of squares on the scale drawing as on the real chessboard.

Exercise S5.3

1 The scale on a drawing is 1 cm represents 5 cm.
Calculate the distance represented by

a 2 cm **b** 4 cm
c 5 cm **d** 10 cm
e 20 cm

2 The scale on a drawing is 1 cm represents 10 cm.
Calculate the distance represented by

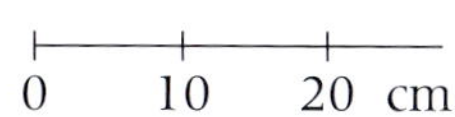

a 2 cm **b** 5 cm
c 2.5 cm **d** 0.5 cm
e 4.1 cm

3 The scale on a drawing is 1 cm represents 50 cm.
Calculate the distance represented by

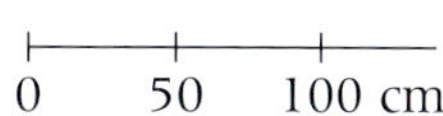

a 4 cm **b** 5 cm
c 10 cm **d** 0.5 cm
e 1.5 cm

4 The scale on a drawing is 1 cm represents 1 km.
Calculate the distance represented by

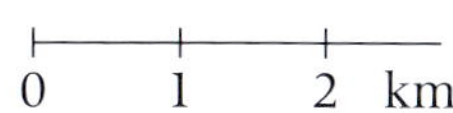

a 2 cm **b** 2.5 cm
c 3.2 cm **d** 10 cm
e 0.25 cm

5 The scale on a drawing is 1 cm represents 2 m.
Calculate the distance represented by

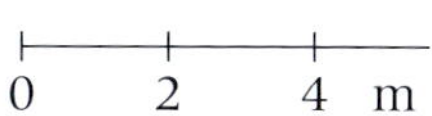

a 8 cm **b** 7.5 cm
c 3.6 cm **d** 10.8 cm
e 0.45 cm

6 Using this scale drawing of the Eiffel Tower, calculate

a the height
b the width of the base.

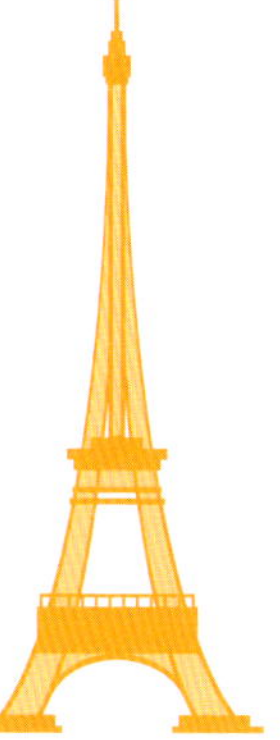

Scale: 1 cm represents 60 m

S5.4 Bearings

This spread will show you how to:

- Understand angle measure using the associated language
- Give bearings accurately

Keywords

Bearing
Direction
Scale
Three-figure bearing

You can give a **direction** using the compass points to describe P.

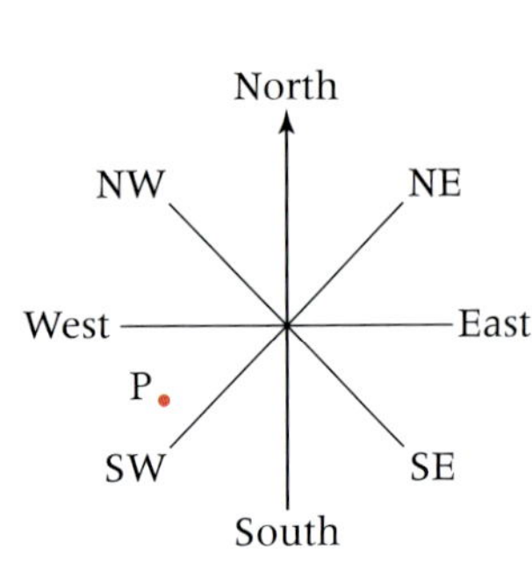

This is not very accurate.
P is between SW and W.

You use a 360° **scale** or a **bearing** to give a direction accurately.

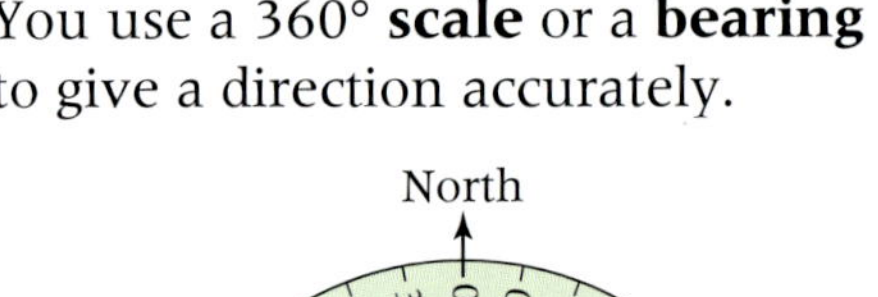

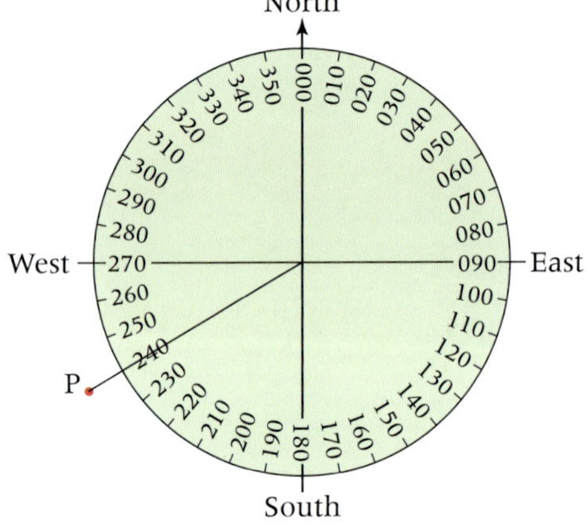

The bearing of P is 240°.
This is very accurate.

000° = North
090° = East
180° = South
270° = West

- To give a bearing accurately you
  - measure from North
  - measure clockwise
  - use three figures.

Example

a Write the bearing of Lincoln from Sheffield.

b Mark the position of Manchester on a bearing of 284° at a distance of 40 miles from Sheffield.

Scale: 1 cm represents 10 miles

N
Manchester
Sheffield
Lincoln

From Sheffield means centre the protractor *at* Sheffield.

a Place the centre on Sheffield and 0° at North. Read off the direction of Lincoln as 115°.

b 1 cm represents 10 miles. 4 cm represents 40 miles.

Exercise S5.4

1 **a** Use a protractor to draw an accurate diagram of the compass points.

b On your diagram, give the **three-figure bearings** of each of the compass points.

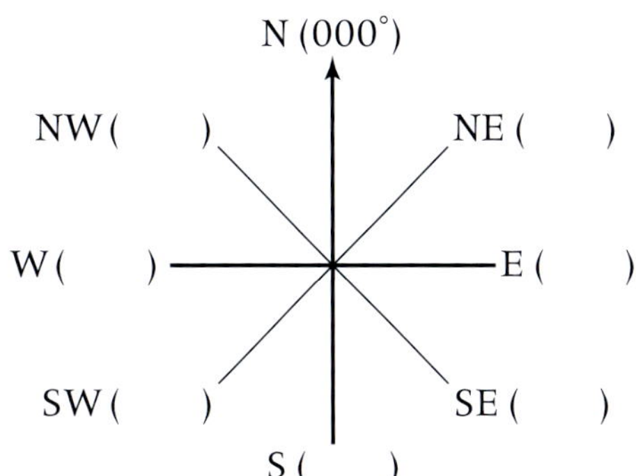

2 Measure the bearing of these places from the Lookout point.

a Battlefield
b Tower
c Church
d Castle
e Buoy
f Yacht
g Needles rocks
h Lighthouse

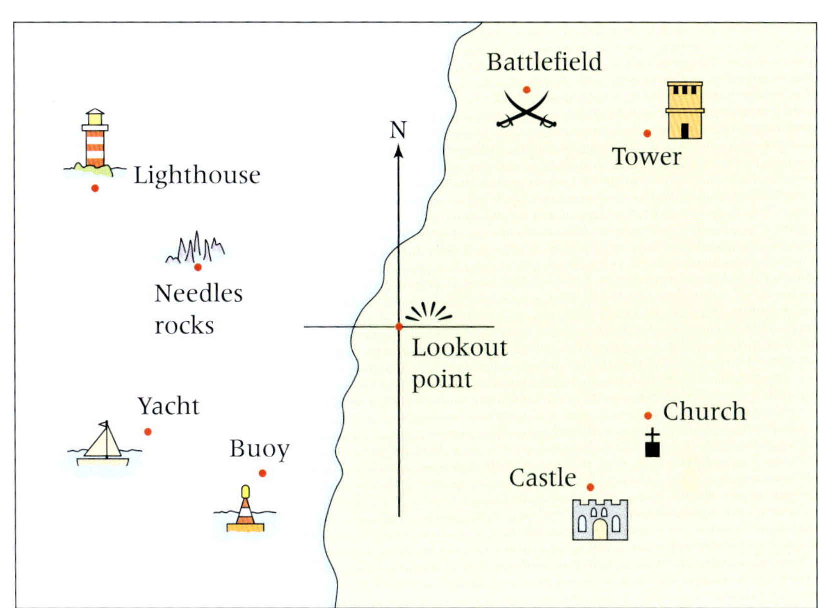

3 For each question, put a cross anywhere on your page.
Plot the points and join them to form a quadrilateral.
Name the shape, then measure and calculate the perimeter.

a

Bearing from the cross	000°	050°	180°	310°
Distance from the cross	5 cm	5 cm	5 cm	5 cm

b

Bearing from the cross	055°	145°	235°	325°
Distance from the cross	5 cm	5 cm	5 cm	5 cm

c

Bearing from the cross	000°	090°	180°	270°
Distance from the cross	5 cm	2.5 cm	5 cm	2.5 cm

d

Bearing from the cross	000°	135°	180°	225°
Distance from the cross	5 cm	5 cm	5 cm	5 cm

e

Bearing from the cross	040°	120°	240°	320°
Distance from the cross	4 cm	5 cm	5 cm	4 cm

S5.5 Measuring instruments

This spread will show you how to:

- Interpret scales on a range of measuring instruments, including those for time and mass
- Recognise the possible inaccuracy of measurements

Keywords
Scale
Temperature
Time

Measuring instruments use **scales** to show measurements.

The scales are divided into small divisions.

You need to work out what one division stands for.

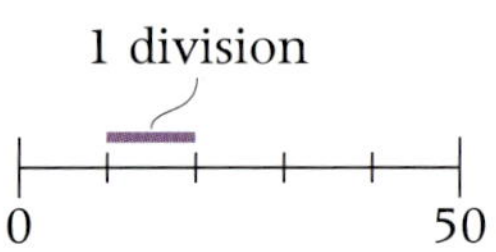

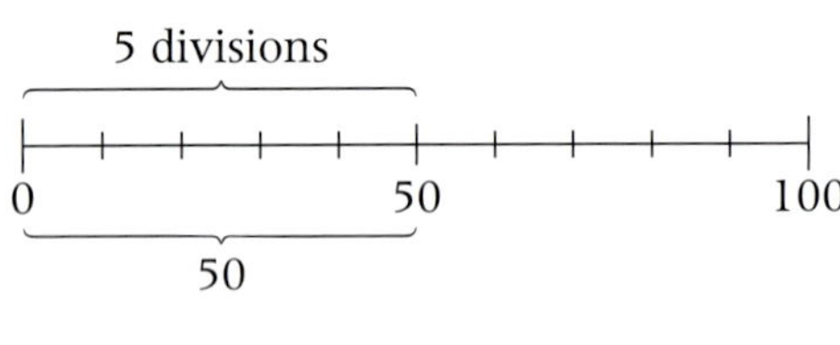

There are 5 divisions from 0 to 50.

$50 \div 5 = 10$

So each division stands for 10.

Example

Write the readings shown on the scales.

a

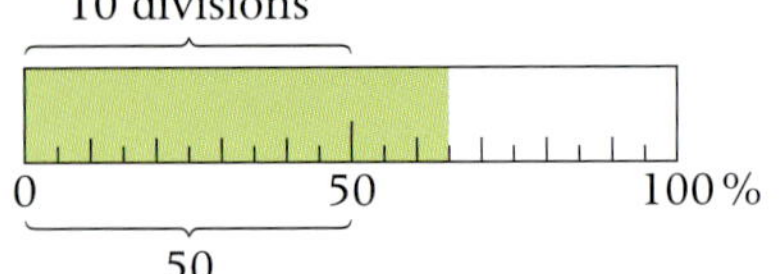

b

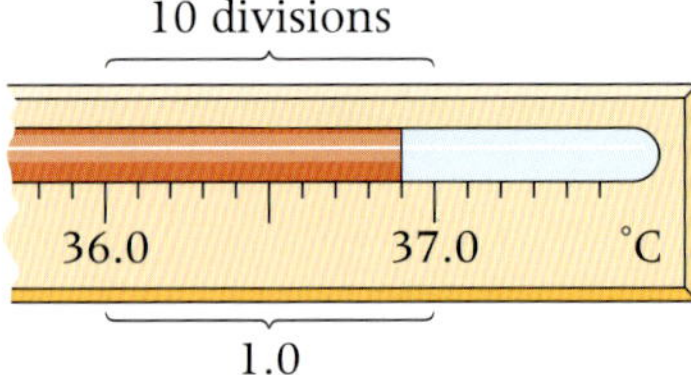

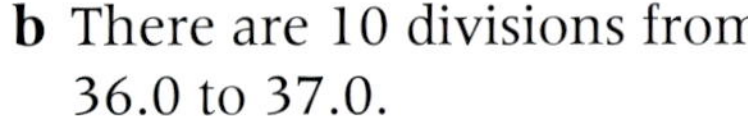

a There are 10 divisions from 0 to 50.

$50 \div 10 = 5$

Each division stands for 5%.
Reading shows
$50 + 5 + 5 + 5 = 65\%$.

b There are 10 divisions from 36.0 to 37.0.

$1.0 \div 10 = 0.1$

Each division stands for 0.1.
Reading shows
$36.0 + 0.9 = 36.9$ °C.

The reading on a clock is only accurate to the nearest minute. The reading is not exact.

The reading shows 2 minutes past 12.

However, the line could be anywhere between 1.5 and 2.5 minutes past. The reading may be inaccurate by up to 1 minute.

Two minutes past

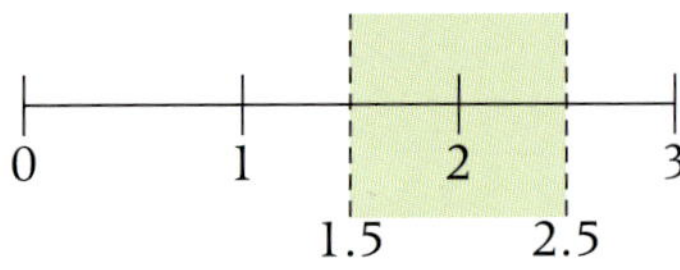

Exercise S5.5

1 For each scale, write what each division represents and the readings shown.

a

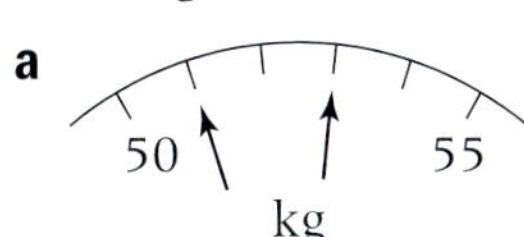

b

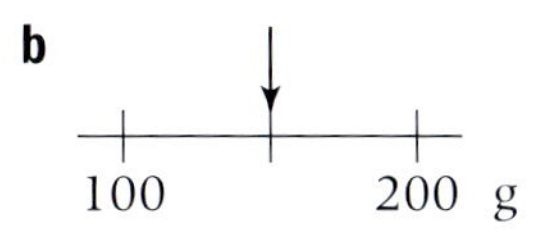

c

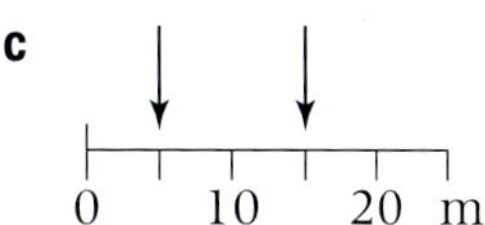

d

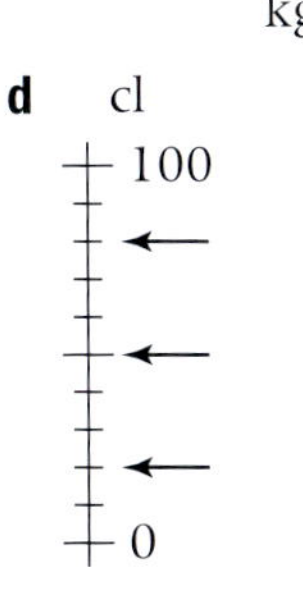

e

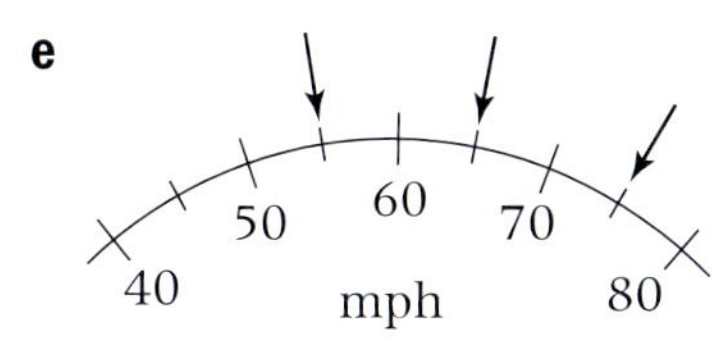

2 For each scale, write what each division represents and the readings shown.

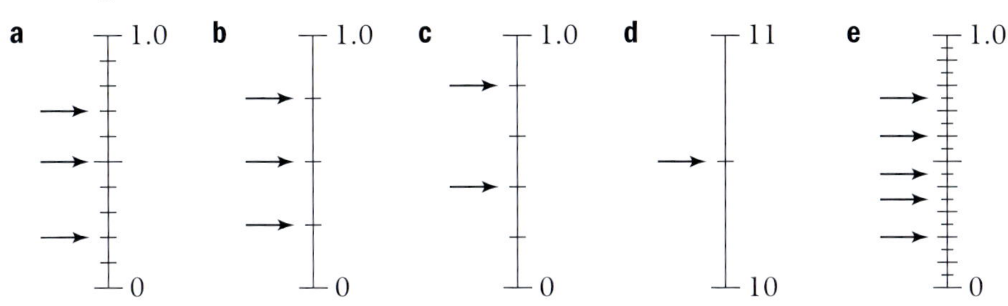

3 This scale shows how to convert from stones to kilograms.

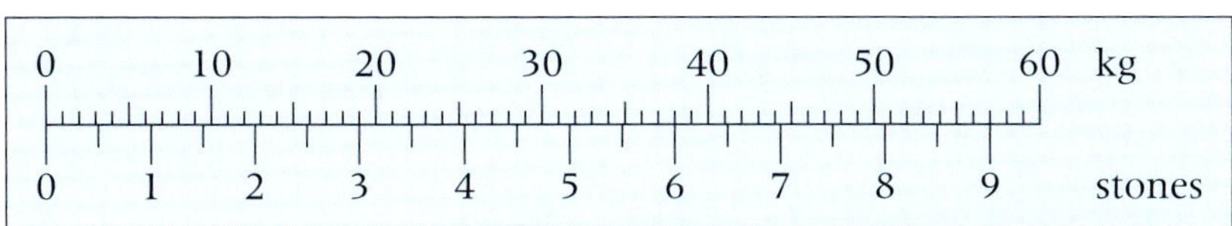

Use the scale to convert these weights in stones to kilograms.

a 8 stones **b** 3 stones **c** 2.5 stones

d 5.5 stones **e** 8.5 stones

DID YOU KNOW?

The stone weight was established as 14 lb in 1352 during the reign of Edward III and hasn't changed since!

4 **a** Give the reading on the thermometer.

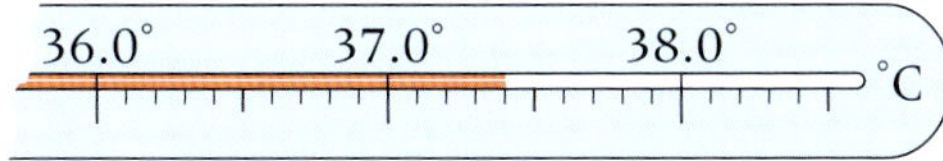

b Calculate how much the temperature is above 36.9 °C.

5 These measurements are given to the nearest centimetre.
Give the lowest and highest measurements they could represent.

a 5 cm **b** 8 cm **c** 1 cm

d 10 cm **e** 20 cm

S5 Exam review

Key objectives

- Interpret scales on a range of measuring instruments, including those for time and mass
- Convert measurements from one unit to another
- Measure and draw lines to the nearest millimetre, and angles to the nearest degree
- Understand angle measure using the associated language
- Draw triangles and other 2-D shapes using a ruler and protractor, given information about their side lengths and angles

1 **a** Measure the bearing of A from B. (1)

b Draw a similar diagram to show a point C with a bearing of 98° from a point D. (2)

2 **a** Draw a line 12 cm long. (1)

b Find the point that is halfway along the line you have drawn. Mark it with a cross (✗). (1)

Draw a grid of centimetre squares.

c On your grid, draw a rectangle that has length 6 cm and width 4 cm. (1)

(Edexcel Ltd., 2003)

Using fractions and percentages

This unit will show you how to

- Express one number as a fraction of another number
- Calculate a given fraction of an amount
- Use mental, written and calculator methods to calculate with fractions
- Convert between fractions, percentages and decimals
- Use fraction and percentage as operators
- Calculate percentages of amounts using a variety of methods
- Calculate percentage increase and decrease
- Solve percentage increase and decrease problems
- Calculate simple interest
- Use checking procedures involving estimation

Before you start ...

You should be able to answer these questions.

	Review
1 Calculate $\frac{1}{5}$ of £30.	Unit N3
2 Calculate $\frac{1}{4} \times 84$.	Unit N3
3 Calculate each of these, clearly showing the method you have used. **a** 50% of £240 **b** 10% of \$60	Unit N3
4 **a** Use your calculator to work out 34% of £250. **b** Increase £190 by 10%.	Unit N3
5 Stuart buys a signed football shirt for £140. At the end of the year the shirt has risen in value by 25%. Calculate the new value of the shirt.	Unit N3

N7.1 Fraction of a quantity

This spread will show you how to:

- Express one number as a fraction of another number
- Calculate a given fraction of an amount
- Use mental, written and calculator methods to calculate with fractions

Keywords
Fraction

You can express one number as a **fraction** of another number.

Example

In a class there are 30 students. 18 of the students are boys. What fraction of the class are boys? Give your answer in its simplest form.

There are 30 students in the class altogether (the **whole**).
18 of the students are boys.
Fraction of the class who are boys $= \frac{18}{30} = \frac{3}{5}$

You can calculate a fraction of a number or quantity in several ways.

Example

Calculate using a mental method.

a $\frac{1}{4}$ of £20 **b** $\frac{3}{5}$ of €200

a

$\frac{1}{4}$ of £20 = 20 ÷ 4 = £5

b

$\frac{1}{5}$ of €200 = 200 ÷ 5 = €40
$\frac{3}{5}$ of €200 = 3 × 40 = €120

Finding $\frac{1}{5}$ of something is the same as dividing by 5.

Example

Calculate $\frac{3}{8}$ of 56 kg using a written method.

$$\frac{3}{8} \text{ of } 56 \text{ kg} = \frac{3}{8} \times 56 \text{ kg}$$
$$= 3 \times \frac{1}{8} \times 56 \text{ kg}$$
$$= \frac{3 \times 56}{8}$$
$$= \frac{168}{8}$$
$$= 21 \text{ kg}$$

To work out a fraction of an amount, multiply the fraction by the amount.

Multiplying by $\frac{1}{8}$ is the same as dividing by 8.

You could also work out $3 \times (56 \div 8) = 3 \times 7 = 21$

Example

Calculate $\frac{7}{8}$ of £93 using a calculator.

Decimal equivalent for $\frac{7}{8} = 7 \div 8 = 0.875$

$\frac{7}{8}$ of £93 $= \frac{7}{8} \times$ £93 $= 0.875 \times 93$
$= 81.375 =$ £81.38

Exercise N7.1

1 Write the fraction of each of these shapes that is shaded, leaving your answer as a fraction in its simplest form.

a 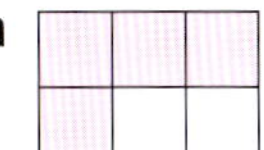 **b** 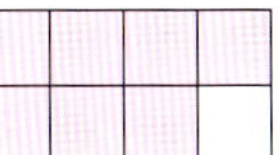 **c** 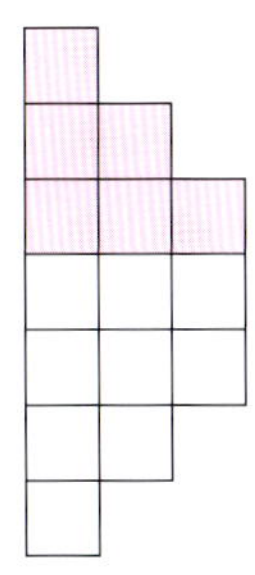 **d** 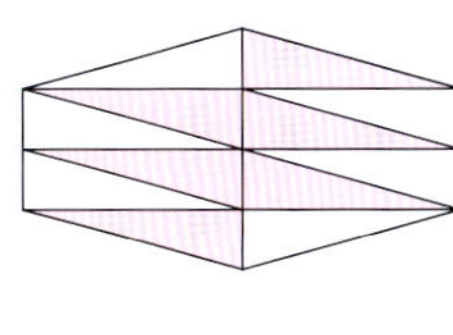

2 Use a mental method to calculate each of these amounts.

a $\frac{1}{2}$ of 40 sheep **b** $\frac{1}{3}$ of 15 apples **c** $\frac{1}{5}$ of 25 shops **d** $\frac{1}{4}$ of 48 marks

3 For each answer state the fraction in its simplest form.

a There are 30 students in a class. 20 are boys and 10 are girls. What fraction of the class are

i boys **ii** girls?

b Horace earns £500 a week. He pays £150 of his money each week in tax. He saves £120 each week. What fraction of his weekly wage does Horace

i pay in tax **ii** save?

c Michael and Shafique share £200. Michael has £80 and Shafique has £120. What fraction of the £200 belongs

i to Michael **ii** to Shafique?

4 Use a mental or written method to work out these.

a $\frac{8}{10}$ of €200

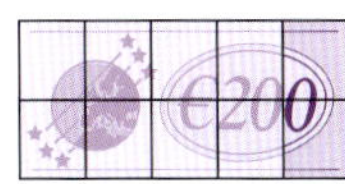

$\frac{1}{10}$ of €200 = 200 ÷ 10 = _____

$\frac{8}{10}$ of €200 = 8 × _____ = _____

b $\frac{2}{5}$ of £40 **c** $\frac{3}{4}$ of 60 minutes **d** $\frac{4}{7}$ of 77 pencils

e $\frac{3}{10}$ of £84 **f** $\frac{5}{6}$ of 42 apples **g** $\frac{7}{12}$ of 36p

5 Use a suitable method to calculate each of these. Where appropriate round your answer to two decimal places.

a $\frac{7}{15}$ of 375 **b** $\frac{9}{10}$ of $450 **c** $\frac{4}{7}$ of 800 kg

d $\frac{5}{9}$ of 234 m **e** $\frac{17}{18}$ of 400 tonnes **f** $\frac{4}{5}$ of 360°

g $\frac{12}{25}$ of 750 marbles **h** $\frac{11}{15}$ of £255 **i** $\frac{4}{15}$ of £345

j $\frac{1}{6}$ of 546 hours

N7.2 Fraction of a fraction

This spread will show you how to:

- Use a fraction as an operator
- Multiply fractions
- Calculate a fraction of a fraction

Keywords
Denominator
Fraction
Numerator

Multiplying by $\frac{1}{5}$ is the same as dividing by 5.

For example, $3 \times \frac{1}{5} = \frac{3}{5} \quad 3 \div 5 = \frac{3}{5}$

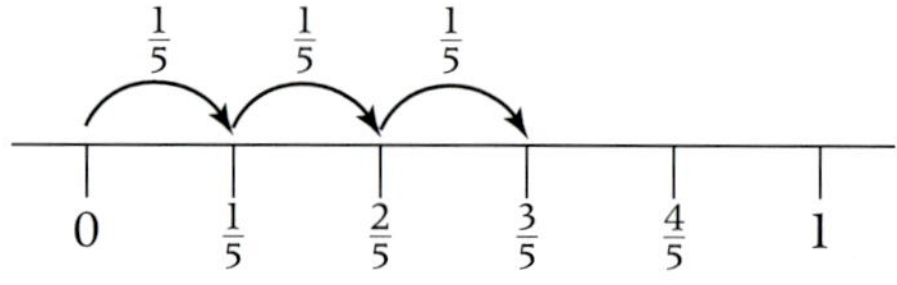

Multiplying by $\frac{1}{10}$ is the same as dividing by 10.

For example, $7 \times \frac{1}{10} = \frac{7}{10} \quad 7 \div 10 = \frac{7}{10}$

- **You can multiply any fraction by a whole number using unit fractions.**

Example

Calculate

a $\frac{2}{3} \times 8$ **b** $\frac{3}{5}$ of £220

a $\frac{2}{3} \times 8 = 8 \times 2 \times \frac{1}{3}$
$= 16 \times \frac{1}{3}$
$= \frac{16}{3}$
$= 5\frac{1}{3}$

b $\frac{3}{5}$ of £220 $= \frac{3}{5} \times 220$
$= 3 \times \frac{1}{5} \times 220$
$= 3 \times 220 \times \frac{1}{5}$
$= 660 \times \frac{1}{5}$
$= \frac{660}{5} = 132$

So $\frac{3}{5}$ of £220 is £132.

Remember: You can find a fraction of an amount by multiplying.

- **You can multiply a fraction by another fraction by multiplying the numerators together and multiplying the denominators together.** $\frac{3}{4} \times \frac{7}{10} = \frac{3 \times 7}{4 \times 10} = \frac{21}{40}$

Example

Sarah saves $\frac{1}{3}$ of her weekly wage.
She puts $\frac{3}{4}$ of the money she saves into a pension. What fraction of her weekly wage does Sarah put into a pension?

Sarah's pension is $\frac{3}{4}$ of the money she saves; she saves $\frac{1}{3}$ of her weekly wage.

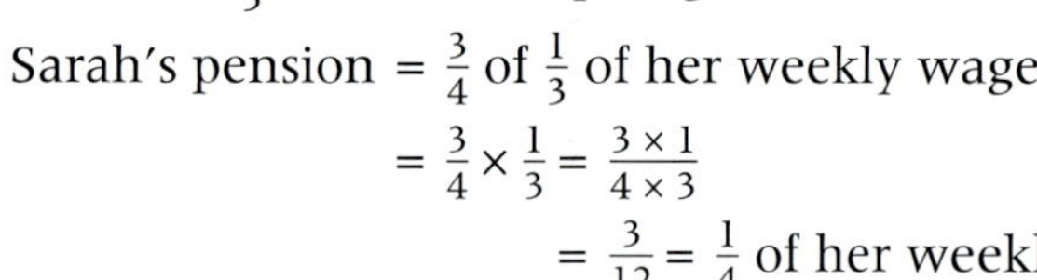

Sarah's pension $= \frac{3}{4}$ of $\frac{1}{3}$ of her weekly wage
$= \frac{3}{4} \times \frac{1}{3} = \frac{3 \times 1}{4 \times 3}$
$= \frac{3}{12} = \frac{1}{4}$ of her weekly wage

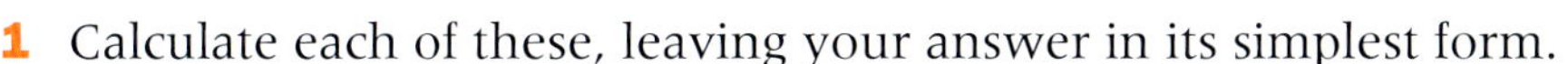

Exercise N7.2

1 Calculate each of these, leaving your answer in its simplest form.

a $3 \times \frac{1}{5}$ **b** $4 \times \frac{1}{7}$ **c** $11 \times \frac{1}{13}$

d $18 \times \frac{1}{3}$ **e** $\frac{1}{7} \times 14$ **f** $\frac{1}{5} \times 20$

2 Calculate each of these, leaving your answer in its simplest form.

a $4 \times \frac{1}{9}$ **b** $8 \times \frac{1}{3}$ **c** $9 \times \frac{1}{2}$

d $10 \times \frac{1}{8}$ **e** $12 \times \frac{1}{6}$ **f** $\frac{1}{8} \times 13$

3 Calculate each of these amounts.

a $\frac{1}{2}$ of 24 apples **b** $\frac{1}{3}$ of 21 shops **c** $\frac{1}{5}$ of 45 texts **d** $\frac{1}{6}$ of 36 cups

4 Use an appropriate method for each of these calculations. The first question has been started for you using the multiplication method.

a $\frac{3}{10}$ of €500 $= 3 \times \frac{1}{10} \times 500$

$= 3 \times 500 \times \frac{1}{10}$

$= 1500 \times \frac{1}{10}$ ($\times\frac{1}{10}$ is the same as $\div 10$)

$= \frac{\_\_}{10}$

= €____

b $\frac{3}{5}$ of £60 **c** $\frac{2}{3}$ of 120 kg **d** $\frac{4}{5}$ of 250p

e $\frac{7}{10}$ of 40 m **f** $\frac{5}{6}$ of 54 cards **g** $\frac{5}{12}$ of \$84

5 Calculate each of these, leaving your answer in its simplest form.

a $6 \times \frac{2}{3}$ **b** $9 \times \frac{1}{3}$ **c** $4 \times \frac{2}{3}$ **d** $\frac{2}{3} \times 12$

e $4 \times \frac{3}{5}$ **f** $\frac{4}{5} \times 3$ **g** $\frac{2}{5} \times 10$ **h** $\frac{3}{8} \times 4$

6 Use a suitable method to calculate each of these amounts.
Where appropriate round your answer to two decimal places.

a $\frac{4}{15}$ of £390 **b** $\frac{7}{10}$ of 4500 m **c** $\frac{3}{7}$ of 50 g **d** $\frac{4}{9}$ of 639 mm

e $\frac{7}{18}$ of 200 kg **f** $\frac{3}{17}$ of 360° **g** $\frac{2}{25}$ of 775 miles **h** $\frac{7}{15}$ of 345 m^2

i $\frac{14}{15}$ of £70 **j** $\frac{5}{6}$ of 18 hours

7 Calculate each of these, leaving your answer in its simplest form.

a $\frac{1}{5} \times \frac{2}{3}$ **b** $\frac{2}{5} \times \frac{3}{4}$ **c** $\frac{2}{7} \times \frac{3}{4}$ **d** $\frac{2}{7} \times \frac{2}{5}$

e $\frac{5}{6} \times \frac{3}{4}$ **f** $\frac{3}{8} \times \frac{5}{9}$ **g** $\frac{3}{5} \times \frac{4}{9}$ **h** $\frac{5}{6} \times \frac{2}{5}$

i $\frac{3}{7} \times \frac{5}{6}$ **j** $(\frac{2}{5})^2$ **k** $(\frac{2}{3})^2$ **l** $(\frac{4}{9})^2$

N7.3 Percentage of a quantity

This spread will show you how to:

- Use percentage as an operator
- Calculate a percentage of an amount using mental and written methods

Keyword

Percentage

Percentages, fractions and decimals are ways of writing the same thing.

$$25\% = \frac{25}{100} = \frac{1}{4} = 0.25$$

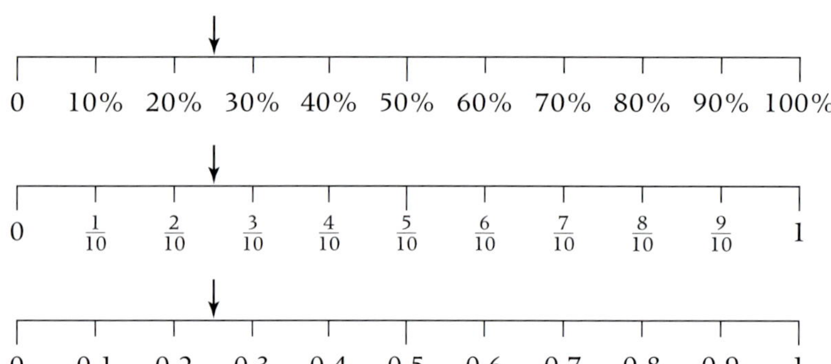

Some useful equivalents to remember.

$1\% = \frac{1}{100} = 0.01$
$10\% = \frac{10}{100} = \frac{1}{10} = 0.1$
$20\% = \frac{20}{100} = \frac{1}{5} = 0.2$
$25\% = \frac{25}{100} = \frac{1}{4} = 0.25$
$50\% = \frac{50}{100} = \frac{1}{2} = 0.5$

You can calculate simple percentages of amounts in your head using equivalent fractions.

Example

Calculate

a 10% of £41

b 50% of 84 m

a $\frac{1}{10}$ of £41 $\quad (10\% = \frac{1}{10})$
$= \frac{1}{10} \times £41$
$= £41 \div 10$
$= £4.10$

b $\frac{1}{2}$ of 84 m $\quad (50\% = \frac{1}{2})$
$= \frac{1}{2} \times 84$ m
$= 84$ m $\div 2$
$= 42$ m

- **To calculate a percentage of an amount using a written method, change the percentage to its equivalent fraction and multiply by the amount.**

Example

Calculate 15% of £80.

15% of £80 $= \frac{15}{100} \times £80$
$= 15 \times £80 \times \frac{1}{100}$
$= \frac{15 \times £80}{100}$
$= \frac{£1200}{100}$
$= £12$

Alternatively you could use a mental method

10% of £80 $= \frac{1}{10}$ of 80
$= 80 \div 10 = 8$
So 5% of 80 $= 8 \div 2 = 4$
So 15% of 80 $= 8 + 4 = 12$
15% of £80 is £12.

15% = 10% + 5%

Exercise N7.3

1 **a** What fraction of this shape is shaded?

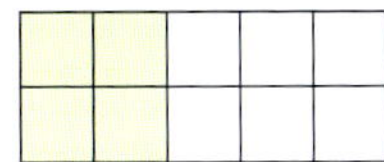

b What percentage of the shape is shaded?

c Copy the shape and shade in more squares so that 70% of the shape is shaded.

2 Calculate these percentages without using a calculator.

a 50% of £60 **b** 50% of 40 **c** 50% of 272p **d** 10% of 40
e 10% of 370p **f** 1% of £700 **g** 50% of 12 kg **h** 50% of £31
i 1% of 200 **j** 1% of 420 m

3 Calculate these percentages using a mental method.

a 10% of £40 **b** 10% of £340 **c** 5% of £120 **d** 20% of 210 km
e 20% of $530 **f** 25% of £300 **g** 20% of £32 **h** 5% of 28 m

4 Calculate these percentages using a mental method.
The first one is started for you.

a 60% of 40

50% of 40 $= \frac{1}{2}$ of 40 = 20
10% of 40 $= \frac{1}{10}$ of 40 = ?
60% of 40 = ?

Use percentages which you can work out quickly in your head.

b 60% of £70 **c** 40% of £30 **d** 40% of 45 m
e 2.5% of £80 **f** 70% of 150 m **g** 15% of 120 kg
h 35% of 400 mm **i** 7.5% of 300 km

5 Write how you would use a mental method to calculate

a 15% of anything **b** 5% of anything **c** 20% of anything
d 90% of anything **e** 11% of anything **f** 17.5% of anything.

17.5% is the VAT rate.

6 Calculate these percentages using a mental or written method.
Show all the steps of your working out.

a 11% of £18 **b** 60% of 7300 km **c** 8% of £30
d 2% of €3000 **e** 7% of 60 m **f** 13% of 40 cm
g 75% of 48 m **h** 3% of £70

7 Rearrange these cards to make three correct statements.

25% of	£450	= £128
30% of	£640	= £130
20% of	£520	= £135

N7.4 Percentage increase and decrease

This spread will show you how to:

- Calculate a percentage of an amount
- Calculate percentage increase and decrease

Keywords
Decrease
Increase
Percentage

You can already calculate a **percentage** of an amount.

See N7.3.

Example

Calculate

a 12% of £50 **b** 12.6% of 320 m

a 12% of £50 $= \frac{12}{100} \times £50$
$= \frac{12 \times £50}{100}$
$= \frac{£600}{100}$
$= £6$

b 12.6% of 320 m $= \frac{12.6}{100} \times 320$
$= 0.126 \times 320$
$= 40.32$ m

This topic is extended to harder percentage increase and decrease on page 368.

Percentages are used in real life to show how much an amount has increased or decreased.

- To calculate a percentage **increase**, work out the increase and add it to the original amount.
- To calculate a percentage **decrease**, work out the decrease and subtract it from the original amount.

Example

a Karen is paid £800 a month. Her employer increases her wage by 6%. Calculate the new wage Karen is paid each month.
b In January a car costs £3400. In February the price is reduced by 17%. What is the new price of the car?

a Increase in wage $= 6\%$ of £800 $= \frac{6}{100} \times £800$
$= \frac{6 \times £800}{100}$
$= \frac{£4800}{100}$
Increase in wage = £48
Karen's new wage = £800 + £48
= £848

- Calculate 6% of Karen's monthly wage.
- Add this to her monthly wage.

b Price reduction $= 17\%$ of £3400 $= \frac{17}{100} \times £3400$
$= 0.17 \times £3400$
Price reduction = £578
New price of car = £3400 − £578
= £2822

Exercise N7.4

1 Calculate these amounts without using a calculator.

a 10% of £400 **b** 10% of 2600 cm **c** 5% of 64 kg
d 25% of 80 m **e** 50% of 380p **f** 5% of £700
g 25% of 12 kg **h** 20% of £31 **i** 15% of 360
j 1% of 720 m **k** 30% of £25 **l** 25% of 444

2 Calculate these percentages, giving your answer to two decimal places where appropriate.

a 45% of 723 kg **b** 25% of $480 **c** 23% of 45 kg
d 21% of 28 kg **e** 17.5% of £124 **f** 34% of 230 m

3 Calculate these percentages, giving your answer to two decimal places where appropriate.

a 4.5% of £320 **b** 2.5% of $4300 **c** 3.6% of 54 kg
d 13.2% of 220 m^2 **e** 4.8% of 245 litres **f** 5.1% of 2050 hectares

4 **a** There are 62 million people living in the UK. 23% of the population are under 18. How many people are under 18?

b The price of a pair of trainers is normally £85. In a sale the price is reduced by 30%. How much cheaper are the trainers in the sale?

c The recommended daily allowance (RDA) of iron is 14 mg. A bowl of cereal provides 45% of the RDA of iron. How much iron is there in a bowl of cereal?

d Joanne earns £78 450 a year. She pays 36% of her earnings in tax. How much money does she pay in tax?

5 Calculate these.

a Increase £450 by 10% **b** Decrease 840 kg by 20%
c Increase £720 by 5% **d** Decrease 560 km by 30%
e Increase £560 by 17.5% **f** Decrease 320 m by 20%

6 **a** A drink can contains 330 ml. The size is increased by 15%. How much drink does it now contain?

b The price of a coat was £95. The price is reduced by 20% in a sale. What is the sale price of the coat?

c A holiday package is advertised in the brochure at a price of £2400. The travel agent reduces the price by 8%. What is the new price of the holiday package?

d A house is bought for £190 000. During the next two years the house increases in price by 23%. What is the new value of the house?

N7.5 Simple interest

This spread will show you how to:

- Solve percentage increase and decrease problems
- Calculate simple interest
- Use checking procedures involving estimation

Keywords
Decrease
Increase
Percentage
Simple interest

Banks and building societies pay interest on money in an account.
The interest is always written as a **percentage**.
For example, 4.5% interest means that the bank pays you an extra 4.5% of the money you put into the bank account.

Example

Jamal puts £750 into a bank account. The bank pays interest of 4.5% on any money he keeps in the account for one year. Calculate the interest Jamal receives at the end of the year.

Estimate: 4.5% of £750 ≈ 5% of £700
= (10% of 700) ÷ 2
= 70 ÷ 2
= £35

Interest = 4.5% of £750 = $\frac{4.5}{100} \times 750$
= 0.045 × 750
= 33.75
= £33.75

Always estimate your answers when working with a calculator.

People sometimes choose to have the interest they earn at the end of each year paid out of their bank account. This is called **simple interest**.

- **To calculate simple interest you multiply the interest earned at the end of the year by the number of years.**

This topic is extended to compound interest on page 368.

Example

Calculate the simple interest on £3950 for

a 4 years at an interest rate of 5%
b 10 years at an interest rate of 8%.

a Interest each year = 5% of £3950
= $\frac{5}{100} \times 3950$
= 0.05 × 3950
= £197.50

Total amount of simple interest after 4 years = 4 × £197.50
= £790

b Interest each year = 8% of £3950
= $\frac{8}{100} \times 3950$
= 0.08 × 3950
= £316

Total amount of simple interest after 10 years = 10 × £316
= £3160

Don't forget to estimate.
5% of £3950
≈ 5% of £4000
= 10% of £4000 ÷ 2
= £400 ÷ 2
= £200

Exercise N7.5

1 Calculate these amounts without using a calculator.

a 10% of £5000 **b** 10% of $260 **c** 5% of £3900

d 25% of €800 **e** 50% of £3.60 **f** 5% of $7500

g 25% of £1240 **h** 20% of £780 **i** 15% of $8430

j 1% of £560 **k** 15% of £230 **l** 25% of €1250

2 Calculate these percentages, giving your answer to two decimal places where appropriate.

a 7% of £3200 **b** 12% of £3210 **c** 27% of €5400

d 3.5% of £2200 **e** 0.3% of €4450 **f** 3.7% of £12 590

3 **a** Jane puts £3700 into a bank account. The bank pays interest of 4% on any money she keeps in the account for one year. Calculate the interest received by Jane at the end of the year.

b Majid puts £30 000 into a savings account. The account pays interest of 8.1% on any money he keeps in the account for one year. Calculate the interest received by Majid at the end of the year.

4 Calculate the simple interest paid on £4580

a at an interest rate of 4% for 3 years

b at an interest rate of 11% for 5 years

c at an interest rate of 4.6% for 4 years

d at an interest rate of 8.5% for 3 years.

5 Calculate the simple interest paid on these.

a An amount of £4500 at an interest rate of 5% for 3 years

b An amount of £8500 at an interest rate of 7% for 5 years

c An amount of £320 at an interest rate of 2.5% for 8 years

d An amount of £3900 at an interest rate of 7.5% for 11 years

6 Calculate these.

a Increase £250 by 10% **b** Decrease £2830 by 20%

c Increase £17 200 by 5% **d** Decrease £3600 by 30%

e Increase £3.60 by 17.5% **f** Decrease £2500 by 20%

7 Calculate these. Give your answer as appropriate to two decimal places.

a Increase £740 by 8% **b** Decrease £39 450 by 32%

c Increase £107.80 by 5% **d** Decrease $8230 by 34%

e Increase $5900 by 4.5%. **f** Decrease £2950 by 17.5%

N7 Exam review

Key objectives

- Calculate a given fraction of a given quantity
- Express a given number as a fraction of another
- Multiply a fraction by an integer, or a unit fraction
- Convert simple fractions of a whole to percentages of a whole and vice versa
- Understand the multiplicative nature of percentages as operators
- Solve simple percentage problems, including increase and decrease

1 Daniel wants to buy a CD.
He finds the CD he is looking for in a shop.
The CD usually costs £12.
In a sale, the CD is now being sold at $\frac{5}{8}$ the original price.

a What is the sale price of the CD? (2)

b By what percentage has the price of the CD been reduced? (2)

2 There are 800 students at Prestfield School.
144 of these students were absent from school on Wednesday.

a Work out how many students were not absent on Wednesday. (2)

Trudy says that more than 25% of the 800 students were absent on Wednesday.

b Is Trudy correct? Explain your answer. (2)

45% of these 800 students are girls.

c Work out 45% of 800. (2)

There are 176 students in Year 10.

d Write 176 out of 800 as a percentage. (2)

(Edexcel Ltd., 2004)

Interpreting diagrams and charts

This unit will show you how to

- Understand the difference between discrete and continuous data
- Interpret data from pictograms and bar charts
- Calculate the modal category of a data set
- Use the shape of comparative bar charts to compare two data sets
- Interpret data from a pie chart
- Interpret data from a stem-and-leaf diagram
- Calculate the mean, mode, median and range from a stem-and-leaf diagram
- Interpret data and trends from time series graphs
- Look at data to find patterns and exceptions

Before you start …

You should be able to answer these questions.

Review

1 Find the value of the angle x. — Unit S2

a

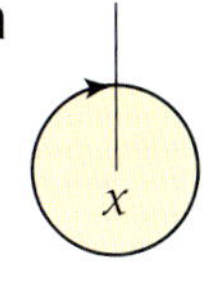

b **c**

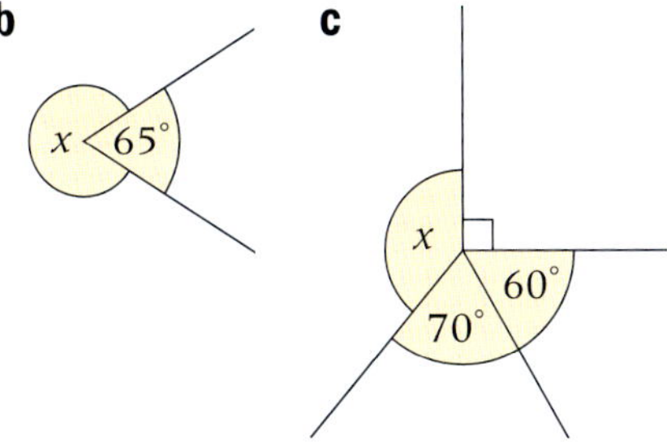

2 Calculate each of these. — Unit N2

a $360 \div 12$ **b** $360 \div 9$

c $360 \div 10$ **d** $360 \div 180$

e $360 \div 90$

3 Find the mean, mode, median and range for each of these sets of numbers. — Unit D4

a 3, 3, 6, 7, 16

b 0, 0, 0, 2, 8, 10, 15

c 3, 3, 4, 6

D5.1 Pictograms and bar charts

This spread will show you how to:

- Understand the difference between discrete and continuous data
- Interpret data from pictograms and bar charts
- Calculate the modal category of a data set

Keywords

Bar chart
Category
Discrete
Modal
Pictogram

- **Discrete** data can only take exact values (usually collected by counting). For example, the number of pets in a home.
- You can interpret information on categories and discrete data from a **pictogram** and a **bar chart**.

The size of each **category** gives the frequency.

Continuous data can have any value, for example, your height.

Example

The pictogram shows the number of students attending school in a week. The class has 20 students.

Monday	○ ○ ○ ○
Tuesday	○ ○ ◖
Wednesday	○ ○ ○
Thursday	○ ◔
Friday	○ ○ ◜

Key: ○ represents 4 students

Calculate

a the number of students that attended on Tuesday

b the number of students that were absent on Thursday.

a $4 + 4 + 2 = 10$ students

b $20 - (4 + 3) = 20 - 7 = 13$ students

Example

The bar chart shows the types of sweets in an assorted box.

Calculate

a the number of chocolate sweets

b the total number of sweets in the box.

c Find the **modal** type of sweet.

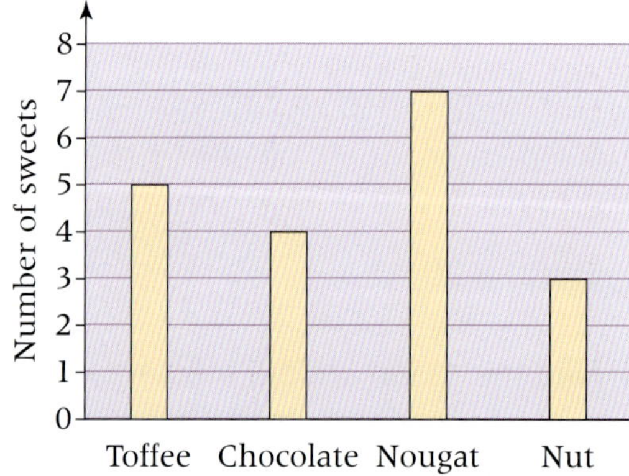

Modal means the most frequent type.

a 4

b $5 + 4 + 7 + 3 = 19$ sweets

c Nougat

Exercise D5.1

1 The numbers of buses that stop at a village through the week are shown on the bar chart.

a How many buses stop at the village on Thursday?

b On which days is there no bus service?

c Calculate the total number of buses that stop at the village throughout the week.

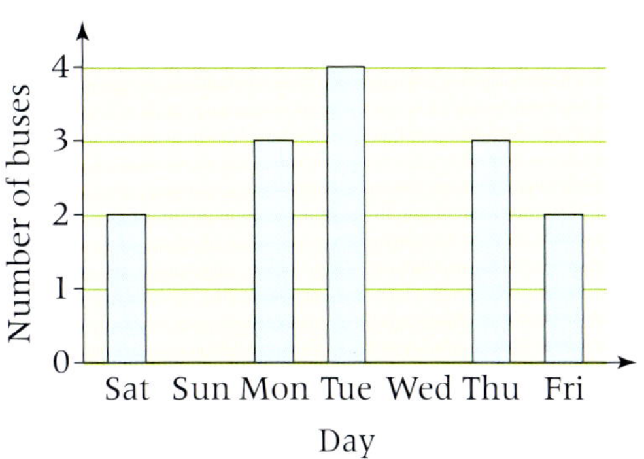

2 The number of foreign language teachers at a school is shown in the pictogram.

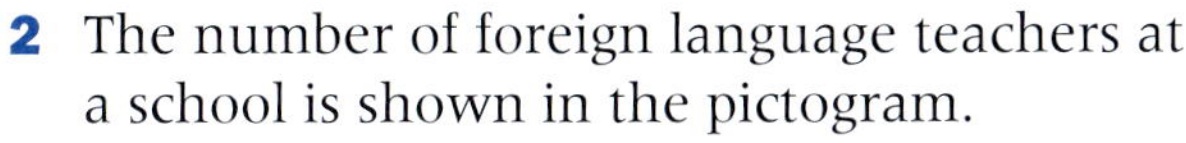

a Which subject has only one teacher?

b How many French teachers teach at the school?

c Calculate the total number of foreign language teachers at the school.

French	
Spanish	
Russian	
German	
Italian	

Key: represents 2 teachers

3 The number of times each racing driver has been the Formula 1 World Champion is shown on the bar chart.

a Who has been the F1 World Champion the most times?

b How many times has Alaïn Prost been the F1 World Champion?

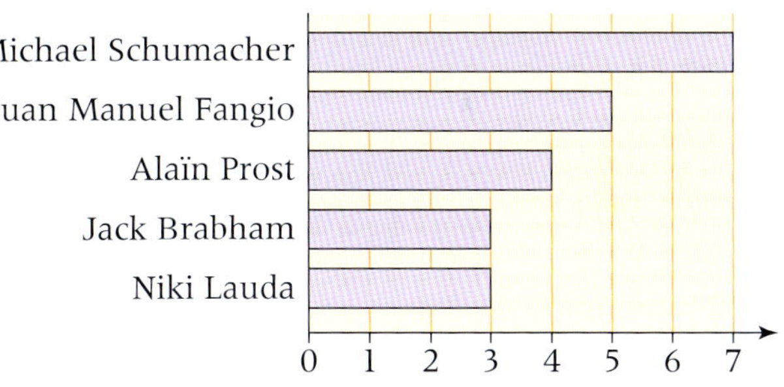

4 Four friends played a series of games. The number of wins for each person is shown in the pictogram. Angela won six games.

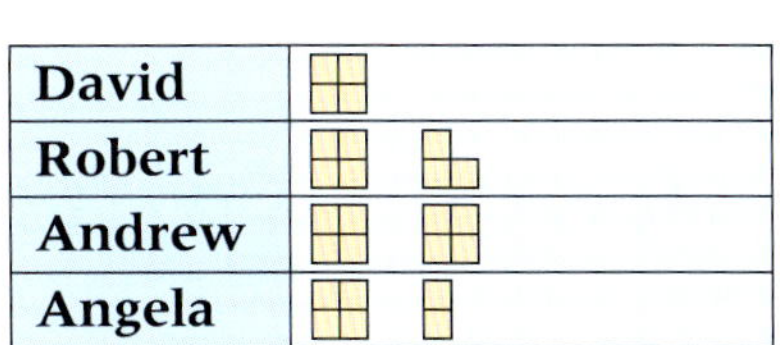

Key: represents ?

a What does ⊞ represent?

b How many games did Robert win?

c Who won the most games and how many did that person win?

d Who won the least games and how many did that person win?

e Calculate the total number of games played.

D5.2 Comparative bar charts

This spread will show you how to:

- Use the shape of comparative bar charts to compare two data sets

Keywords
Bar chart
Comparative bar chart

You can interpret information on categories and discrete data from a **bar chart**.

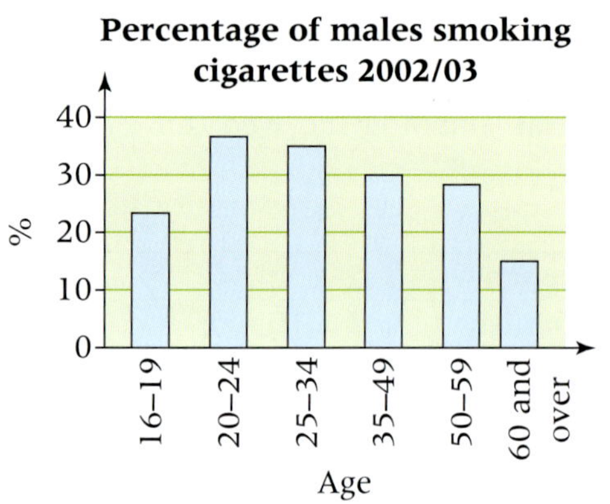

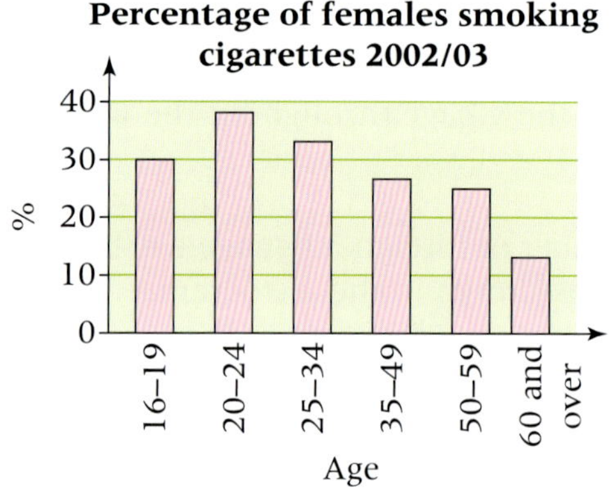

- You can combine two bar charts to create a **comparative bar chart**, which helps you to compare two sets of data.

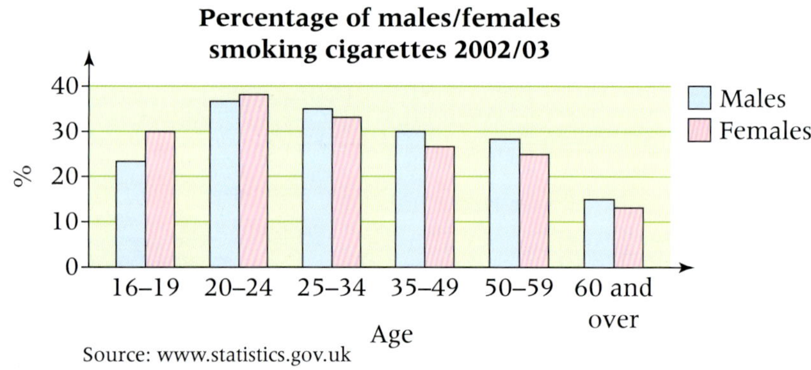

Source: www.statistics.gov.uk

The sets of data you compare must be the same type, for example, Boy/Girl.

Example

The chart shows the sales of pork pies one weekend at two rival butchers.

Calculate

a the number of pork pies that Billy sold on Sunday

b the total number of pork pies that were sold by both butchers.

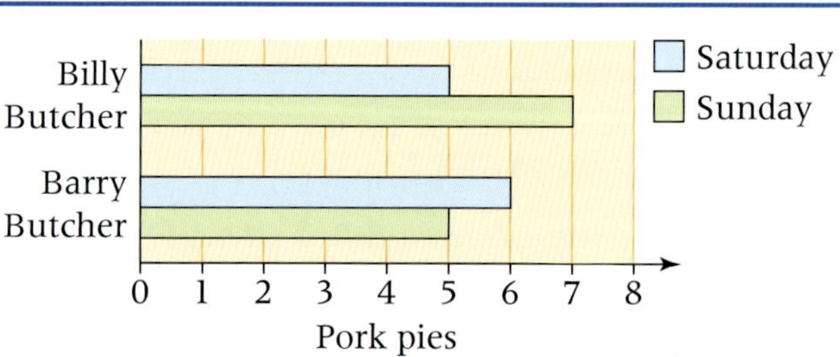

a 7 pork pies

b $5 + 7 + 6 + 5 = 23$ pork pies

Exercise D5.2

1 The number of letters delivered one week to No 10 and No 12 on the same street is shown on the bar chart.

a State the number of letters that were delivered to

i No 10 on Tuesday

ii No 12 on Friday.

b Which day did

i No 12 receive no letters

ii No 10 receive 2 letters?

c Which was the only day that No 12 received more post than No 10?

d Calculate the total number of letters delivered in the week to

i No 10

ii No 12.

e Which day did the postman deliver the most letters to No 10 and No 12 combined?

2 The number of visits to the swimming baths over two weeks for Daniel and Emma is shown on the bar chart.

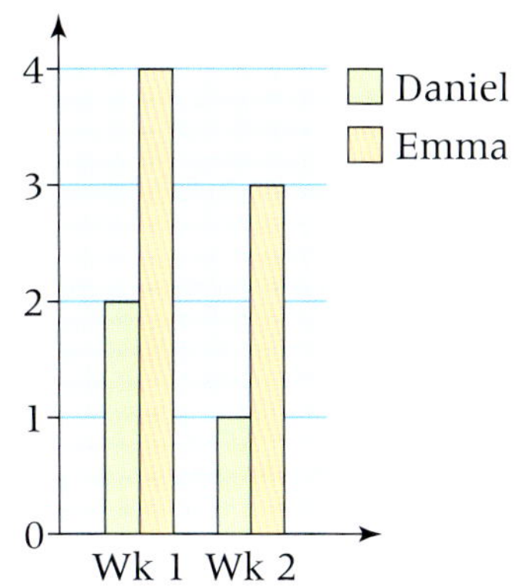

a State the number of visits for

i Emma in Week 1

ii Daniel in Week 2.

b Calculate the total number of visits for

i Daniel

ii Emma.

3 The bar chart shows the number and flavour of milkshakes that Mark and Peter drank.

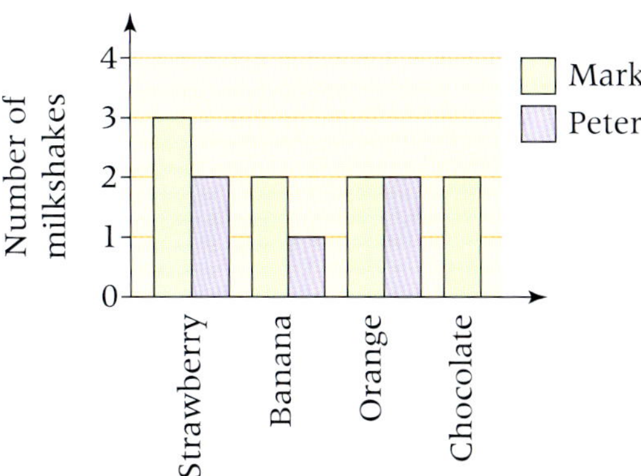

a State the number of

i banana milkshakes drunk by Mark

ii orange milkshakes drunk by Peter.

b Which flavour milkshake was not drunk by Mark?

c Which flavour milkshake did Mark and Peter drink the same number of?

d Calculate the total number of milkshakes drunk by

i Mark

ii Peter.

D5.3 Pie charts

This spread will show you how to:

- Interpret data from a pie chart
- Calculate the modal category of a data set

Keywords
Category
Modal
Pie chart
Proportion
Sector

- You can interpret data from a **pie chart**.

A pie chart does not show the actual number of items in each **category**.

A pie chart does show the size of each category compared to the total number of items.

Drinks sold at a machine

Cappuccino is the biggest **sector**.

The pie chart shows the **proportion** of drinks bought at a vending machine.
More cappuccinos were sold compared to the other categories, but we do not know how many drinks were sold.

Example

60 vehicles are shown on the pie chart.

a Calculate the numbers of cars, vans, buses and lorries.

b State the modal type of vehicle.

Vehicles Parked in the High St.

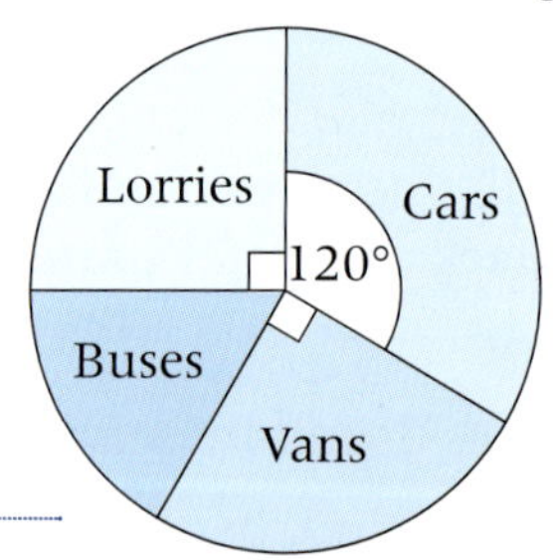

a 60 vehicles = 360°
1 vehicle = 6°
The angle for Buses is 360° − (90° + 90° + 120°) = 60°

Cars	120 ÷ 6 = 20 cars
Vans	90 ÷ 6 = 15 vans
Buses	60 ÷ 6 = 10 buses
Lorries	90 ÷ 6 = 15 lorries

Check: 20 + 15 + 10 + 15 = 60 vehicles

b The modal type of vehicle is Car.

The angles at a point add to 360°.

Modal means the most frequent type.

Exercise D5.3

1 The survey results for the favourite band of 100 people are shown.

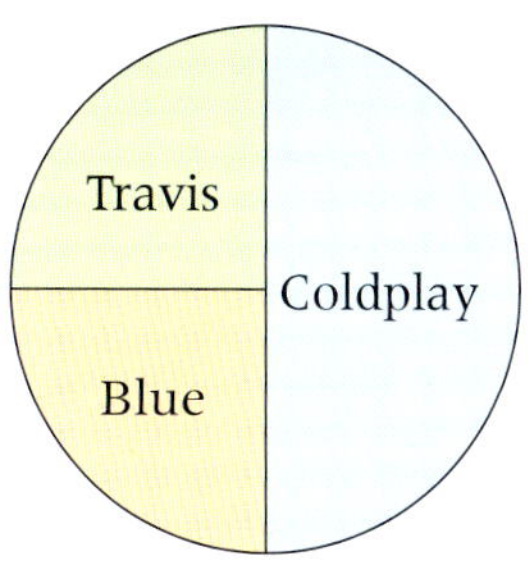

a What fraction of the circle represents

i Coldplay **ii** Travis **iii** Blue?

b Calculate the number of people who voted for

i Coldplay **ii** Travis **iii** Blue.

c State the modal band.

2 Six drinks of soup are represented on the pie chart.

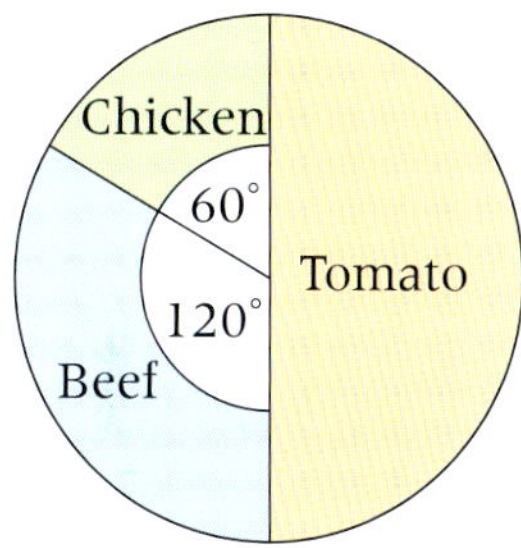

a Copy and complete 6 drinks = 360°
1 drink = ___°

b Calculate the number of drinks that are

i tomato **ii** chicken **iii** beef.

c State the modal type of drink.

3 A shop sells 12 loaves of three different types: organic, wholegrain and white.

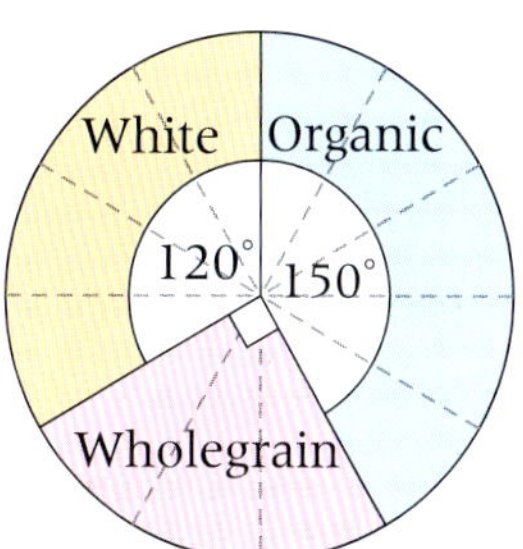

a Copy and complete 12 loaves = 360°
1 loaf = ___°

b Calculate the number of loaves that are

i organic **ii** wholegrain **iii** white.

c State the modal type of loaf.

4 Nine newspapers delivered on a street are shown on the pie chart.

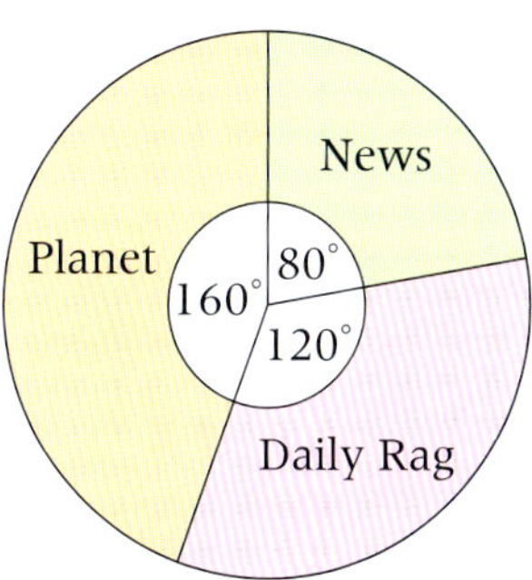

a Copy and complete 9 newspapers = 360°
1 newspaper = ___°

b Calculate the number of newspapers delivered that are

i the Planet **ii** the News **iii** the Daily Rag.

c State the modal newspaper.

5 A car dealer sells 18 cars in one week of three different types: diesel, petrol and electric.

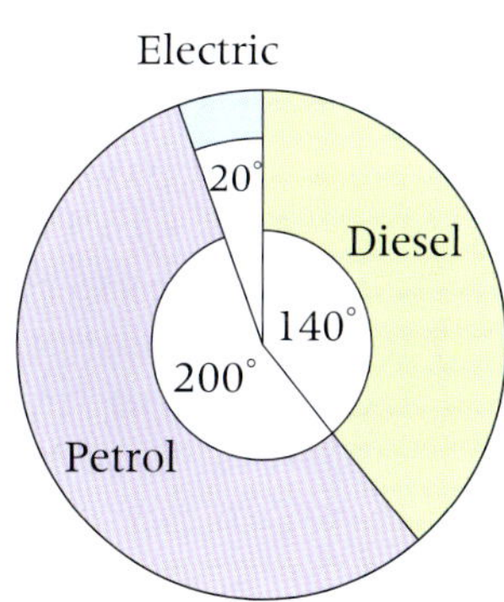

a Calculate the angle that represents one car.

b Calculate the number of cars sold that are

i diesel **ii** petrol **iii** electric.

c State the modal type of car sold.

D5.4 Stem-and-leaf diagrams

This spread will show you how to:

- Interpret data from a stem-and-leaf diagram
- Calculate the mean, mode, median and range from a stem-and-leaf diagram

Keywords
Ordered
Stem-and-leaf diagram

- You can interpret numerical data from a **stem-and-leaf diagram**.

Hannah recorded the lengths in mm of the leaves that fell off an oak tree one October day.

The numbers in her stem-and-leaf diagram are

104, 105, 108, 112, 112, 114, 115, 120, 124, 127, 132, 138, 141, 142, 147.

stem	leaf
140	1 2 7
130	2 8
120	0 4 7
110	2 2 4 5
100	4 5 8

This means 147.

This means 120.

Key: 110 | 4 means 114 mm

Always give the key.

Her stem-and-leaf diagram is **ordered** as the data is in numerical order.

- **You can calculate the mean, mode, median and range from the stem-and-leaf diagram.**

Example

The ages of nine people in a judo club are shown in the diagram.

Calculate

a the mean
b the mode
c the median
d the range.

40	6
30	2 4 5 5
20	7 8
10	1 3

Key: 20 | 7 means 27

a Mean = (11 + 13 + 27 + 28 + 32 + 34 + 35 + 35 + 46) ÷ 9
= 261 ÷ 9 = 29
b Mode = 35, the most common age
c Median = 32, the middle age when arranged in order
d Range = 46 − 11=35 years (highest age minus lowest age)

Exercise D5.4

1 Emma takes a spelling test every week. Her marks are shown in the stem-and-leaf diagram.

0	9
10	1 2 3 3 4 4 5 7 7
20	0 0 0

Key: | 10 | 5 | means 15 marks

a Write out her 13 scores in numerical order, smallest first.

b Calculate

i the mean **ii** the mode **iii** the median **iv** the range.

2 The cost, in pence, of various types of sweets are

38 44 48 20 29 37 29 39 40

a Copy and complete the stem-and-leaf diagram.

20	
30	
40	

Key: | 20 | 9 | means 29p

b Redraw your table to give an ordered stem-and-leaf diagram.

c Calculate

i the mean **ii** the mode **iii** the median **iv** the range.

3 The weights, in kilograms, of seven students are shown in the stem-and-leaf diagram.

40	3 5 5
50	0 6
60	3 9

Key: | 40 | 3 | means 43 kg

Calculate

a the mean **b** the mode **c** the median **d** the range.

4 The reaction times, in seconds, for 15 people are shown in the stem-and-leaf diagram.

9	9
10	0 1 1 1 2 3 5 7
11	8 9
12	0 7 7

Key: | 10 | 5 | means 10.5 seconds

Calculate

a the mean **b** the mode **c** the median **d** the range.

D5.5 Time series graphs

This spread will show you how to:

- Interpret data and trends from time series graphs
- Look at data to find patterns and exceptions

Keywords
Horizontal
Line graph
Time series graph
Trend

- You can interpret numerical data from a **time series graph** as it changes with time.

The cost of renting a cottage, month-by-month, is shown in this **line graph**.

Time is always on the **horizontal** axis.

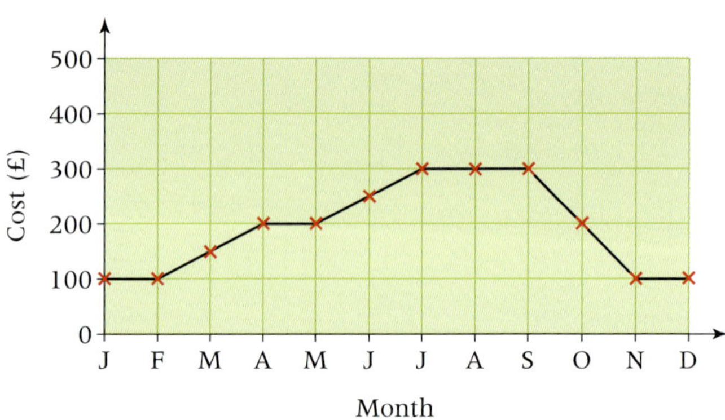

- You can see the **trend** from the graph.

  Prices increase to a peak in July, August and September, then decrease in the winter months.

Example

The average attendance, to the nearest 1000, at UK motorbike speedway matches is shown in the line graph.

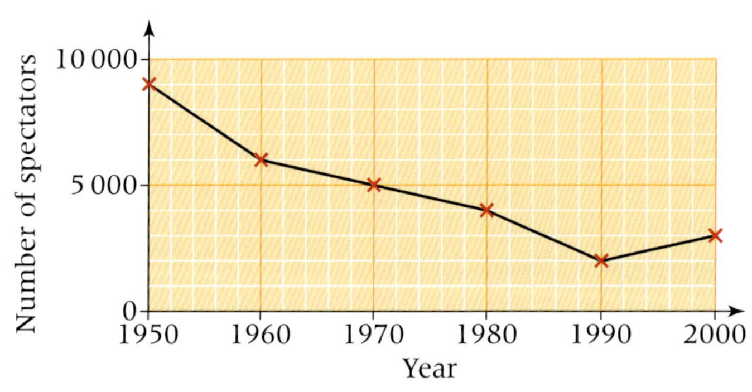

a What was the maximum average attendance?
b When was the lowest average attendance?
c What was the average attendance in 1970?
d Describe the trend of the attendances.

a 9000 spectators
b 1990
c 5000 spectators
d The trend is down until 1990, when there is an increase.

Exercise D5.5

1 The line graph shows the temperature measured at midday daily for one week.

a State the day when the midday temperature was

i 18 °C **ii** 22 °C **iii** 14 °C.

b Which day was the midday temperature

i the coldest **ii** the hottest?

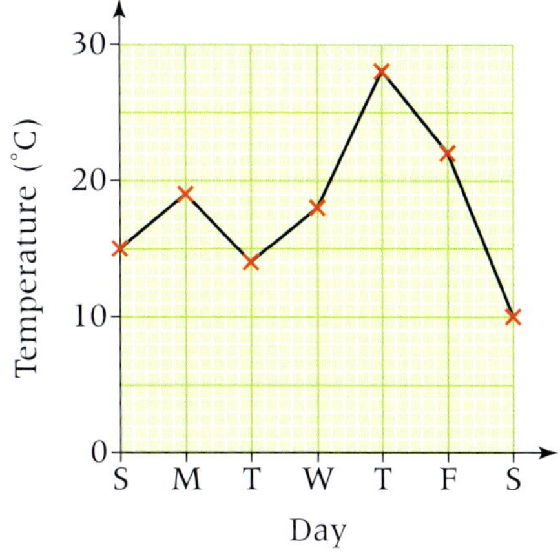

2 The weight, in grams, of an octopus is shown on the graph.

a Give the weight of the octopus after one month in

i grams **ii** kilograms.

b State the weight of the octopus after three months.

c When does the octopus weigh 1.5 kg?

d When is the steepest increase in weight?

e When does the octopus first reach its greatest weight?

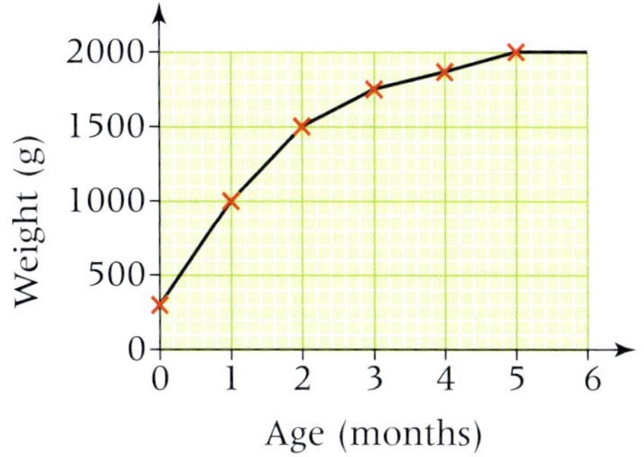

3 The graph shows the temperature of an oven.

a State the maximum temperature of the oven.

b How many minutes did it take to reach the maximum temperature?

c What was the temperature after 10 minutes?

d How long did it take for the oven temperature to reach 100 °C?

e Calculate the rise in temperature from the 5th to the 10th minute.

f In what time interval was the steepest rise in temperature?

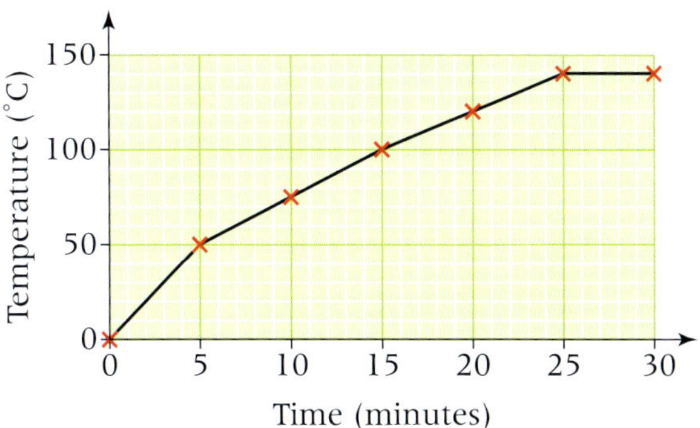

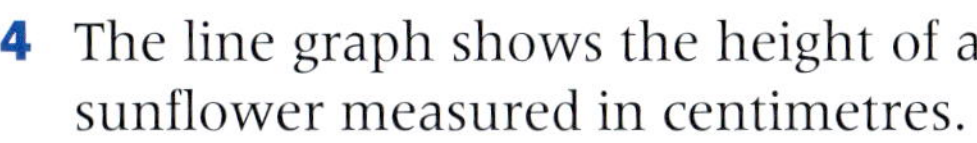

4 The line graph shows the height of a sunflower measured in centimetres.

a State the height after

i Week 3 **ii** Week 4.

b When was the height 70 cm?

c What do you think happened during Week 6?

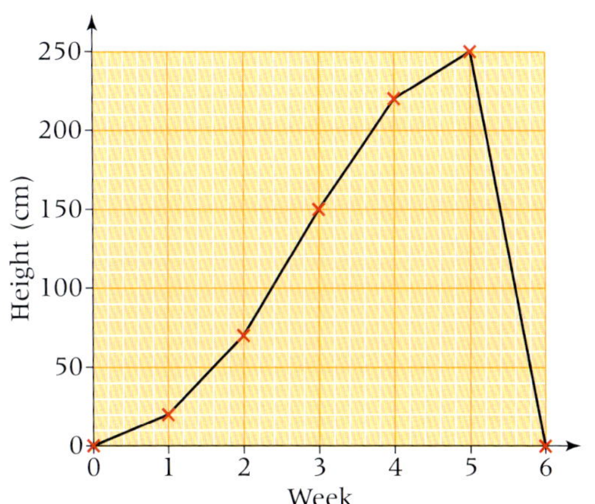

D5 Exam review

Key objectives

- Calculate mean, range and median of small data sets with discrete data
- Interpret a wide range of graphs and diagrams and draw conclusions
- Look at data to find patterns and exceptions
- Compare discrete distributions and make inferences, using the shapes of distributions and measures of average and range

1 Lucy does a survey to find out the most popular flavour of ice cream among the students in her class.
The table shows the results:

Flavour	Frequency
Strawberry	6
Chocolate	10
Vanilla	6
Mint	3
Other	5

a What is the modal flavour? (1)

b Draw an accurate pie chart to represent Lucy's data. (4)

2 Shirin recorded the number of students late for school each day for 21 days.
The stem and leaf diagram shows this information.

Number of students late

1	4	5	7	8	8	9				
2	2	2	5	6	6	7	7	9	9	9
3	0	1	3	4	6					

Key: | 1 | 4 | means 14 students late

a Find the median number of students late for school. (1)

b Work out the range of the number of students late for school. (1)

(Edexcel Ltd., 2003)

S6 Perimeter, area and volume

This unit will show you how to

- Use the vocabulary associated with circles
- Calculate the circumference and area of a circle
- Calculate the area of rectangles and triangles and shapes made from rectangles and triangles
- Find the surface area of cuboids
- Find the length of a side of a cuboid, given the surface area.
- Calculate the volume of cuboids
- Understand units of measure and the dimensions they represent

Before you start ...

You should be able to answer these questions.

1 Give the mathematical name for these distances in a circle. — Review: Unit S4

a

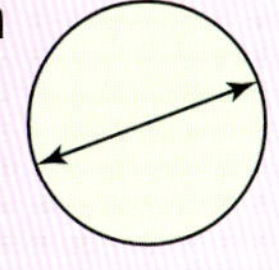

b

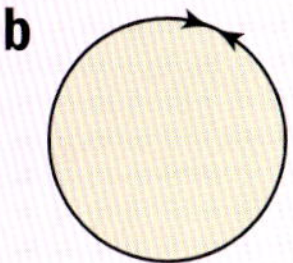

c

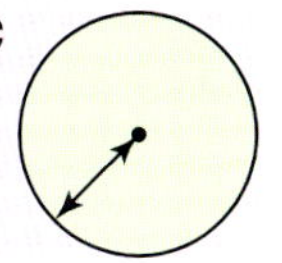

2 Calculate the area of this rectangle. State the units of your answer. — Review: Unit S1

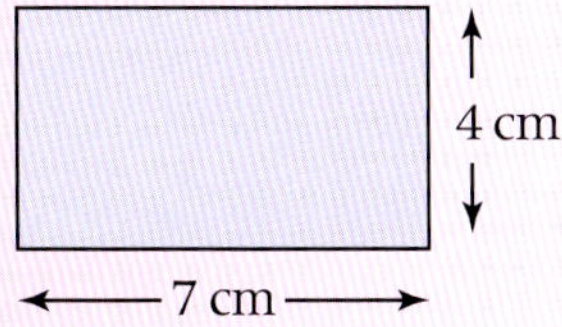

3 Match each quantity with its units: — Review: Unit S1

a Length   **i** cm^2

b Area   **ii** cm^3

c Volume   **iii** cm

S6.1 Circumference and area of a circle

This spread will show you how to:

- Use the vocabulary associated with circles
- Calculate the circumference and area of a circle

Keywords
Centre
Circle
Circumference
Diameter
Pi (π)
Radius

In a **circle**:

- the **radius** is r
- the **diameter** is d
- the **circumference** is C.

C, d and r are all measures of length.

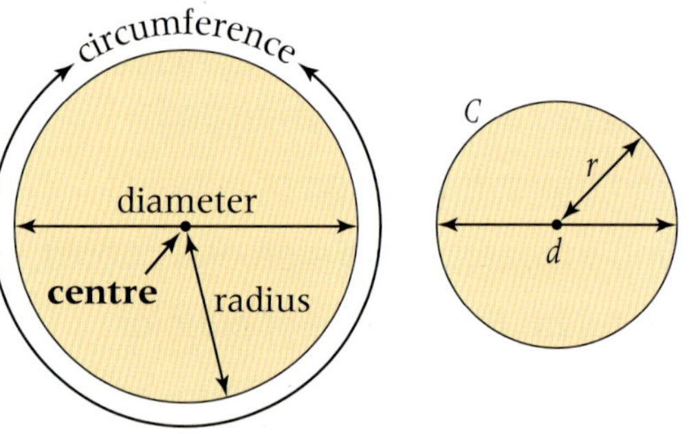

The perimeter of a circle is called the circumference.

- Diameter = 2 × radius
- $C = \pi \times \text{diameter} = \pi d = 2\pi r$

$d = 2 \times r$

$\pi = 3.14 \ldots$

Example

Calculate the circumference of this circle.

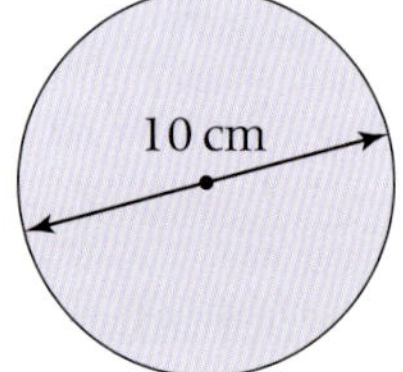

$C = \pi \times d$
$= 3.14 \times 10$
$= 31.4$ cm   Remember to state the units.

- Area of a circle = π × radius × radius
  $= \pi \times r \times r$ or πr^2

r^2 means $r \times r$

Example

A circular lawn has radius 3 metres.

a Calculate the area of the lawn. State the units of your answer.

b Calculate the length of edging stones needed to fit all round the edge of the lawn.
Give your answer to a suitable degree of accuracy.

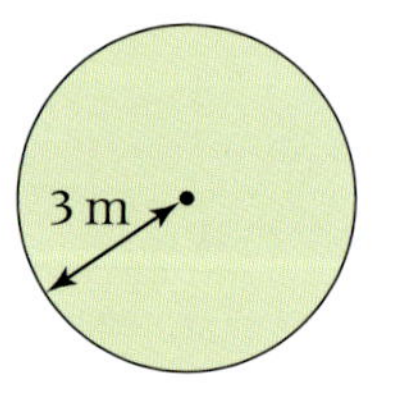

a Area $= \pi r^2$
$= 3.14 \times 3 \times 3$
$= 3.14 \times 9$
$= 28.26\ \text{m}^2$

b Circumference $= \pi d$
$= 3.14 \times 6$
$= 18.84$

So 19 m of edging stones are needed.

Area is measured in square units.

Exercise S6.1

Take $\pi = 3.14$ for all questions on this page.

1 Calculate the circumferences of these circles. State the units of your answers.

a diameter = 10 cm

b diameter = 8 m

c diameter = 12 cm

d 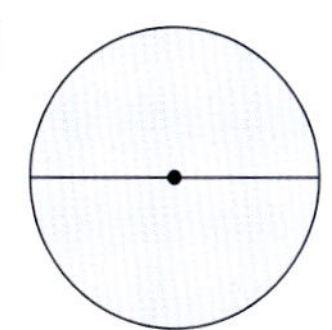

diameter = 20 m

e 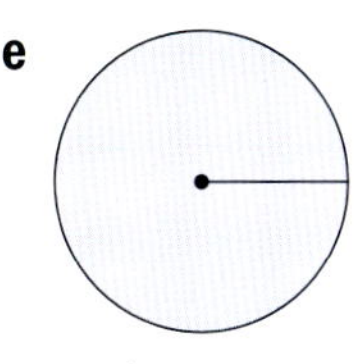

radius = 2 m

f 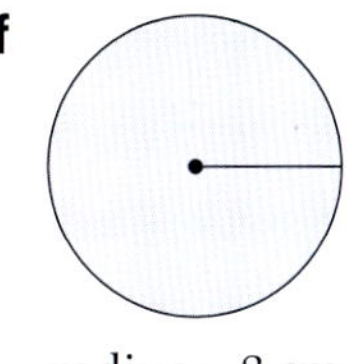

radius = 8 cm

g 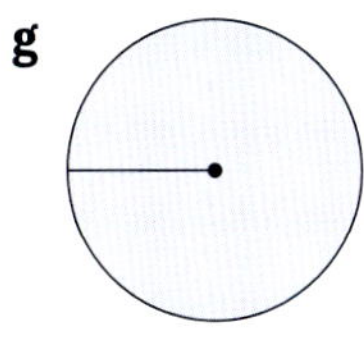

radius = 1.5 m

h 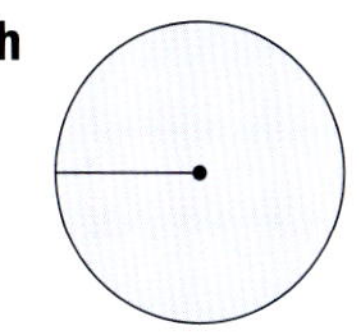

radius = 3.5 cm

2 Calculate the diameter of a circle, if its circumference is

a 18.84 cm **b** 15.7 m **c** 28.26 cm **d** 47.1 m **e** 314 cm

DID YOU KNOW?

The world's largest tyre, in Michigan, USA, has a diameter of 24.4 m. That's a 76.6 m circumference!

3 Calculate the areas of these circles. State the units of your answers.

a radius = 7 cm

b 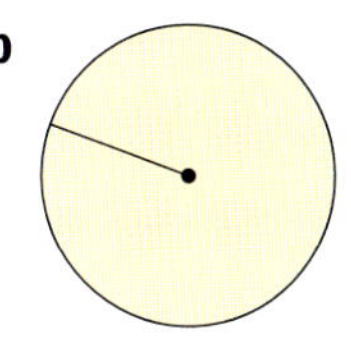

radius = 5 m

c 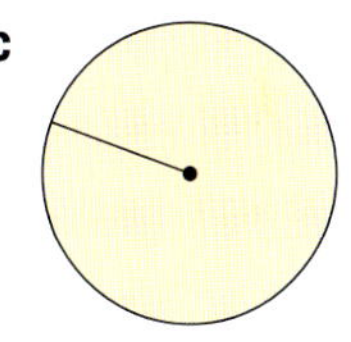

radius = 4 cm

d 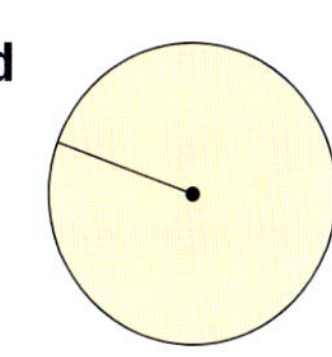

radius = 3 m

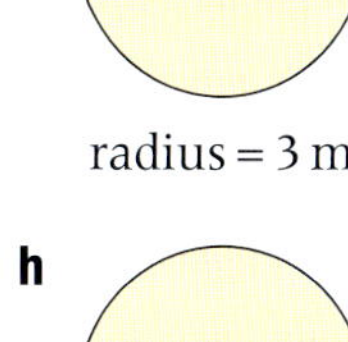

e 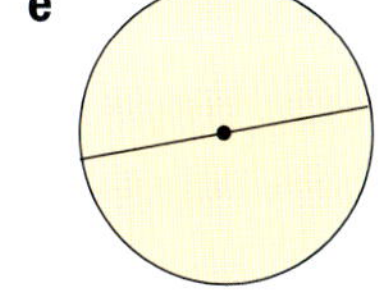

diameter = 20 m

f 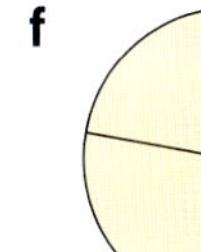

diameter = 16 cm

g 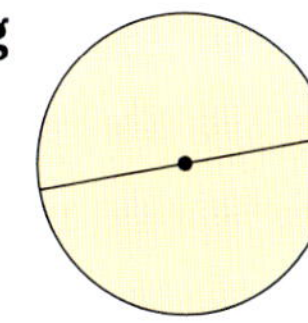

diameter = 12 mm

h 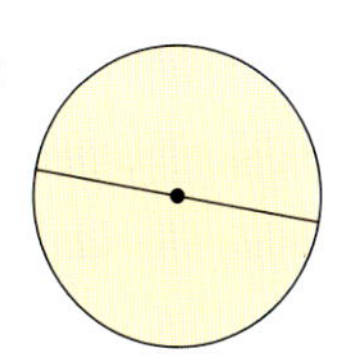

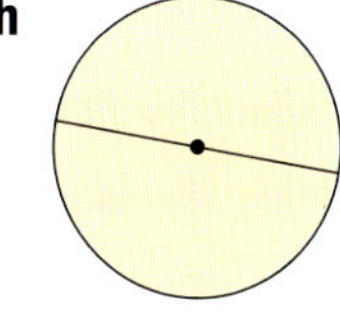

diameter = 18 cm

4 A garden pond is circular.
The radius of the pond is 1.5 m.

a Calculate the diameter of the pond.

b Calculate the circumference of the pond.

c Calculate the area of the pond.

Give your answers to a suitable degree of accuracy.

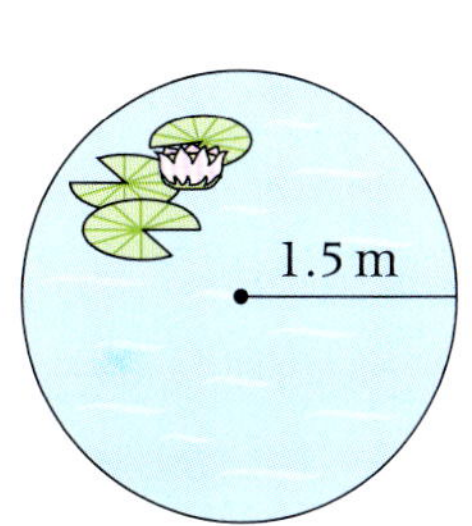

S6.2 Area of a rectangle and a triangle

This spread will show you how to:

- Calculate the area of rectangles and triangles and shapes made from rectangles and triangles

Keywords
Area
Diagonal
Perimeter
Rectangle
Right-angled triangle
Square centimetre

- **The area is the amount of surface a shape covers.**

You can find the area of a **rectangle** using the formula

- **Area of rectangle = length × width**

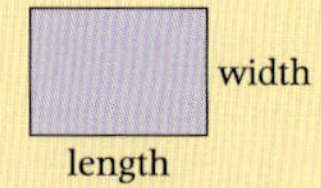

This formula also works for a square.

You can find the area of a **right-angled triangle** in several ways

Area = 6 squares and 4 half squares
$= 6 + 2$
$= 8 \text{ cm}^2$

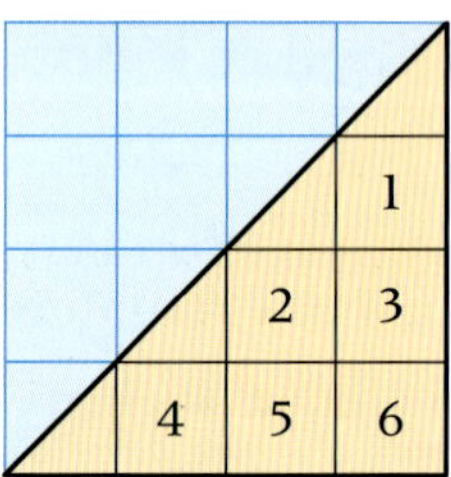

The **diagonal** line splits the square into two halves.

Area $= \frac{1}{2}$ of the area of the square
$= \frac{1}{2}$ of 16 cm^2
$= 8 \text{ cm}^2$

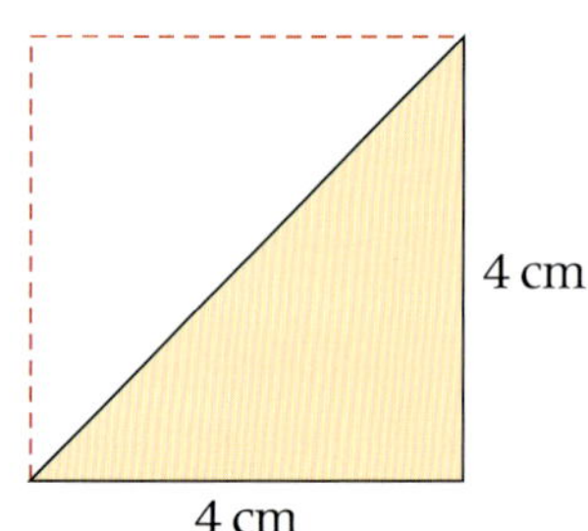

Area of the triangle is half the area of the square.

- **Area of a right-angled triangle = $\frac{1}{2}$ × base × height**

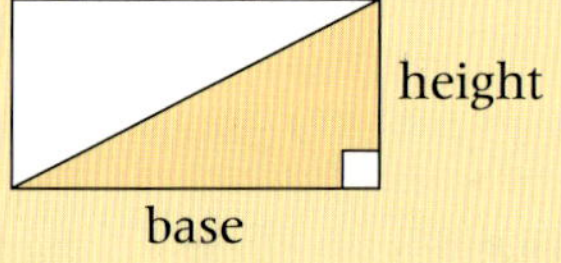

The height is perpendicular (at right angles) to the base.

Example

Calculate the area of the triangle.
State the units of your answer.

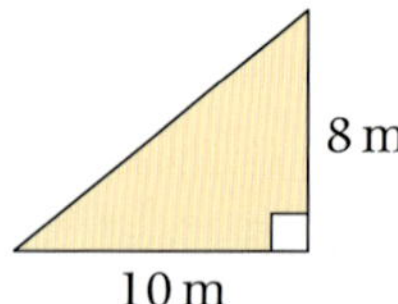

This topic is extended to the area of a parallelogram and trapezium on page 382.

Area $= \frac{1}{2} \times 10 \times 8$
$= 40 \text{ m}^2$

The units are square metres or m^2.

Exercise S6.2

1 Find the area of each of these triangles. Each square represents one square centimetre.

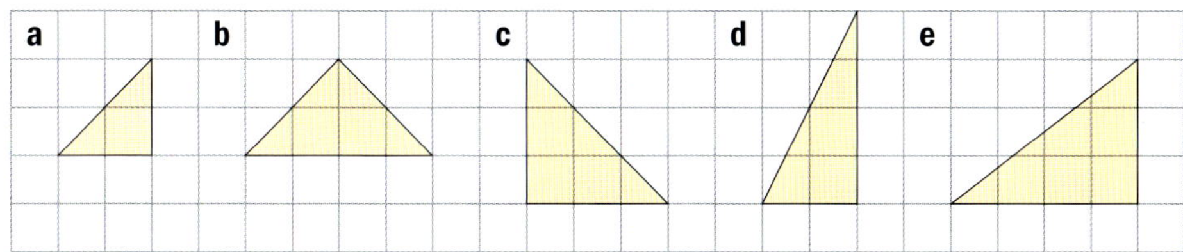

2 Calculate the area of each of these right-angled triangles. State the units of your answers.

a 4 cm, 8 cm
b 10 m, 20 m
c 3 cm, 6 cm
d 8 m, 16 m
e 10 cm, 20 cm

3 Calculate the area of each of these triangles. State the units of your answers.

a 10 cm, 8 cm
b 4 m, 8 m
c 6 cm, 6 cm
d 8 mm, 9 mm
e 7 m, 7 m

4 **a** Calculate the perimeter of this triangle.

b Calculate the area of the triangle.

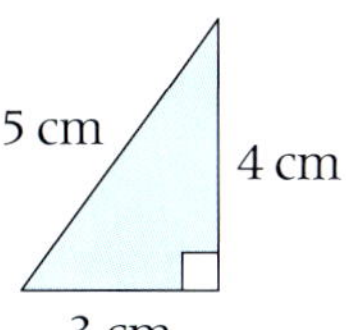

c Two of these triangles are placed together to form five different shapes.

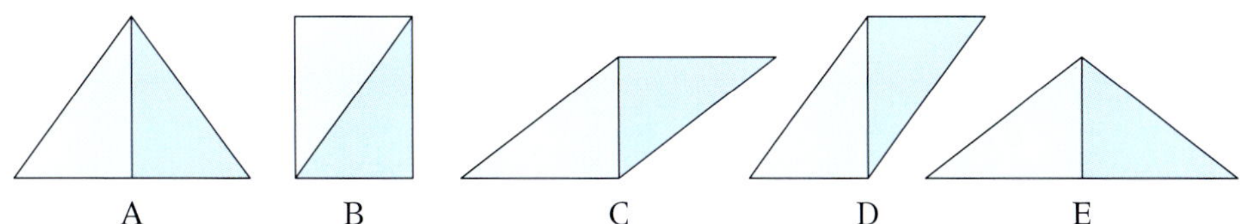

Copy and complete this table.

Shape	Perimeter (cm)	Area (cm^2)
A		
B		
C		
D		
E		

S6.3 Surface area

This spread will show you how to:

- Find the surface area of cuboids
- Find the length of a side of a cuboid, given the surface area

Keywords
Cuboid
Faces
Net
Surface area

A **cuboid** has six rectangular **faces**.

A cereal box is a cuboid.

You can unfold the cereal box to see its net.

When you unfold the cuboid, the six rectangles form the **net**.
The area of the net gives you the **surface area** of the cuboid.

- **The surface area of a cuboid is the total area of its faces.**

Example

Calculate the surface area of this cuboid.
State the units of your answer.

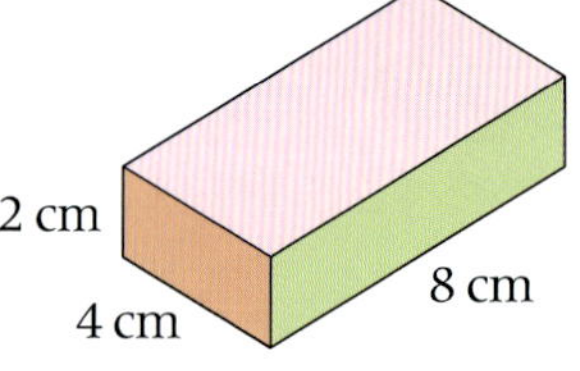

There are two green rectangles, two red rectangles and two pink rectangles.

Area of one red rectangle $= 2 \times 4 \quad = 8\ \text{cm}^2$
Area of one pink rectangle $= 4 \times 8 = 32\ \text{cm}^2$
Area of one green rectangle $= 2 \times 8 = 16\ \text{cm}^2$
$56\ \text{cm}^2$

Total surface area $= 56 \times 2$
$= 112\ \text{cm}^2$

Units of area are cm^2.

- **You can find the length of a side, given the surface area of a cube.**

Example

The surface area of a cube is $150\ \text{cm}^2$.
Calculate the length of one side of the cube.

A cube has six square faces.

The area of one square $= 150 \div 6$
$= 25\ \text{cm}^2$

Length of one side $= \sqrt{25}$
$= 5\ \text{cm}$

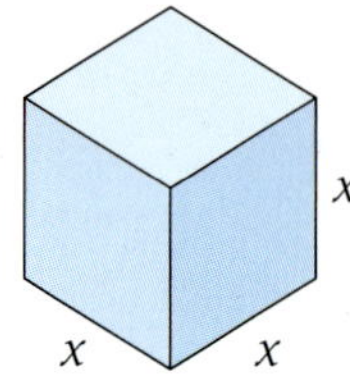

Area $25\ \text{cm}^2$, sides x, x

Exercise S6.3

1 These nets make cuboids. Each square represents a 1 cm square.

a **b** **c** **d**

Calculate the surface area of each cuboid.
State the units of your answers.

2 A 3 cm by 4 cm by 5 cm cuboid is shown. Calculate

a the area of the red rectangle

b the area of the orange rectangle

c the area of the green rectangle

d the surface area of the cuboid.

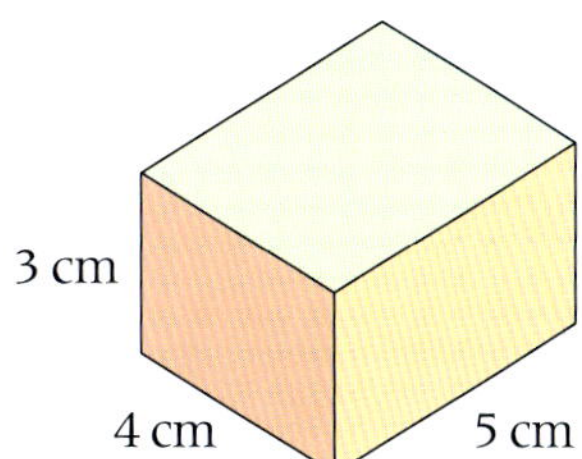

3 Calculate the surface area of each of these cuboids.
State the units of your answers.

a

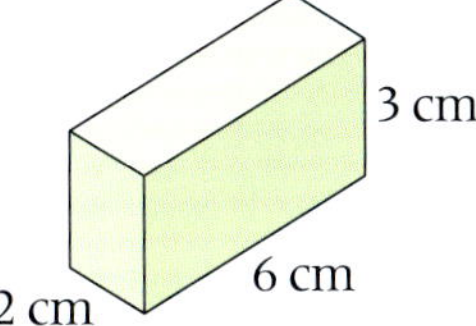

b

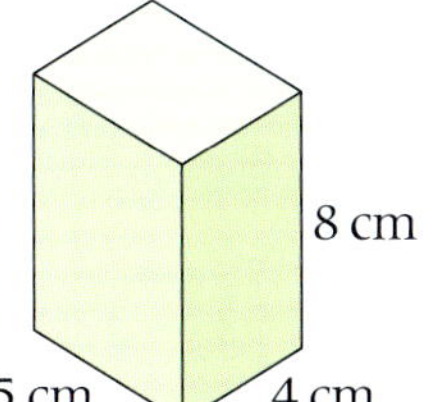

c

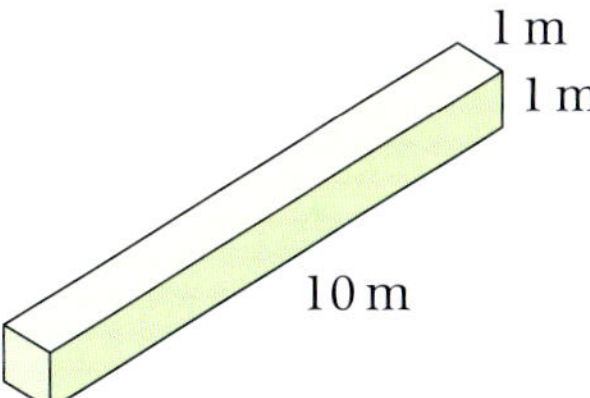

d

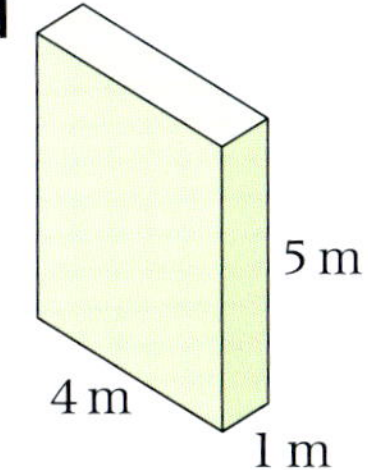

e

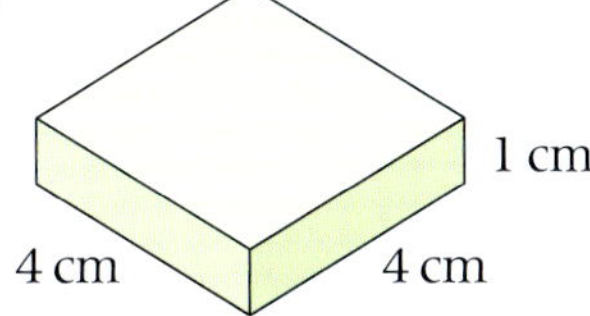

f

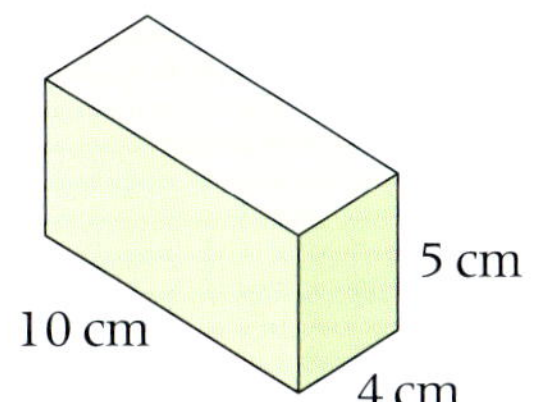

4 Calculate the surface area of each of these cubes.
State the units of your answers.

a length 5 cm **b** length 8 m **c** length 2.5 cm

d length 15 mm **e** length 0.5 m

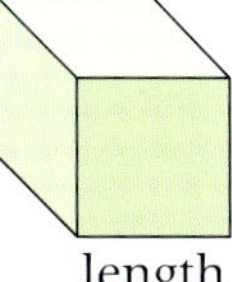

5 Calculate the length of one side of a cube if the surface area of the cube is

a 600 cm^2 **b** 54 cm^2 **c** 294 cm^2

d 9600 cm^2 **e** 37.5 cm^2

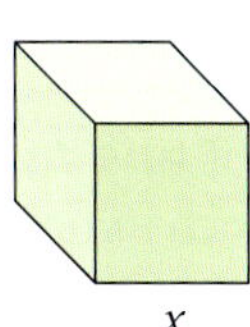

S6.4 Volume

This spread will show you how to:

- Calculate the volume of cuboids

Keywords
Cube
Cubic centimetre (cm^3)
Cubic metre (m^3)
Cuboid
Volume

- **The volume of a 3-D shape is the amount of space it takes up.**

You measure volume using **cubes**.

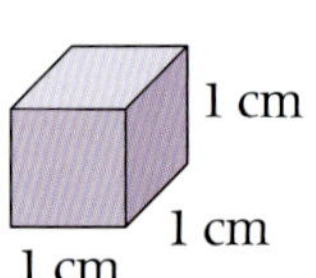

One **cubic centimetre** is 1 cm^3.

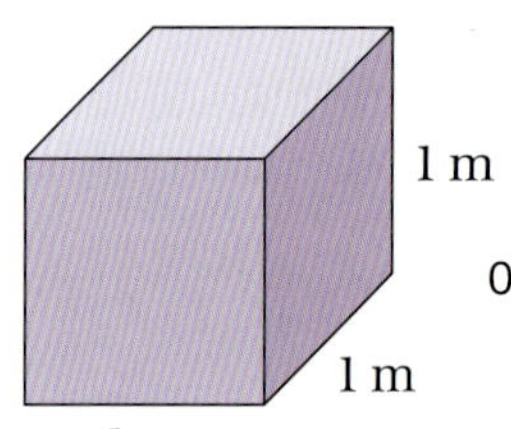

One **cubic metre** is 1 m^3.

The ^{3} in cm^3 shows there are 3 dimensions.

Example

Find the volume of these shapes made from centimetre cubes.
State the units of your answers.

a

b

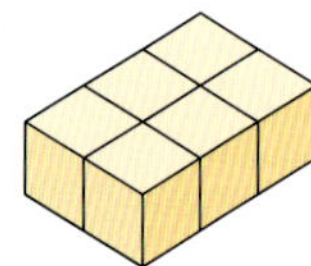

a Volume $= 5\ cm^3$

b Volume $= 6\ cm^3$

The volume of a **cuboid** can be found by counting the number of layers.

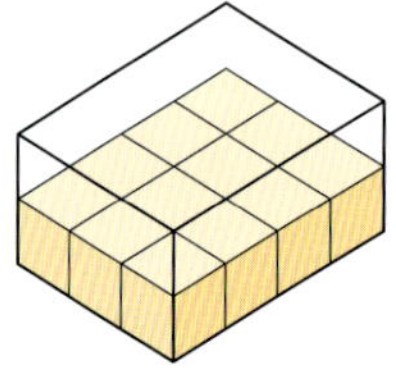

On the bottom layer, there are $3 \times 4 = 12$ cubes.

For 2 layers, there are $2 \times 12 = 24$ cubes.

- **Volume of a cuboid = length × width × height**

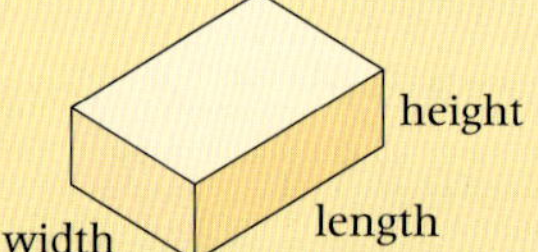

Example

Calculate the volume of this cuboid.
State the units of your answer.

Volume $= 2 \times 4 \times 10$
$= 80\ m^3$

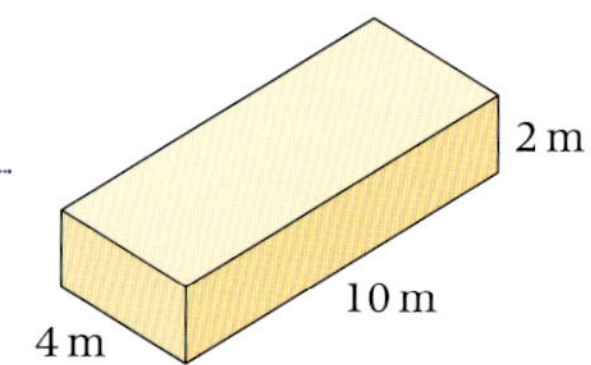

The units are cubic metres or m^3.

Exercise S6.4

1 **a** Calculate the volume, for each solid.
Each cube represents 1 cm³.

i 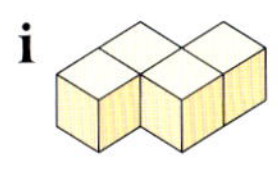
ii 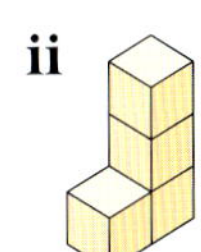
iii 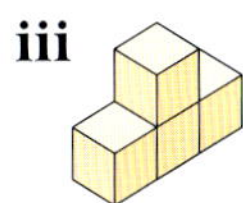
iv 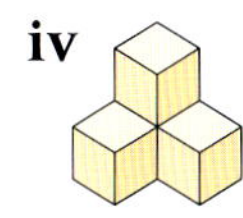
v 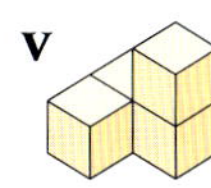
vi 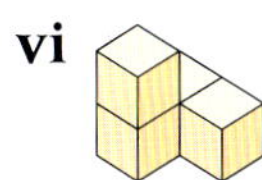
vii 

b All these solids fit together to make a cube.
Find the volume of the cube.

c What are the dimensions of the cube?

2 Calculate the volume, in cm³, of each cuboid.

a

b 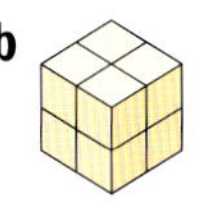

c 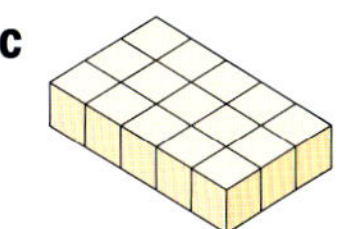

d 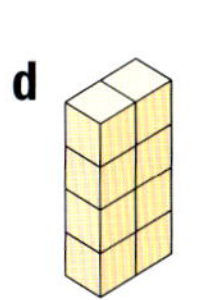

e 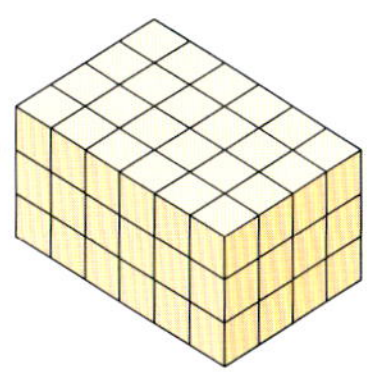

f 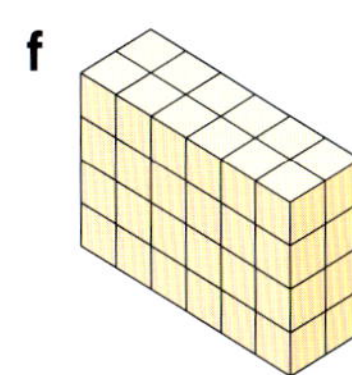

g 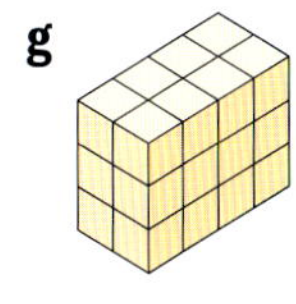

h 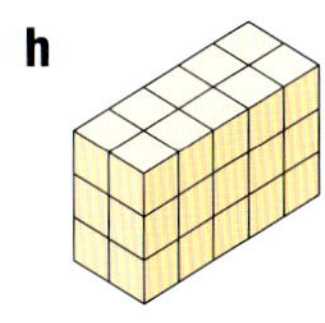

i 

j 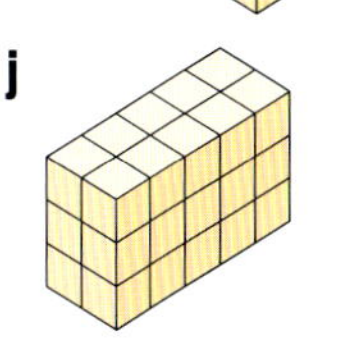

3 Calculate the volume of each cuboid. State the units of your answers.

a 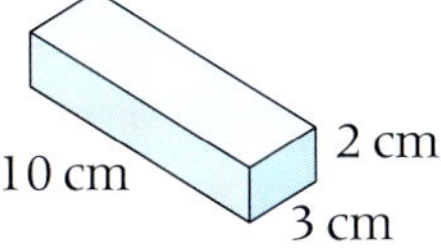

b 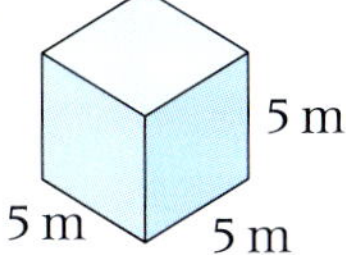

c 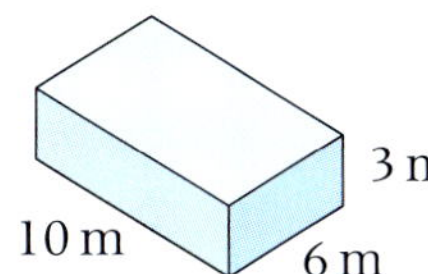

d 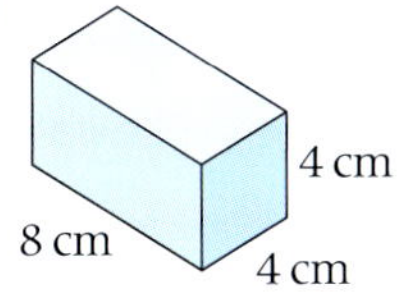

e 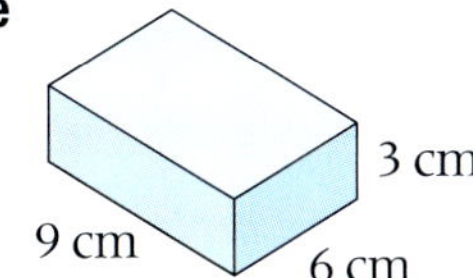

f 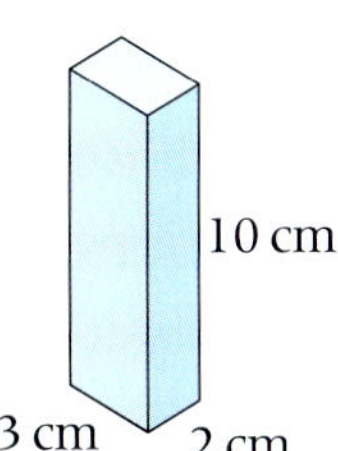

4 A **prism** has a constant cross-section.
The cross-section of this prism is an L-shape.

- **Volume of a prism = area of cross-section × length**

Calculate the volume of this prism.

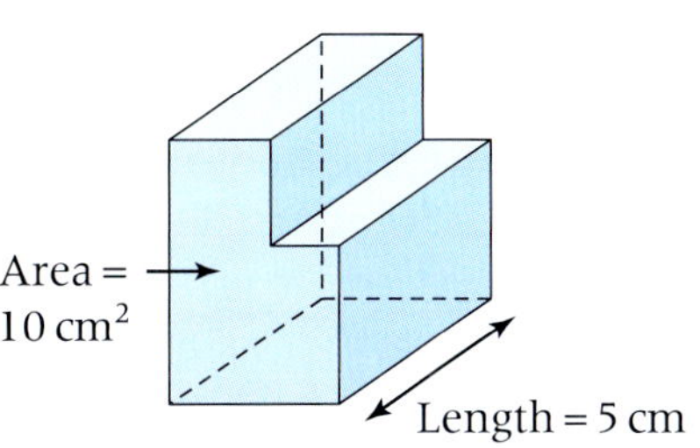

S6.5 Units and dimensions

This spread will show you how to:

- Understand units of measure and the dimensions they represent

Keywords
Area
Dimensions
Length
Units
Volume

- The **length** AB is the distance between two points A and B.

A ——————— B

You measure a distance in **units** of length (cm, m, km).

Length has 1 **dimension** or 1-D.

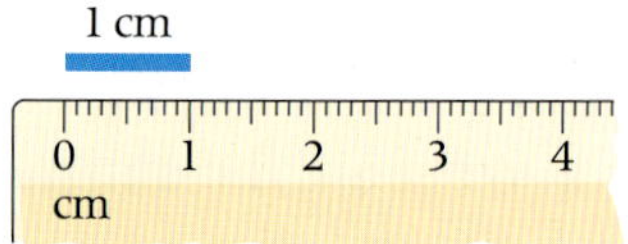

- The **area** is the amount of surface a shape covers.

You measure an area in squares (cm^2, m^2, km^2).

Area has 2 dimensions or 2-D.

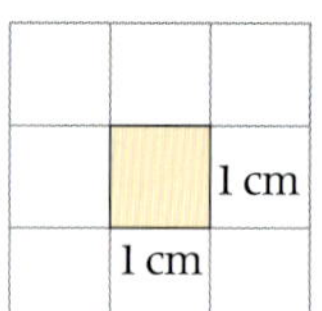

- The **volume** of a shape is the amount of space it takes up.

You measure a volume in cubes (cm^3, m^3, km^3).

Volume has 3 dimensions or 3-D.

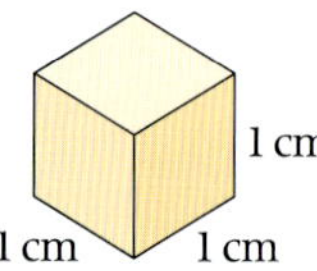

Capacity and volume both measure 3-D space and use the same units.

You can use a letter, such as L, to link length, area and volume.

length
L

L area = L^2
L

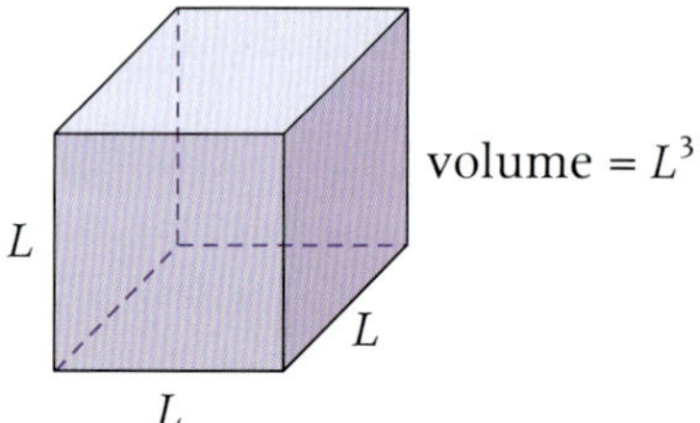

Example

Give the dimensions and a suitable unit of measurement for each of these. Choose either length (L), area (L^2), volume (L^3), or none of these dimensions (N).

a The surface covered by a carpet
b The distance a piece of string stretches
c The space inside the boot of a car
d The value of a 50 pence piece

a Area (L^2), 2-D, m^2
b Length (L), 1-D, cm
c Volume (L^3), 3-D, m^3
d None of these (N)

Exercise S6.5

Choose one of length (L), area (L^2), volume (L^3), or none of these dimensions (N) for each of these.

1 Surface covered by a lake
2 8 litres
3 Distance from Sheffield to Leeds
4 The amount of paper needed to cover a parcel
5 2 cm^3
6 Your height
7 4 kilograms
8 The thickness of a book
9 Your weight
10 The radius of a circle
11 The space inside a classroom
12 One minute
13 7
14 The amount of lemonade in a bottle
15 60 miles per hour
16 The distance from you to the nearest door
17 The circumference of a circle
18 The surface of a cylinder
19 £10
20 The width of a car
21 The amount of petrol/diesel in a car
22 The size of a field
23 The diameter of a circle
24 5 m^3
25 The cost of a train journey
26 The distance travelled by a train
27 The time taken for a train journey
28 The speed of a train
29 The weight of an elephant
30 The surface of a cuboid
31 The height of the Eiffel Tower
32 3 cm^2
33 The weight of a laptop computer
34 The height of a tree
35 9 kilometres
36 The weight of a cat
37 π
38 The altitude of a hot air balloon
39 The length of a dog's tail
40 The weight of a burger
41 The distance a rope will stretch
42 The surface covered by a wall
43 The amount of glass in a window
44 A right angle
45 The space inside a train carriage
46 The weight of a train
47 The perimeter of a field
48 2 inches
49 The amount of paint inside a tin
50 The surface covered by a tin of paint
51 length × width
52 length + width + length + width

width
length

53 π × diameter

diameter

54 π × radius × radius

radius

55 length × width × height

height
length
width

S6 Exam review

Key objectives

- Recall the definition of a circle and the meaning of related terms, including centre, radius, diameter, circumference
- Find circumferences of circles, recalling relevant formulae
- Calculate perimeters and areas of shapes made from triangles and rectangles
- Find the surface area of simple shapes by using the formulae for the areas of triangles and rectangles
- Find volumes of cuboids, recalling the formula and understanding the connection to counting cubes and how it extends this approach

1 Work out the circumference of this circle.
Give your answer correct to one decimal place.

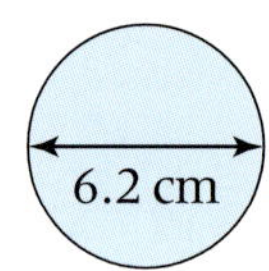

(2)

2 A shaded shape has been drawn on the centimetre grid.

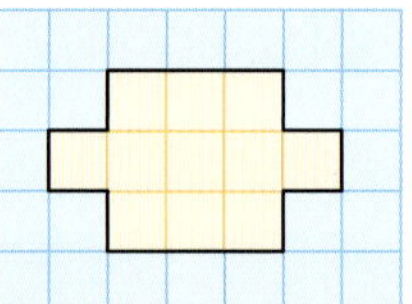

a i Find the area of the shaded shape.
ii Find the perimeter of the shaded shape. (2)

The shaded shape has two lines of symmetry.

b Copy the diagram and draw the two lines of symmetry on the shaded shape. (2)

c Find the volume of this prism.

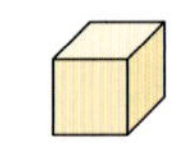

represents 1 cm^3

Diagram **NOT** accurately drawn

(2)

(Edexcel Ltd., 2005)

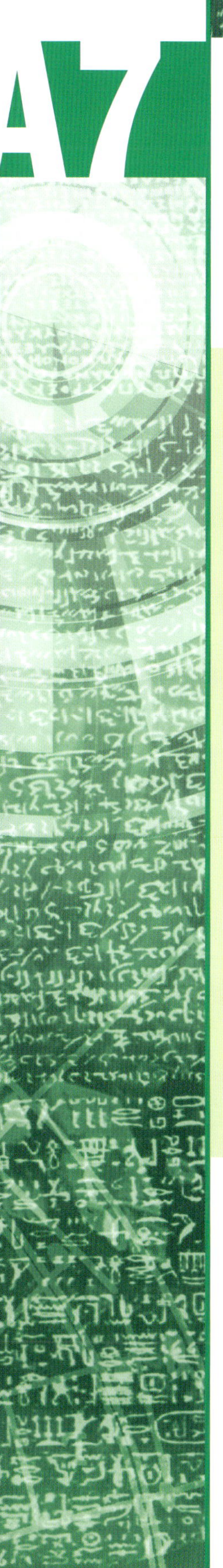

A7 More sequences

This unit will show you how to

- Generate and describe sequences derived from patterns
- Use term-to-term rules to work out the next pattern or term in a sequence
- Use position-to-term rules to work out any pattern or term in a sequence

Before you start ...

You should be able to answer these questions.

	Review
1 Write the first five multiples of each number. **a** 2 **b** 4 **c** 3 **d** 5	Unit A3
2 The first five terms of a sequence are 4 7 10 13 16 Work out the next two terms in the sequence.	Unit A3
3 Work out the value of each expression when $n = 4$. **a** $n + 7$ **b** $2n - 3$ **c** $n^2 + 1$ **d** $2n^2 - 1$	Unit A1
4 Solve these equations. **a** $4x + 3 = 27$ **b** $5y - 11 = 4$	Unit A5

A7.1 Sequence patterns

This spread will show you how to:

- Generate and describe sequences derived from patterns

Keywords
Sequence
Term-to-term rule

Here is a **sequence** of patterns made from dots.

Pattern 1 Pattern 2 Pattern 3

The next two patterns are:

Pattern 4 Pattern 5

Examiner's tip
The techniques in this unit are useful for your coursework.

The numbers of dots make a sequence.

Pattern number	1	2	3	4	5
Number of dots	5	7	9	11	13

The **term-to-term rule** is 'add 2'.

- **You can use the term-to-term rule to work out the number of dots in the next pattern.**

Example

Here is a sequence of patterns made from stars.

a Draw the next two patterns in the sequence.
b Describe how the patterns grow.
c Complete the table for the sequence.

Pattern 1 Pattern 2 Pattern 3

Pattern number	1	2	3	4	5
Number of stars					

d Work out the number of stars in the 6th and 7th patterns.
e Work out the number of stars in the 10th pattern.

a

Pattern 4 Pattern 5

b Add one star to each side
So add three stars each time.

c

Pattern number	1	2	3	4	5
Number of stars	3	6	9	12	15

d 6th pattern: $15 + 3 = 18$ stars.
7th pattern: $18 + 3 = 21$ stars.

e Number of stars = 3 × pattern number.
Number of stars in pattern 10 $= 3 \times 10$
$= 30$.

The relationship between the number of dots and the pattern number is the **position-to-term rule**.

The pattern number is the pattern's position in the sequence.

- **You can use the position-to-term rule to work out the number of dots in any pattern.**

Exercise A7.1

1 For each sequence

- draw pattern number 4 and pattern number 5
- copy and complete the table.

a

Pattern 1 Pattern 2 Pattern 3

Pattern number	1	2	3	4	5
Number of dots					

b

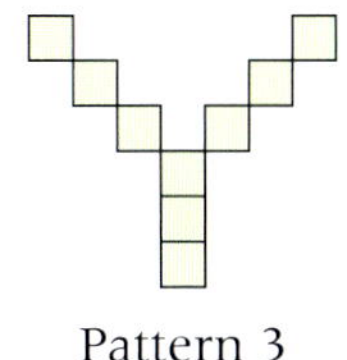

Pattern 1 Pattern 2 Pattern 3

Pattern number	1	2	3	4	5
Number of square tiles					

2 For this sequence

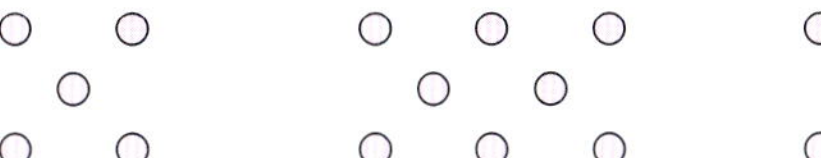

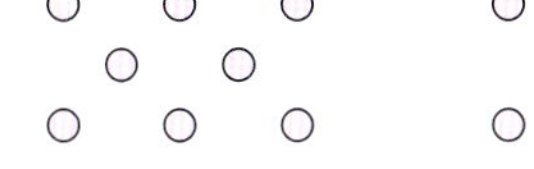

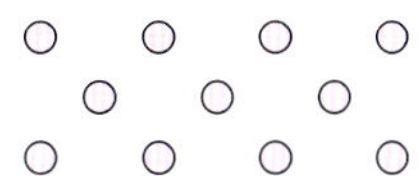

Pattern 1 Pattern 2 Pattern 3

- draw pattern number 4 and pattern number 5
- describe how the pattern grows
- copy and complete the table.

Pattern number	1	2	3	4	5
Number of dots					

- work out the number of dots in pattern number 6 and pattern number 7.

3 These patterns are made from square tiles.

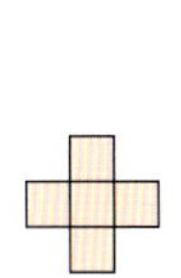

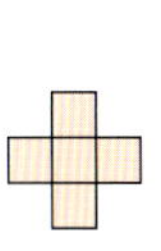

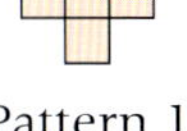

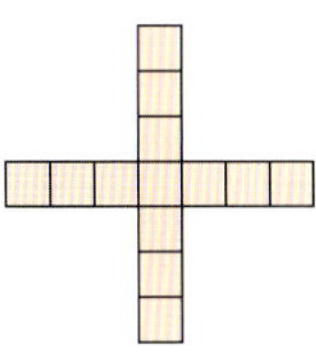

Pattern 1 Pattern 2 Pattern 3

a Draw the 4th and 5th patterns in the sequence.

b Describe how the patterns grow.

c Copy and complete the table.

Pattern number	1	2	3	4	5
Number of tiles					

d Copy and complete:

Number of tiles = ____ × pattern number

e Use your formula from **d** to work out the number of tiles in pattern number 10.

A7.2 More patterns

This spread will show you how to:

- Generate and describe sequences derived from patterns

Keywords
Sequence

Here is a pattern made from matches.

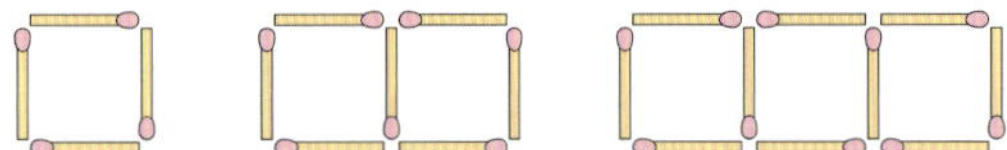

To make the next pattern, you add three matches.

Pattern number	1	2	3	4	5
Number of matches	4	7	10	13	16

The term-to-term rule is 'add 3'.

You can show how the pattern grows in steps of three.

Pattern 1		$1 + 3 = 1 + 1 \times 3$ matches
Pattern 2		$1 + 3 + 3 = 1 + 2 \times 3$ matches
Pattern 3		$1 + 3 + 3 + 3 = 1 + 3 \times 3$ matches
Pattern 4		$1 + 3 + 3 + 3 + 3 = 1 + 4 \times 3$ matches

You can work out the number of matches in the 10th pattern.

$$\text{Pattern } 10 \rightarrow 1 + 10 \times 3 \text{ matches}$$
$$= 1 + 30 = 31 \text{ matches}$$

You can write a rule.

Number of matches in pattern = 1 + pattern number × 3

Example

For the pattern shown above

a Work out the number of matches in the 6th pattern.
b Check your answer to **a** by drawing the 6th pattern.
c Is there a pattern with 26 matches?

a Number of matches in 6th pattern $= 1 + 6 \times 3$
$= 1 + 18 = 19$

b The 6th pattern will have 6 squares of matches.
Number of matches = 19

c The numbers of matches go up in 3s.

7th pattern	8th pattern	9th pattern
22	25	28

No pattern has 26 matches.

Exercise A7.2

1 For each sequence

- draw the next two patterns
- copy and complete the table.

Pattern number					
Number of matches					

a

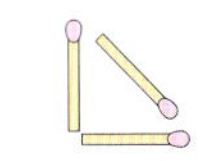

Pattern 1

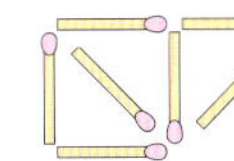

Pattern 2

Pattern 3

b

Pattern 1

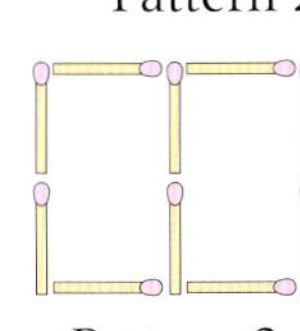

Pattern 2

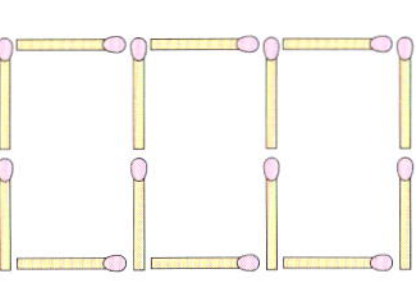

Pattern 3

2 Here is a sequence of patterns made of matches.

Pattern 1

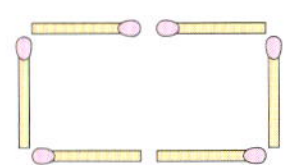

Pattern 2

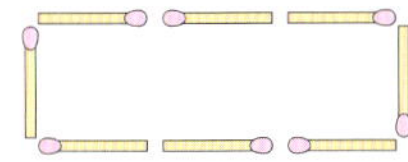

Pattern 3

a Draw the next two patterns in the sequence.

b Copy and complete the table.

Pattern number					
Number of matches					

c Describe how the pattern grows.

3 Here is a sequence of patterns made of matches.

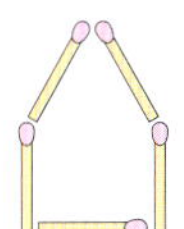

Pattern 1

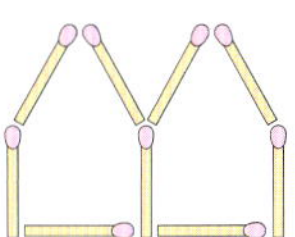

Pattern 2

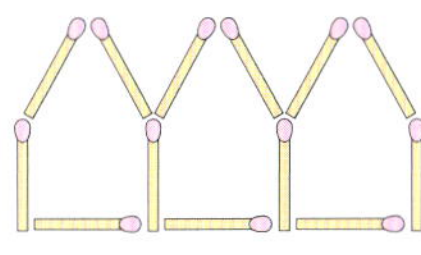

Pattern 3

a Draw the next two patterns in the sequence.

b Copy and complete the table.

Pattern number					
Number of matches					

c Describe how the pattern grows.

d Copy and complete:

Pattern 1	1 hut	$1 + 1 \times 4$ matches
Pattern 2	__ huts	$1 + 2 \times 4$ matches
Pattern 3	__ huts	$1 +$ __ $\times 4$ matches
Pattern 4	__ huts	$1 +$ __ $\times$ __ matches
Pattern 6	__ huts	$1 +$ __ $\times$ __ matches
Pattern 10	__ huts	$1 +$ __ $\times$ __ matches

The shapes formed by the matches look like huts.

e Draw pattern 6 to check your answer to part **d**.

f Is there a pattern with 33 matches? Explain how you know.

A7.3 Sequences of numbers

This spread will show you how to:

- Use term-to-term rules to work out the next pattern or term in a sequence

Keywords
Consecutive
Term
Term-to-term rule

The **terms** in a number sequence follow a pattern.

You can see how the pattern grows by looking at the differences between **consecutive** terms.

Consecutive terms are next to each other.

Here are the first five terms of a number sequence.

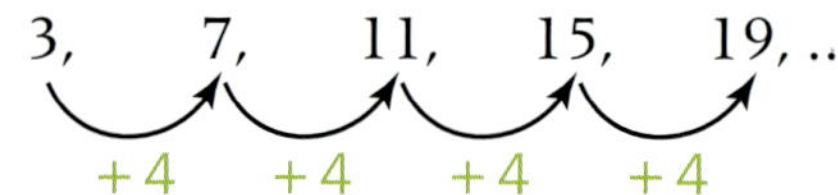

Difference

The **term-to-term rule** is 'add 4'.

- **You can use the term-to-term rule to work out the next term in a sequence.**

Example

Here is a number sequence.

20 17 14 11 8

a Write the next two terms.
b Explain how you worked them out.

---

a

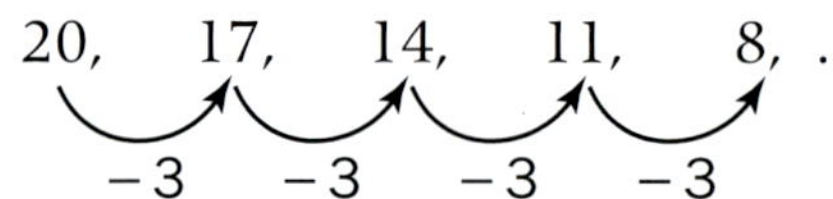

Difference

The next two terms are 5 and 2.

b To get from one term to the next, you subtract 3.

Example

Here is a number sequence.

3 8 13 18 23

a Write the next two terms in this sequence.
b Write the term-to-term rule for the sequence.
c Is 100 a term in this sequence?
Explain how you know.

---

a

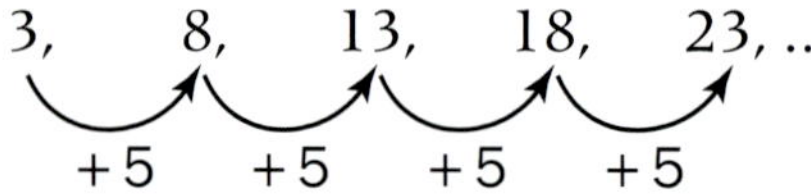

Difference

The next two terms are 28 and 33.

b The term-to-term rule is 'add 5'.
c All the terms in the sequence end in 3 or 8.
So 100 is not a term in the sequence.

Exercise A7.3

1 For each sequence

- write the next two terms
- explain how you worked them out.

a 5, 11, 17, 23, 29, ...

b 30, 26, 22, 18, 14, ...

2 **a** Write the next two terms of this sequence.

11 22 33 44 55

b Is 87 a term of this sequence? Explain how you know.

c What is the 9th term of the sequence?

3 **a** Write the next two terms of this sequence.

15 20 25 30 35

b Is 105 a term in this sequence? Explain how you know.

c Write down the first term in the sequence greater than 1000.

4 For each sequence

- write the next two terms
- write the term-to-term rule.

a −5, −3, −1, 1, 3, ...

b 10, 6, 2, −2, −6, ...

c 26, 20, 14, 8, 2, ...

5 The table shows some rows of a number pattern.

Row 1	1	1×1
Row 2	$1 + 3$	2×2
Row 3	$1 + 3 + 5$	3×3
Row 4	$1 + 3 + 5 + 7$	
Row 5		
Row 6		
Row 7		

a Copy the table. Complete it by following the patterns in the rows.

b Copy and complete these statements.

The sum of the first 3 odd numbers = 3×3

The sum of the first 4 odd numbers = __ × __

The sum of the first 5 odd numbers = __ × __

c Work out the sum of the first 10 odd numbers.
(Do not add them up!)

d Can 72 be written as a sum of consecutive odd numbers starting from 1?
Explain how you know.

A7.4 More sequences of numbers

This spread will show you how to:

- Use term-to-term rules to work out the next pattern or term in a sequence

Keywords
Consecutive
Position-to-term rule
Term

Here is a sequence of squares made from dots.

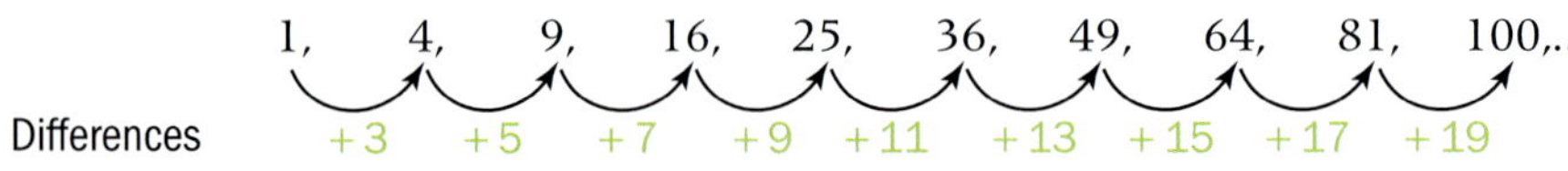

Pattern 1	Pattern 2	Pattern 3
$1 \times 1 = 1^2 = 1$	$2 \times 2 = 2^2 = 4$	$3 \times 3 = 3^2 = 9$

The numbers of dots in the patterns make the sequence of square numbers.

The position-to-term rule for this sequence is

Term = term number × term number = (term number)^2

The sequence up to the 10th term is:

1, 4, 9, 16, 25, 36, 49, 64, 81, 100,...

Differences +3 +5 +7 +9 +11 +13 +15 +17 +19

The differences follow the pattern 3, 5, 7, 9, 11, ...
They are the consecutive odd numbers, starting at 3.

- **The differences between consecutive terms in a sequence may follow a pattern.**

Example

For each sequence of numbers

- work out the differences between consecutive terms
- describe the pattern in the differences
- work out the next two terms.

a 4 6 10 16 24
b 1 2 6 13 23

a 4, 6, 10, 16, 24, ...

Differences +2 +4 +6 +8

The differences are consecutive even numbers, starting at 2.
The next two differences are +10 and +12.
The next two terms are 24 + 10 = 34 and 34 + 12 = 46.

6th term = 5th term + 10
7th term = 6th term + 12

b 1, 2, 6, 13, 23, ...

Differences +1 +4 +7 +10

The differences are the sequence with start number 1, term-to-term rule 'add 3'.
The next two differences are 10 + 3 = 13 and 13 + 3 = 16.
The next two terms are 23 + 13 = 36 and 36 + 16 = 52.

Exercise A7.4

1 For each sequence

- work out the differences between consecutive terms
- describe the pattern in the differences
- work out the next two terms.

a 2 4 7 11 16 **b** 5 7 11 17 25

c 52 50 46 40 32 **d** 5 6 9 14 21

2 Here are the first three terms in a number sequence.

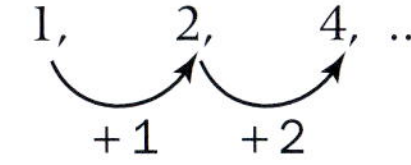

Mick says, 'The differences are doubling each time.'

a Work out the next three terms in Mick's series.

Cora says, 'The differences are consecutive whole numbers.'

b Work out the next three terms in Cora's sequence.

3 Here are the first three terms in a number sequence.

5 6 10

Callum says, 'The next term in the sequence is 17.'
Zoe says, 'The next term in the sequence is 19.'

a Explain why they could both be correct.

b Work out the next term in Callum's sequence.

c Work out the next term in Zoe's sequence.

4 Here are the first five terms in a sequence.

2 5 10 17 26

a Work out the differences.

b Work out the next two terms in the sequence.

c Copy and complete this table to compare the sequence with the sequence of square numbers.

Term number	1	2	3	4	5	6	7
Sequence of square numbers	1 ↓ +1	4 ↓ +1	9 ↓	16 ↓	25 ↓	36 ↓	49 ↓
Sequence	2	5					

d Copy and complete the position-to-term rule for this sequence:

term = (term number)^2 + ____

5 Repeat question **1** for these sequences.

a –2 –4 –7 –11 –16 **b** 10 8 4 –2 –10

A7.5 Generating sequences from an nth term

This spread will show you how to:

- Use position-to-term rules to work out any pattern or term in a sequence

Keywords
General term
nth term

A position-to-term rule links a term to its position in the sequence.

- **The position-to-term rule is often called the nth term.**

Another name for the nth term is the **general term**.

- **You can work out the terms of a sequence by substituting the position numbers 1, 2, 3, 4, ... into the nth term.**

The 1st term is in position 1.
The 2nd term is in position 2.
The nth term is in position n.

Example

Generate the first five terms of the sequences with these nth terms.

a $3n + 2$ **b** $n^2 + 3$

This topic is extended to finding the general term on page 380.

a $3n + 2$

1st term substitute $n = 1$: $3 \times 1 + 2 = 3 + 2 = 5$
2nd term substitute $n = 2$: $3 \times 2 + 2 = 6 + 2 = 8$
3rd term substitute $n = 3$: $3 \times 3 + 2 = 9 + 2 = 11$
4th term substitute $n = 4$: $3 \times 4 + 2 = 12 + 2 = 14$
5th term substitute $n = 5$: $3 \times 5 + 2 = 15 + 2 = 17$
The first five terms are 5, 8, 11, 14, 17.

b $n^2 + 3$

1st term: $1^2 + 3 = 4$
2nd term: $2^2 + 3 = 4 + 3 = 7$
3rd term: $3^2 + 3 = 9 + 3 = 12$
4th term: $4^2 + 3 = 16 + 3 = 19$
5th term: $5^2 + 3 = 25 + 3 = 28$
The first five terms are 4, 7, 12, 19, 28.

Example

Here is a sequence of patterns made from counters.

The general term for the sequence is

$C = 3n - 1$

where C is the number of counters and n is the pattern number.
Which pattern in the sequence has 89 counters?

Pattern 1 Pattern 2 Pattern 3

For the pattern with 89 counters

$89 = 3n - 1$
$89 + 1 = 3n - 1 + 1$
$90 = 3n$
$90 \div 3 = 3n \div 3$
$30 = n$

The 30th pattern has 89 counters.

Substitute 89 into the formula.
Solve the equation to find n.

Add 1 to both sides.
Divide both sides by 3.

Exercise A7.5

1 Generate the first five terms of the sequences with these nth terms.

a $3n - 1$ **b** $2n + 3$ **c** $2n - 5$ **d** $6n + 2$

e $4n + 5$ **f** $3 - 2n$ **g** $4n - 2$ **h** $-3n + 8$

2 Write the first five terms of the sequences with these nth terms.

a $n^2 + 1$ **b** $n^2 + 7$ **c** $n^2 - 4$ **d** $n^2 - 1$

e $n^2 + 10$ **f** $3 - n^2$ **g** $n^2 - 2$ **h** $-n^2 + 1$

3 Here is a sequence of beach hut patterns made from matchsticks. The formula for the number of matchsticks in a pattern is

$$M = 4n + 1$$

where M is the number of matchsticks and n is the pattern number.

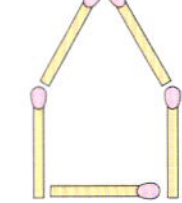
Pattern 1

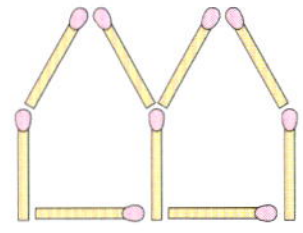
Pattern 2

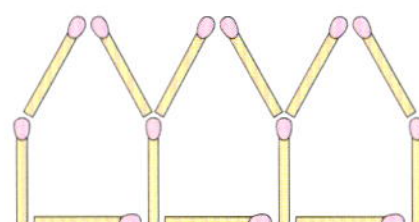
Pattern 3

a Work out the number of matchsticks in pattern number 9.

b Work out the number of matchsticks in pattern number 10.

Trina has 40 matchsticks.

c What is the pattern number of the largest pattern she can make?

d How many matchsticks will she have left over?

4 Here is a sequence of patterns made from tiles.

The general term for the sequence is

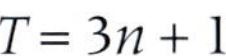

$$T = 3n + 1$$

where T is the number of tiles and n is the pattern number.

Pattern 1

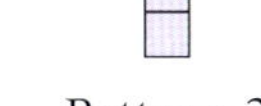
Pattern 2

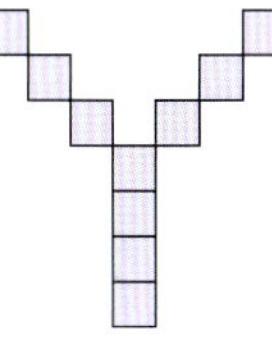
Pattern 3

a Work out the number of tiles in pattern number 15.

b Which pattern has 28 tiles?

c Is there a pattern with 39 tiles? Explain your answer.

d Which is the largest pattern you could make with 39 tiles? How many tiles would you have left over?

5 Here is a sequence of patterns made from 2p coins.

The formula for this sequence is

$$C = 3n + 1$$

where C is the number of 2p coins and n is the pattern number.

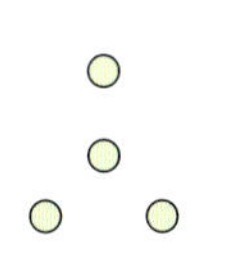
Pattern 1

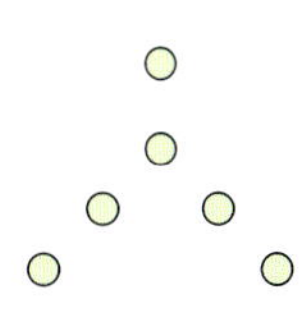
Pattern 2

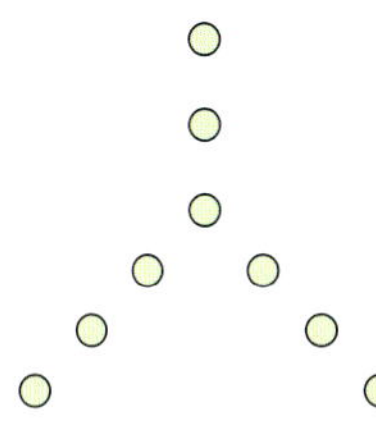
Pattern 3

a How many 2p coins are there in the 5th pattern?

b What is the value of the 5th pattern?

c Which pattern has twenty-two 2p coins?

d What is the value of the pattern in part **c**?

e Which pattern has value 68p?

A7 Exam review

Key objectives

- Generate terms of a sequence using term-to-term and position-to-term definitions of the sequence
- Use linear expressions to describe the *n*th term of an arithmetic sequence, justifying its form by reference to the activity or context from which it was generated

1 The first five terms of a number sequence are given below.

2 7 12 17 22

a Write down the next two terms of the sequence. (2)

b Explain the rule you used to find your answer to part **a**. (1)

c Why is the number 125 not a term in the sequence? (1)

2 Here are some patterns made up of dots.

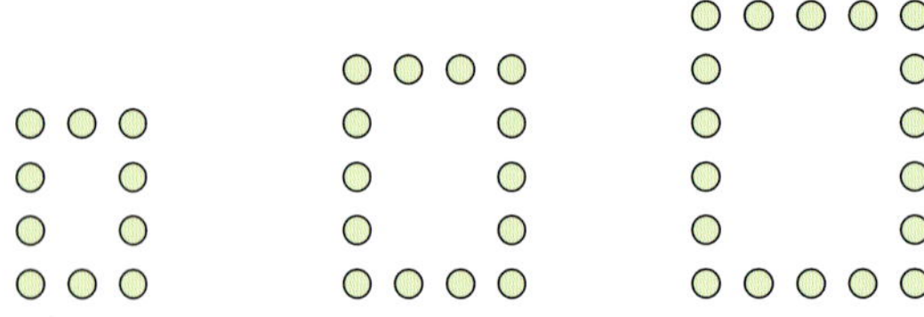

a Draw pattern number 4. (1)

b Copy and complete the table: (1)

Pattern number	1	2	3	4	5
Number of dots	10	14	18		

c How many dots are used in pattern number 10? (1)

(Edexcel Ltd., 2004)

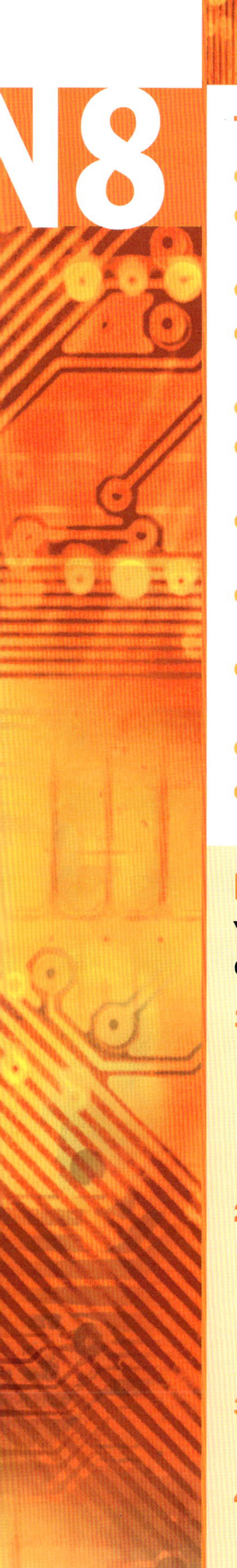

Number problems

This unit will show you how to

- Use a ratio to compare the size of two objects
- Use ratio notation and scales and express a ratio in its simplest form
- Divide a quantity in a given ratio
- Use a range of mental and written methods for whole number and decimal calculations
- Use approximation to estimate the answers to calculations
- Solve word problems, making sure you know what the question is asking you to find out
- Use correct units in calculations and express answers to a given degree of accuracy
- Calculate a fraction or a percentage of an amount using a variety of methods
- Check if an answer is of the correct order of magnitude and use inverse operations to check solutions
- Carry out more complex calculations using a calculator
- Interpret the display on a calculator

Before you start ...

You should be able to answer these questions.

	Review
1 A car is 240 cm long. A motorbike is 160 cm long. What is the ratio of the length of the car compared to the length of the motorbike? Express your answer in its simplest form.	Unit N4
2 Steve buys 2 CDs for £6.50 each 2 pens for £0.40 each. He pays with a £20 note. How much change should he get from the £20 note?	Unit N3
3 Calculate these without a calculator. **a** $\frac{3}{7}$ of £56 **b** 15% of 80 kg	Unit N3
4 Calculate these using your calculator. **a** $3.2 \times 4.3 - 2.6$ **b** 3.44^2 **c** $\sqrt[3]{3375}$	Unit N3, N5

N8.1 Solving ratio problems

This spread will show you how to:

- Use a ratio to compare the size of two objects
- Use ratio notation and scales and express a ratio in its simplest form
- Divide a quantity in a given ratio

Keywords
Ratio
Scale

You can compare the size of two objects using a **ratio**.

This topic is extended to dividing with harder ratios on page 370.

Example

a Morrissey the kitten is 25 cm tall. Honey the puppy is 75 cm tall. What is the ratio of Morrissey's height compared to Honey's height? Express your answer in its simplest form.

b Loopy the dog is 100 cm (1 m) tall. Compare his height to Honey the puppy. Express your answer in its simplest form.

a Morrissey's height : Honey's height
= 25 cm : 75 cm
= 25 : 75
= 1 : 3
Honey is 3 times taller than Morrissey.

b Honey's height: Loopy's height
= 75 : 100
= 3 : 4
Honey's height is $\frac{3}{4}$ of Loopy's height.
Loopy's height is $\frac{4}{3}$ × Honey's height.

You can simplify a ratio by dividing both parts of the ratio by the same number.

You can divide a quantity in a given ratio.

Example

Anne and Parvez share £200 in the ratio 3 : 7.
How much money do they each receive?

Anne receives 3 parts for every 7 parts that Parvez receives.

Total number of parts = 3 + 7 = 10 parts
Each part = £200 ÷ 10 = £20

Anne will receive 3 parts = 3 × £20 = £60
Parvez will receive 7 parts = 7 × £20 = £140

£200 is divided into 10 parts.

- A **scale** is a ratio expressed in the form 1 : *n*.

Example

A map has a scale of 1 : 20 000. A distance on the map is 4.5 cm.
What is this distance in real life?

Using the scale you can say
Distance in real life = 20 000 × distance on the map
= 20 000 × 4.5 cm
= 90 000 cm
= 900 m

Exercise N8.1

1 Write each ratio in its simplest form.

a 4 : 6	**b** 6 : 10	**c** 10 : 25	**d** 16 : 24
e 25 : 45	**f** 50 : 60	**g** 46 : 58	**h** 200 : 250

2 Express these pairs of objects as ratios in their simplest form.

a There are 40 boys and 55 girls in Year 11.
What is the ratio of boys to girls in Year 11?

b In a batch of apples there are 8 bad apples and 52 good apples.
What is the ratio of good apples to bad apples?

c In a crowd of football supporters there are 12 000 men and 8000 women. What is the ratio of women to men?

d Hugh has 50p. Gwen has £3. What is the ratio of Hugh's money to Gwen's money?

3 Solve these problems.

a In a batch of concrete the ratio of sand to cement is 5 : 2.
How much sand is needed to mix with 10 kg of cement?

Hint for part **a**:
Amount of sand =
$\frac{5}{2}$ × amount of cement

b In a school the ratio of teachers to students is 2 : 25. If there are 500 students at the school, how many teachers are there?

c In a metal alloy the ratio of aluminium to zinc is 3 : 4. How much aluminium is needed to mix with 20 kg of zinc?

d For his vegetable beds, Paul mixes some sand and compost in the ratio 3 : 4. How much compost does he mix with 72 kg of sand?

4 Use the scales to work out the measurements in each of these calculations.

a A map has a scale of 1 : 500.

i What is the distance in real life of a measurement of 10 cm on the map?

ii What is the distance on the map of a measurement of 20 m in real life?

b A map has a scale of 1 : 5000.

i What is the distance in real life of a measurement of 4 cm on the map?

ii What is the distance on the map of a measurement of 600 m in real life?

5 Solve these problems.

a Divide £30 in the ratio 3 : 7.

b Divide 250 kg in the ratio 7 : 3.

c Divide 40 tonnes in the ratio 5 : 3.

d Divide 135 litres in the ratio 5 : 4.

N8.2 Solving word problems

This spread will show you how to:

- Use a range of mental and written methods for whole number and decimal calculations
- Use approximation to estimate the answers to calculations
- Solve word problems, making sure you know what the question is asking you to find out

Keywords
Divisor
Estimate
Grid method
Standard method
Whole number

It is important to identify the information that is necessary to solve a problem. Once you have read the problem, highlight or underline the key words in the question.

Example

Brian puts 23 boxes into his van. Each box weighs 21.7 kg. Work out the total weight of the 23 boxes.

Highlight the key words and numbers.

Estimate first. $23 \times 21.7 \approx 20 \times 20$
$= 400$ kg

Rewrite the calculation as

$$23 \times 21.7 = 23 \times 217 \div 10$$

Work out 23×217 using the **grid method** or **standard method**.

×	200	10	7
20	$20 \times 200 = 4000$	$20 \times 10 = 200$	$20 \times 7 = 140$
3	$3 \times 200 = 600$	$3 \times 10 = 30$	$3 \times 7 = 21$

$$23 \times 217 = 4000 + 600 + 200 + 30 + 140 + 21$$
$$= 4991$$

Total weight: $23 \times 21.7 = 23 \times 217 \div 10$
$= 4991 \div 10 = 499.1$ kg

Identify the operations required, in this case multiplication.
Always make an **estimate**.
Perform the calculation.

Read the problem and check that you know exactly what the question is asking you to find out. Make sure you answer using the correct units.

Example

Veronica sells 54 identical doll's houses for £675. She sells each doll's house for the same price. Work out the price at which Veronica sold each doll's house.

You are asked to find a price.
So your answer will be an amount of money.

Estimate first. £675 ÷ 54 ≈ 650 ÷ 50
= £13

```
54)675
  - 540    54 × 10
    135
  - 108    54 × 2
     27
  -  27    54 × 0.5
      0
```

10 + 2 + 0.5 = 12.5
675 ÷ 54 = **12.5**

Veronica sells each doll's house for £12.50.

Exercise N8.2

1 Calculate each of these using an appropriate written method. Remember to do a mental approximation first.

a 14×18 **b** 28×13 **c** 43×59

d 126×19 **e** 37×59 **f** 148×17

2 Use an appropriate method of calculation to work out each of these. Where appropriate leave your answer in remainder form.

a $62 \div 6$ **b** $84 \div 9$ **c** $136 \div 8$

d $181 \div 7$ **e** $192 \div 13$ **f** $213 \div 3$

3 **a** Boris plants 12 seeds in each pot. He has 154 seeds. How many pots will he fill with seeds? How many seeds will there be left?

b Jamal runs a ski lift. Each lift carriage when full can hold nine passengers. He has 112 people in his queue. How many carriages can he fill with the people in the queue?

c Dewi packs eggs into boxes. He packs 12 eggs into each box. He has 103 eggs. How many boxes will he be able to fill? How many eggs will he have left over?

d Karen packs rice into sacks. Each sack holds 15 kg. How many sacks can she fill with 208 kg of rice? How much rice is left over?

e The diagram shows the measuring scale on a petrol tank.

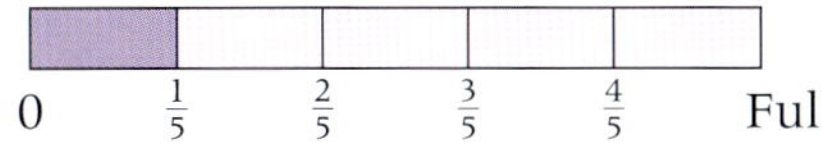

i What fraction of the petrol tank is empty?

ii The petrol tank when full holds 40 litres. A litre of petrol costs 91p. What is the cost of the petrol required to fill the tank?

4 Use an appropriate method of calculation to work out each of these.

a 18×2.3 **b** 13.7×8 **c** 5.24×13

d $26.6 \div 7$ **e** $33.6 \div 6$ **f** $67.2 \div 8$

5 **a** Tina buys 13 bags of flour. Each bag costs £1.12. How much does this cost in total?

b Lemonade costs 96 pence per litre. How much does 4.3 litres of lemonade cost?

c Uri carries 27 boxes from a van. Each box weighs 2.84 kg. What is the total weight of boxes carried by Uri?

d A box will hold 24 glasses. Christine has 672 glasses. Calculate the number of boxes she will need to pack all the glasses.

e Heather is paid £25.20 for six hours work. How much does she get paid each hour?

f Sam has 50.4 kg of sugar. She packs the same amount of sugar into 21 bags. How much sugar is there in each bag?

N8.3 Calculating with fractions and percentages

This spread will show you how to:

- Calculate a fraction or a percentage of an amount using a variety of methods

Keywords
Fraction
Percentage

You can calculate a **fraction** or a **percentage** of something using different methods.

Mental methods

Example

Calculate **a** $\frac{2}{3}$ of £240 **b** 20% of €104

a Find $\frac{1}{3}$ of £240 = 240 ÷ 3 = £80
Calculate $\frac{2}{3}$ of £240 = 2 × 80 = £160

b 20% of €104
Find 10% of €104 = $\frac{1}{10}$ of €104
= $\frac{1}{10}$ × €104
= €104 ÷ 10
= €10.4
Calculate 20% of €104 = 2 × 10.4 = €20.8

Written methods

Example

Calculate **a** $\frac{4}{7}$ of £63 **b** 45% of 320 kg

a $\frac{4}{7}$ of £63 = $\frac{4}{7}$ × £63
= 4 × $\frac{1}{7}$ × £63
= $\frac{4 \times 63}{7}$
= $\frac{252}{7}$
= £36

Multiplying by $\frac{1}{7}$ is the same as dividing by 7.

b 45% of 320 kg = $\frac{45}{100}$ × 320
= $\frac{45 \times 1 \times 320}{100}$
= $\frac{45 \times 320}{100}$
= $\frac{14\,400}{100}$
= 144 kg

This is the same as multiplying the amount by 45 and dividing by 100, that is, you work out 45 × 320 ÷ 100

Calculator methods

Example

Calculate

a $\frac{7}{16}$ of 130 m (to one decimal place) **b** 43% of £75

a Decimal equivalent of $\frac{7}{16}$ = 7 ÷ 16 = 0.4375
$\frac{7}{16}$ of 130 m = $\frac{7}{16}$ × 130 m
= 0.4375 × 130 m
= 56.875 m
= 56.9 m (1 dp)

b Decimal equivalent of 43% = 43 ÷ 100 = 0.43
43% of £75 = 0.43 × £75 = £32.25

Another method of finding a percentage of an amount is to find 1%, and then multiply by the percentage. To find 12% of 3500 m

1% of 3500 = 3500 ÷ 100 = 35

12% of 3500 = 12 × 35 = 420 m

Exercise N8.3

1 Use a mental method to calculate each of these amounts.

a $\frac{1}{4}$ of 60 carrots **b** $\frac{1}{3}$ of 24 rulers **c** $\frac{1}{8}$ of 32 windows

2 Use a mental method to calculate each of these amounts.

Try to use the equivalent fractions.

a 50% of £90 **b** 1% of 600 m^2 **c** 10% of 48 kg

d 25% of 60 kg **e** 20% of 70p **f** 1% of £65

3 Calculate these fractions of amounts without using a calculator.

a $\frac{3}{4}$ of €40 **b** $\frac{2}{3}$ of 60p **c** $\frac{5}{8}$ of 48 g

d $\frac{3}{10}$ of 80p **e** $\frac{1}{7}$ of £84 **f** $\frac{5}{9}$ of $45

4 Calculate these percentages without using a calculator.

a 25% of £60 **b** 5% of 40 **c** 10% of 272p

d 75% of 40 **e** 70% of £40 **f** 30% of £50

g 15% of 35 m **h** 2.5% of £26 **i** 45% of 400 mm

5 Calculate these fractions of amounts without using a calculator.

a A jacket normally costs £130. In a sale the jacket is priced at $\frac{4}{5}$ of its normal selling price. What is the new price of the jacket?

b Hector rents out a holiday flat. His flat is available for 45 weeks of the year. He rents the flat out to tourists for $\frac{7}{9}$ of the time it is available. For how many weeks is Hector's flat occupied by tourists?

6 Calculate these percentages without using a calculator.

a Kelvin collects models. He owns 170 models. He has painted 20% of the models. How many of the models has he painted?

b A barrel can hold 70 litres. Water is poured into the barrel until it is 80% full. How much water is there in the barrel?

7 Use a suitable method to calculate each of these. Where appropriate round your answer to two decimal places.

a $\frac{7}{12}$ of 450 **b** $\frac{9}{10}$ of 360 m **c** $\frac{2}{7}$ of 400 kg **d** $\frac{7}{9}$ of 250 mm

e $\frac{3}{4}$ of 9 tonnes **f** $\frac{4}{11}$ of 2365 m **g** $\frac{7}{25}$ of 43 000 **h** $\frac{3}{7}$ of £345

8 Calculate these using an appropriate method.

a Increase £450 by 10%. **b** Decrease 76 kg by 5%.

c Increase $990 by $\frac{1}{3}$. **d** Decrease 620 km by $\frac{1}{100}$.

N8.4 Checking procedures for solving problems

This spread will show you how to:

- Use correct units in calculations and express answers to a given degree of accuracy
- Check if an answer is of the correct order of magnitude and use inverse operations to check solutions

Keywords
Approximate
Estimate

You need to know how to check that your answer to a calculation is sensible.

Example

Estimate the answer to **a** $18.3(9.8 + 21.9)$ **b** $\frac{6.93 \times 8.09}{1.79}$

a You can round each of these numbers to the nearest 10.

$$18.3(9.8 + 21.9) \approx 20\ (10 + 20)$$
$$= 20 \times 30 = 600$$

Check: the exact answer is 580.11. The estimate is close so this looks right.

b You can round each of these numbers to the nearest whole number.

$$\frac{6.93 \times 8.09}{1.79} \approx \frac{7 \times 8}{2}$$
$$= \frac{56}{2} = 28$$

Check: the exact answer is 31.3 (1 dp). The estimate is close so this looks right.

Rounding is also called **approximation**.

You should check to see if your answer is reasonable – it is about the size you would expect and in the right sort of units.

Example

a Juan is working out the calculation 12.3×0.987 on his calculator. His answer is 13.5683. Without calculating, explain how you know that his answer is wrong.

b Matt has been working out a calculation to find the perimeter of a triangle. His answer is 152.1 mm^2. Identify and correct the two mistakes Matt has made.

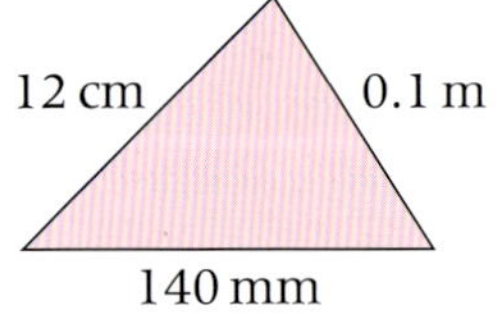

a When you multiply 12.3×0.987 the answer should be less than 12.3. Juan's answer is more than 12.3, so his answer is wrong.

b i Matt has given the answer in mm^2. The measurement for perimeter is in m, cm, mm. He has used the wrong units.

ii Matt has not changed the measurements into the same unit. He has just added the numbers together. His calculation is incorrect.

Perimeter of triangle = 12 cm + 0.1 m + 140 mm
= 12 cm + 10 cm + 14 cm = 36 cm

When you multiply by a number less than 1, the answer should be smaller.

Converting units:

1 m = 100 cm

0.1 m = 0.1 × 100 cm = 10 cm

10 mm = 1 cm

140 mm = 140 ÷ 10 cm = 14 cm

Exercise N8.4

1 Write a suitable estimate for each of these calculations.

a $18.3 + 35.8 \approx 18 + \_\_\_$
$= \_\_\_\_$

b $18.76 + 9.8 - 7.12 \approx \_\_\_ + \_\_\_ - \_\_\_$
$= \_\_\_\_$

c $494 + 319$ **d** $6.89 - 3.11$ **e** $84.2 + 39.76 - 53.612$

2 Write a suitable estimate for each of these calculations. In each case clearly show how you estimated your answer.

a $3.08 \times 8.72 \approx 3 \times \_\_\_$ (rounding to the nearest whole number)
$= \_\_\_\_$

b $19.26 \div 3.76 \approx \_\_\_ \div 4$ (rounding to the nearest whole number)
$= \_\_\_\_$

c 63.47×2.01 **d** $29.49 \div 19.71$ **e** $\frac{7.9 \times 3.21}{5.86}$ **f** $\frac{49 \times 3.8}{19.8}$

3 For each calculation read the question and the answer carefully. Use an appropriate method to check each of the answers.

i Identify if the answer is right or wrong.

ii Explain your method of checking.

iii Correct any mistakes.

a Q. Inzimam and Gary share £300 in the ratio 3 : 2. Calculate how much money each person receives.
A. *Inzimam gets £100; Gary gets £150.*

b Q. Write down all the factor pairs of 80.
A. *The factor pairs are* 1×80; 2×40; 3×25; 4×20; 5×16; 6×15; 8×10.

c Q. Mr. Homes works out how many units of electricity he used last year.
At the start of the year, the electricity meter reading was 6919 units.
At the end of the year, the electricity meter reading was 7102 units.
How many units of electricity did he use during the year?
A. *221 units were used.*

d Q. A shuttlecock costs 65p. Sean has £5. He buys as many as he can.

i Work out the number of shuttlecocks Sean can buy with £5.

ii Work out the amount of change Sean should get from £5.

A. **i** 0.076 923 **ii** *0 pence. He has spent it all on shuttlecocks.*

e Q. Will buys 57 digital memory packs at £32.50 each. Work out the total amount spent by Will.
A. *£18 525*

N8.5 Calculator methods

This spread will show you how to:

- Carry out more complex calculations using a calculator
- Interpret the display on a calculator

Keywords
Brackets
Fraction key
Memory
Sign change key

You should learn to use the function keys of your own calculator. Always carry out an estimate before you use the calculator.

Example

Calculate $4(3.29)^2$.

Estimate: $4(3.29)^2 \approx 4 \times 3^2 = 4 \times 9 = 36$

You type 4 (3 · 2 9) x^2 =

The calculator should display 4(3.29)2 43.2964

Answer = 43.2964 = 43.30 (2 decimal places)

$4(3.29)^2$ means $4 \times (3.29)^2$.

The calculator used here is a Casio FX-85. Your calculator may have a different key sequence.

You can store an answer in the memory of your calculator using the STO and RCL keys.

Example

Calculate $\dfrac{12.4}{4.23 - 1.9}$

You type

Answer = 5.321 888 412 = 5.32 (2 decimal places)

You can enter fractions using your calculator.
You use the (−) key to enter a negative number into a calculator.

Example

Calculate $\frac{3}{8} \times -\frac{2}{5}$.

You type

The calculator should display 3⌟8×-2⌟5 -3⌟20.

Answer = $-\frac{3}{20}$

You can calculate with square roots and cube roots.

Example

Calculate the value of $\sqrt[3]{2197}$.

You type

The calculator should display

Answer = 13

$\sqrt[3]{2197}$ is the cube root of 2197.
$\sqrt[3]{2197} = 13$ because
$13^3 = 13 \times 13 \times 13$
$= 2197$

Exercise N8.5

1 Use your calculator to work out each of these calculations. Write all the figures on your calculator display.

a $3(4.3 + 1.6)$ **b** $3(2.4^2 + 1.03)$ **c** $2(9.3 - 5.7)^2$

d $25(7.2 - 2.4)^2$ **e** $4(2.8^2 - 2.04)^2$ **f** $33.17 \div (4.13 + 2.3^2)$

2 Calculate each of these, giving your answer to one decimal place. Use the bracket keys and/or memory keys of your calculator.

a $\frac{6.3 + 4.7}{5.4 - 3.2}$ **b** $\frac{5.7 \times 3.2}{3.7 + 2.5}$ **c** $\frac{4.6 - 2.13}{5.2 - 1.59}$ **d** $\frac{3.7 \times 2.84}{3.8 + 2.02}$

3 Use your calculator to work out each of these. Write all the figures on your calculator.

a $\frac{352.5 \times 28.4}{(1.6 + 3.7)^2}$ **b** $\frac{(22.4 + 13.2) \times 6.3^2}{185.9 \times (9.5 - 3.87)}$

c $\frac{19.28 \times (27.3 - 8.64)}{(26.3 + 14.7) \times 5.3}$ **d** $\frac{5.44^2 \times 3.16 + 2.8}{4.61 - 2.79}$

4 Calculate these using the appropriate keys on your calculator. Give your answer as appropriate either as a whole number, a fraction or a decimal to two decimal places.

a $3 \times -2 + 6$ **b** $\frac{1}{3} + \frac{2}{5}$ **c** $\sqrt[3]{120}$

d $-7 - -3.5$ **e** $\frac{3}{5} \times 125$ **f** $4.3^2 - \sqrt{20}$

g $\sqrt{(12.3 + 17.29)}$ **h** $1.2(-3.4 + 1.76)^2$

5 **a** David shares £3000 equally between his seven grandchildren. How much does each child receive? Give your answer to an appropriate degree of accuracy.

b Littleton FC want to send 795 fans on a trip to watch their team play at Bigfield FC. Each coach can hold 52 people. How many coaches do the club need to order for the trip?

c Brad is packing cricket balls into boxes. Each box can hold 48 cricket balls. Each week he packs a total of 25 000 cricket balls.

i How many boxes will he fill with cricket balls each week?

ii How many cricket balls will be left over each week?

6 Hugh is working out how much he saves each year. Each week he saves £19.75. There are 52 weeks in one year.

a Hugh calculates his answer as £10 270. Explain why Hugh's answer is wrong.

b Calculate the correct answer.

7 Jeff needs 550 tiles to tile a bathroom. The tiles are sold in boxes of 15.

a How many boxes of tiles does Jeff need to buy?

b A box of tiles costs £19.99. How much does Jeff spend on tiles?

DID YOU KNOW?

Ada Lovelace, the daughter of the poet Lord Byron, worked closely with Charles Babbage to design the first computer. She also wrote the first computer program.

N8 Exam review

Key objectives

- Use ratio notation, including reduction to its simplest form and its various links to fraction notation
- Calculate a given fraction of a given quantity
- Convert fractions of a whole to percentages of a whole and vice versa
- Divide a quantity in a given ratio
- Estimate answers to problems involving decimals, round to a given number of significant figures
- Solve word problems about ratio and proportion
- Use calculators effectively and efficiently, knowing how to enter complex calculations

1 This is a recipe for lemon biscuits:

60 g caster sugar	25 g ground almonds	
1 tsp lemon zest	80 g butter	50 g flour

a Calculate the ratio of grams of sugar to grams of butter in the recipe. (2)
Express your answer in its simplest form.

b The above recipe makes 8 biscuits.
Work out the amounts of the ingredients needed to make 12 biscuits. (3)

2 Jade made a train journey. Her train should have arrived at 14 40.
It arrived 1 hour 50 minutes late.

a At what time did her train arrive? (1)

The railway company gave Jade some money back, because her train was late.
The company used this rule to work out the amount of money.

Find $\frac{1}{4}$ of the ticket price
Then round up this answer to the next whole number of pounds

Jade's ticket price was £33.56.

b i Work out $\frac{1}{4}$ of £33.56.

ii Round up your answer in part **i** to the next whole number of pounds. (3)

(Edexcel Ltd., 2005)

S7 Further transformations

This unit will show you how to

- Recognise and visualise reflections, rotations, translations and enlargements
- Understand that reflections are specified by a mirror line
- Understand that rotations are specified by a centre of rotation and an angle and direction of turn
- Understand that translations are specified by a distance and direction
- Understand that enlargements are specified by a centre of enlargement and a scale factor
- Transform triangles and other 2-D shapes by reflection, rotation, translation and enlargement
- Understand the effect of enlargement on the angles and lengths of shapes and how this relates to the scale factor of the enlargement
- Recognise similar shapes

Before you start ...

You should be able to answer these questions.

1 State the equations of lines **a** and **b**. — Review: Unit A4

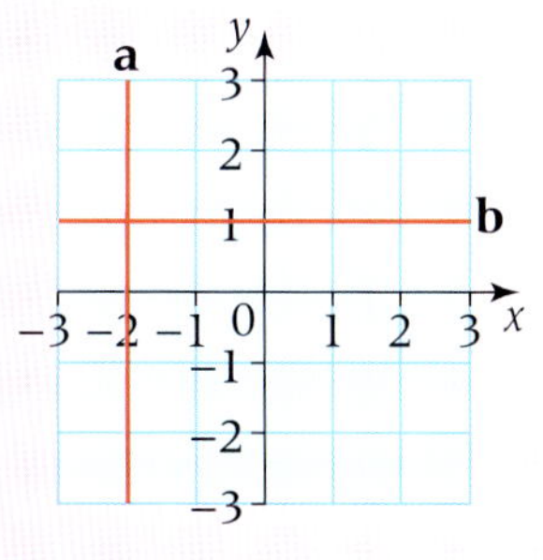

2 State the value of these angles. — Review: Unit S2

a

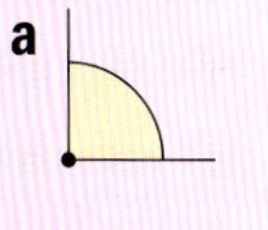

b

c

3 Evaluate each of these. — Review: Unit N3

a 3.2×2 **b** 3.8×2 **c** 4.7×2

d 3.8×3 **e** 4.6×4

S7.1 More reflections

This spread will show you how to:

- Recognise and visualise reflections
- Understand that reflections are specified by a mirror line

Keywords
Congruent
Equidistant
Mirror line
Reflection
Transformation

A **transformation** changes the position of a shape.

- **A reflection flips the shape over.**

You specify the **mirror line** or reflection line.

The two shapes are **congruent** – they are exactly the same size and the same shape.

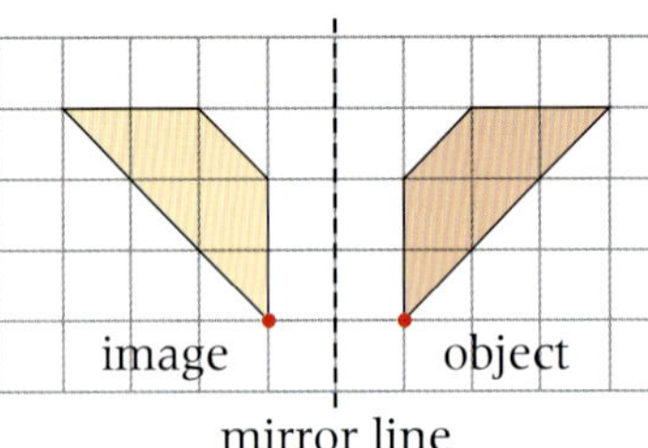

The object and image are **equidistant** from the mirror line.

Each dot is 1 unit from the mirror line.

Example

a Draw the mirror line so that shape B is a reflection of shape A.

b Give the equation of the mirror line.

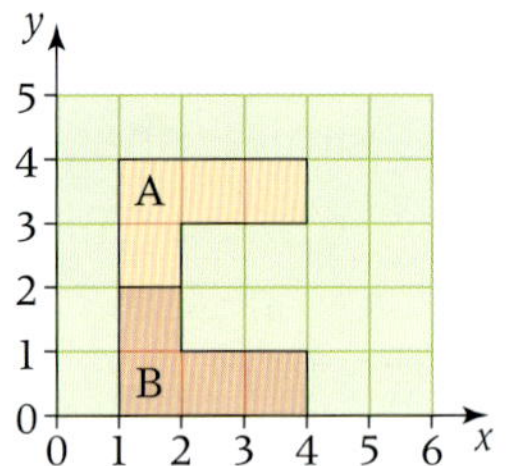

a

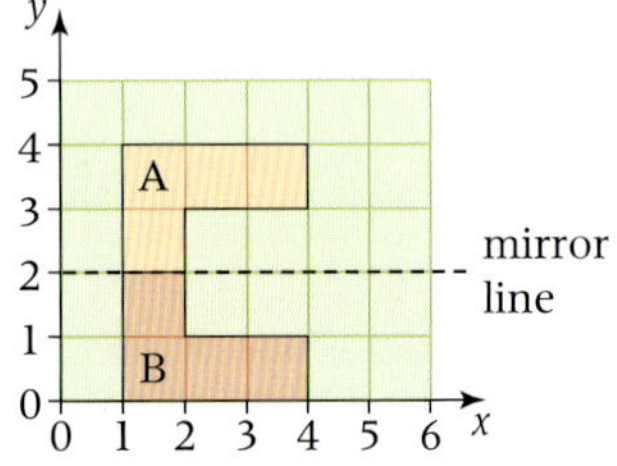

b Each point on the mirror line has y-coordinate 2.
$y = 2$

You can rotate the page to make the mirror line vertical.

To reflect a shape, you choose a point on the object and find the position of the corresponding point in the image.

Example

Reflect this pattern using the y-axis as the mirror line.

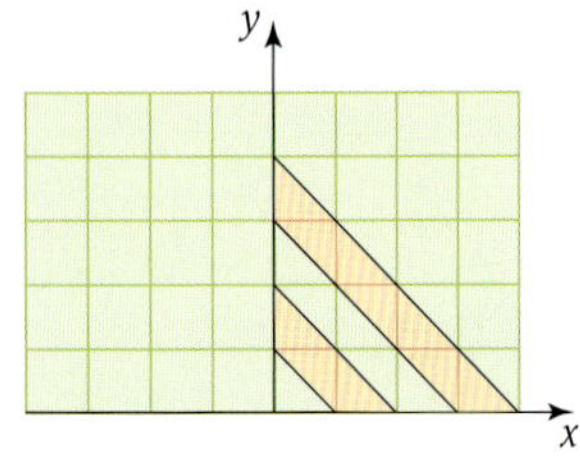

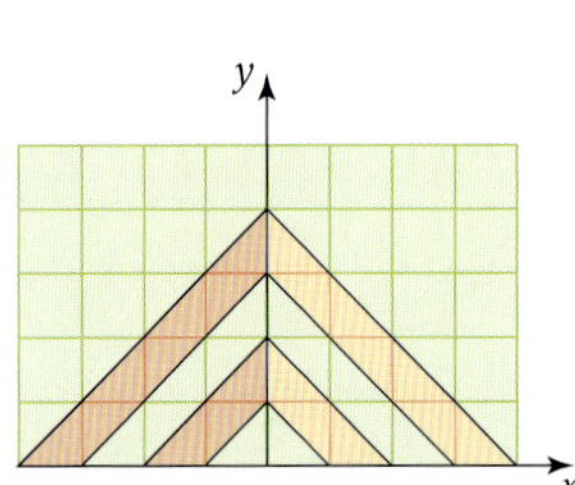

The equation of the y-axis is $x = 0$.

Exercise S7.1

1 Copy the diagrams on square grid paper. Reflect each shape in the mirror line. Give the mathematical name of each new shape you have made.

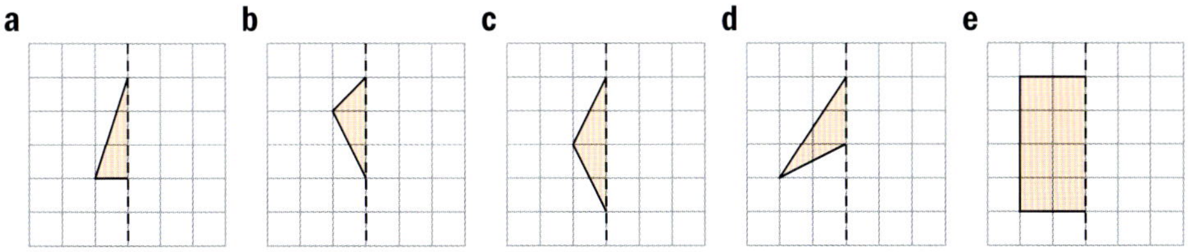

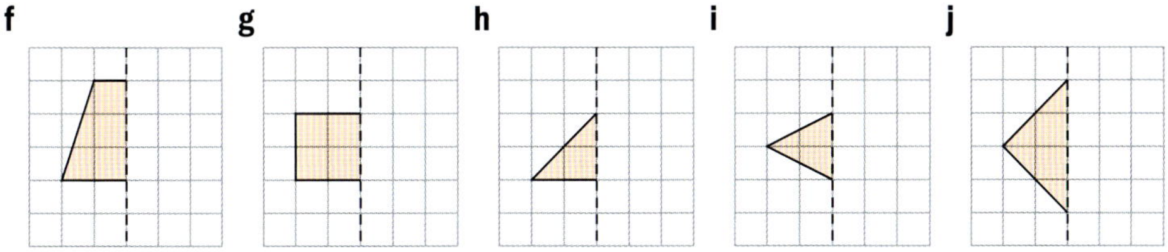

a **b** **c** **d** **e**

f **g** **h** **i** **j**

Explain why it is impossible to draw a parallelogram using this method.

2 Copy and complete the diagrams to show the reflections in both the x-axis and the y-axis.

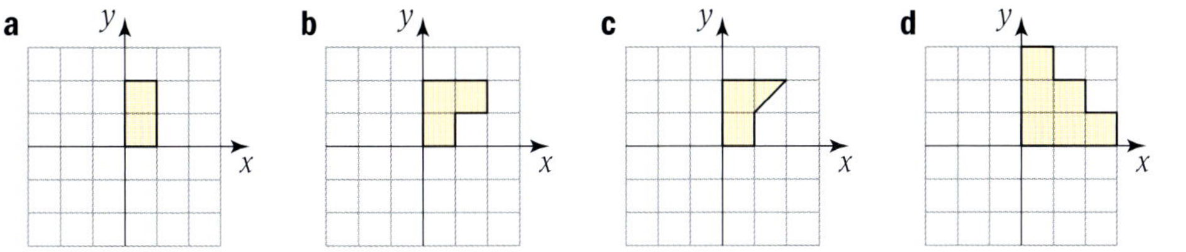

a **b** **c** **d**

3 Copy and complete the diagrams to show the reflection of each triangle in the mirror line $x = 3$.

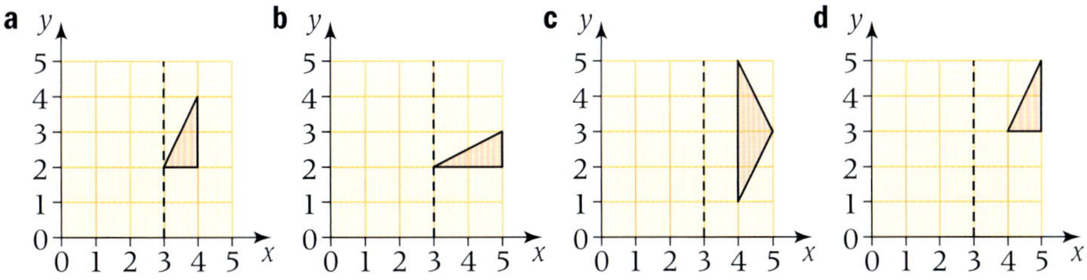

a **b** **c** **d**

4 Give the equation of the mirror line for each reflection.

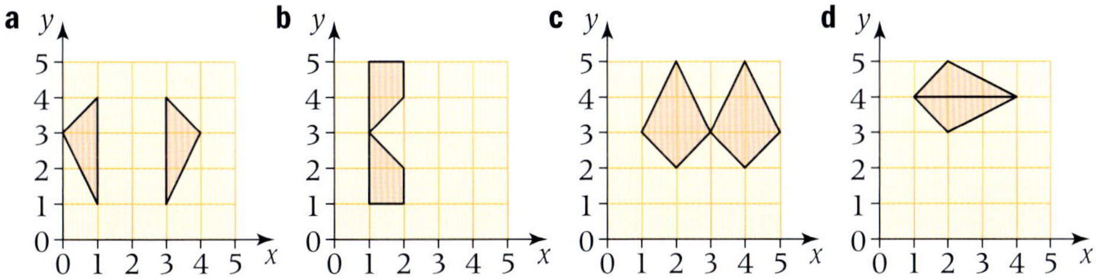

a **b** **c** **d**

S7.2 Rotating shapes

This spread will show you how to:

- Recognise and visualise rotations
- Understand that rotations are specified by a centre of rotation and an angle and direction of turn

Keywords
Anticlockwise
Centre of rotation
Clockwise
Congruent
Origin
Rotation
Transformation

A **rotation** turns a shape.

The turn is 90°. The direction is **clockwise**.

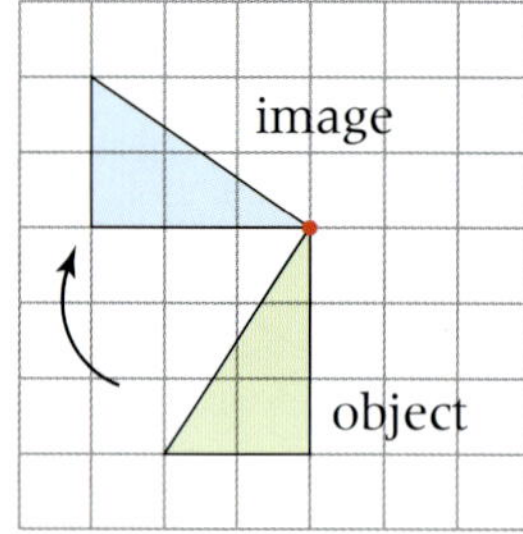

The dot is the **centre of rotation**.

The two shapes are **congruent**.

Example

Draw the pentagon after a rotation of 180° about the dot.

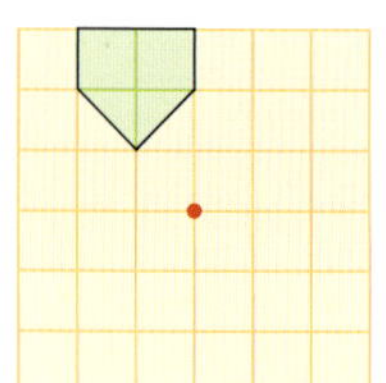

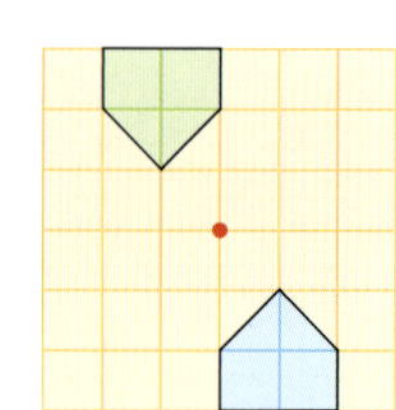

A rotation of 180° can be clockwise or anticlockwise.

You can use tracing paper to help rotate shapes.

- **To describe a rotation you give**
  - **the centre of rotation – the point about which it turns**
  - **the angle of turn**
  - **the direction of turn – either clockwise or anticlockwise.**

Example

a Give the mathematical name of the green shape.
b Draw the position of the green shape after a rotation of 90° clockwise about the origin.

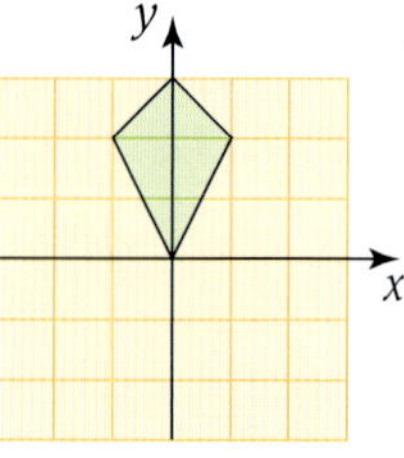

The **origin** is the point (0, 0).

a Kite

b

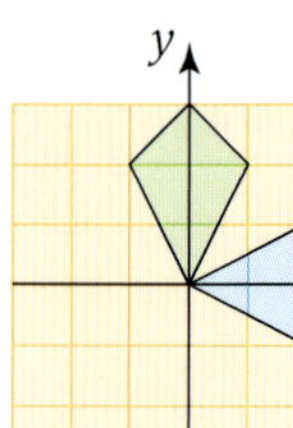

Exercise S7.2

1 State the angle and direction of turn for each of these rotations, green shape to blue shape.

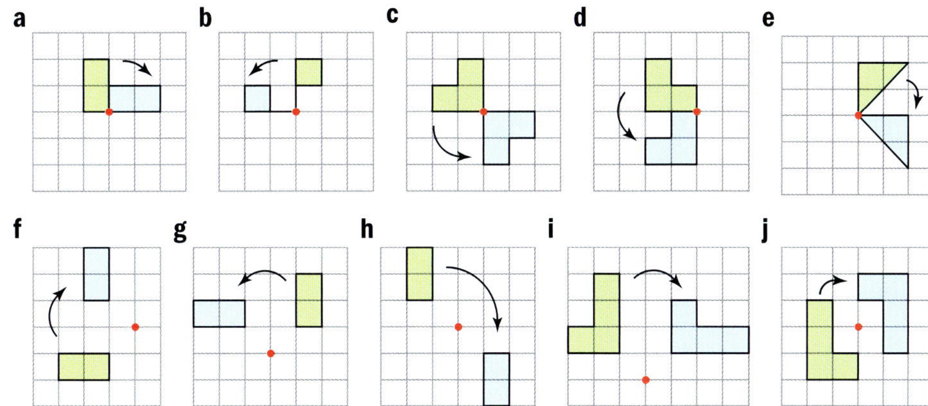

2 Copy these shapes on square grid paper. Rotate each shape through the given angle and direction about the dot (•).

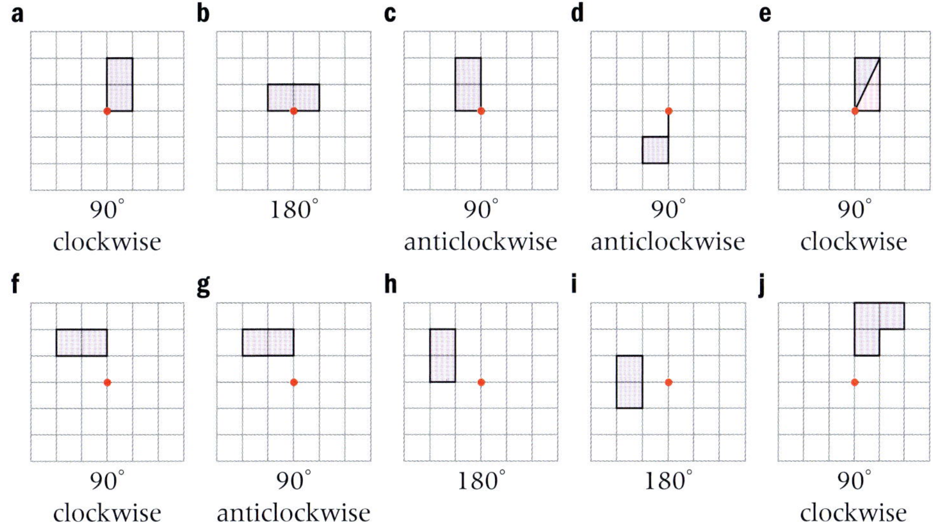

3 **a** Plot and join the points (0, 0) (1, 2) (0, 4) and (−1, 2) on a copy of this grid.

b Give the mathematical name of this shape.

c Rotate the shape through 90° anticlockwise about the origin.

d Give the coordinates of the rotated points.

e Are the two shapes congruent?

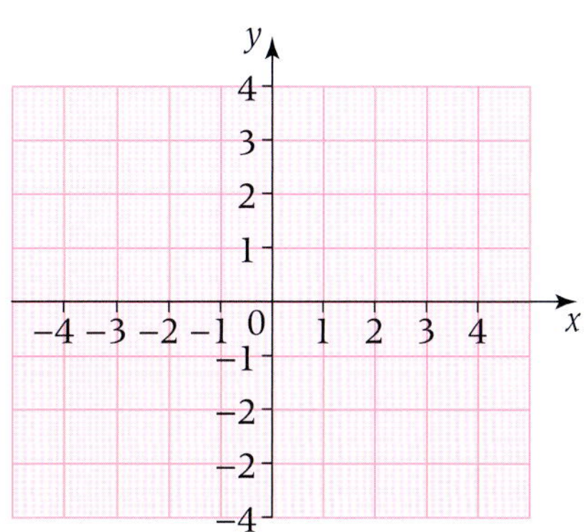

S7.3 More translations

This spread will show you how to:

- Recognise and visualise translations
- Understand that translations are specified by a distance and direction

Keywords
Congruent
Slide
Translation

- **A translation is a sliding movement.**

You specify the distance moved

- right or left, then
- up or down.

Translate the object 5 units right and 1 unit up $\begin{pmatrix}5\\1\end{pmatrix}$.

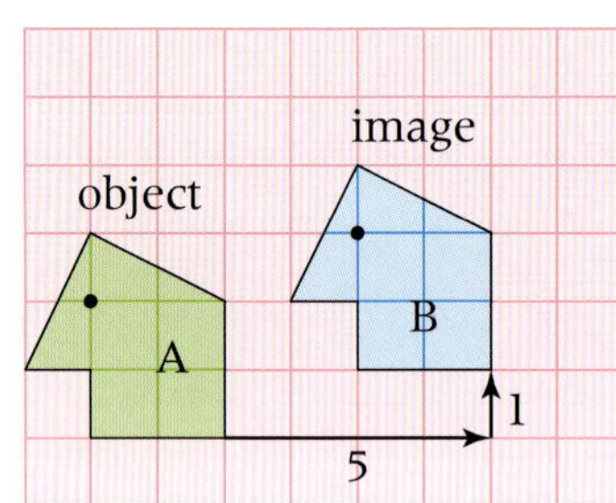

The two shapes are congruent.

You can write the translation in a column like this $\begin{pmatrix}\text{right}\\\text{up}\end{pmatrix}$.

$\begin{pmatrix}5\\1\end{pmatrix}$ means 5 right and 1 up.

Left and down are negative directions.

Shape B to shape A is $\begin{pmatrix}\text{5 left}\\\text{1 down}\end{pmatrix}$ or $\begin{pmatrix}-5\\-1\end{pmatrix}$.

Here are some other examples of translations.

2 units right and 1 unit down

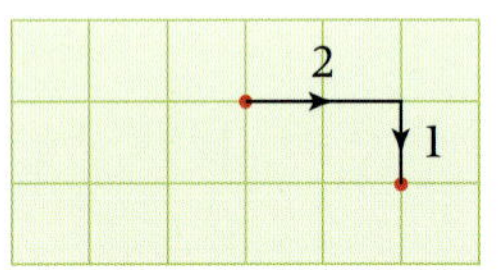

$\begin{pmatrix}2\\-1\end{pmatrix}$

4 units left and 1 unit down

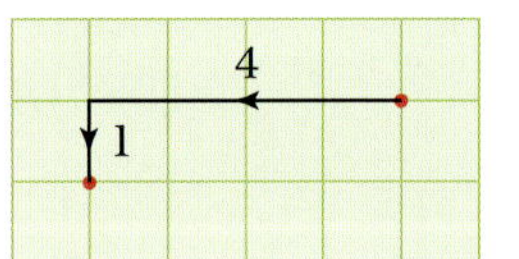

$\begin{pmatrix}-4\\-1\end{pmatrix}$

Example

Describe fully the transformation that moves the shaded triangle to

a shape A
b shape B.

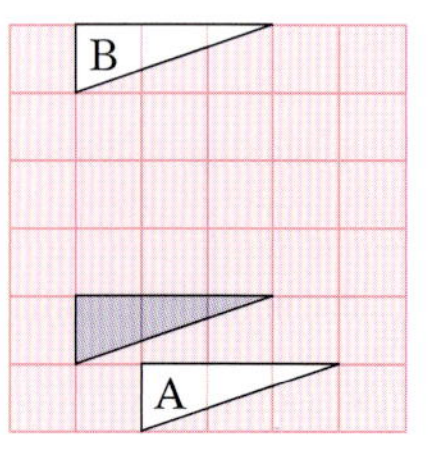

a A translation of 1 unit right and 1 unit down = $\begin{pmatrix}1\\-1\end{pmatrix}$.

b A translation of 0 units right and 4 units up = $\begin{pmatrix}0\\4\end{pmatrix}$.

Exercise S7.3

1 Which of these shapes are translations of the green shape?

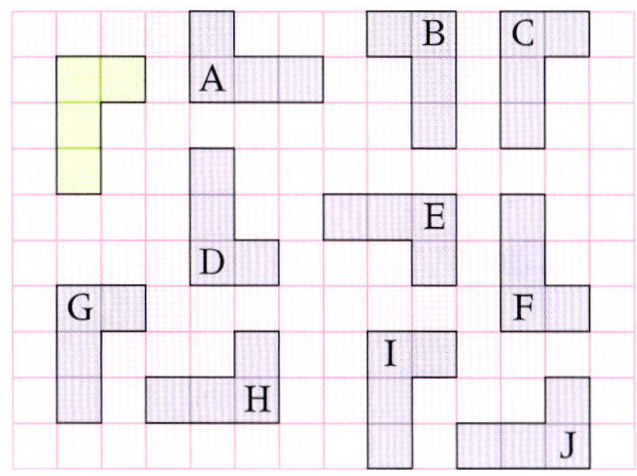

2 Describe these translations.

a A to B **b** A to C
c A to D **d** A to E
e B to C **f** B to E
g C to E **h** C to A
i D to C **j** D to B
k B to A **l** B to D

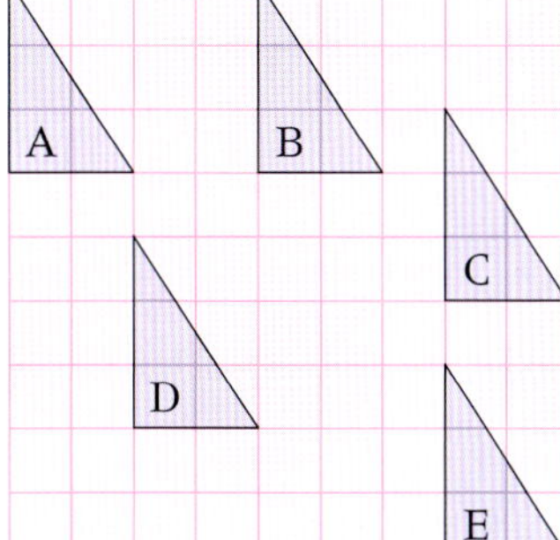

3 **a** Give the coordinates of the point A.
b What is the mathematical name of the shape?
c Copy the diagram. Draw the shape after a translation of 4 units left and 3 units down.
d State whether the two shapes are congruent.
e Give the coordinates of the point A after the translation.

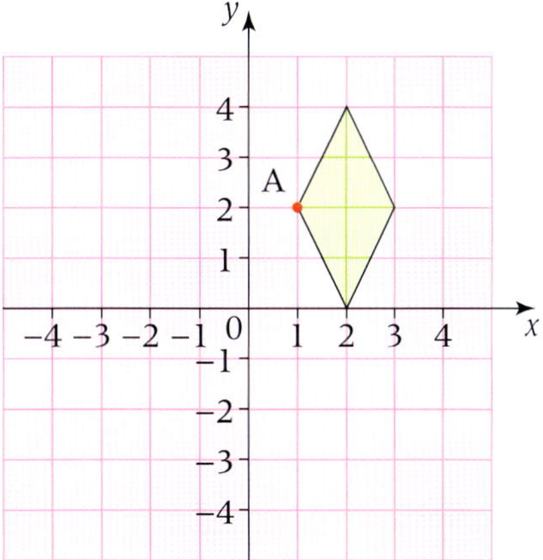

4 Copy these quadrilaterals onto square grid paper. Translate the shapes and give the mathematical name of the new shape you have made.

a

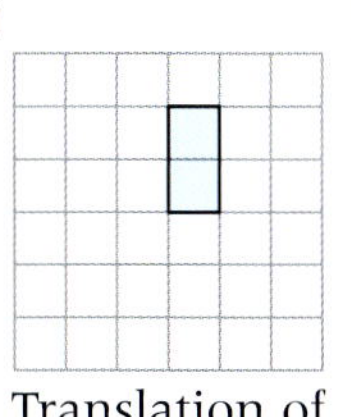

Translation of $\begin{pmatrix} 1 \\ 0 \end{pmatrix}$

b

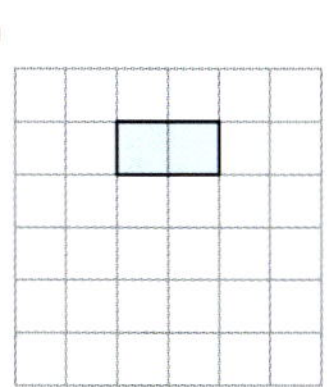

Translation of $\begin{pmatrix} 0 \\ -1 \end{pmatrix}$

c

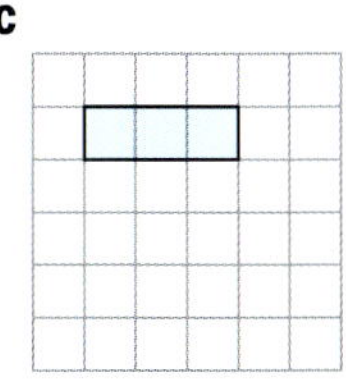

Translation of $\begin{pmatrix} 0 \\ -1 \end{pmatrix}$

d

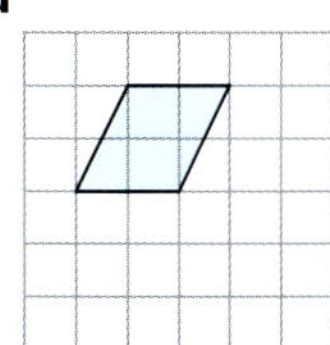

Translation of $\begin{pmatrix} 2 \\ 0 \end{pmatrix}$

e

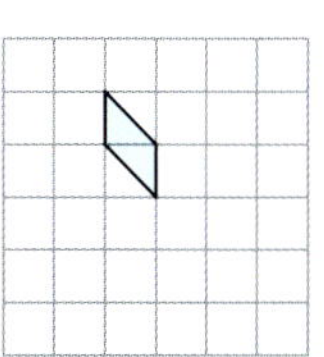

Translation of $\begin{pmatrix} 1 \\ -1 \end{pmatrix}$

S7.4 Enlargements

This spread will show you how to:

- Understand that enlargements are specified by a centre of enlargement and a scale factor

Keywords
Enlargement
Multiplier
Proportion
Scale factor
Similar

In an **enlargement** the lengths change by the same **scale factor**.

- **The scale factor is the multiplier in an enlargement.**

The two bees are **similar** – the same shape but different sizes.

The bee has been enlarged by a scale factor of 2.

3 cm × 2 = 6 cm

The green rectangle is an enlargement of the yellow rectangle.

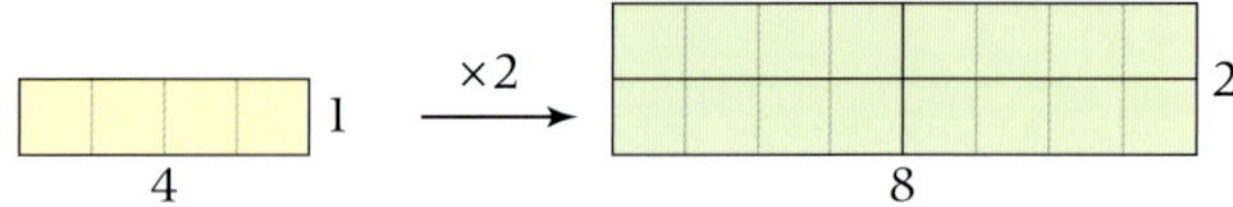

Corresponding lengths are multiplied by 2:

$4 \times 2 = 8 \qquad 1 \times 2 = 2$

The scale factor of this enlargement is 2.

- **In an enlargement**
  - **the angles stay the same**
  - **the lengths increase in proportion.**

Example

a Decide if these triangles are enlargements of the purple triangle. If so, calculate the scale factor.

b List the triangles that are similar to the purple triangle.

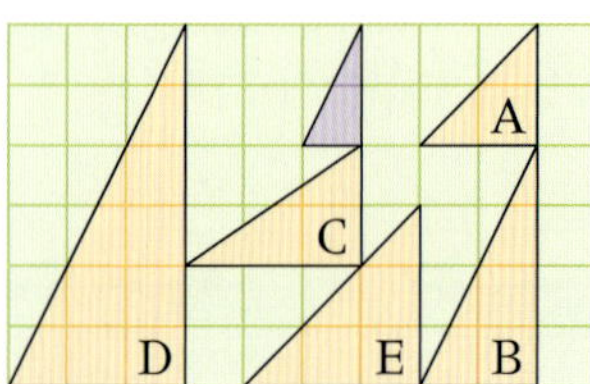

a A No, not an enlargement, as different shape.
B Yes, enlargement. Each length is multiplied by 2. Scale factor 2.
C No, not an enlargement, as a different shape.
D Yes, enlargement. Each length is multiplied by 3. Scale factor 3.
E No, not an enlargement, as a different shape.

b B and D are similar to the purple triangle.

Exercise S7.4

1 **a** Decide if these rectangles are enlargements of the yellow rectangle. If so, calculate the scale factor.

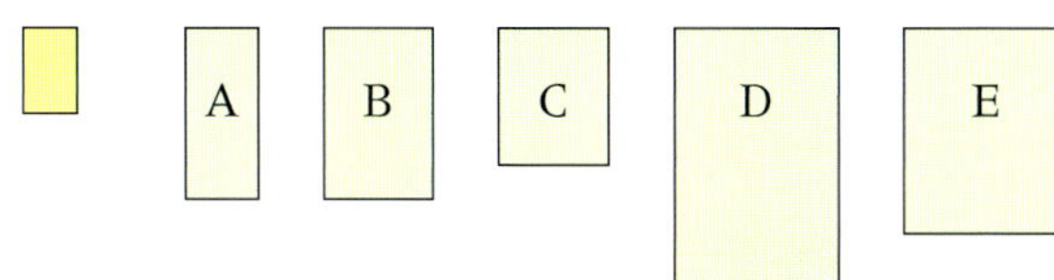

b List the rectangles that are similar to the yellow rectangle.

2 **a** Decide if these triangles are enlargements of the yellow triangle. If so, calculate the scale factor.

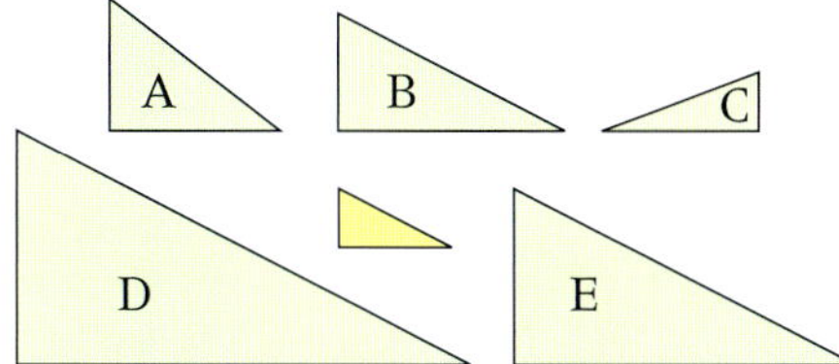

b List the triangles that are similar to the yellow triangle.

3 List the kites that are similar to the yellow kite.

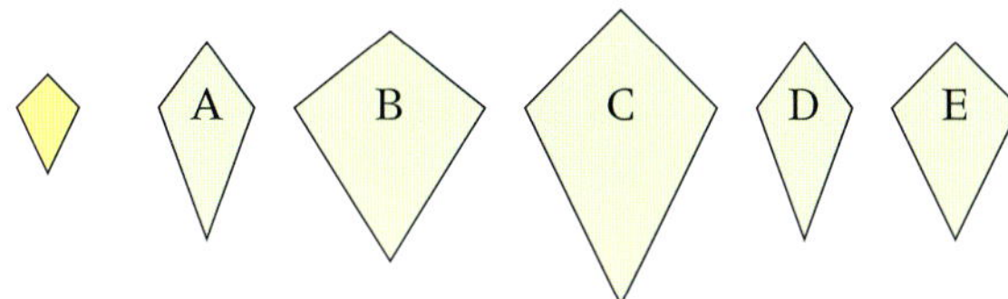

4 One rhombus is an enlargement of the yellow rhombus. State the letter of this shape and calculate the scale factor of the enlargement.

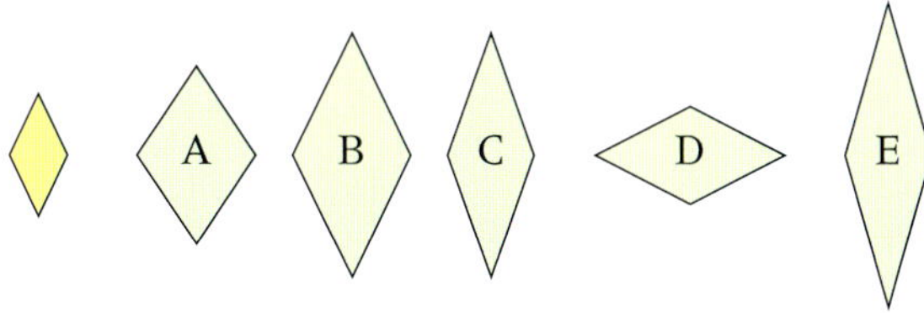

5 **a** Draw these triangles by plotting the coordinates on a copy of this grid.

Yellow	(5, 5)	(7, 5)	(5, 6)
A	(6, 9)	(9, 9)	(6, 10)
B	(2, 7)	(8, 7)	(2, 10)
C	(8, 4)	(10, 4)	(8, 6)
D	(2, 2)	(10, 2)	(2, 6)
E	(1, 0)	(5, 0)	(1, 2)

b State the triangles that are enlargements of the yellow triangle. Calculate the scale factor of the enlargement in each case.

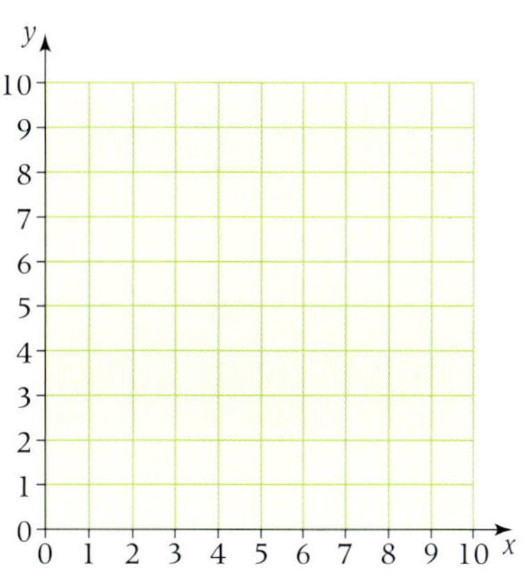

S7.5 Enlarging shapes

This spread will show you how to:

- Understand the effect of enlargement on the angles and lengths of shapes and how this relates to the scale factor of the enlargement
- Recognise similar shapes

Keywords
Enlargement
Multiplier
Scale factor
Similar

- In an **enlargement**
  - the angles stay the same
  - the lengths increase in proportion.

The green triangle is an enlargement of the yellow triangle.

Corresponding lengths are multiplied by 3:
$2 \times 3 = 6$

The **scale factor** of this enlargement is 3.

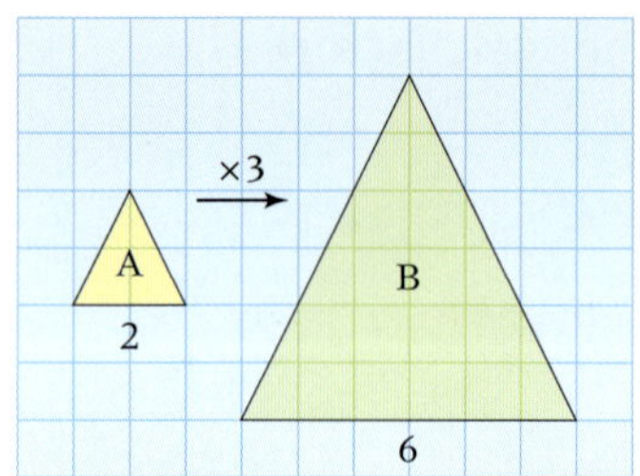

The two triangles are **similar** – the same shape but different sizes.

The scale factor of enlargement from B to A is $\frac{1}{3}$. You divide all the lengths by 3: $6 \div 3 = 2$.

- The scale factor is the **multiplier** in an enlargement.

Example

a Draw an enlargement of the yellow arrowhead with scale factor 2.
b Measure the length x. Measure the corresponding length y in your enlargement.
c Measure the angle at A. Measure the corresponding angle in your enlargement.

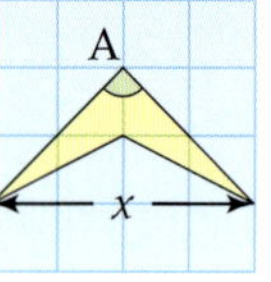

a

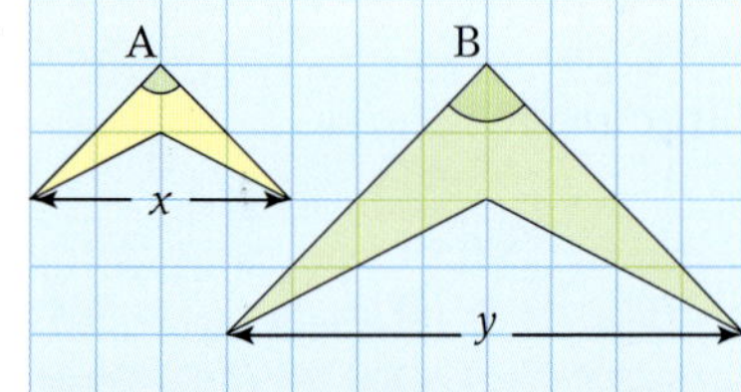

b $x = 4$ units
$y = 8$ units

c Angle A = 90°
Angle B = 90°

Check:
Scale factor is 2
$4 \times 2 = 8$.

Angles stay the same in enlargements.

Example

On the isometric paper draw an enlargement of the cuboid with scale factor 2.

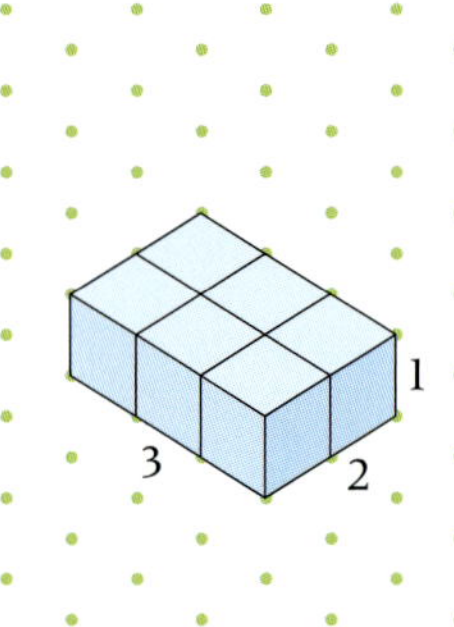

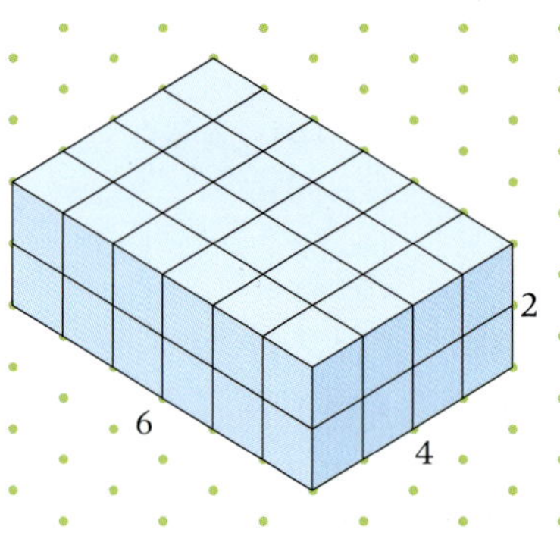

Check:
Scale factor is 2
$3 \times 2 = 6$
$2 \times 2 = 4$
$1 \times 2 = 2$

This topic is extended to similar shapes on page 386.

Exercise S7.5

1 Copy each diagram onto square grid paper. Enlarge each shape by the given scale factor.

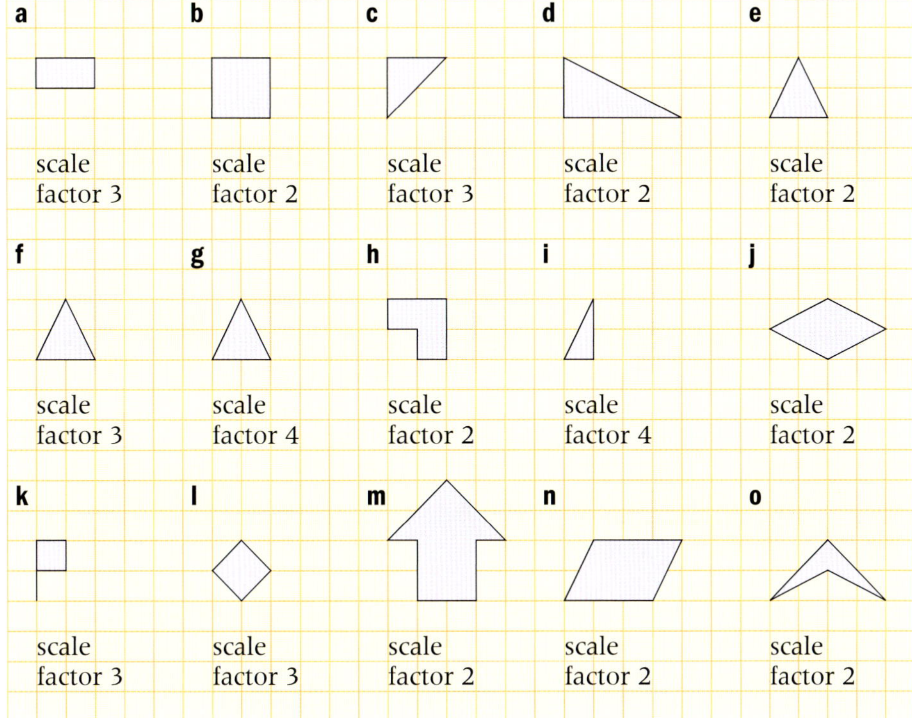

2 **a** Copy the 'L' shape onto square grid paper.

b Calculate the perimeter of the shape.

c Draw the shape after an enlargement of scale factor 3.

d Calculate the perimeter of the enlarged shape.

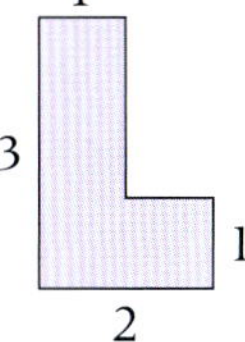

3 **a** Copy the triangle onto square grid paper.

b Measure the shaded angle.

c Draw the triangle after an enlargement of scale factor 2.

d Measure the corresponding shaded angle in the enlargement.

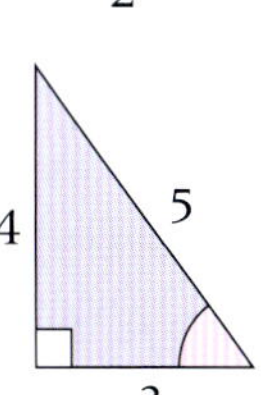

4 Copy each diagram onto isometric paper.
Enlarge each shape by the given scale factor.

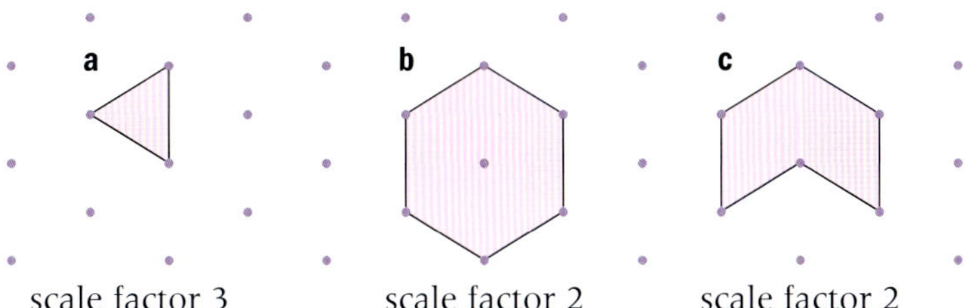

S7 Exam review

Key objectives

- Understand that rotations are specified by a centre and an angle
- Understand that translations are specified by giving a distance and direction, and enlargements by a centre and a positive scale factor
- Understand that reflections are specified by a mirror line
- Recognise and visualise rotations, translations and reflections
- Transform triangles and other 2-D shapes by reflection, rotation and translation, recognising that these transformations preserve length and angle, so that any figure is congruent to its image under any of these transformations
- Recognise, visualise and construct enlargements of objects using positive integer scale factors
- Recognise that enlargements preserve angle but not length

1 Rectangle B is an enlargement of rectangle A.
What is the scale factor of the enlargement? (1)

2.5 cm A 4 cm

5 cm B 8 cm

2 a Copy the grid and rotate triangle A 180° about O.
Label your new triangle B. (2)

b On your grid, enlarge triangle A by scale factor 2, centre O.
Label your new triangle C. (3)

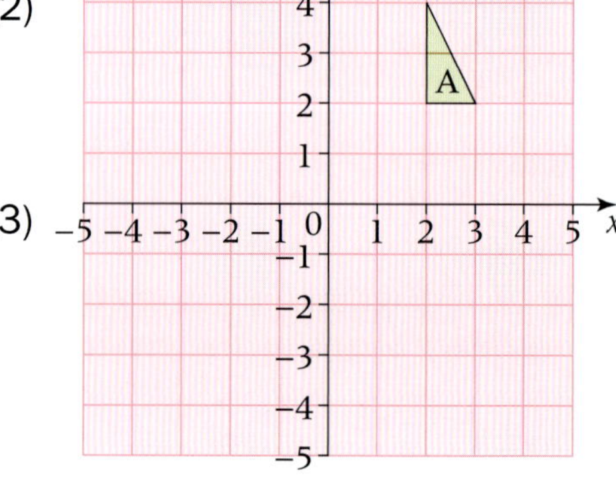

(Edexcel Ltd., 2003)

Further probability

This unit will show you how to

- Understand and use the probability scale
- Calculate probabilities of outcomes, including equally likely outcomes, giving answers in their simplest form
- List all outcomes for single events
- Use and interpret two-way tables for discrete data
- Calculate the expected frequency of an outcome
- Understand and use relative frequency
- List all outcomes for two successive events in a systematic way

Before you start ...

You should be able to answer these questions.

1 Choose a number from the rectangle that is

a prime

b square

c triangular

d a multiple of 4

e a factor of 10.

1	7	9	
4	5	2	8
10	3		6

Review: Unit N5

2 Cancel these fractions to their simplest form.

a $\frac{12}{18}$ **b** $\frac{8}{10}$ **c** $\frac{5}{20}$ **d** $\frac{15}{25}$ **e** $\frac{10}{10}$

Review: Unit N3

3 Work out each of these.

a $1 - \frac{1}{4}$ **b** $1 - \frac{7}{10}$ **c** $1 - \frac{3}{5}$

Review: Unit N3

4 Work out each of these.

a $\frac{1}{5} \times 50$ **b** $\frac{2}{5} \times 100$ **c** $\frac{2}{3} \times 300$

Review: Unit N3

D6.1 Probability revision

This spread will show you how to:

- Understand and use the probability scale
- Calculate probabilities of outcomes, including equally likely outcomes, giving answers in their simplest form
- List all outcomes for single events

Keywords
Event
Outcome
Probability
Probability scale

An **event** is an activity, for example, spinning a spinner.

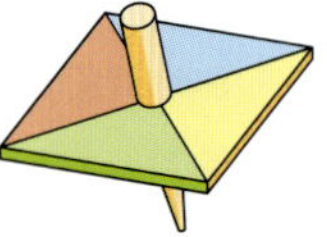

The possible **outcomes** are Blue, Yellow, Green, Red.

The **probability** is a number that measures how likely it is that an outcome will happen.

All probabilities have a value between 0 and 1.

0 means impossible.
1 means certain.

- You can show a probability on a **probability scale**.

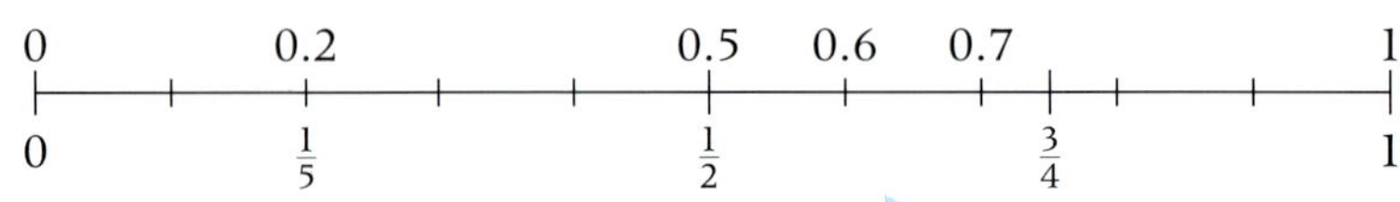

more and more likely to happen

You can use decimals, fractions or percentages.

- **You can calculate the probability using these formulae:**

Probability of an outcome happening $= \dfrac{\textbf{number of ways the outcome can happen}}{\textbf{total number of all possible outcomes}}$

Probability of an outcome not happening = 1 – probability of the outcome happening

Probability of an outcome can be written as P(outcome).

Example

20 counters are put into a bag. There are 1 red, 3 blue, 4 yellow and 12 green counters. Edward takes out a counter without looking. Calculate, giving your answers in simplest form, the probability that Edward takes out

a a red counter
b any colour that is not red
c a green or yellow counter
d a purple counter.

'Simplest form' means cancel the fraction.

a There is one red counter. There are 20 possible outcomes. P(red) = $\frac{1}{20}$
b $1 - \frac{1}{20} = \frac{19}{20}$ P(not red) = $\frac{19}{20}$
c There are 16 green or yellow counters. P(green or yellow) = $\frac{16}{20} = \frac{4}{5}$
d There are 0 purple counters. P(purple) = $\frac{0}{20} = 0$

Exercise D6.1

For all these questions, the outcomes are equally likely.

1 A bag contains four red and six blue balls. One ball is taken out of the bag.

a Calculate the probability that the ball is red.

b Calculate the probability that the ball is blue.

c Draw a probability scale as shown.

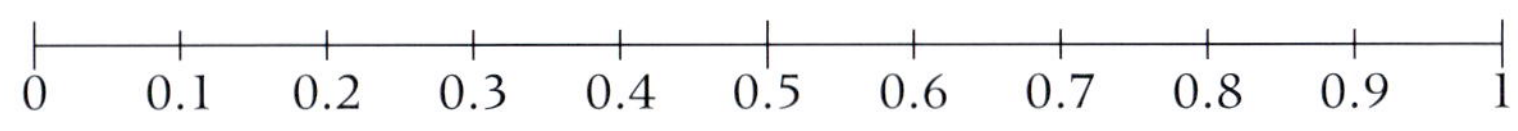

On your scale mark the positions of P(red) and P(blue).

d Which colour ball is most likely to be taken out?

e Which colour ball is least likely to be taken out?

2 This spinner is spun. Calculate the probability that the spinner lands on

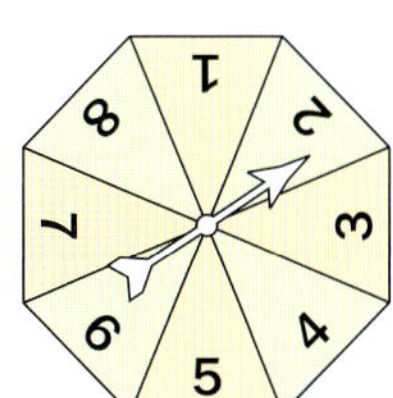

a a 3

b an even number

c a number greater than 6

d a number less than 6

e a square number

f a multiple of 3

g a multiple of 4

h a prime number.

3 A raffle has only one prize. 200 raffle tickets are sold. Calculate the probability of winning the prize if you buy

a one ticket

b two tickets.

4 There are 25 students in a class. Each student is given a different number from 1 to 25. The 25 numbers are put into a bag, and one is taken out. Calculate the probability that the number is

a odd **b** not odd **c** a multiple of 5

d a multiple of 10 **e** a square number

f a prime number **g** greater than 18

h less than 10 **i** not less than 10.

D6.2 Two-way tables

This spread will show you how to:

- Use and interpret two-way tables for discrete data

Keywords
Equally likely
Random
Two-way table

When someone or something is chosen at **random**, each person or item must be **equally likely** to be chosen.

For example, picking counters from a bag, provided the counters are replaced and the bag is well shaken.

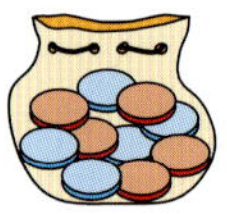

- A **two-way table** links two types of information.

100 students were asked whether they preferred swimming or running for exercise.

	Boys	Girls
Swimming	9	28
Running	31	32

9 boys preferred swimming.

32 girls preferred running.

$9 + 28 = 37$ students preferred swimming.
$31 + 32 = 63$ students preferred running.
$9 + 31 = 40$ students were boys.
$28 + 32 = 60$ students were girls.

You can calculate the probability of randomly selecting the different groups of students using the two-way table.

Example

The two-way table gives the number of males/females and adults/children in a room.
One person is selected at random.

	Adults	Children
Male	18	5
Female	15	12

Calculate the probability that the person selected is

a a female

b a child

c an adult male.

---

a $15 + 12 = 27$ female

P(female) $= \frac{27}{50}$

The total number of people is $18 + 5 + 15 + 12 = 50$.

b $5 + 12 = 17$ children

P(child) $= \frac{17}{50}$

c There are 18 adult males.

P(adult male) $= \frac{18}{50} = \frac{9}{25}$

Exercise D6.2

1 The speeds of vehicles passing a school are measured. The first 50 vehicles are shown in the two-way table.
Calculate the probability that a vehicle passing the school is travelling at

	Number of vehicles
30 mph or under	15
Over 30 mph	35

a 30 mph or under

b over 30 mph.

The speed limit outside the school is 30 mph.

c What is the probability that a vehicle passing the school is breaking the law?

2 The numbers of students in a class who are right-handed or left-handed are shown in the two-way table.

	Boys	Girls
Right-handed	12	13
Left-handed	3	2

a How many students are in the class altogether?

A student is chosen at random.
What is the probability that the student is

b a right-handed boy   **c** a right-handed girl

d a left-handed boy   **e** a left-handed girl?

3 A building set consists of red and yellow bricks.
Each brick is either a cube or a cuboid.
The two-way table shows the number of each type of brick in the set.

	Cubes	Cuboids
Red	8	5
Yellow	1	6

a How many bricks are in the building set altogether?

b How many yellow bricks are in the building set altogether?

c How many cubes are in the building set altogether?

A brick is selected at random. Calculate the probability that the brick is

d yellow   **e** a cube.

4 The numbers of people in a library are shown in the two-way table.

	Males	Females
Under 18	5	10
18 or over	20	15

a How many people are in the library altogether?

b How many males are in the library?

c How many under 18s are in the library?

One person is selected at random. Calculate the probability that the person is

d a male   **e** an under 18   **f** a female who is under 18.

D6.3 Expected frequency

This spread will show you how to:

- Calculate the expected frequency of an outcome

Keywords
Expect
Expected frequency
Trial

The probability is a number that measures how likely it is that an outcome will happen.

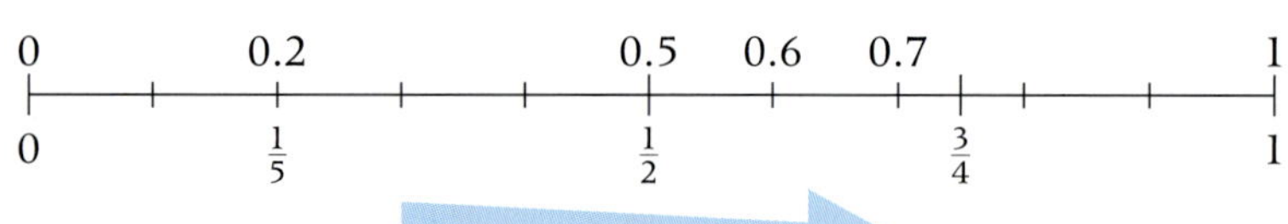

0 means impossible.
1 means certain.

If you know the probability of an outcome, you can calculate how many times you **expect** the outcome to happen.

Example

The probability of getting a Tail when you spin a coin is $\frac{1}{2}$.
How many Tails would you expect if the coin was spun 100 times?

$\frac{1}{2}$ of 100 = 50 Tails

- The **expected frequency** is the number of times you expect the outcome to happen.

- **Expected frequency = probability × number of trials**

Each spin of the coin is called a **trial**.

Example

Red and yellow counters are put in a bag.
The probability of taking out a red counter is $\frac{3}{5}$.

a Calculate the probability of taking out a yellow counter.
b If a counter is taken out and replaced 20 times, how many yellow counters would you expect to be taken out?

a $1 - \frac{3}{5} = \frac{2}{5}$

P(yellow counter) = $\frac{2}{5}$

b Expected frequency = probability × number of trials

$= \frac{2}{5} \times 20$

= 8 yellow counters

The expected frequency does not guarantee the outcome.

For example, if you spin a coin 100 times you may not always get the outcome of 50 Tails.

Exercise D6.3

1 A coin is spun.

a State the probability of spinning a Head.

b If the coin is spun 50 times, how many Heads would you expect?

2 An ordinary dice is rolled.

a Find the probability of rolling a 3.

b If the dice is rolled 60 times, how many 3s would you expect?

3 The spinner is made from a regular pentagon.

a Calculate the probability of spinning an even number.

b If the spinner is spun 100 times, how many even numbers would you expect?

4 The probability of sun on a day in June on the Costa del Sunny is $\frac{2}{3}$. Calculate the number of days in June on which you would expect sun on the Costa del Sunny.

30 days has September, April, June and November.

5 A bag contains 3 red balls and 7 green balls. One ball is taken out and then replaced back in the bag.

a Calculate the probability that the ball is red.

b Calculate the probability that the ball is green.

If a ball is taken out and replaced 100 times, how many of the balls would you expect to be

c red **d** green?

6 The probability that a fuchsia plant will survive after a severe ground frost is $\frac{4}{5}$.

a Calculate the probability that the fuchsia will not survive after a severe ground frost.

A gardener has 50 of these fuchsia plants.

b How many of the plants should he expect to die after a severe ground frost?

7 The probability of seeing a red car is $\frac{7}{20}$.

a Calculate the probability of not seeing a red car.

b If 100 cars go past you, how many of these would you expect not to be red?

8 The probability that a seed germinates is $\frac{9}{10}$. If 60 seeds are planted, how many seeds would you expect to germinate?

D6.4 Relative frequency

This spread will show you how to:

- Understand and use relative frequency

Keywords
Biased
Estimate
Fair
Relative frequency
Trial

- The probability is a number that measures how likely it is that an outcome will happen.

You can calculate a **theoretical** probability for objects such as coins and dice.

It is not always possible to calculate the theoretical probability, for example the probability of a car accident on a stretch of road.

You can, however, **estimate** the probability from experiments.

This topic is extended on page 398.

- **The estimated probability is called the relative frequency.**

Example

James spins a square spinner 50 times.
The results are shown in the data collection sheet.

Colour	Tally	Frequency
Red	𝍸 𝍸	10
Blue	𝍸 𝍸 IIII	14
Yellow	𝍸 IIII	9
Green	𝍸 𝍸 𝍸 II	17

a Estimate the probability of getting green on the spinner.
b Do you think the spinner is biased?
Explain your answer.

The spinner is **biased** if the colours are NOT all equally likely.

a The spinner was green on 17 out of 50 occasions.
Estimated probability of getting green = $\frac{17}{50}$

b You would expect each frequency to be about the same for a fair spinner.
The spinner could be biased as there are many more green than yellow.
However, James needs to spin the spinner many more times before he can make the decision.

The spinner is **fair** if the colours are all equally likely.

The estimated probability becomes more and more reliable the greater the number of trials.

Each spin of the spinner is called a **trial**.

Exercise D6.4

1 A tetrahedron dice is rolled 50 times. The scores are shown.

4	3	2	2	1	4	2	3	1	4
3	2	1	4	4	3	2	1	1	2
4	2	2	3	1	1	2	4	4	3
3	3	2	1	4	3	4	2	2	1
2	1	4	2	4	3	4	2	1	1

a Copy and complete the frequency chart to show the 50 scores.

Score	Tally	Frequency
1		
2		
3		
4		

b State the modal score.

c Estimate the probability of rolling a

i 1 **ii** 2 **iii** 3 **iv** 4

2 A spinner is made from a regular octagon.
The spinner is spun 40 times and the colour is recorded.

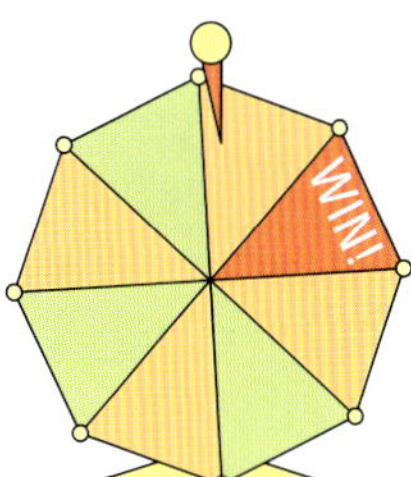

Orange	Red	Orange	Green	Orange	Orange
Orange	Green	Red	Green	Orange	Orange
Orange	Green	Orange	Orange	Orange	Orange
Orange	Green	Orange	Orange	Red	Green
Green	Green	Orange	Green	Green	Green
Orange	Orange	Orange	Red	Green	Green
Green	Orange	Green	Red		

a Copy and complete the frequency chart to show the 40 colours.

Colour	Tally	Frequency
Orange		
Green		
Red		

b State the modal colour.

c Estimate the probability of spinning

i orange **ii** green **iii** red.

3 A coin is spun 100 times. The outcomes are shown in the frequency table.

	Frequency
Head	45
Tail	55

a Calculate the estimated probability of spinning a Head.

b Calculate the estimate probability of spinning a Tail.

c Do you think the coin is biased? Explain your answer.

D6.5 Two events

This spread will show you how to:

- List all outcomes for two successive events in a systematic way

Keywords
Event
Outcome
Successive
Systematic

An **event** is an activity.

- You can list the possible **outcomes** for two successive events.

Successive means following on, for example, 6, 7, 8.

Example

A restaurant decides to offer a two-course meal for £5.99.

The meal must be one starter and one main course.

List the different choices that are possible.

Onion Soup – Beef Steak
Onion Soup – Grilled Salmon
Onion Soup – Veggie Pasta
Caesar Salad – Beef Steak
Caesar Salad – Grilled Salmon
Caesar Salad – Veggie Pasta

This list is **systematic**. It is in order.

Example

A spinner has colours red, yellow, blue and green.
A coin has two faces, Heads or Tails.

Matthew spins the spinner and the coin.

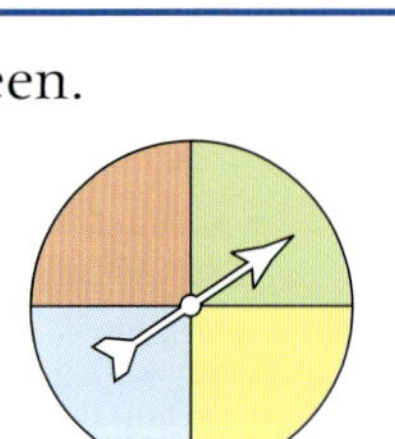

a List all the possible outcomes.
b What is the probability that he gets green and a Head?

a

Colour on spinner	Red	Red	Yellow	Yellow	Blue	Blue	Green	Green
Head/Tail on coin	Head	Tail	Head	Tail	Head	Tail	Head	Tail

b Green and Head occurs once.
There are 8 possible outcomes.

P(Green and a Head) = $\frac{1}{8}$

Exercise D6.5

1 At a sports club, there are three activities, but only two sessions. You have to choose one different activity for each session. Copy and complete the table to show the possible choices.

Choose 2 from
Tennis
Badminton
Squash

Session 1	Session 2
Tennis	Badminton
Tennis	

2 Four people, Arthur, Ben, Chris and Darryl, enter a competition.

a List the four possible winners of the competition.

It is decided to give another award, for 'Most improved player'.

b Copy and complete the table to show the 12 possible prize winners.

Winner	Most improved player
A	B
A	

3 A spinner is labelled 1, 2, 3. Another spinner is labelled A, B, C. Both spinners are spun.

a List the nine possible outcomes.

b Calculate the probability of getting a 3 and a C.

4 A dice is numbered from 1 to 6. A coin has Head or Tail. The dice is rolled and the coin is spun.

a List the 12 possible outcomes.

b Calculate the probability of getting a 3 and a Head.

c Calculate the probability of getting an even number and a Tail.

5 Three tracksuit tops are in a drawer. Another drawer has three tracksuit bottoms. One top and one bottom are randomly taken out of the drawers.

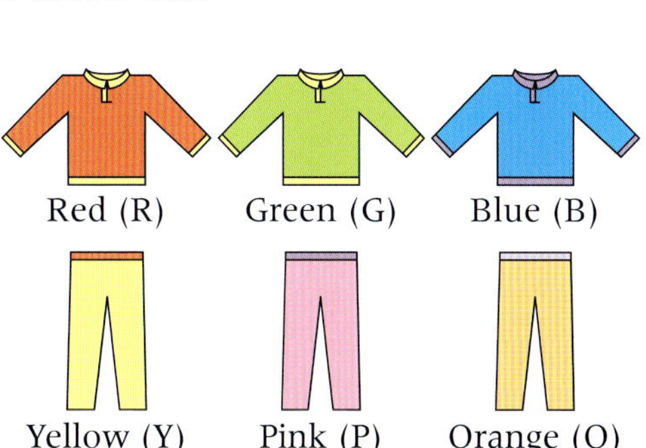

a List the nine possible outcomes.

b Calculate the probability of getting blue and yellow.

c Calculate the probability of not getting blue and yellow.

d Calculate the probability of getting blue and orange.

e Calculate the probability of not getting blue and orange.

D6 Exam review

Key objectives

- Use two-way tables for discrete data
- Understand and use the probability scale
- Understand and use estimates or measures of probability from theoretical models (including equally likely outcomes)
- List all outcomes for single events, and for two successive events, in a systematic way

1 Thomas rolls a fair six-sided dice, numbered 1 to 6, and then tosses a fair coin.

a Copy and complete the table to list the possible outcomes. (4)

Number on dice	Head/Tail on coin

b What is the probability that he gets an even number and a Head? (2)

2 Mark throws a fair coin.
He gets a Head.
Mark's sister then throws the same coin.

a What is the probability that she will get a Head? (1)

Mark throws the coin 30 times.

b Explain why he may not get exactly 15 Heads and 15 Tails. (1)

(Edexcel Ltd., 2004)

A8 Real-life graphs

This unit will show you how to

- Use and interpret conversion graphs
- Read values from graphs
- Draw conversion graphs
- Draw, use and interpret distance–time graphs
- Work out the average speed for a journey from a distance–time graph

Before you start ...

You should be able to answer these questions.

	Review
1 Draw a coordinate grid with x- and y-axes from 0 to 10. **a** Plot these points on your grid. (0, 4) (2, 6) (4, 4) (2, −4) **b** Connect the points in order and state the shape you have formed.	Unit A4, S4
2 Work out each of these. **a** 2.5 × 10 **b** 0.7 × 10 **c** 1.6 × 15 **d** 48.8 ÷ 8	Unit N6
3 Convert these distances in kilometres to metres. **a** 2 km **b** 60 km **c** 0.5 km **d** 3.5 km	Unit S1
4 If £1 = €1.45, work out these prices in euros. **a** Bananas £1 per bunch! **b** Ferry to Calais £15 **c** Flights to Berlin now only £34.50	Unit N8

A8.1 Conversion graphs

This spread will show you how to:

- Use and interpret conversion graphs
- Read values from graphs

Keywords
Conversion graph
Convert
Scale
Unit

Weights can be measured in kilograms or pounds.

You can use a **conversion graph** to

- **convert** a weight in pounds to a weight in kg
- convert a weight in kg to a weight in pounds.

kg is the metric **unit**, pound is the imperial unit.

- **You can use a conversion graph to convert between units of measurement.**

Example

Here is a conversion graph for converting pounds to kg and kg to pounds.
Use the graph to convert

a 12 pounds to kilograms **b** 7 kilograms to pounds.

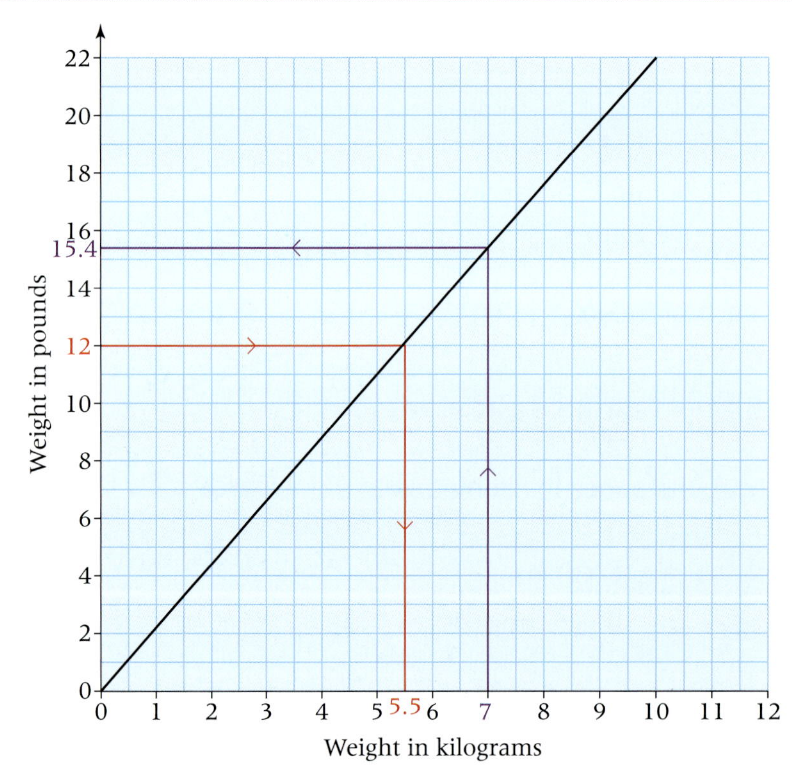

The vertical scale goes up in 2s.

The horizontal scale goes up in 1s.

a i Find 12 pounds on the pounds (vertical) axis.
ii Draw a line across to the graph.
iii Draw a line down from the graph to the kilograms (horizontal) axis.
iv Read the value off the axis.

12 pounds = 5.5 kilograms

b i Find 7 kilograms on the kilograms (horizontal) axis.
ii Draw a line up to the graph.
iii Draw a line across from the graph to the pounds (vertical) axis.
iv Read the value off the axis.
Use the scale to estimate the value.

7 kilograms = 15.4 pounds

Exercise A8.1

1 Use the conversion graph on page 338 to convert

a 22 pounds to kilograms
b 11 pounds to kilograms
c 3 kilograms to pounds
d 9 kilograms to pounds
e 8.5 pounds to kilograms
f 3.5 pounds to kilograms
g 7.5 kilograms to pounds
h 2.5 kilograms to pounds.

2 Here is a conversion graph for

- euros to pounds
- pounds to euros.

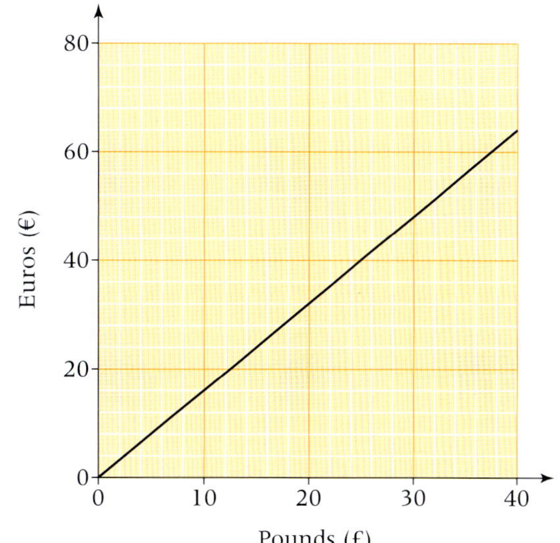

a Use the graph to convert

i €32 to pounds
ii €48 to pounds
iii £40 to euros
iv £25 to euros
v €50 to pounds
vi €35 to pounds

b Which is worth more: £1 or €1? Explain how you know.

3 Here is a conversion graph for

- miles to kilometres
- kilometres to miles.

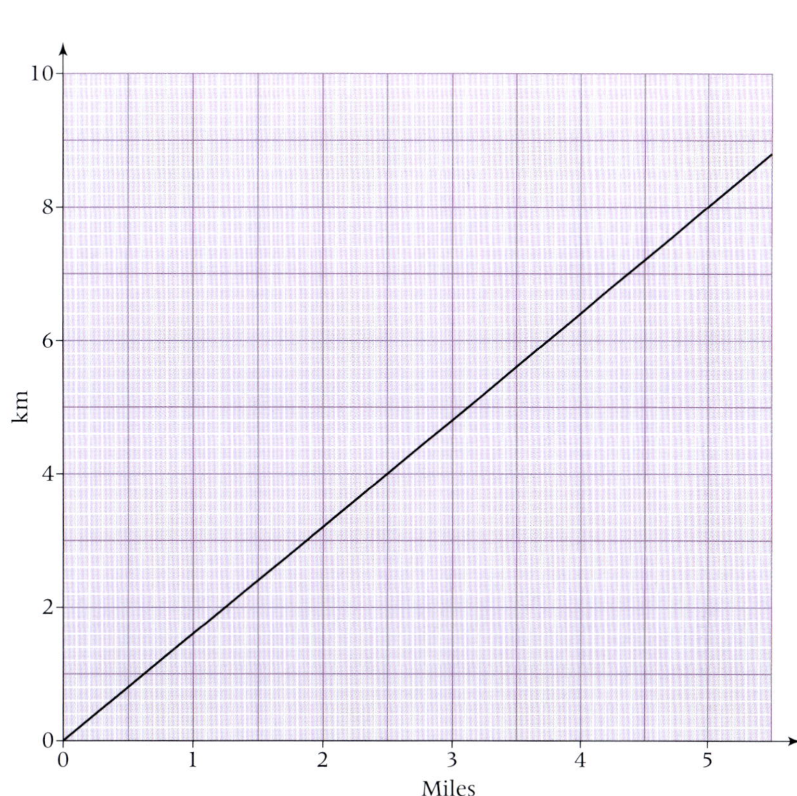

a Use the graph to convert

i 4 km to miles
ii 5 miles to km
iii 1 mile to km
iv 1 km to miles.

b Which is longer, 1 mile or 1 km? Explain how you know.

A8.2 More conversion graphs

This spread will show you how to:

- Use and interpret conversion graphs
- Read values from graphs

Keywords
Degrees Celsius (°C)
Degrees Fahrenheit (°F)

You can use this conversion graph to convert

- temperatures in °C to °F
- temperatures in °F to °C.

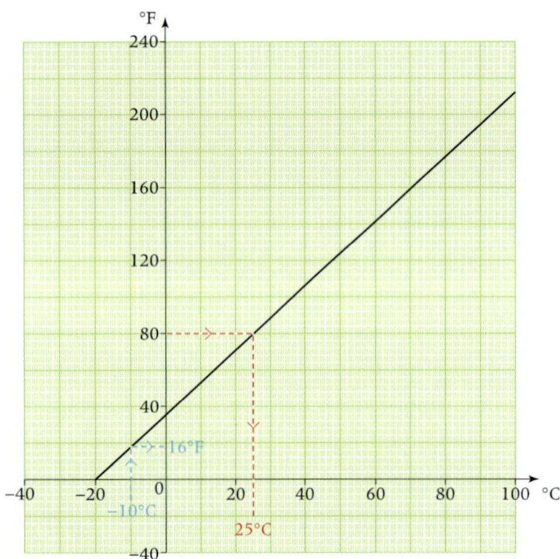

A weather forecast often gives temperatures in degrees Fahrenheit (°F) and degrees Celsius (°C).

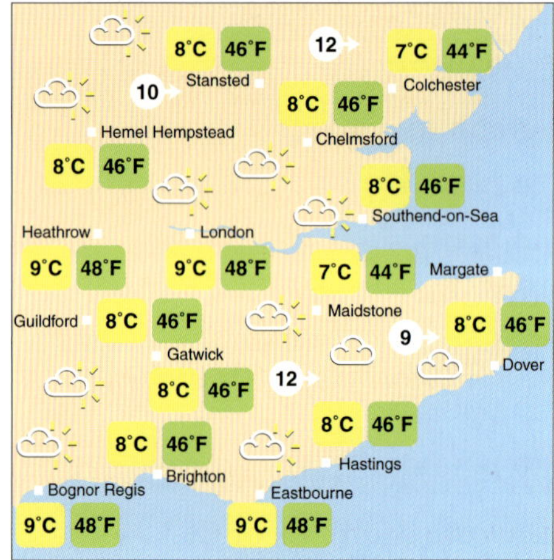

Example

Use the conversion graph to convert

a 80 °F to °C **b** −10 °C to °F.

a To convert 80 °F to °C

- **i** Find 80 °F on the °F (vertical) axis. (This scale goes up in 40s.)
- **ii** Draw a line across to the graph.
- **iii** Draw a line down from the graph to the °C (horizontal) axis.
- **iv** Read the value off the axis.

80 °F = 25 °C

b To convert −10 °C to °F

- **i** Find −10 °C on the °C (horizontal) axis.
- **ii** Draw a line up to the graph.
- **iii** Draw a line across from the graph to the °F (vertical) axis.
- **iv** Read the value off the axis.

−10 °C = 16 °F

Example

Use the conversion graph to help you write these temperatures in order, from coldest to hottest.

10 °F 60 °F 10 °C 80 °C 160 °F

To compare temperatures, they need to be measured in the same units.

From the graph, 10 °C = 50 °F

80 °C = 176 °F.

So in order the temperatures are 10°F 50°F 60°F 160°F 176°F

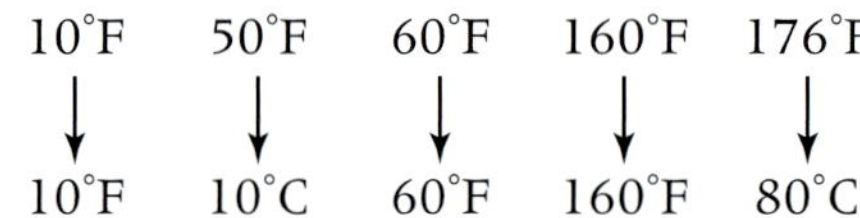

10°F 10°C 60°F 160°F 80°C

Use the original values in the answer.

Exercise A8.2

Use the conversion graph for °F to °C on page 340 to answer questions **1** to **3**.

1 Convert these temperatures.

a 120 °F to °C **b** 60 °F to °C **c** 40 °C to °F **d** 65 °C to °F

e 75 °F to °C **f** 8 °F to °C **g** −5 °C to °F **h** −20 °C to °F

2 Which is hotter?

a 20 °C or 20 °F **b** 150 °F or 40 °C **c** 18 °C or 60 °F

d 90 °F or 30 °C **e** −8 °C or 8 °F **f** −30 °F or −30 °C

3 **a** The melting point of ice is 32 °F. What is this in °C?

b The boiling point of water is 100 °C. What is this in °F?

c Average body temperature for humans is 37 °C. What is this in °F?

4 Here is a conversion graph for inches to centimetres.

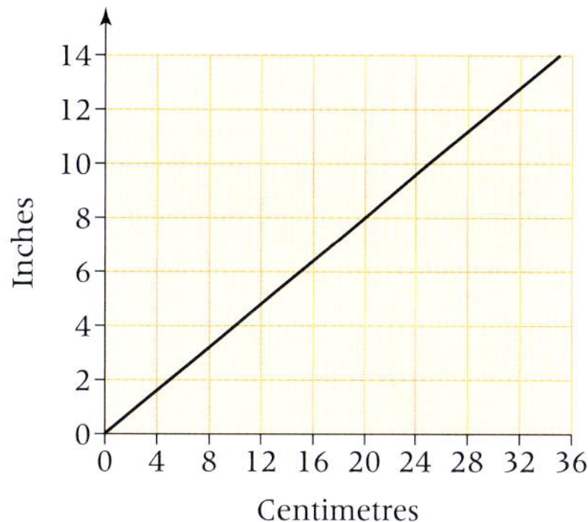

a Use the conversion graph to convert

i 6 inches to cm **ii** 10 cm to inches

iii 4 cm to inches **iv** 10 inches to cm.

b Use the graph to help you write these sets of lengths in order.

i 2 inches 12 cm 13 inches 9 cm 4 inches

ii 23 cm 1 inch 16 cm 5 inches 3 cm

c For a handling data project, a group of students measured the lengths of their feet.

Some measured in inches instead of centimetres.

i Convert all the measurements in inches to centimetres.

ii Write the students' feet in order of size.

Name	Length of foot
Nadia	19 cm
Mel	10 inches
Jonathan	13 inches
Omar	23 cm
Shelley	10.5 inches
Dustin	12 inches
Jake	29 cm

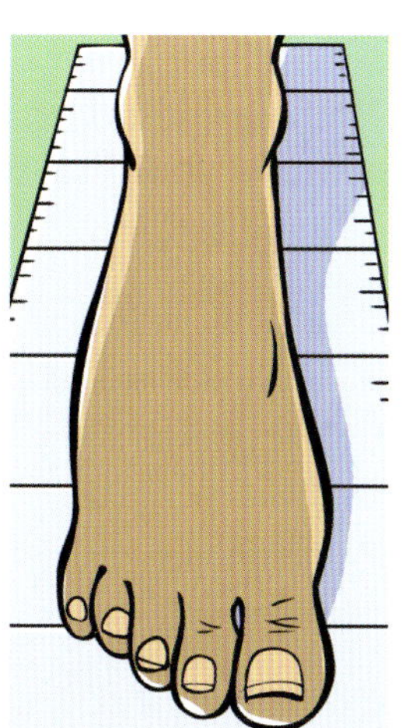

A8.3 Drawing conversion graphs

This spread will show you how to:

- Use and interpret conversion graphs
- Draw conversion graphs

Keywords
Conversion
Table of values

Joe is comparing data on heights of trees.
Some of the data is in feet and some is in metres.
He draws a conversion graph to help him convert from feet to metres easily.

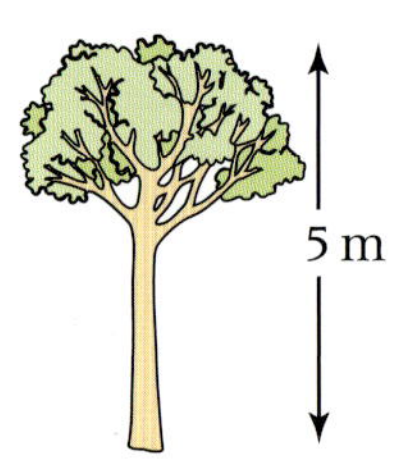

From a ruler, Joe sees that 1 foot ≈ 30 cm = 0.3 m.

1 foot is approximately equal to 30 cm.

He uses this **conversion** to draw a **table of values**.

Metres	0	0.3	3.0
Feet	0	1	10

0 feet = 0 metres

1 foot = 0.3 m
↓ ×10 ↓ ×10
10 feet = 3 m

He writes the coordinate pairs from the table. (0, 0) (0.3, 1) (3, 10)

To plot a straight line you only need to plot two points.

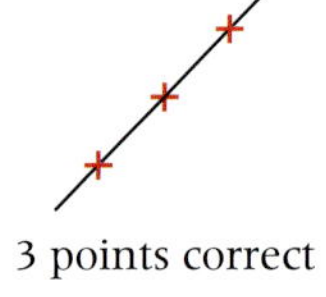

If you plot three, you can tell if you have made a mistake.

3 points correct

1 point must be wrong

He plots the points on a coordinate grid and joins them with a straight line.

- To draw a conversion graph
  - draw up a table of values with at least three values
  - plot the points from the table on a coordinate grid
  - join the points with a straight line
  - extend your line to the edges of the grid.

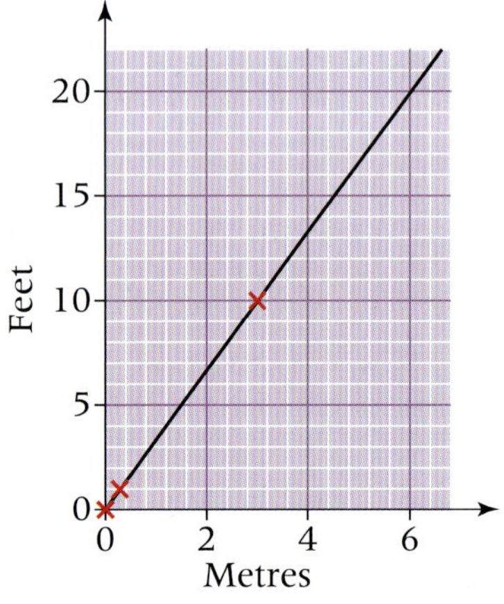

Example

Complete the table of values for a conversion graph for UK pounds to US dollars.

Pounds	0	1	10
US dollars		1.7	

Pounds	0	1	10
US dollars	0	1.7	17

0 pounds = 0 dollars

1 pound = 1.7 dollars
↓ ×10 ↓ ×10
10 pounds = 17 dollars

Exercise A8.3

1 **a** Copy and complete this table of values for a grams-to-ounces conversion graph.

Ounces	0	1	10
Grams		28	

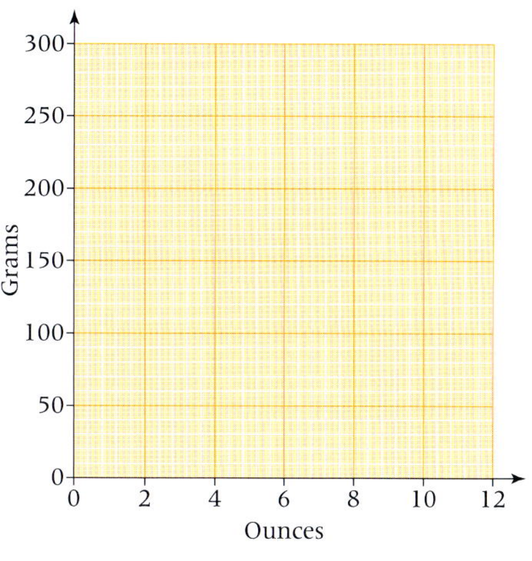

b Write three pairs of coordinates for the conversion graph.
Write them in this order
(number of ounces, number of grams)

c Copy these axes onto graph paper.

d Plot the points on the coordinate grid.
Join the points with a straight line.
Extend your line to the edge of the grid.

e Use your graph to convert

i 4 ounces to grams **ii** 200 g to ounces.

2 **a** Copy and complete this table of values for a kilometres-to-miles conversion graph.

Miles		5	10
Kilometres	0	8	

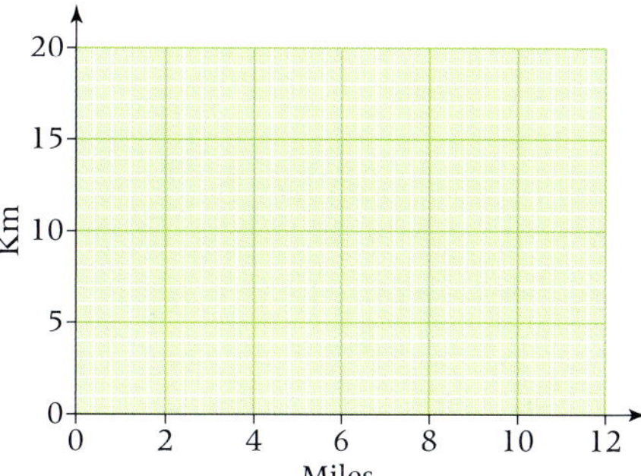

b Write three pairs of coordinates for the conversion graph.
Write them in this order
(number of miles, number of kilometres)

c Copy these axes onto graph paper.

d Plot the points on the coordinate grid and join the points with a straight line. Extend your line to the edge of the grid.

e Use your graph to convert

i 6 miles to kilometres **ii** 5000 metres to miles.

3 **a** Copy and complete this table of values for pounds to New Zealand dollars.

Pounds	0	1	10
NZ dollars		2.4	

b Write three pairs of coordinates for the conversion graph.

c Draw the graph by copying the axes and plotting the points.

d Use your graph to convert

i £8 to NZ dollars **ii** 12 NZ dollars to pounds.

e Use your graph to work out which is more

i £6 or 12 NZ dollars? **ii** 18 NZ dollars or £7?

A8.4 Distance–time graphs

This spread will show you how to:

- Draw, use and interpret distance–time graphs

Keywords

Distance
Horizontal
Time
Vertical

You can plot a graph for a journey.

- **You plot time on the horizontal axis and distance on the vertical axis.**

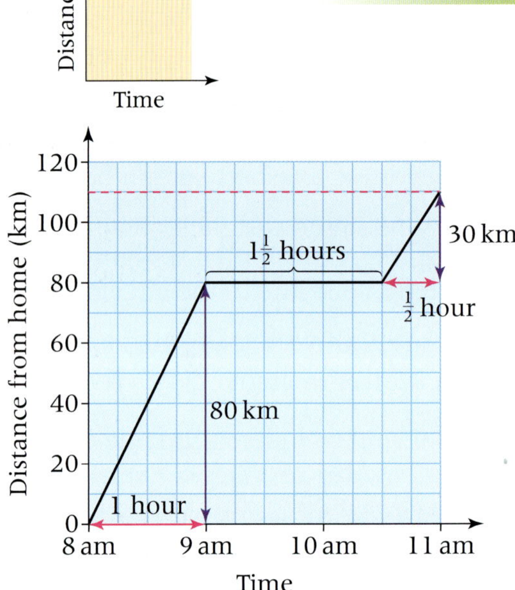

Shaun is a salesman.

- He leaves home at 8 am.
- He drives 80 km to his first meeting. This takes 1 hour.
- He arrives at 9 am.
- His meeting lasts $1\frac{1}{2}$ hours.
- At 10.30 am he sets off again.
- He drives 30 km to his next appointment. This takes $\frac{1}{2}$ hour.
- At 11 am he is 110 km from home.

While Shaun is in a meeting, he is not travelling. His distance from home does not change. The line on the graph is **horizontal**.

- **A horizontal line on a distance–time graph shows a break in the journey.**

Example

The graph shows Tristan's trip to the cinema and back again.

a What time did Tristan leave home?
b How far is the cinema from his home?
c How long was he at the cinema for?
d How long did he spend travelling home?

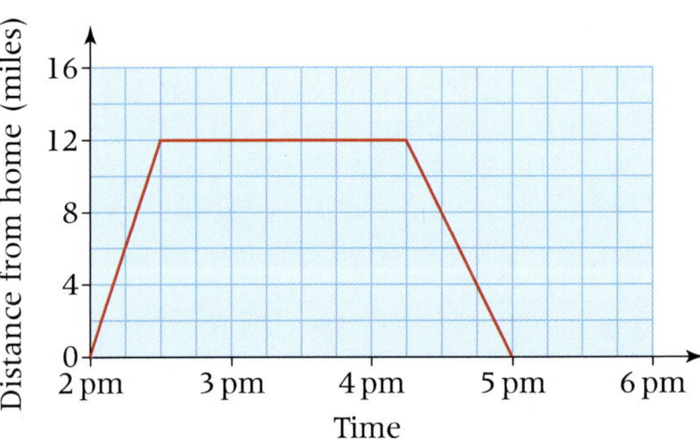

a He left home at 2 pm.
b 12 miles
c From 2.30 to 4.15 = $1\frac{3}{4}$ hours
d $\frac{3}{4}$ hour

On a journey going away from home, your distance from home is increasing. The graph slopes up.

On a journey back home, your distance from home is decreasing. The graph slopes down.

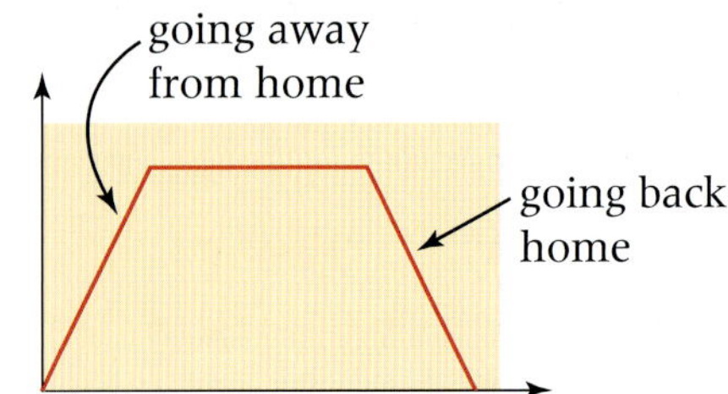

Exercise A8.4

1 Shona runs a corner shop.
The graph shows her trip from the shop to the cash and carry.

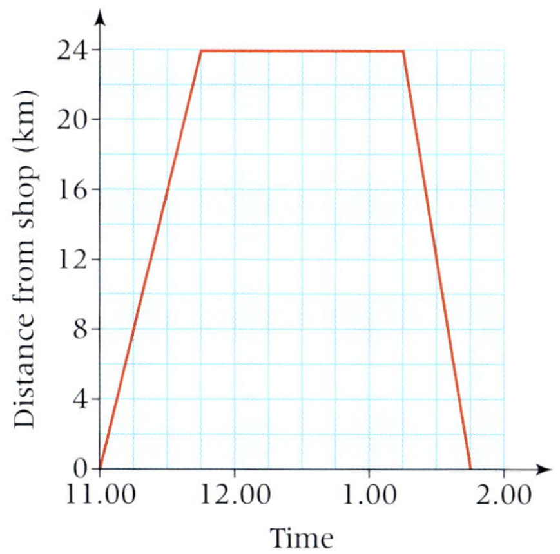

a What time did she set off to the cash and carry?

b What time did she arrive at the cash and carry?

c How long did the journey to the cash and carry take?

d How long did she spend at the cash and carry?

e How long did the journey home from the cash and carry take?

f How far is the cash and carry from Shona's shop?

g How many kilometres did she travel in total?

2 The graph shows a train journey.

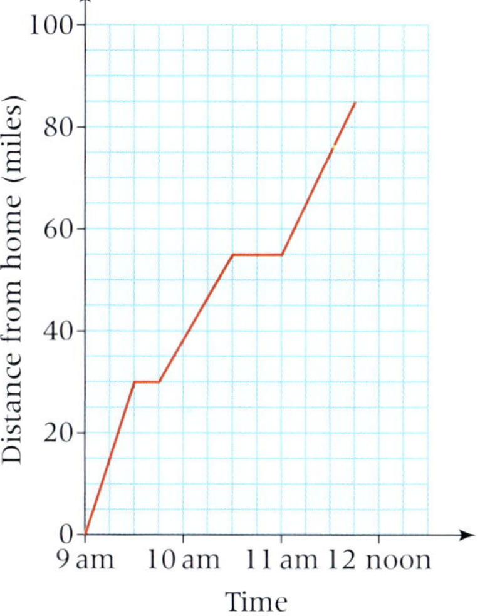

a How many stations did the train stop at?

b How far did the train travel before it stopped for the first time?

c How far did the train travel in total?

d How long did the whole journey take?

e At one station stop, the train had to wait for a connection.

Which stop do you think this was?

How long did it wait for?

3 The graph shows Luke's trip to the shop.

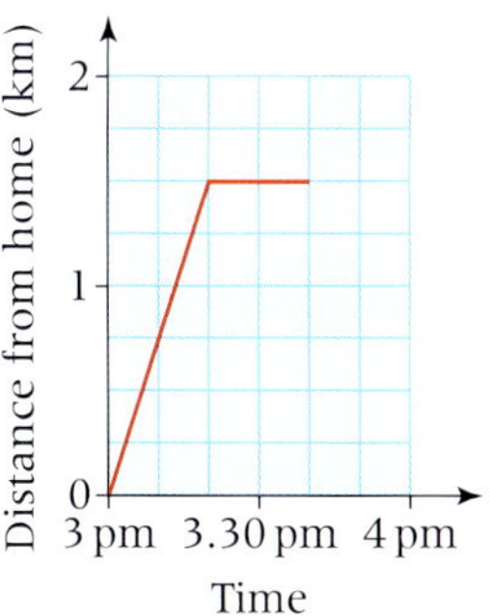

a How far is the shop from Luke's home?

b How many minutes did it take Luke to walk to the shop?

c How many minutes did he spend in the shop?

d What time did he leave the shop?

e Luke bought a magazine, and read it as he walked home.

It took him 30 minutes to walk home.

What time did he arrive home?

f Copy the graph onto graph paper.

Follow these steps to complete the graph for Luke's trip.

i Mark the time that Luke arrived home on the time (horizontal) axis.

ii Join this point to the graph where he left the shop.

Speed from a distance–time graph

This spread will show you how to:

- Draw, use and interpret distance–time graphs
- Work out the average speed for a journey from a distance–time graph

Keywords
Average speed
Speed

You can work out speeds from a distance–time graph.

You use the formula **speed** $= \dfrac{\text{distance}}{\text{time}}$

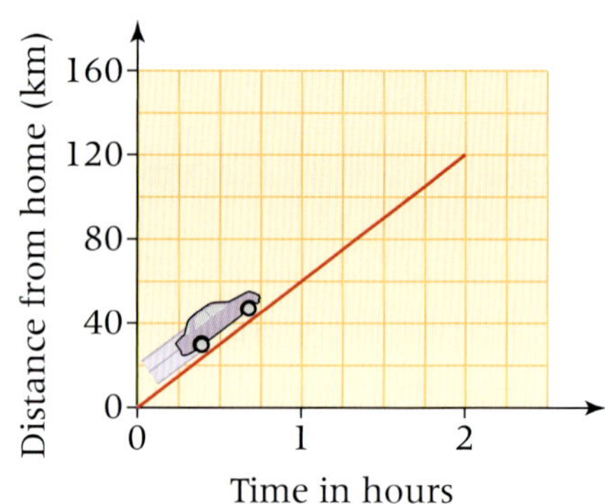

This graph shows a car journey.

The car travels 120 km (distance) in 2 hours (time).

speed $= \dfrac{\text{distance}}{\text{time}} = \dfrac{120\text{ km}}{2\text{ hours}} = 60$ km per hour

In real life, a car does not travel at exactly the same speed for 2 hours. It may have to slow down for junctions or traffic. The **average speed** is 60 km per hour.

- **Average speed** $= \dfrac{\textbf{total distance}}{\textbf{total time}}$

Example

The graph shows a train journey.

a Explain what could have happened at 2 pm.
b Work out the average speed for the first part of the journey.
c Work out the average speed for the second part of the journey.
d For which part of the journey was the train travelling fastest?
e Work out the average speed for the whole journey.

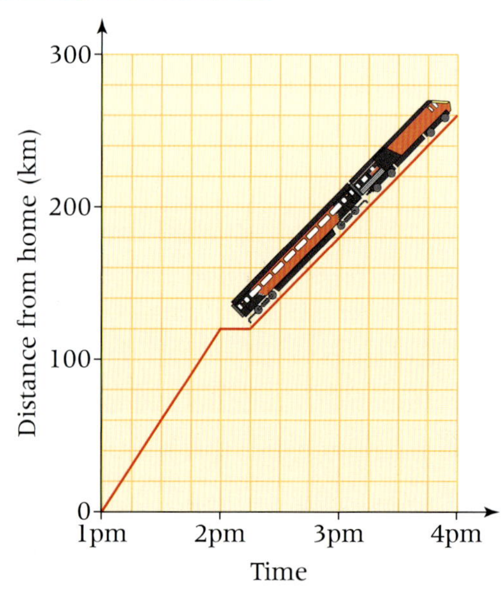

a Train stopped, probably at a station.

b Distance 120 km, time 1 hour.
Average speed = distance ÷ time
= 120 ÷ 1 = 120 km per hour

c Distance 140 km, time 1.75 hours.
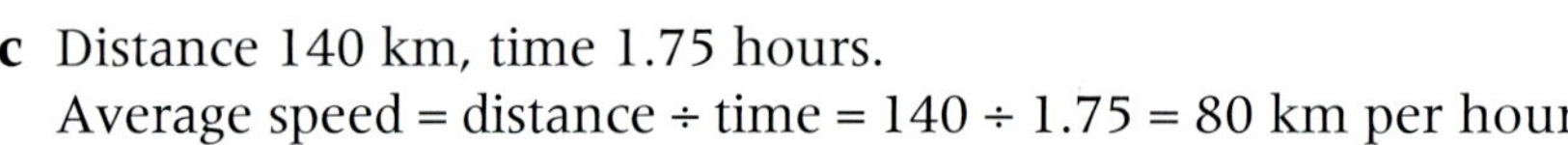
Average speed = distance ÷ time = 140 ÷ 1.75 = 80 km per hour

For speed in km per hour, the distance must be in km and the time must be in hours.

d Train travelled fastest for first part.

e Total distance is 260 km, total time is 3 hours.
Average speed for whole journey

$= \dfrac{\text{total distance}}{\text{total time}} = \dfrac{260}{3} = 86.6666$ km per hour

$= 87$ km per hour (to the nearest km)

Exercise A8.5

1 For each graph work out the average speed.

a

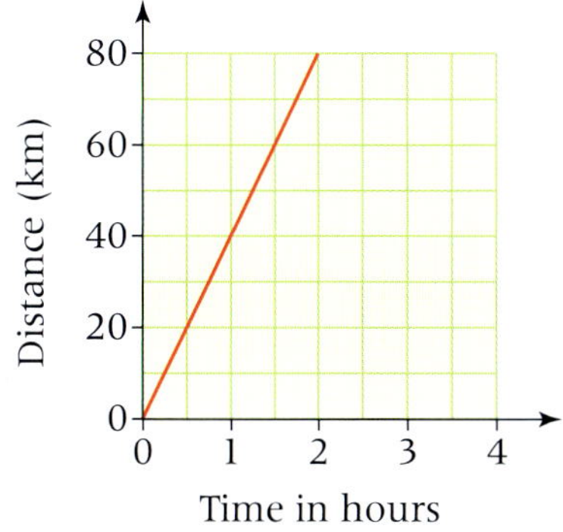

b

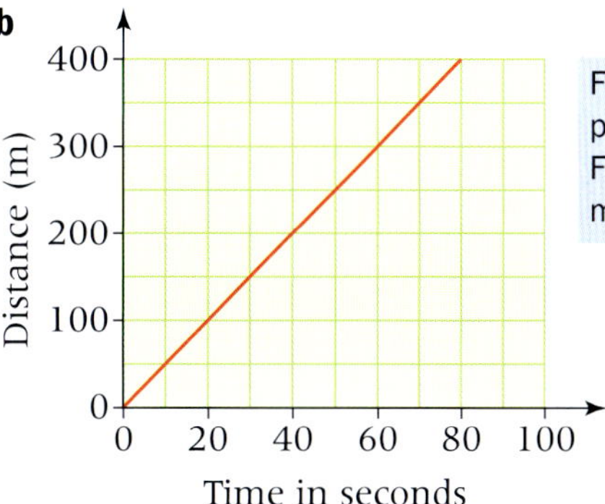

For part **a**, use km per hour.
For part **b**, use metres per second.

2 For each graph, work out

- the average speed for each part of the journey
- the average speed for the whole journey.

a

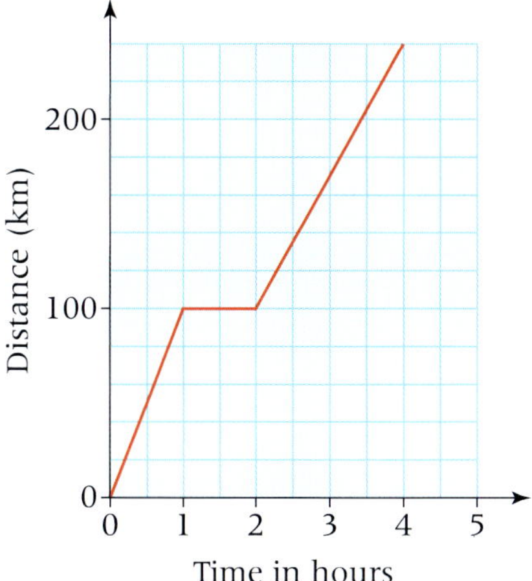

b

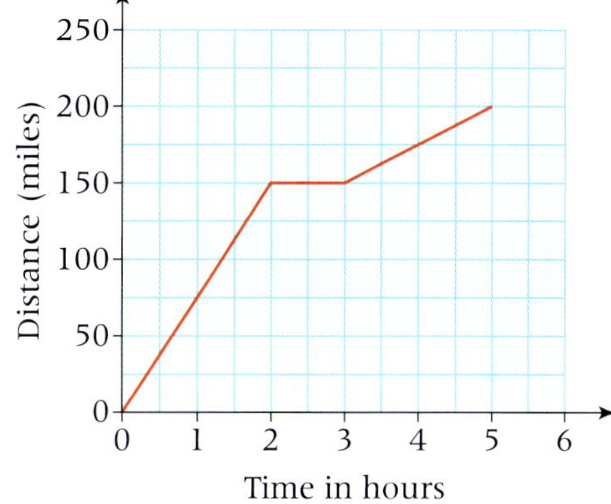

3 The graph shows Amit's car journey to Wales.

a What distance does he travel in the first part of the journey?

b What is the time taken for the first part of the journey?
Write your answer in hours, as a decimal.

c Use your answers from **a** and **b** to work out his average speed in km per hour for the first part of the journey.

d Work out Amit's average speed in km per hour for the second part of the journey.

e Work out his average speed in km per hour for the whole journey.

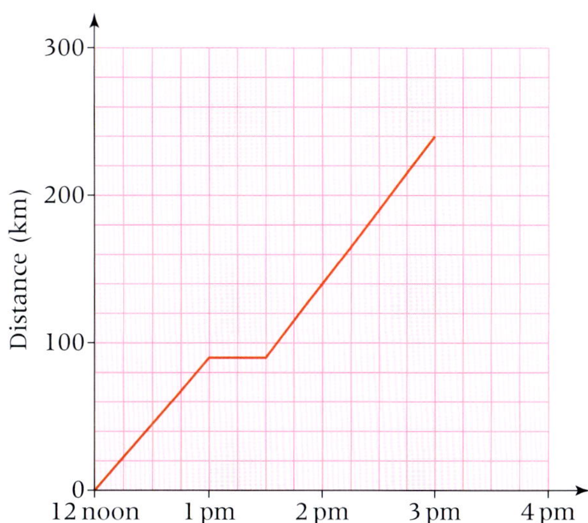

A8 Exam review

Key objectives

- Construct linear functions from real-life problems and plot their corresponding graphs
- Discuss and interpret graphs arising from real situations

1 **a** Copy and complete this table for a kilometre to mile conversion graph. (1)

Miles		5	10
Kilometres	0		16

b Use your table to draw a kilometre to mile conversion graph. (4)

c Use your graph to change (2)

i 4 miles to kilometres **ii** 12 kilometres to miles.

2 A man left home at 12 noon to go for a cycle ride. The travel graph represents part of the man's journey.

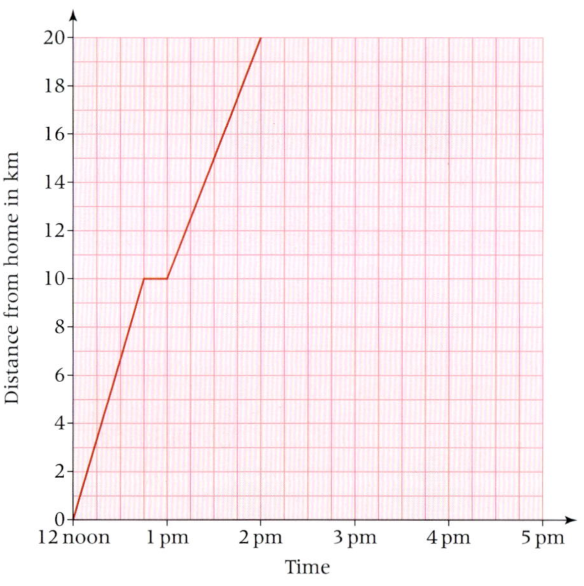

At 12.45 pm the man stopped for a rest.

a For how many minutes did he rest? (1)

b Find his distance from home at 1.30 pm. (1)

The man stopped for another rest at 2 pm.
He rested for one hour.
Then he cycled home at a steady speed. It took him 2 hours.

c Copy and complete the travel graph. (2)

(Edexcel Ltd., 2005)

S8 Properties of shapes

This unit will show you how to

- Use angle properties of equilateral, isosceles and right-angled triangles
- Use angle properties of quadrilaterals
- Find exterior angles in triangles and quadrilaterals
- Know the names of general polygons
- Understand that regular polygons have equal sides and equal angles
- Understand congruence and recognise congruent shapes
- Understand and create tessellations
- Use 2-D representations of 3-D shapes
- Calculate the surface area and volume of cuboids
- Use nets to construct cuboids from given information

Before you start ...

You should be able to answer these questions.

Review

1 State the sum of the three angles *x*, *y*, *z*. — Unit S2

a

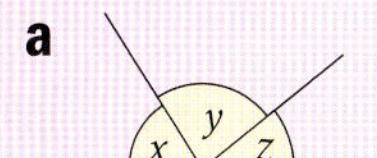

b

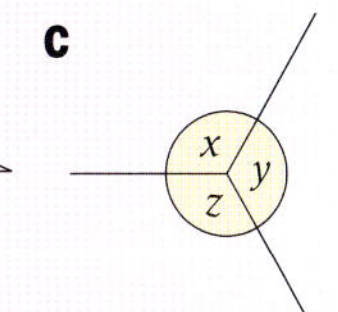

c

x y z

2 Find the value of each unknown angle. — Unit S2

a

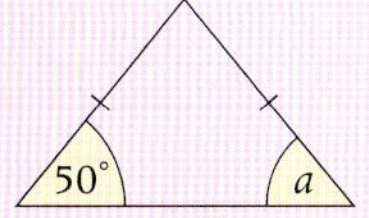

b

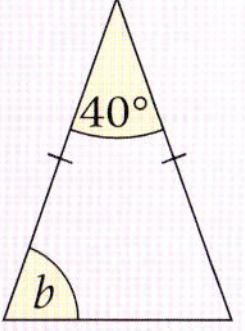

3 Calculate the area of the rectangle.
State the units of your answer. — Unit S6

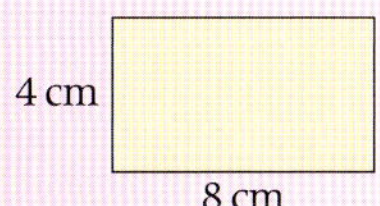

4 Find the volume of each solid. — Unit S6

a

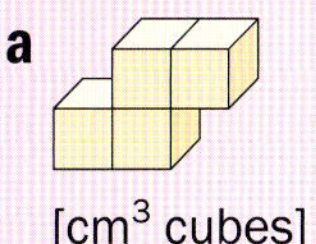

[cm^3 cubes]

b

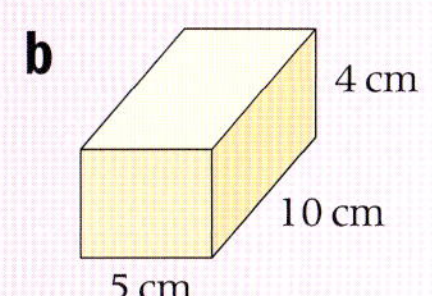

S8.1 Properties of triangles and quadrilaterals

This spread will show you how to:

- Use angle properties of equilateral, isosceles and right-angled triangles
- Use angle properties of quadrilaterals
- Find exterior angles in triangles and quadrilaterals

Keywords
Diagonal
Exterior
Interior
Quadrilateral
Triangle

- **The angles in a triangle add to 180°.**

A **quadrilateral** is a 2-D shape with 4 sides and 4 angles.

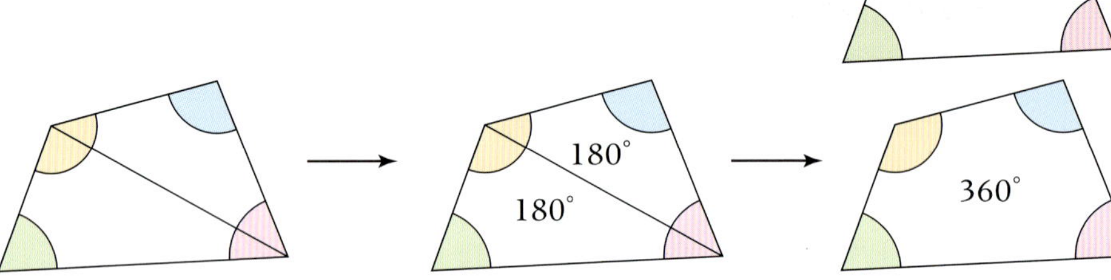

You can draw a **diagonal** on a quadrilateral to form 2 triangles.

The angles in each triangle add to 180°.

$2 \times 180 = 360°$

This works for any quadrilateral.

- **The angles in a quadrilateral add to 360°.**

The angles inside a shape are called **interior** angles.

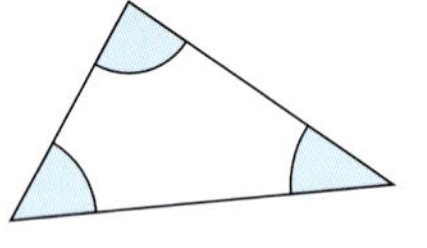

You find the **exterior** angles by extending each side of the shape in the same direction.

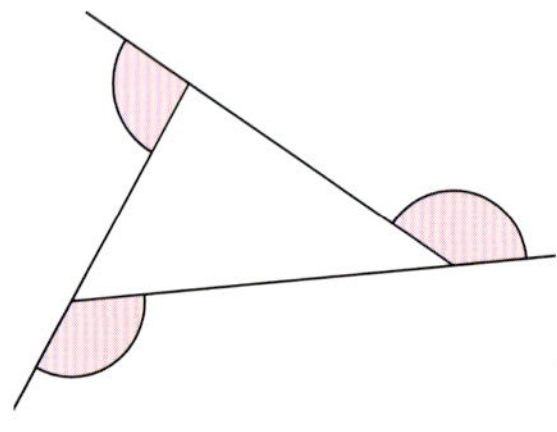

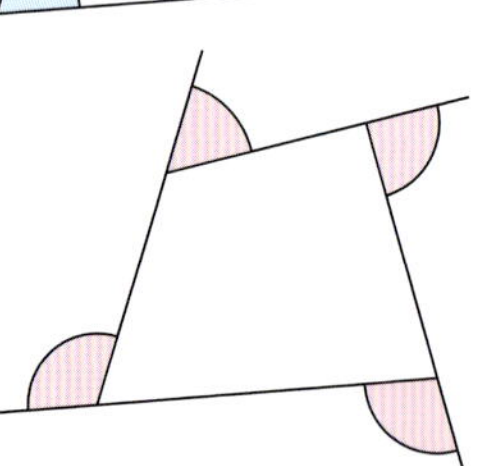

- **Interior angle + exterior angle = 180°**
  (angles on a straight line add to 180°)

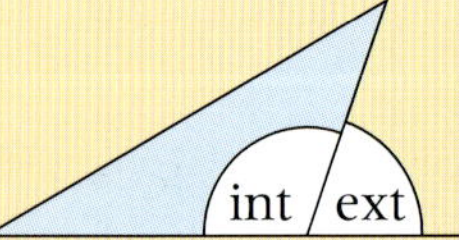

Example

Calculate the value of x and y.
Give a reason for each answer.

$125° + 110° + 58° = 293°$

$360° - 293° = 67°$

$x = 67°$ (angles in a quadrilateral add to 360°)

$180° - 67° = 113°$

$y = 113°$ (angles on a straight line add to 180°)

Exercise S8.1

1 Give the mathematical name of each coloured shape in the regular hexagon.

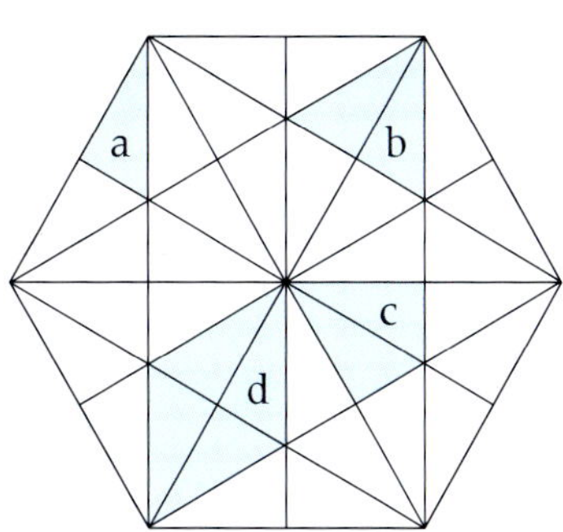

2 Calculate the value of the unknown angles.

a

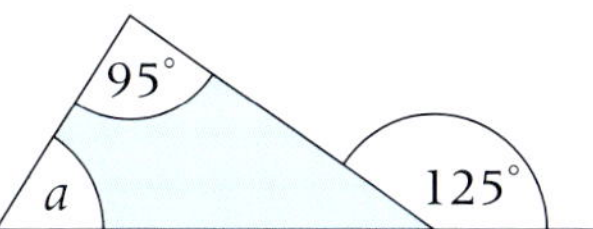

b

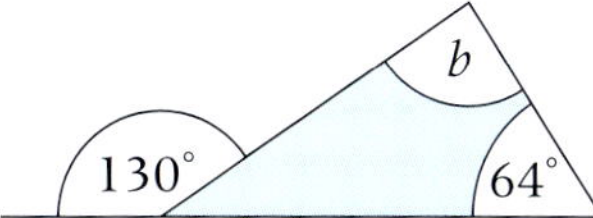

c

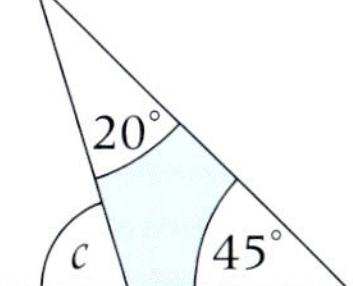

3 Calculate the value of the unknown angles.

a

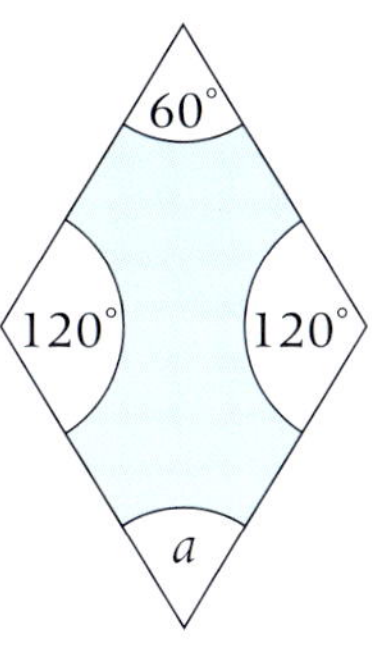

b

c

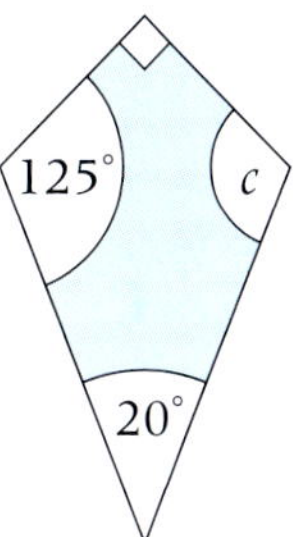

d

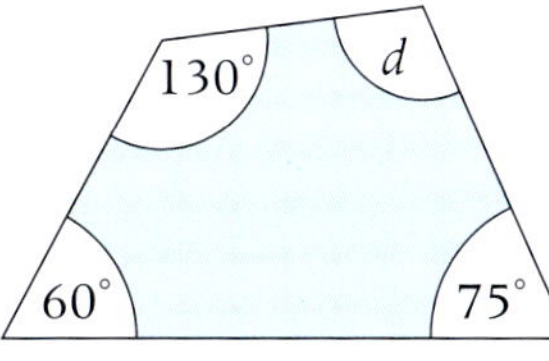

e

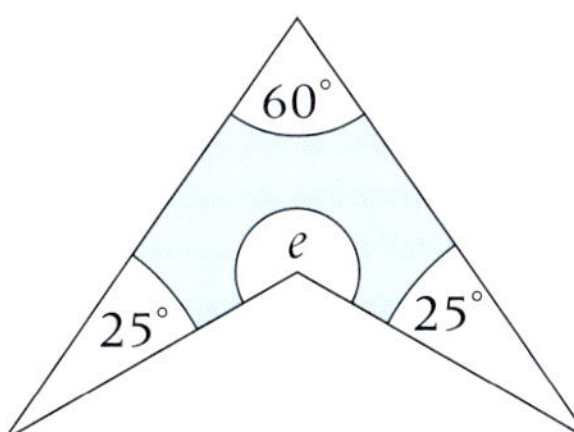

4 Calculate the value of the angles marked by a letter.

a

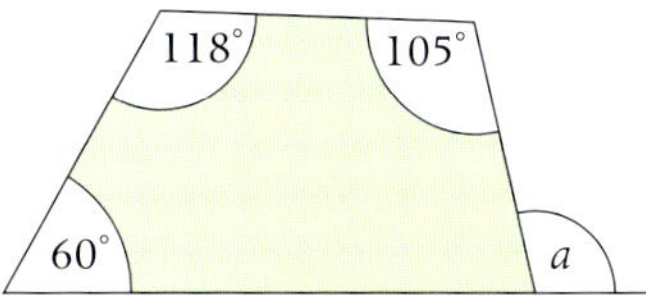

b

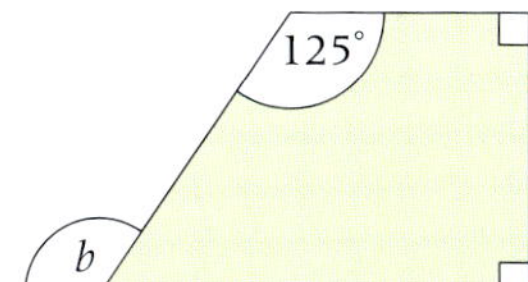

c

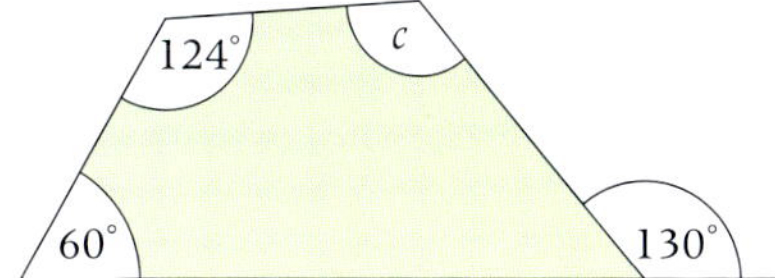

S8.2 Polygons

This spread will show you how to:

- Find exterior angles in triangles and quadrilaterals
- Know the names of general polygons
- Understand that regular polygons have equal sides and equal angles

Keywords
Isosceles
Polygon
Regular

- **A polygon** is a 2-D shape with many sides and many angles.

You should know the names of these polygons.

Sides	Name
3	triangle
4	quadrilateral
5	pentagon
6	hexagon
7	heptagon
8	octagon
9	nonagon
10	decagon

- **A regular** shape has equal sides and equal angles.

A regular pentagon has 5 equal sides and 5 equal angles.

Example

A regular octagon is drawn inside a circle.
There are 8 **isosceles** triangles.
Calculate the values of x and y.

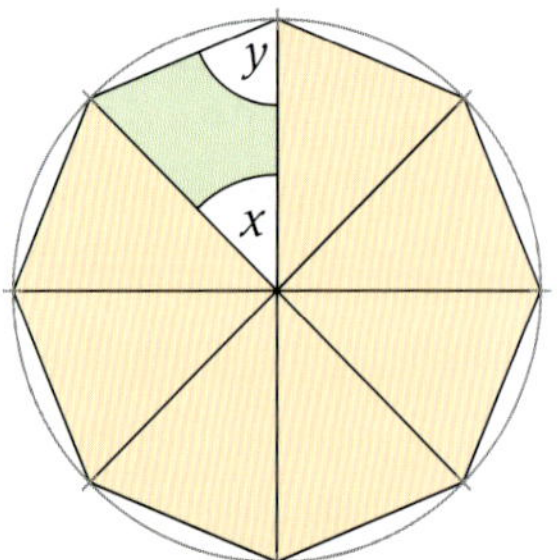

$360° \div 8 = 45°$

$x = 45°$ (angles at a point add to 360°)

$180° - 45° = 135°$

$2y = 135°$

$y = 67\frac{1}{2}°$ (angles in a triangle add to 180°)

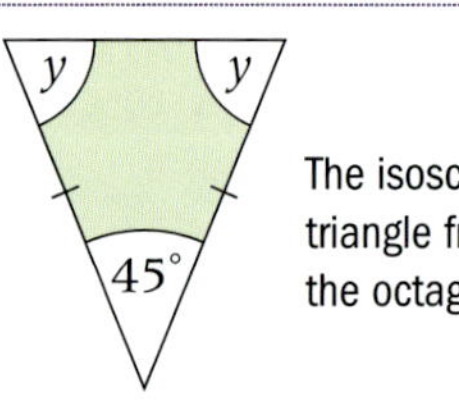

The isosceles triangle from the octagon.

- For any polygon
  - sum of exterior angles = 360°
  - interior angle + exterior angle = 180°

You can use the example to find the interior and exterior angles of an octagon.

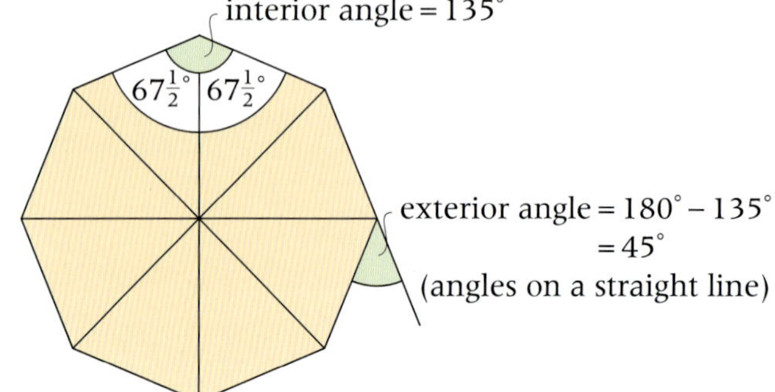

Exercise S8.2

1 **a** Use a protractor and a ruler to draw a regular hexagon.

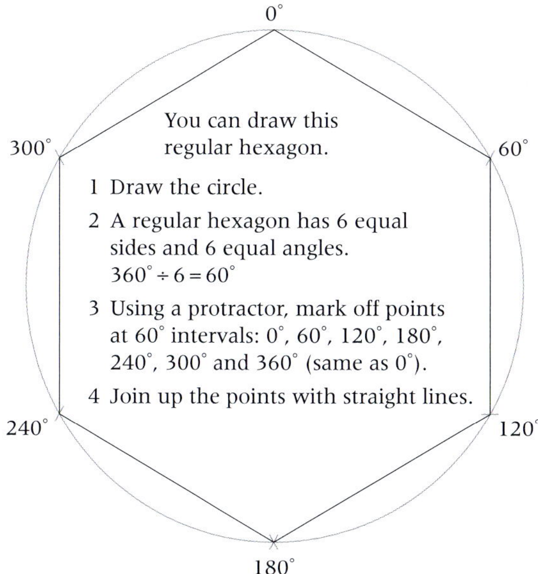

2 A regular pentagon is made from five isosceles triangles.

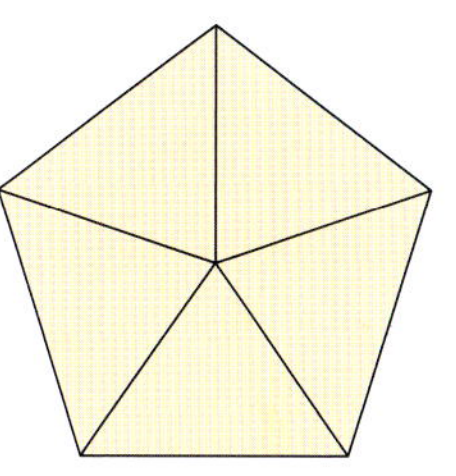

a Calculate the three angles in an isosceles triangle for a pentagon.

b Use your protractor to measure and check these angles.

c Use your results to work out the interior angle of a regular pentagon.

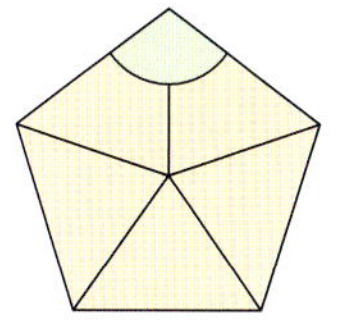

d Work out the exterior angle of a regular pentagon.

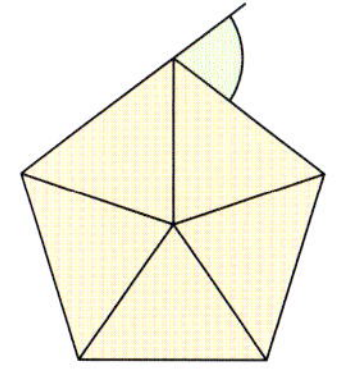

3 You can split a quadrilateral into two triangles by drawing a diagonal.

Angle sum of quadrilateral $= 2 \times 180°$

$= 360°$

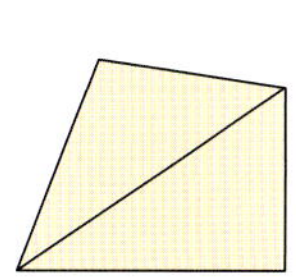

You can split a pentagon into triangles by drawing diagonals.

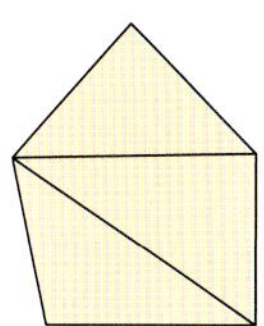

a How many triangles are formed?

b What is the angle sum of a pentagon?

c Find the angle sum of a hexagon in a similar way.

S8.3 Tessellations

This spread will show you how to:

- Understand congruence and recognise congruent shapes
- Understand and create tessellations

Keywords
Congruent
Quadrilateral
Regular
Rotate
Tessellation
Translate
Triangle

- **Congruent** shapes are exactly the same shape and size.

You can make patterns with an L shape like this.

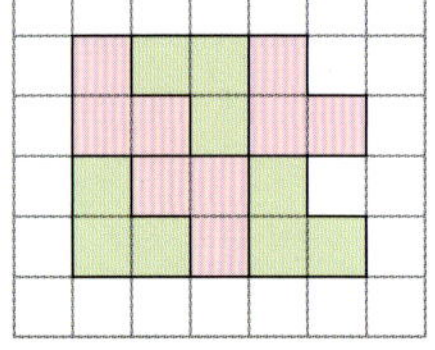

The L shapes fit together so that there are no gaps or overlaps. The L shape tessellates.

- A **tessellation** is a tiling pattern with no gaps.

Example

Draw at least five more trapezium shapes on the grid to show how the shape tessellates.

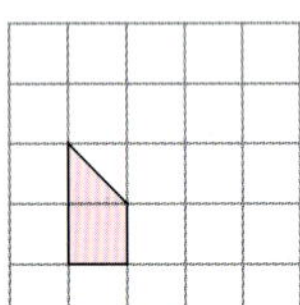

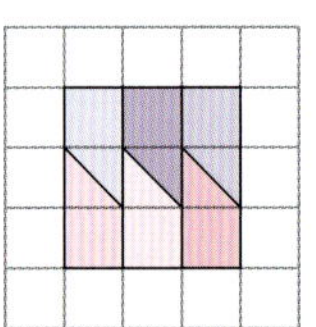

You are allowed to **rotate** or **translate** the shape.

- Any **triangle** tessellates.

Rotate the triangle about the midpoint of the sides.

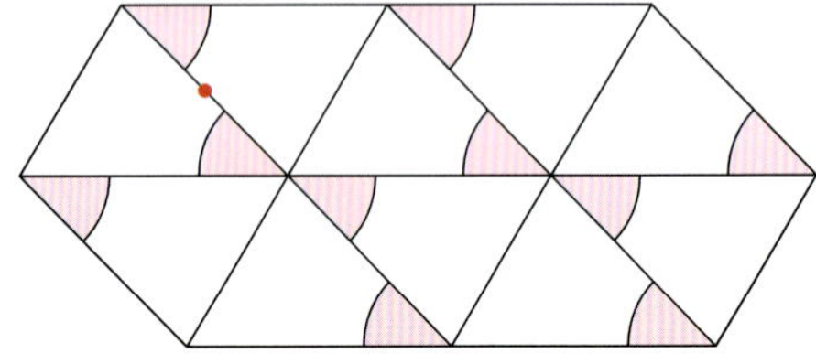

Even a scalene triangle tessellates.

- Any **quadrilateral** tessellates.

Rotate the quadrilateral about the midpoint of the sides.

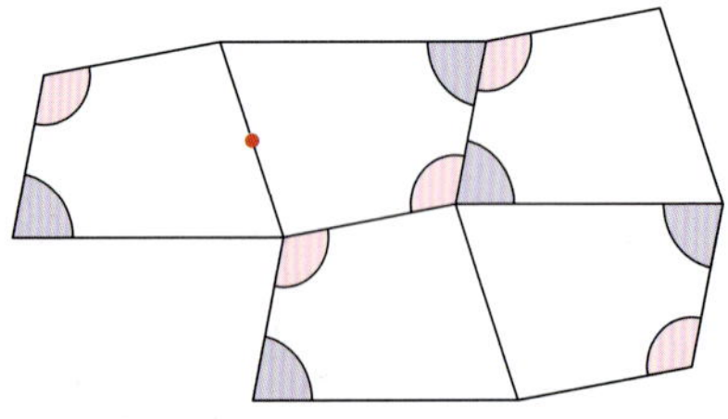

This means that all the special quadrilaterals also tessellate.

Only three regular shapes tessellate.

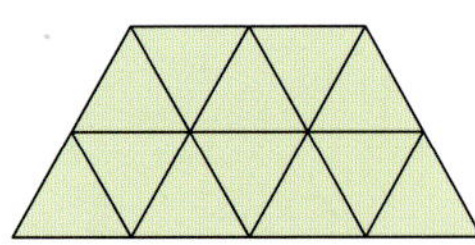

Equilateral triangles

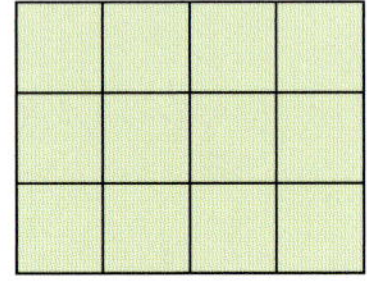

Squares

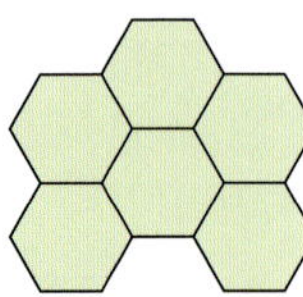

Hexagons

A **regular** shape has equal sides and equal angles.

Exercise S8.3

1 Copy these shapes onto square grid paper.
Show how each shape tessellates, using rotations and translations.

a b c d

e f g h i

j k l

m n o p

2 Draw a 4 by 2 rectangle on square grid paper.

Remove the triangle and translate it to the new position as shown.

Remove a second triangle and translate it to the new position as shown.

Cut out the shape and tessellate it using translations.

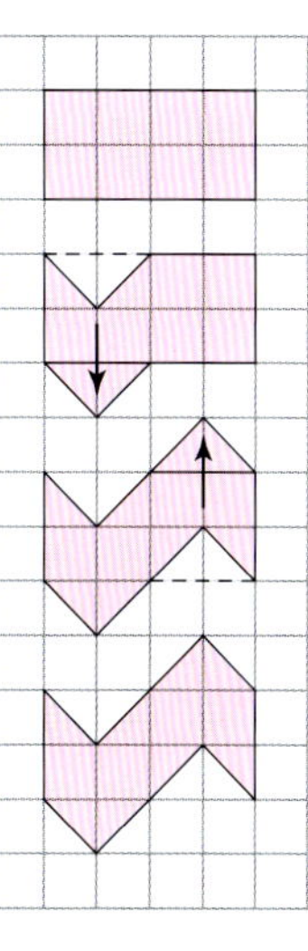

DID YOU KNOW?

Quilters make tessellating patterns out of fabric and assemble highly technical artwork.

S8.4 3-D shapes and isometric paper

This spread will show you how to:

- Use 2-D representations of 3-D shapes
- Calculate the surface area and volume of cuboids

Keywords
3-D
Congruent
Cube
Cuboid
Isometric
Surface area
Volume

You can draw some **3-D** shapes using **isometric** paper.

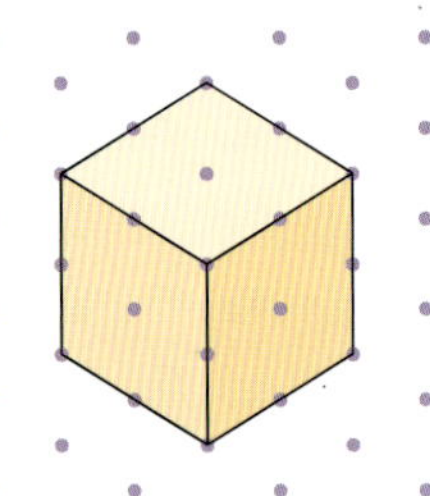

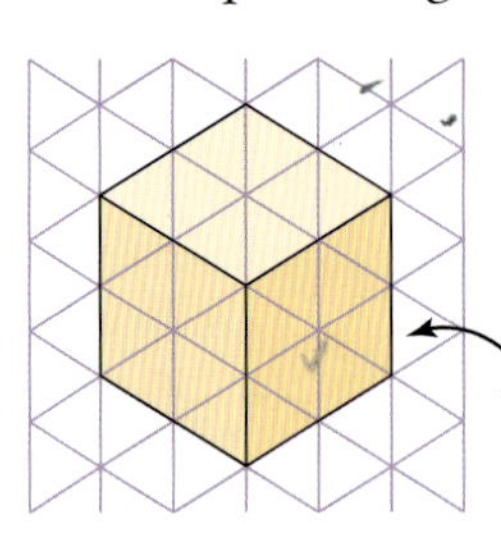

The equal edges of the cube are shown by lines of equal length.

3-D means 3-dimensional.

The paper must be this way round.

Example

Add one cube to shape B so that it is congruent to shape A.

A

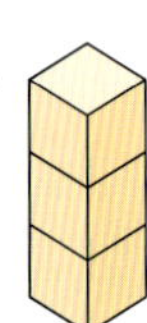

B

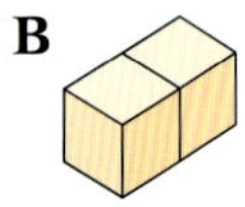

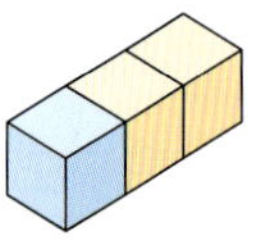

Congruent shapes are identical in size and shape.

- **Surface area of a cuboid is the total area of its faces.**
- **Volume of cuboid = length × width × height**

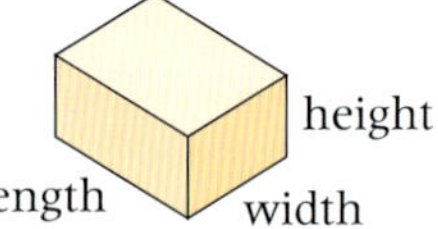

Example

a On isometric paper, draw a cuboid with dimensions 2 cm by 3 cm by 4 cm.
b Calculate the surface area of the cuboid. State the units of your answer.
c Calculate the volume of the cuboid. State the units of your answer.

a

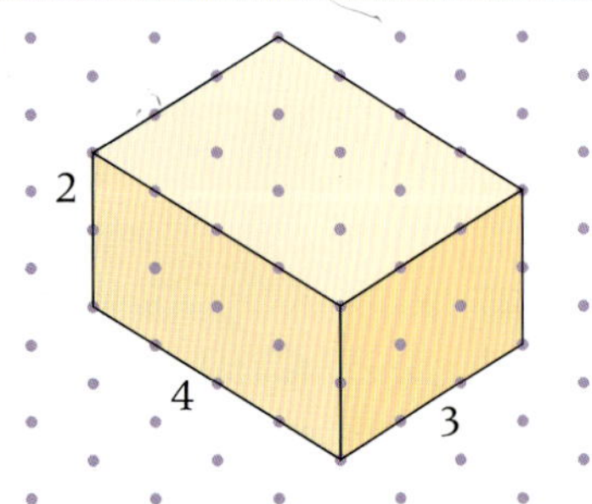

b Area of the 4 cm by 3 cm rectangle = $4 \times 3 = 12\ \text{cm}^2$
Area of the 3 cm by 2 cm rectangle = $3 \times 2 = 6\ \text{cm}^2$
Area of the 2 cm by 4 cm rectangle = $2 \times 4 = 8\ \text{cm}^2$
$26\ \text{cm}^2$

Surface area = $26 \times 2 = 52\ \text{cm}^2$

c Volume = $2 \times 3 \times 4 = 24\ \text{cm}^3$

There are three hidden faces you must account for when finding the total surface area. Opposite faces are congruent.

Exercise S8.4

1 Draw these solids *after* the shaded cube is removed.

a

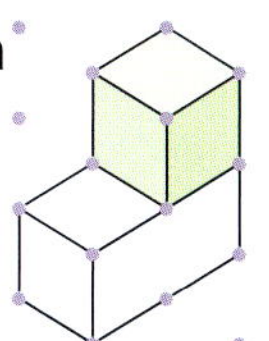

b

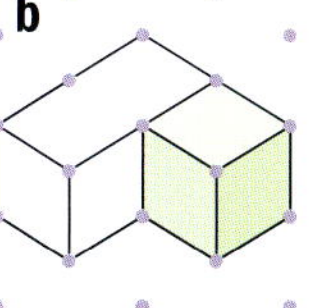

c

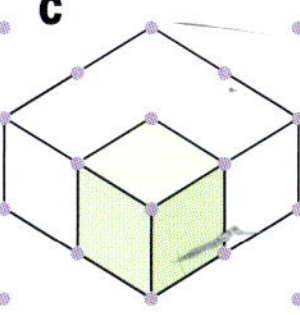

d

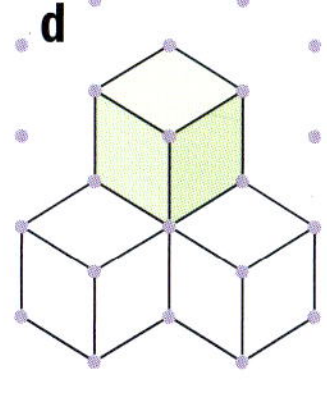

e

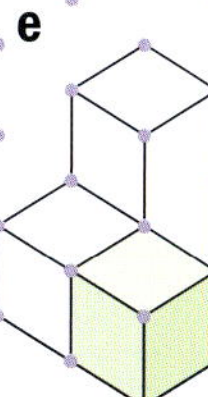

f

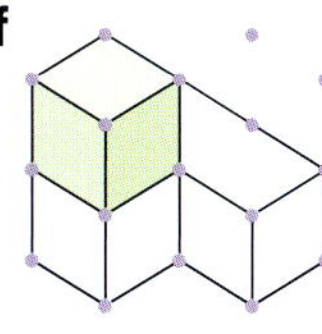

g

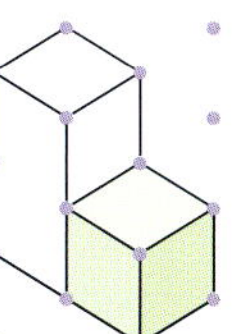

h

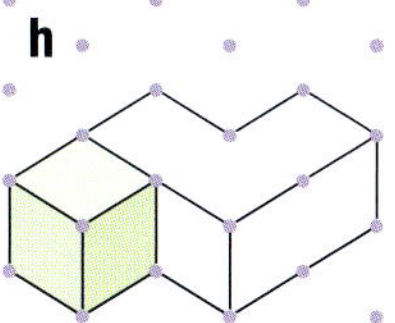

i

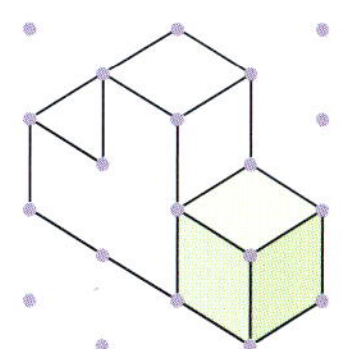

j

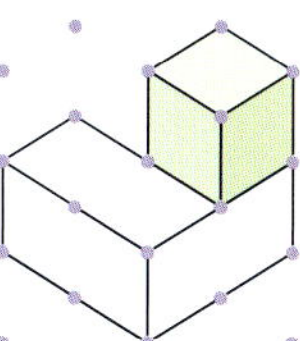

2 **a** State the dimensions of this cuboid.

b Calculate the surface area of the cuboid.

c Calculate the volume of the cuboid.

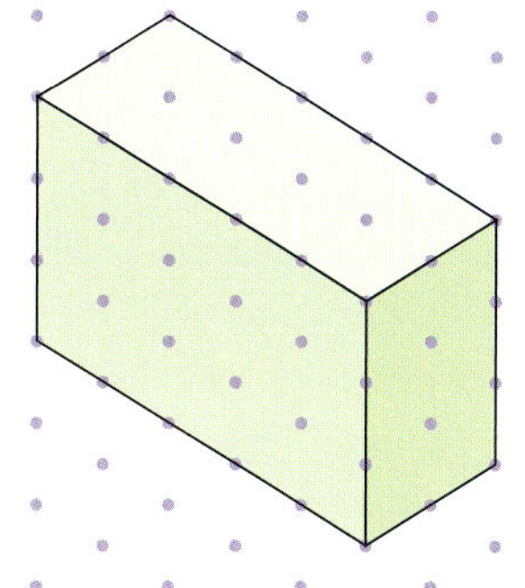

3 **a** Copy and complete the drawing of the 1 cm by 2 cm by 5 cm cuboid on isometric paper.

b Calculate the surface area of the cuboid.

c Calculate the volume of the cuboid.

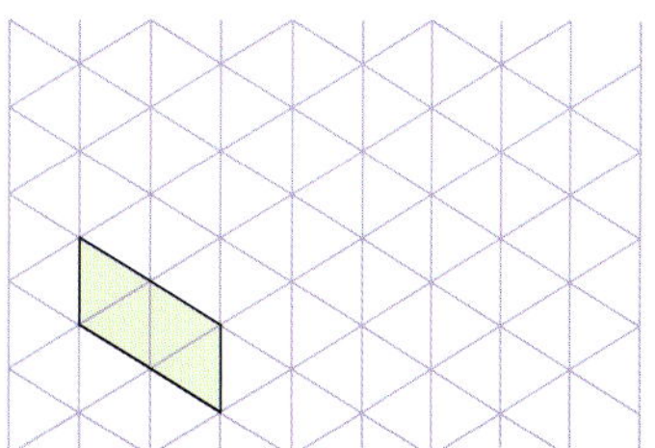

4 **a** On isometric paper, draw a cuboid with dimensions 1 cm by 3 cm by 4 cm.

b Calculate the surface area of the cuboid. State the units of your answer.

c Calculate the volume of the cuboid. State the units of your answer.

5 On isometric paper, draw a cube with a volume of 27 cm^3.

S8.5 Cubes and cuboids

This spread will show you how to:

- Use nets to construct cuboids from given information
- Use 2-D representations of 3-D shapes

Keywords
Cube
Cuboid
Edge
Face
Net
Solid
Surface area
Three-dimensional (3-D)
Vertex
Volume

A **solid** is a **three-dimensional (3-D)** shape.

- A **net** is a 2-D arrangement that can be folded to form a solid shape.

The net of a **cube** has 6 squares.

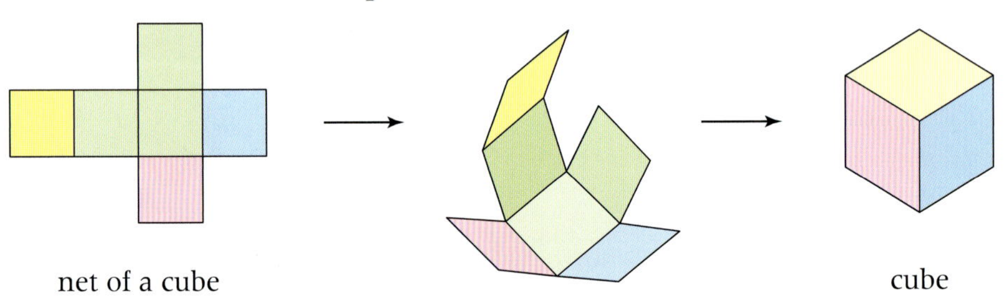

There are many other possible nets of a cube.

The net of a **cuboid** has 6 rectangles.

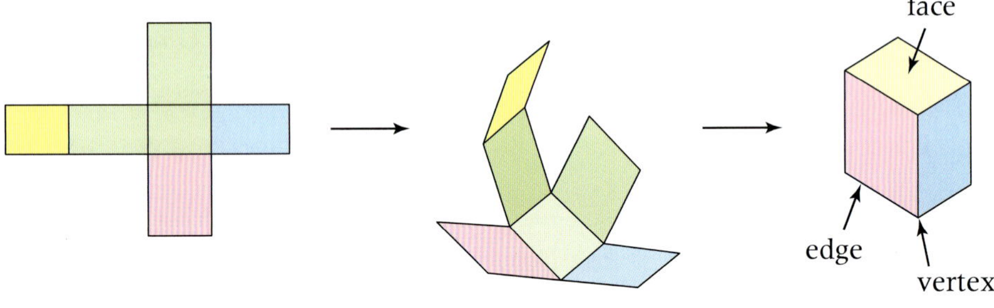

A cuboid has 6 **faces**, 12 **edges** and 8 vertices.

The plural of **vertex** is vertices.

Example

Here is the net of a cuboid.

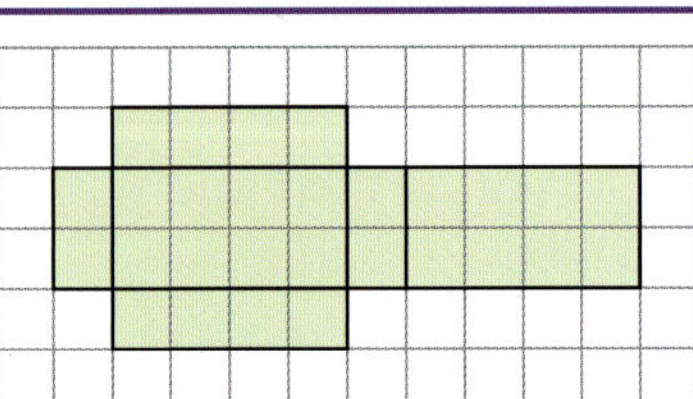

a State the dimensions of the cuboid.
b Calculate the **surface area** of the cuboid.
c Calculate the **volume** of the cuboid.

---

a 1 cm by 2 cm by 4 cm
b Surface area = $(4 + 8 + 2) \times 2$
$= 28 \text{ cm}^2$
c Volume = $2 \times 4 \times 1$
$= 8 \text{ cm}^3$

Exercise S8.5

1 State whether each arrangement of squares is a net of a cube.

a 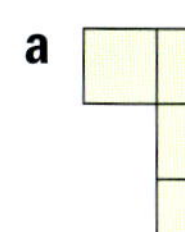 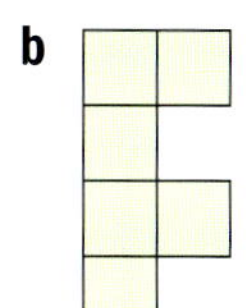

b

c 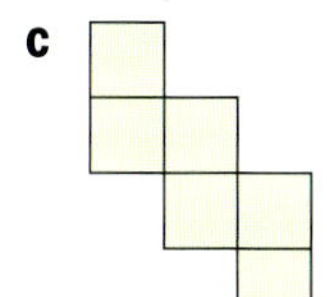

d 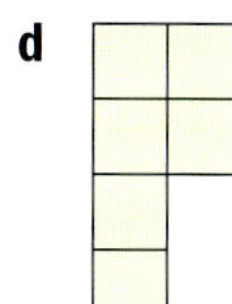

e 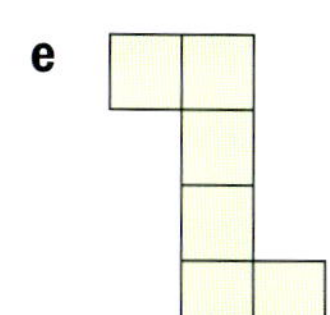

2 These nets make cuboids. Copy the nets onto centimetre squared paper and cut them out to make the cuboids. Write the dimensions and calculate the surface area and the volume of each cuboid.

a 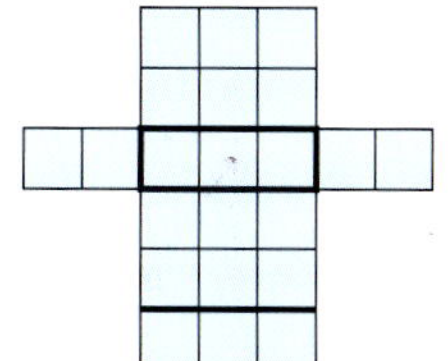

b 

c 

3 The volume of a cube is 27 cm^3.
On square grid paper, draw the net of this cube.

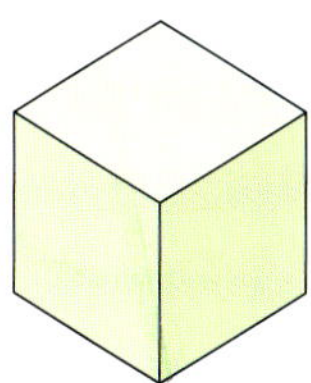

4 On square grid paper, draw the net for each cuboid.

a

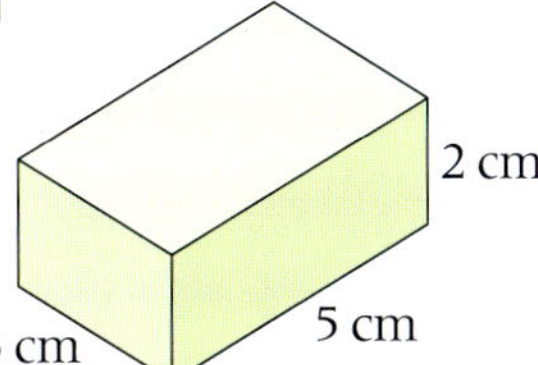

b

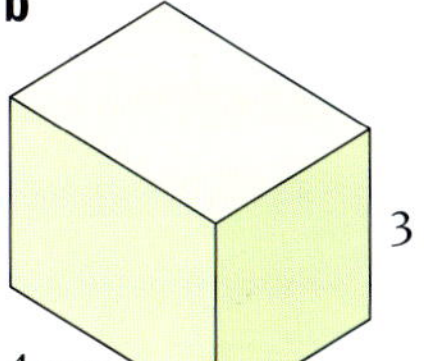

c

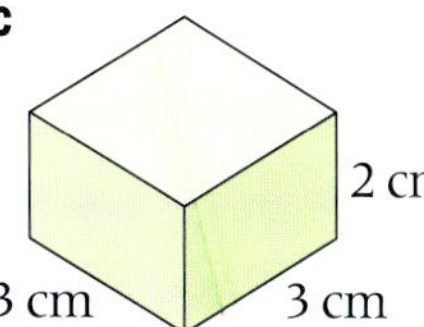

d

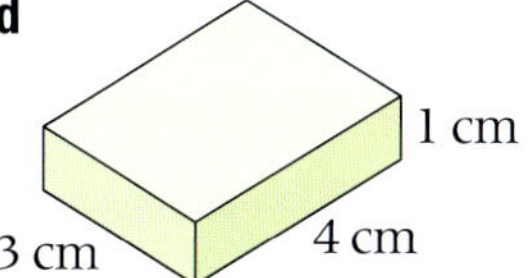

e

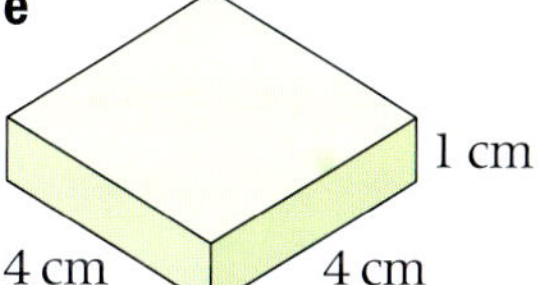

S8 Exam review

Key objectives

- Use angle properties of equilateral, isosceles and right-angled triangles
- Understand congruence
- Recall the essential properties of special types of quadrilaterals
- Calculate and use the sums of the interior and exterior angles of triangles and quadrilaterals
- Calculate and use the angles of regular polygons
- Explore the geometry of cuboids, and shapes made from cuboids
- Use 2-D representations of 3-D shapes and analyse 3-D shapes through 2-D projections and cross-sections, including plan and elevation

1 **a** What is the name of this polygon? (1)

b A B C D E F G H

One of the above polygons is congruent to the polygon in part **a**. Which is it? (1)

c Draw a diagram to show how the shape H in part **b** will tessellate. Draw at least 8 shapes. (2)

2 Here is a cuboid

4 cm

6 cm

3 cm

Diagram **NOT** accurately drawn

a Write

i the number of edges on this cuboid

ii the number of vertices on this cuboid. (2)

b Draw an accurate net for the cuboid. (3)

(Edexcel Ltd., 2003)

Plus section

This section contains material that extends the topics in this book to cover grades D and C.

There are 20 lessons in total.

Number

Algebra

Shape

Data

+1 Rounding decimals

This spread will show you how to:

- Round numbers to a given number of decimal places
- Round numbers to a given number of significant figures
- Use approximation to estimate an answer

Keywords
Approximate
Decimal places
Estimate
Rounding
Significant figures

You can round a decimal number to a given accuracy.

To round 718.394 to 2 **decimal places**, look at the **thousandths** digit.

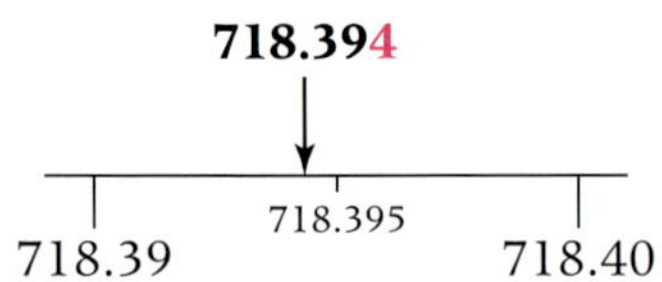

The **thousandths** digit is **4**, so round down to 718.39.

718.394 ≈ 718.39 (to 2 decimal places).

You can also round numbers to a given number of **significant figures**.

- **The first non-zero digit in a number is called the 1st significant figure – it has the highest value in the number.**

You only need to know how to round to 1 significant figure for your exam.

Example

Round 54.76 to 2 **significant figures**:

Look at the 3rd significant figure.

Tens	Units	•	tenths	hundredths
5	4	•	7	6

The **3rd significant** figure is **7**, so the number is rounded up to 55.

54.76 ≈ 55 (to 2 significant figures).

- **When rounding numbers to a given degree of accuracy, look at the next digit. If it is 5 or more then round up, otherwise round down.**

You can **estimate** the answer to a calculation by rounding the numbers.

Example

Estimate the answer to $\frac{6.23 \times 9.89}{18.7}$.

You can round each of the numbers to 1 significant figure.

$$\frac{6.23 \times 9.89}{18.7} \approx \frac{6 \times 10}{20} = \frac{60}{20} = 3$$

Exercise +1

1 Round each of these numbers to the

i nearest 10 **ii** nearest 100 **iii** nearest 1000.

a 3487 **b** 3389 **c** 14 853 m

d £57 792 **e** 92 638 kg **f** £86 193

g 3438.9 **h** 74 899.36

2 Round each of these numbers to the nearest whole number.

a 3.738 **b** 28.77 **c** 468.63

d 369.29 **e** 19.93 **f** 26.9992

g 100.501 **h** 0.001

3 Round each of these numbers to the nearest

i 3 dp **ii** 2 dp **iii** 1 dp.

a 3.4472 **b** 8.9482 **c** 0.1284

d 28.3872 **e** 17.9989 **f** 9.9999

g 0.003 987 **h** 2785.5555

4 Round each of these numbers to the nearest

i 3 sf **ii** 2 sf **iii** 1 sf.

a 8.3728 **b** 18.82 **c** 35.84

d 278.72 **e** 1.3949 **f** 3894.79

g 0.008 372 **h** 2399.9 **i** 8.9858

j 14.0306 **k** 1403.06 **l** 140 306

5 Write a suitable estimate for each of these calculations.
In each case, clearly show how you estimated your answer.

a 4.98×6.12 **b** $17.89 + 21.91$

c $\dfrac{5.799 \times 3.1}{8.86}$ **d** $34.8183 - 9.8$

e $\dfrac{32.91 \times 4.8}{3.1}$ **f** $\{9.8^2 + (9.2 - 0.438)\}^2$

+2 Adding and subtracting fractions

This spread will show you how to:

- Add and subtract fractions

Keywords
Cancel
Common denominator
Equivalent
Fraction

It is easy to add or subtract **fractions** when they have the same denominator.

+

=

$\frac{3}{8} + \frac{1}{8} = \frac{4}{8}$

- **You can add or subtract fractions with different denominators by first writing them as equivalent fractions with the same denominator.**

Example

Calculate **a** $\frac{3}{5}+\frac{1}{3}$ **b** $1\frac{3}{4}-\frac{5}{7}$

a $\frac{3}{5}+\frac{1}{3}$

$\frac{3}{5}+\frac{1}{3} = \frac{9}{15}+\frac{5}{15}$

$= \frac{9+5}{15}$

$= \frac{14}{15}$

$\frac{3}{5} \xrightarrow{\times 3} \frac{9}{15}$ (numerator × 3, denominator × 3) $\frac{1}{3} \xrightarrow{\times 5} \frac{5}{15}$ (numerator × 5, denominator × 5)

The lowest **common denominator** is the least common multiple of 5 and 3, which is 15.

b $1\frac{3}{4}-\frac{5}{7}$

Change the mixed number to an improper fraction:

$1\frac{3}{4} = \frac{7}{4}$

$1\frac{3}{4}-\frac{5}{7} = \frac{7}{4}-\frac{5}{7}$

$= \frac{49}{28}-\frac{20}{28}$

$= \frac{49-20}{28}$

$= \frac{29}{28} = 1\frac{1}{28}$

$\frac{7}{4} \xrightarrow{\times 7} \frac{49}{28}$ (numerator × 7, denominator × 7) $\frac{5}{7} \xrightarrow{\times 4} \frac{20}{28}$ (numerator × 4, denominator × 4)

The lowest common denominator is the least common multiple of 4 and 7, which is 28.

- **You can compare and order fractions by writing them as equivalent fractions with the same denominator.**

Example

Which is bigger: $\frac{3}{7}$ or $\frac{4}{9}$?

You need an equivalent fraction for both $\frac{3}{7}$ and $\frac{4}{9}$.

$\frac{3}{7} \xrightarrow{\times 9} \frac{27}{63}$ (numerator × 9, denominator × 9) $\frac{4}{9} \xrightarrow{\times 7} \frac{28}{63}$ (numerator × 7, denominator × 7)

$\frac{27}{63} < \frac{28}{63}$ so $\frac{3}{7} < \frac{4}{9}$ $\frac{4}{9}$ is bigger.

The common denominator of these equivalent fractions will be $7 \times 9 = 63$.

Exercise +2

1 Work out these.

a $\frac{1}{3}+\frac{1}{3}$ **b** $\frac{3}{8}+\frac{2}{8}$ **c** $\frac{8}{11}-\frac{3}{11}$

d $\frac{8}{17}+\frac{5}{17}$ **e** $\frac{14}{23}-\frac{11}{23}$ **f** $\frac{5}{27}+\frac{8}{27}$

2 Work out each of these, leaving your answer in its simplest form.

a $\frac{2}{3}+\frac{1}{3}$ **b** $\frac{8}{9}-\frac{2}{9}$ **c** $\frac{8}{11}+\frac{5}{11}$ **d** $\frac{15}{13}-\frac{8}{13}$

e $\frac{14}{9}+\frac{1}{9}$ **f** $\frac{17}{12}-\frac{9}{12}$ **g** $1\frac{2}{3}+\frac{2}{3}$ **h** $4\frac{2}{7}-\frac{5}{7}$

3 Work out these.

a $\frac{1}{3}+\frac{1}{2}$ **b** $\frac{1}{4}+\frac{3}{5}$ **c** $\frac{3}{5}-\frac{1}{3}$ **d** $\frac{4}{5}-\frac{2}{7}$

e $\frac{5}{8}+\frac{1}{3}$ **f** $\frac{4}{9}+\frac{2}{5}$ **g** $\frac{7}{9}-\frac{2}{11}$ **h** $\frac{7}{15}+\frac{3}{7}$

Write both fractions as equivalent fractions with the same denominator.

4 Work out each of these, leaving your answer in its simplest form as appropriate.

a $\frac{2}{5}-\frac{1}{15}$ **b** $\frac{1}{2}-\frac{1}{3}$ **c** $\frac{2}{5}+\frac{7}{20}$ **d** $\frac{1}{2}-\frac{1}{6}$

5 Work out each of these, leaving your answer in its simplest form.

a $\frac{4}{5}+\frac{2}{3}$ **b** $1\frac{1}{2}+\frac{3}{5}$ **c** $1\frac{1}{3}+1\frac{1}{4}$ **d** $1\frac{2}{7}+\frac{3}{5}$

e $2\frac{2}{5}-\frac{1}{3}$ **f** $3\frac{3}{8}-1\frac{1}{2}$ **g** $4\frac{1}{3}-2\frac{3}{4}$ **h** $3\frac{4}{7}-2\frac{8}{9}$

6 Work out each of these, leaving your answer in its simplest form.

a Pete walked $3\frac{2}{3}$ miles before lunch and then a further $2\frac{1}{4}$ miles after lunch. How far did he walk altogether?

b A bag weighs $2\frac{3}{16}$ lb when it is full. When empty the bag weighs $\frac{3}{8}$ lb. What is the weight of the contents of the bag?

c Henry and Paula are eating peanuts. Henry has a full bag weighing $1\frac{3}{16}$ kg. Paula has a bag that weighs $\frac{4}{5}$ kg. What is the total mass of their two bags of peanuts?

d Simon spent $\frac{2}{3}$ of his pocket money on a computer game. He spent $\frac{1}{5}$ of his pocket money on a ticket to the cinema. Work out the fraction of his pocket money that he had left.

7 Write if each of these statement are true or false.

a $\frac{9}{2}<3$ **b** $\frac{17}{24}>\frac{5}{8}$ **c** $\frac{11}{12}>\frac{8}{9}$ **d** $\frac{2}{3}<\frac{5}{7}$

e $\frac{3}{5}>\frac{4}{7}$ **f** $\frac{26}{25}>\frac{16}{15}$ **g** $\frac{5}{4}<\frac{12}{7}$ **h** $\frac{7}{4}>\frac{12}{10}$

< means less than
> means more than.

+3 Prime factor decomposition

This spread will show you how to:

- Express a number as the product of its prime factors
- Recognise and use the HCF and LCM of two numbers

Keywords
Factor
HCF
LCM
Prime factor
Prime number

- A **prime factor** is a prime number that is also a factor of another number.
  Factors of 28 are {1, 2, 4, 7, 14, 28}. Prime factors are {2, 7}.

- Every whole number can be written as the product of its prime factors.

Here are two common methods to find prime factors.

Factor trees

Split the number into a **factor** pair.
Continue splitting until you reach a prime factor.

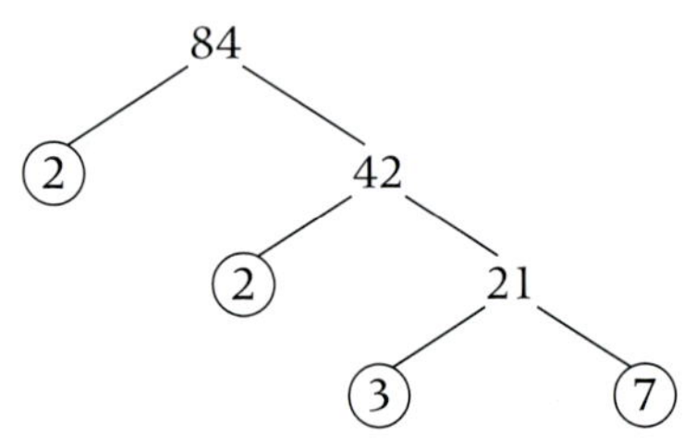

$84 = 2 \times 2 \times 3 \times 7 = 2^2 \times 3 \times 7$

Division by prime numbers

Divide the number by the smallest **prime number**.
Repeat dividing by larger prime numbers until you reach a prime number.

(2)	84
(2)	42
(3)	21
	(7)

$84 = 2 \times 2 \times 3 \times 7 = 2^2 \times 3 \times 7$

- You can find the **highest common factor (HCF)** of a set of numbers by using prime factors.

For example, the HCF of 30 and 135:

$30 = 2 \times 3 \times 5 \quad = 2 \times 3 \times 5$

2	30
3	15
	5

$135 = 3 \times 3 \times 3 \times 5 = 3 \times 3 \times 3 \times 5$

3	135
3	45
3	15
	5

HCF $= 3 \times 5 = 15$

- Write each number as the product of its prime factors.
- Pick out the common factors 3 and 5.
- Multiply these together to get the HCF.

- You can find the **least common multiple (LCM)** of a set of numbers by using prime factors.

For example, the LCM of 28 and 126:

$28 = 2^2 \times 7 = 2 \times 2 \times 7$

2	28
2	14
	7

$126 = 2 \times 3^2 \times 7 = 2 \times 3 \times 3 \times 7$

2	126
3	63
3	21
	7

HCF $= 2 \times 7 = 14$
LCM $= 2 \times 3 \times 3 \times 14 = 252$

- Write each number as the product of its prime factors.
- Pick out the common factors 2 and 7.
- Multiply these together to get the HCF – 14.
- Multiply the HCF by the remaining factors – the remaining factors are 2, 3 and 3.

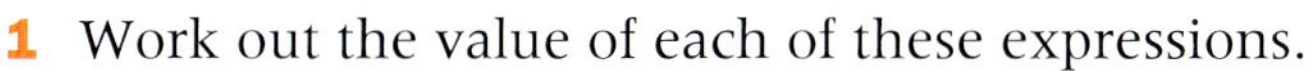

Exercise +3

1 Work out the value of each of these expressions.

a 3×5^2 **b** $2^3 \times 5$ **c** $3^2 \times 7$ **d** $2^2 \times 3^2 \times 5$ **e** $3^2 \times 7^2$

2 Express these numbers as products of their prime factors.

a 18 **b** 24 **c** 40 **d** 39

e 48 **f** 82 **g** 100 **h** 144

i 180 **j** 315 **k** 444 **l** 1350

3 In each of these questions, Jack has been asked to write each of the numbers as the product of its prime factors.

i Mark his work and identify any errors he has made.

ii Correct any of Jack's mistakes.

a 126

2	126
3	63
3	21
	7

Answer: $126 = 2 \times 3^2$

b 210

2	210
3	105
3	21
	7

Answer: $2 \times 3^2 \times 7$

c 221

221

Answer: 221

4 The number 18 can be written as $2 \times 3 \times 3$.
You can say that 18 has three prime factors.

a Find three numbers with exactly three prime factors.

b Find five numbers with exactly four prime factors.

c Find four numbers between 100 and 300 with exactly five prime factors.

d Find a two-digit number with exactly six prime factors.

5 Find the HCF of

a 9 and 24 **b** 15 and 40 **c** 18 and 24

d 96 and 144 **e** 12, 15 and 18 **f** 425 and 816.

6 Find the LCM of

a 9 and 24 **b** 15 and 40 **c** 18 and 24

d 20 and 30 **e** 12, 15 and 18 **f** 48, 54 and 72.

7 Cancel these fractions to their simplest forms using the HCF of the numerator and denominator to help.

a $\frac{6}{8}$ **b** $\frac{12}{18}$ **c** $\frac{60}{96}$

d $\frac{36}{54}$ **e** $\frac{117}{169}$ **f** $\frac{26}{65}$

+4 Harder percentage increase and decrease

This spread will show you how to:

- Calculate percentage increase and decrease using a range of methods

Keywords
Decrease
Increase
Percentage

Percentages are used in real life to show how much an amount has increased or decreased.

- To calculate a **percentage increase**, work out the increase and add it to the original amount.
- To calculate a **percentage decrease**, work out the decrease and subtract it from the amount.

Example

a Alan is paid £940 a month. His employer increases his wage by 3%. Calculate the new wage Alan is paid each month.
b A new car costs £19 490. After one year the car depreciates in value by 8.7%. What is the new value of the car?

a Calculate 3% of the amount.
Add to the original amount.
Increase in wage = 3% of £940 = $\frac{3}{100} \times$ £940
$= \frac{3 \times 940}{100} = \frac{2820}{100}$
Increase in wage = £28.20 per month
Alan's new wage = £940 + £28.20 = £968.20

The percentage calculation has been worked out using a written method.

b Calculate 8.7% of the amount.
Subtract from the original amount.
Depreciation = 8.7% of £19 490
$= \frac{8.7}{100} \times$ £19 490
$= 0.087 \times$ £19 490
Price reduction = £1695.63
New value of car = £19 490 − £1695.63 = £17 794.37

The percentage calculation has been worked out using a calculator.

Money saved in a bank account usually earns **interest**.

To calculate **compound interest** you work out the amount of money in the bank account at the end of each year. At the end of the next year the interest is paid on **all** the money.

You will not need to know about compound interest for your exam.

Example

Ben puts £1200 into a bank account. Each year the bank pays a rate of interest of 10%. Work out the amount of money in Ben's bank account after 3 years.

Year 1 amount
= (100 + 10)% of £1200
= 110% × £1200
= 1.1 × £1200 = £1320

Year 2 amount
= 110% of £1320
= 1.1 × £1320
= £1452

Year 3 amount
= 110% of £1452
= 1.1 × £1452
= £1597.20

Exercise +4

1 Calculate these amounts using an appropriate method.

a 25% of 18 kg
b 20% of 51 m
c 15% of 360°
d 2% of 37 cm
e 65% of 510 ml
f 17.5% of 360°
g 28% of 65 kg
h 31% of 277 kg
i 3.6% of 154 kg
j 0.3% of 1320 m^2

2 Calculate each of these using a mental or written method.

a Increase £350 by 10%
b Decrease 74 kg by 5%
c Increase £524 by 5%
d Decrease 756 km by 35%
e Increase 960 kg by 17.5%

3 Calculate these. Give your answers to 2 decimal places as appropriate.

a Increase £340 by 17%
b Decrease 905 kg by 42%
c Increase £1680 by 4.7%
d Decrease 605 km by 0.9%
e Increase $2990 by 14.5%

4 These are the weekly wages of five employees at Suits-U clothing store. The manager has decided to increase all the employees' wages by 4%. Calculate the new wage of each employee.

Employee	Original wage	Increase	New wage
Hanif	£350	$350 \times 1.04 = ?$	
Bonny	£285.50		
Wilf	£412.25		
Gary	£209.27		
Marielle	£198.64		

Give your answers to 2 decimal places as appropriate.

5 **a** Patricia puts £8000 into a bank account. Each year the bank pays a compound interest rate of 5%. Work out the amount of money in Patricia's bank account after 2 years.

b Simone puts £12 500 into a savings account. Each year the building society pays a compound interest rate of 6%. Work out the amount of money in Simone's bank account after 2 years.

c Antonio invests £3400 into a Super Saver account. Each year the account pays a compound interest rate of 6%. Work out the amount of money in Antonio's account after 3 years.

Compound interest will not appear in your exam. However it is a good example of maths in everyday life.

+5 Dividing in a given ratio

This spread will show you how to:

- Divide an amount in a given ratio
- Solve multi-step problems involving ratio

Keywords
Ratio
Scale

You can divide a quantity in a given **ratio**.

Example

Sean and Patrick share £348 in the ratio 5 : 7.
How much money do they each receive?

Sean receives 5 parts for every 7 parts that Patrick receives.

Total number of parts = 5 + 7
= 12 parts
Each part = £348 ÷ 12
= £29

Sean will receive 5 parts = 5 × £29
= £145
Patrick will receive 7 parts = 7 × £29
= £203

Check your answer by adding up the two parts.
They should add up to the amount being shared!
£145 + £203 = £348

Some calculations involving ratio and **scale** need to be broken down into smaller steps.

Example

A model is made of a truck.
The length of the model is 28 centimetres.
The length of the real truck is 6.3 metres.
Work out the ratio of the length of the model to the length of the real truck.
Write your answer in the form 1 : n.

Step 1
Express the ratio in equal units.

Length of model : length of truck
28 cm : 6.3 m
28 cm : 630 cm
28 : 630

When a ratio is expressed in different units, convert the measurements to the same unit.

Step 2
Express the ratio in the form 1 : n.

÷28 (28 : 630) ÷28
1 : 22.5

The scale is 1 : 22.5.

Exercise +5

1 Solve each of these problems.

a The ratio of boys to girls in a class is 4 : 5. There are 12 boys in the class. How many girls are there?

b In a metal alloy the ratio of aluminium to tin is 8 : 5. How much aluminium is needed to mix with 55 kg of tin?

c The ratio of the number of purple flowers to the number of white flowers in a garden is 5 : 11. There are 132 white flowers. How many purple flowers are there?

d The ratio of Key Stage 3 students to Key Stage 4 students in a school is 7 : 6. There are 588 Key Stage 3 students. How many Key Stage 4 students are there at the school?

Hint for part **a**:
For every 4 boys, there are 5 girls.
8 boys → 10 girls
12 boys → ? girls

2 **a** A map has a scale of 1 : 400. A distance in real life is 4.8 m. What is this distance on the map?

b In a school the ratio of teachers to students is 1 : 22.5. If there are 990 students at the school, how many teachers are there?

c The model of an aircraft is in the scale 1 : 32. If the real aircraft is 12.48 m long, how long is the model?

3 A map has a scale of 1 : 5000.

a What is the distance in real life of a measurement of 6.5 cm on the map?

b What is the distance on the map of a measurement of 30 m in real life?

4 Solve each of these problems.

a Divide £90 in the ratio 3 : 7.

b Divide 369 kg in the ratio 7 : 2.

c Divide 103.2 tonnes in the ratio 5 : 3.

d Divide 35.1 litres in the ratio 5 : 4.

e Divide £36 in the ratio 1 : 2 : 3.

5 Solve each of these problems. Give your answers to 2 decimal places where appropriate.

a Divide £75 in the ratio 8 : 7.

b Divide £1000 in the ratio 7 : 13.

c Divide 364 days in the ratio 5 : 2.

d Divide 500 g in the ratio 2 : 5.

e Divide 600 m in the ratio 5 : 9.

+6 Expanding and factorising

This spread will show you how to:

- Multiply a single term over a bracket
- Take out common factors

Keywords
Brackets
Expand
Simplify

- **You can multiply out brackets.**
  - **You multiply each term inside the bracket by the term outside.**

$2(x+4) = 2 \times x + 2 \times 4 = 2x + 8$

Example

Expand each expression.

a $4(y+3)$ **b** $2(3x+1)$ **c** $n(n+5)$

a $4(y+3) = 4 \times y + 4 \times 3$
$= 4y + 12$

b $2(3x+1) = 2 \times 3x + 2 \times 1$
$= 6x + 2$

c $n(n+5) = n \times n + 5 \times n$
$= n^2 + 5n$

Expand means 'multiply out'.

$n \times n = n^2$

To **simplify** expressions with brackets, expand the brackets and collect like terms.

Example

Simplify each of these.

a $2(p+2) + 3p$ **b** $m(m+2) + m$ **c** $3(x+1) + 2(4x+2)$

a $2(p+2) + 3p = 2p + 4 + 3p$
$= 2p + 3p + 4$
$= 5p + 4$

b $m(m+2) + m = m^2 + 2m + m$
$= m^2 + 3m$

c $3(x+1) + 2(4x+2)$
$= 3x + 3 + 8x + 4$
$= 11x + 7$

Like terms have the same power of the same letter. m^2 and m are **not** like terms.

- **Factorising is the 'opposite' of expanding brackets.**
- **To factorise an expression, look for a common factor for all the terms.**

A common factor divides into all the terms.

$$2(x+4) \underset{\text{factorise}}{\overset{\text{expand}}{=}} 2x + 8$$

In number...
a factor is a number that exactly divides into another number.

2, 3 and 4 are factors of 12.

In algebra...
a factor is a number or letter that exactly divides into another term.

3 and $2x$ are factors of $6x$.

Exercise +6

1 Expand the brackets in these expressions.

a $3(m+2)$ **b** $4(p+6)$ **c** $2(x+4)$ **d** $5(q+1)$

e $2(6+n)$ **f** $3(2+t)$ **g** $4(3+s)$ **h** $2(4+v)$

2 Expand these.

a $3(2q+1)$ **b** $2(4m+2)$ **c** $3(4x+3)$ **d** $2(3k+1)$

e $5(2+2n)$ **f** $3(4+2p)$ **g** $4(1+3y)$ **h** $2(5+4z)$

3 At a pick-your-own farm, Lucy picks n apples.
Mary picks 5 more apples than Lucy.
a Write down, in terms of n, the number of apples Mary picks.

Nat picks 3 times as many apples as Mary.
b Write down, in terms of n, the number of apples Nat picks.

4 Expand and simplify each of these.

a $3(p+3)+2p$ **b** $2(m+4)+5m$ **c** $4(x+1)-2x$

d $2(5+k)+3k$ **e** $4(2t+3)+t-2$ **f** $3(2r+1)-2r+4$

5 On Monday a shop sells s DVDs.
On Tuesday the shop sells 6 more DVDs than on Monday.
a Write an expression for the number of DVDs it sells on Tuesday.

On Wednesday the shop sells twice as many DVDs as on Tuesday.
b Write an expression for the number of DVDs it sells on Wednesday.

On Thursday the shop sells 7 more DVDs than on Wednesday.
c Write an expression for the number of DVDs it sells on Thursday.

Give your answer in its simplest form.

6 Find all the common factors of

a $2x$ and 6 **b** $4y$ and 12 **c** 10 and $20j$ **d** 6 and $12p$

e 9 and $6q$ **f** $6t$ and 4 **g** $4x$ and 10 **h** $24t$ and 8

Hint for **6a**:
2 and x are factors of $2x$.
1, 2, 3 and 6 are factors of 6.
2 is the common factor of $2x$ and 6.

7 Find the highest common factor of

a $3x$ and 9 **b** $12r$ and 10 **c** $6m$ and 8 **d** 4 and $4z$

8 Find the highest common factor of

a y^2 and y **b** $4s^2$ and s **c** $7m$ and m^3 **d** $2y^2$ and $2y$

9 Factorise these.

a $2x+10$ **b** $3y+15$ **c** $8p-4$ **d** $6+3m$

e $5n+5$ **f** $12-6t$ **g** $14+4k$ **h** $9z-3$

+7 Inequalities

This spread will show you how to:

- Solve linear inequalities in one variable

Keywords
Greater than
Inequality
Less than
Solution set

- In an **equation**, the left-hand side equals the right-hand side.
- In an **inequality**, the left-hand and right-hand sides are not necessarily equal.
  - An inequality usually has a range of values.

You use one of these signs to show the relationship between the two sides of an inequality.

$<$	**less than**	$>$	**greater than**
$\leqslant$	less than or equal to	$\geqslant$	greater than or equal to

You can show inequalities on a number line.

$x<2$	x is less than 2	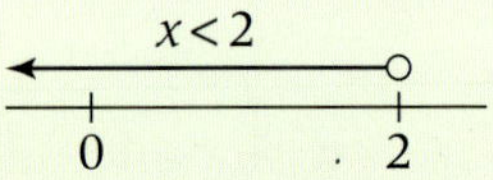
$x>2$	x is greater than 2	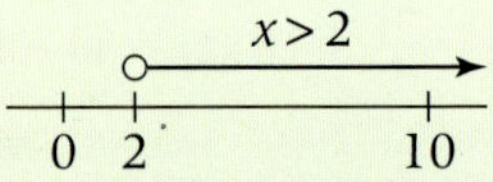
$y\leqslant 4$	y is less than or equal to 4	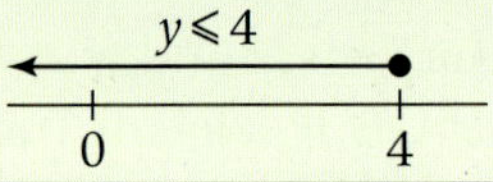
$y\geqslant 4$	y is greater than or equal to 4	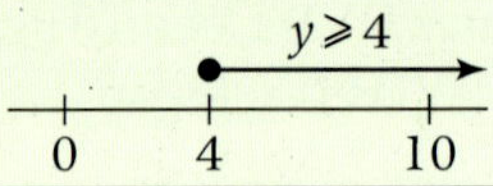

The open circle shows that 2 is not included.

The filled-in circle shows that 4 is included.

Inequalities can have more than one term, for example $3x+4<19$.

You can solve an inequality to find a set of values for x.

Example

a Solve the inequality $3x+4<19$
b Show the solution set on a number line.

a
$$3x+4<19$$
$$3x+4-4<19-4$$
$$3x<15$$
$$x<5$$

b

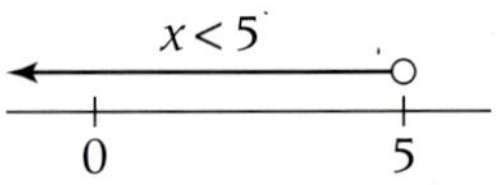

The solution set is $x<5$.

Sometimes a letter is bounded by two inequalities.

Example

List the whole numbers that satisfy the inequality $-3<x\leqslant 2$.

$-3<x$ means '-3 is less than x,' or 'x is greater than -3'.
$x\leqslant 2$ means 'x is less than or equal to 2'.
The whole numbers that satisfy the inequality are $-2, -1, 0, 1, 2$.

Exercise +7

1 Show these inequalities on a number line.

a $x < 1$ **b** $x \geqslant 1$ **c** $x \geqslant 5$ **d** $x < -2$

e $x < 1.5$ **f** $x > -4$ **g** $x \leqslant 3$ **h** $x \leqslant -1.5$

2 **a** If $x > 5$, what can you say about **i** $2x$ **ii** $4x$?

b If $y \leqslant 6$, write an inequality for **i** $3y$ **ii** $5y$.

c If $x \geqslant -4$, write an inequality for $5x$.

d If $m < -3$, write an inequality for $6m$.

3 Solve these inequalities and show the solution sets on number lines.

a $2x \leqslant 4$ **b** $2x < 10$ **c** $3x > -6$ **d** $4x \geqslant -16$

4 Match each inequality to a number line.

a 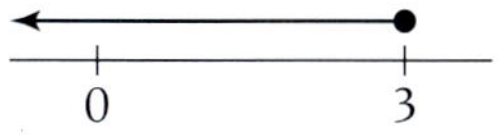

b 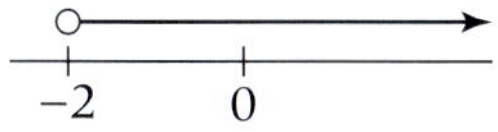

c 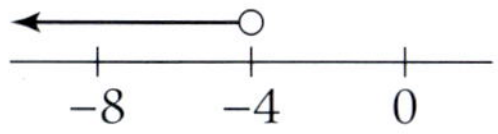

d 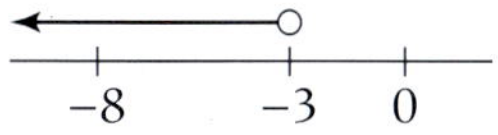

e 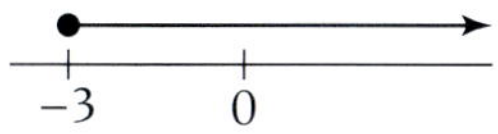

f 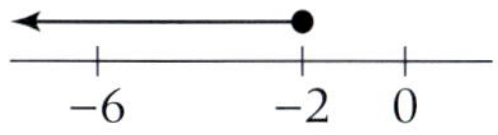

i $x < -3$

ii $x \leqslant 3$

iii $x - 1 \leqslant -3$

iv $x \geqslant -3$

v $x - 1 > -3$

vi $x + 2 < -2$

5 Copy and complete these.

a If $3x + 2 > 11$ then $3x > \square$ and $x > \square$

b If $7x - 4 > 31$ then $7x > \square$ and $x > \square$

c If $2x + 9 \leqslant 11$ then $2x \leqslant \square$ and $x \leqslant \square$

d If $5x - 3 \geqslant 12$ then $5x \geqslant \square$ and $x \geqslant \square$

6 Solve each of these inequalities.
Show each solution on a number line.

a $x + 7 \leqslant 12$ **b** $x - 2 \geqslant 4$ **c** $3x + 5 \geqslant 11$ **d** $2x - 5 < 3$

e $5x + 1 \geqslant -4$ **f** $6x - 2 \leqslant 16$ **g** $3x + 2 > 11$ **h** $2x - 9 \leqslant -5$

7 List the whole numbers that satisfy each of these inequalities.

a $-3 \leqslant x < 2$ **b** $-2 < x \leqslant 3$ **c** $-1 \leqslant x \leqslant 4$

d $0 < x \leqslant 5$ **e** $-3 \leqslant x \leqslant -1$ **f** $0 < x < 2$

+8 Intersection of two lines

This spread will show you how to:

- Plot graphs of functions in which y is given explicitly in terms of x

Keywords
Solution

A straight line graph is made up of an infinite number of points.

- **All the points on a straight line satisfy the equation of the line.**

- **Where two straight lines cross, the coordinates satisfy the equations of both lines.**

The lines $x = 4$ and $y = -1$ are drawn on this graph.

Every point on the line $x = 4$ has x-coordinate 4.
Every point on the line $y = -1$ has y-coordinate -1.

So the point where they cross is $(4, -1)$.

The point P satisfies both equations: $x = 4$ **and** $y = -1$. It has coordinates $(4, -1)$.

P is the **solution** to the equations $x = 4$ and $y = -1$.

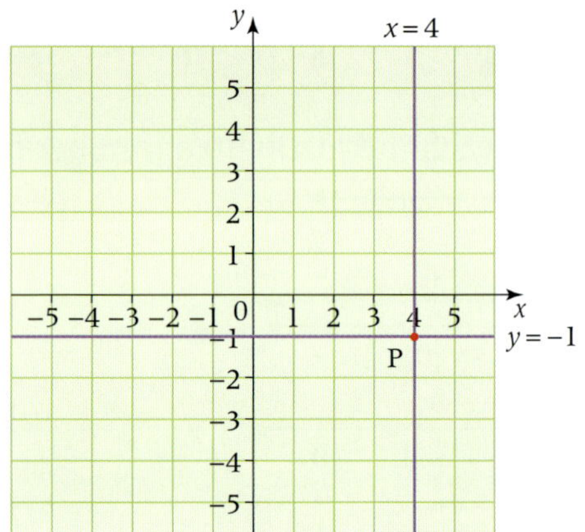

Example

a Draw the graphs of $y = x + 6$ and $y = 2x$ on the same pair of axes.
b Write the coordinates of the point where they cross.
c What can you say about this point?

a $y = x + 6$

x	−2	−1	0	1	2
y	4	5	6	7	8

$y = 2x$

x	−2	−1	0	1	2
y	−4	−2	0	2	4

b Graphs cross at (6, 12)

c (6, 12) satisfies both equations.

$x = 6$ is the solution to $x + 6 = 12$ and $2x = 12$.

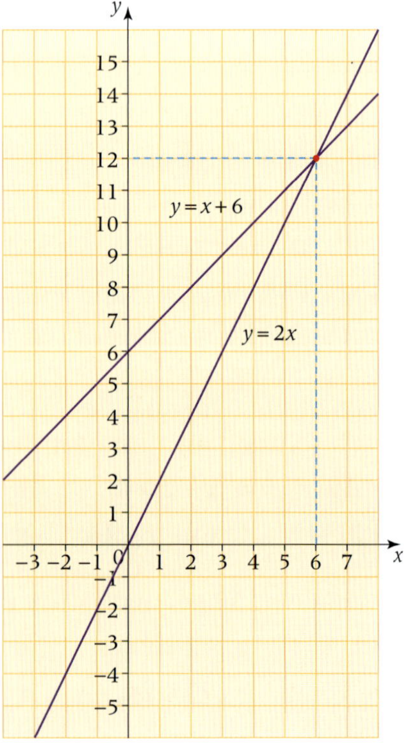

Exercise +8

1 Write the coordinates of the points where these lines cross.
Draw graphs to check your answers.

a $x = 2$ and $y = 3$

b $x = -1$ and $y = -4$

c $x = 3$ and $y = 7$

d $x = -2$ and $y = -4$

e $x = 7$ and $y = -2$

f $y = 1$ and $x = 4$

2 For each pair of equations, decide whether the lines will cross and then draw the graphs on a copy of this grid to check your answers.

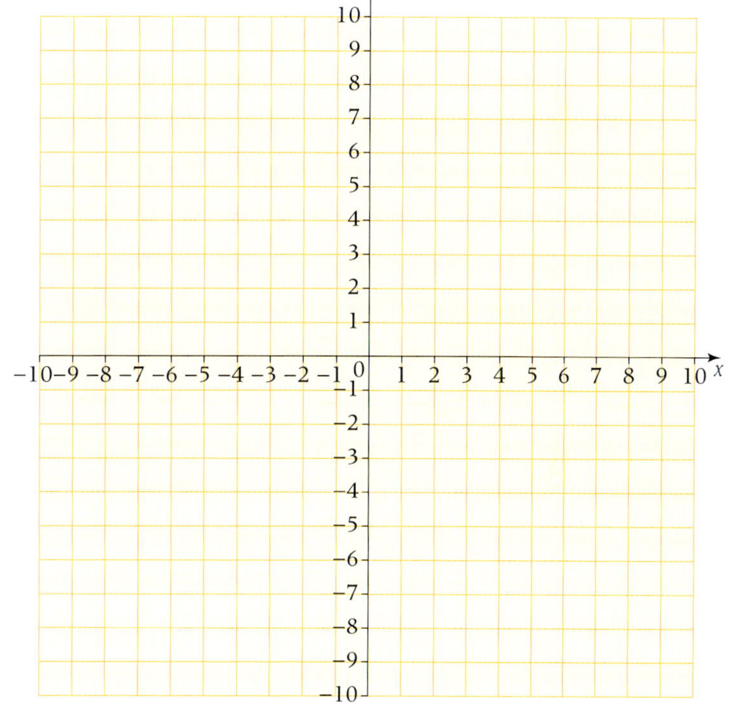

a $y = 2x + 1$ and $y = 4x + 2$

b $y = 3$ and $y = x + 1$

c $y = 3x + 2$ and $y = 3x - 1$

d $y = x$ and $y = -x$

3 For each pair of lines from question **2** that cross, write the coordinates of the point where the lines cross.

4 **a** Draw the graphs of $y = -x + 2$ and $y = 2x - 1$ on the same axes.

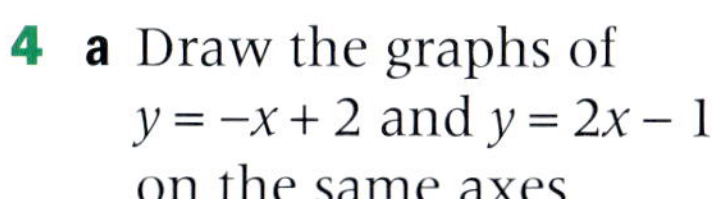

b Write the coordinates of the point where the two lines cross.

c Does the point (1, 2) satisfy both of these equations?
Explain how you know.

5 **a** Draw the graphs of $y = 2x - 4$ and $y = x - 1$ on the same axes.

b Write the coordinates of the point where they cross.

6 **a** Draw the graphs of these two equations on the same axes.

$$x + 2y = 8$$
$$x - y = 2$$

b Where the two lines cross, the x and y values satisfy both these equations.
Write these x and y values.

+9 Equations with brackets and fractions

This spread will show you how to:

- Solve linear equations that require prior simplification of brackets and fractions

Keywords
Brackets
Expand

- **To solve equations with the unknown on both sides and brackets**
  - **Expand the brackets**
  - **Use the balance method.**

Example

Solve each equation.

a $2(y + 4) = 4y$

b $6r - 2 = 4(r + 3)$

a $2(y + 4) = 4y$ — Expand the brackets.

$2y + 8 = 4y$ — Subtract $2y$ from both sides.

$8 = 4y - 2y$

$8 = 2y$ — Divide both sides by 2.

$4 = y$

b $6r - 2 = 4(r + 3)$

$6r - 2 = 4r + 12$ — Subtract $4r$ from both sides.

$6r - 4r - 2 = 4r - 4r + 12$

$2r - 2 = 12$ — Add 2 to both sides.

$2r = 12 + 2$

$2r = 14$ — Divide both sides by 2.

$r = 7$

Example

Solve each equation.

a $\frac{x}{4} = -3$

b $\frac{x + 3}{2} = 5$

a

$$\frac{x}{4} = -3$$

$$4 \times \frac{x}{4} = -3 \times 4$$

$$x = -12$$

b

$$\frac{x + 3}{2} = 5$$

$$2 \times \frac{x + 3}{2} = 5 \times 2$$

$$x + 3 = 10$$

$$x = 10 - 3$$

$$x = 7$$

Exercise +9

1 Solve these equations.

a $2(r+6)=5r$ **b** $6(s-3)=12s$

c $4(2t+8)=24t$ **d** $5(v-1)=6v$

2 Solve these equations.

a $2(a+5)=7a-5$ **b** $3(b-2)=5b-2$

c $2(c+6)=5c-3$ **d** $3d+8=2(d+2)$

3 Solve these equations.

a $3(2x-4)=7x-18$ **b** $2(3y+2)=5y-2$

c $4(2z+1)=6z+15$ **d** $-4(6m+1)=-17m-18$

4 Solve these equations.

a $2(e+3)=4e-1$ **b** $4f+3=2(f+2)$

c $4(2g+1)=6g+1$ **d** $3(2h+3)=5h+8$

5 Solve these.

a $\frac{x}{3}=3$ **b** $\frac{m}{4}=-2$ **c** $\frac{-n}{3}=6$ **d** $\frac{m}{5}=4$

6 Find the value of the unknown in each of these.

a $\frac{s}{3}+5=8$ **b** $4-\frac{t}{2}=1$ **c** $\frac{u}{5}+7=5$ **d** $16=\frac{v}{4}+13$

7 Solve these.

a $\frac{2x}{3}+5=9$ **b** $\frac{3y}{2}-5=4$ **c** $3-\frac{2z}{5}=-3$ **d** $\frac{3q}{2}+5=-7$

8 Solve these equations.

a $\frac{x+5}{3}=2$ **b** $\frac{x-3}{4}=2$ **c** $\frac{x+9}{2}=-4$ **d** $\frac{10-x}{4}=1$

9 I think of a number.
I divide my number by 4 and add 6.

a Write an expression for 'I divide my number by 4 and add 6'.
Use n to represent the number.

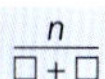

b My answer is 10.
Using your expression from part **a**, write an equation to show this.

Expression **a** = 10

c Solve your equation to find the number, n.

+10 The general term

This spread will show you how to:

- Generate and describe integer sequences using position-to-term definitions

Keywords
General term
*n*th term
Position-to-term

- A **position-to-term** rule links a term with its position in the sequence.

For example, the 4 times table: 4, 8, 12, 16, 20, 24, ... can be written as

1st term = T(1)	2nd term = T(2)	3rd term = T(3)	4th term = T(4)	5th term = T(5)	6th term = T(6)
4	8	12	16	20	24

$T(1) = 1 \times 4$
$T(2) = 2 \times 4$
$T(3) = 3 \times 4$
$T(10) = 10 \times 4$

The **general term** or ***n*th term** of the 4 times table is $T(n) = n \times 4$ or $4n$.

You can generate a sequence from the general term.

Example

Find the first three terms and the 10th term of the sequence with general term $3n + 2$.

1st term $3 \times 1 + 2 \rightarrow 5$
2nd $3 \times 2 + 2 \rightarrow 8$
3rd $3 \times 3 + 2 \rightarrow 11$
10th $3 \times 10 + 2 \rightarrow 32$

- **To find the general term of a linear sequence.**
  - **work out the common difference**
  - **write the common difference as the coefficient of *n***
  - **compare the terms in the sequence to the multiples of *n***

A linear sequence has an equal spacing between terms.

Example

Find the general term of the sequence
5, 8, 11, 14, 17 ...

The common difference is +3.
The *n*th term contains the term $3n$.
Compare the sequence to the multiples of 3:

$3n$	3	6	9	12	15
Term	5	8	11	14	17

Each term is 2 more than a multiple of 3.
The general term is $3n + 2$.

Check:
$n = 1 \rightarrow 3 + 2 = 5$
$n = 2 \rightarrow 6 + 2 = 8$
$n = 3 \rightarrow 9 + 2 = 11$
...

Exercise +10

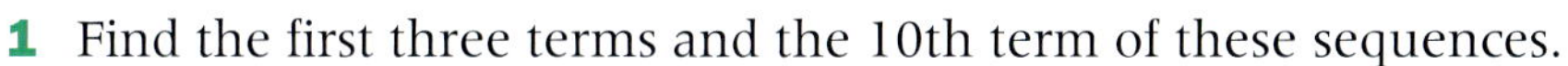

1 Find the first three terms and the 10th term of these sequences.

a $5n+1$ **b** $3n+8$ **c** $8n-4$ **d** $6n-8$

e $24-2n$ **f** $15-5n$ **g** $7n-20$ **h** $4n-6$

2 Copy and complete the table of results for each sequence.

a $3n+8$

Term number	Term
1	11
2	
3	
5	
10	
n	$3n+8$

b $6n-15$

Term number	Term
1	−9
2	
3	
5	
10	
n	$6n-15$

3 Write the first five terms of the sequences with nth term

a n^2+4 **b** n^2-2 **c** $2n^2$ **d** $12-n^2$

4 **a** Find the common difference for the series 5, 9, 13, 17, 21, ...

b Copy and complete this statement:

The nth term contains the term □ n.

c Copy and complete this table to show the sequence and the multiples of n.

Sequence					
□ n					

d Compare the terms in the sequence to the multiples of n and write the general term for the sequence.

5 Follow the steps in question **4** to find the general terms for these sequences.

a 11, 17, 23, 29, 35, ... **b** 1, 10, 19, 28, 37, ...

c 15, 22, 29, 36, 43, ... **d** −10, −6, −2, 2, 6, ...

e 20, 17, 14, 11, 8, ... **f** 15, 11, 7, 3, −1, ...

g 16, 8, 0, −8, −16, ... **h** 31, 23, 15, 7, −1, ...

6 Find the nth term for each of these arithmetic sequences.

a 7, 11, 15, 19, 23, ... **b** −6, −2, 2, 6, 10, ...

c 32, 23, 14, 5, −4, ... **d** 15, 9, 3, −3, 9, ...

+11 Area of a parallelogram and a trapezium

This spread will show you how to:

- Use formulae to find the area of any parallelogram
- Use formulae of rectangles, triangles and parallelograms to find the area of any trapezium

Keywords

Area
Base
Parallelogram
Perpendicular height
Trapezium

You can find the formula for the **area** of any **parallelogram**.

For this parallelogram ...

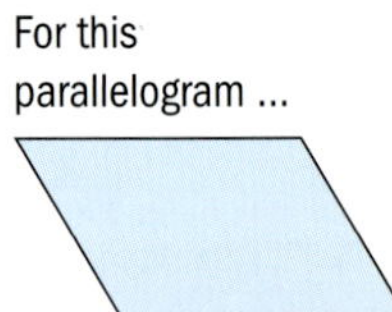

cut off one triangle ...

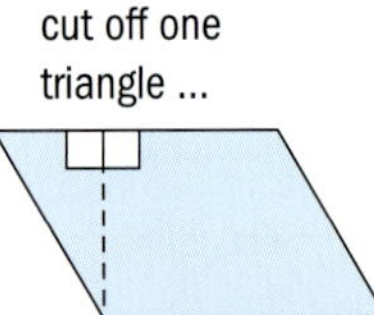

and fit it on the other end ... to make a rectangle.

height

base

- **Area of parallelogram = base × perpendicular height.**

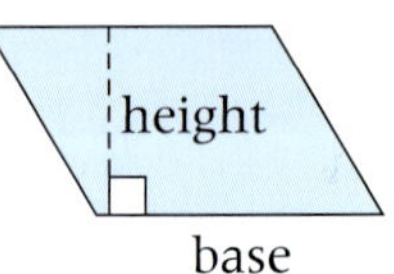

The height must be perpendicular to the base.

You can find the formula for the area of any **trapezium**.

You can fit two **congruent** trapeziums together to make a parallelogram.

Congruent means identical.

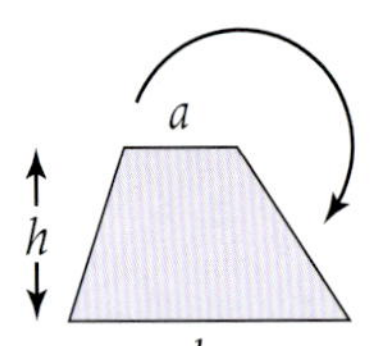

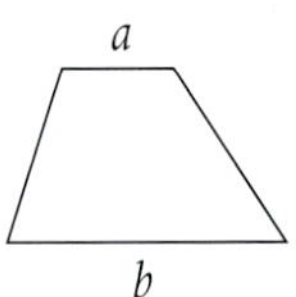

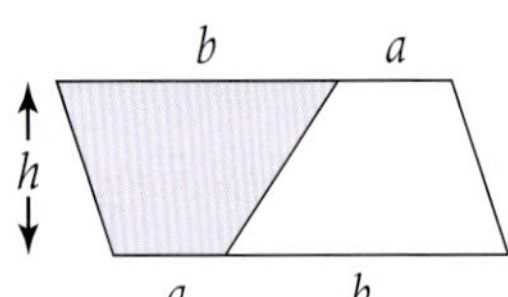

The base of the parallelogram is $a + b$ and the height is h.
Area of parallelogram = $(a + b) \times h$
Area of trapezium = half area of parallelogram.

- **Area of trapezium = $\frac{1}{2} \times (a + b) \times h$**

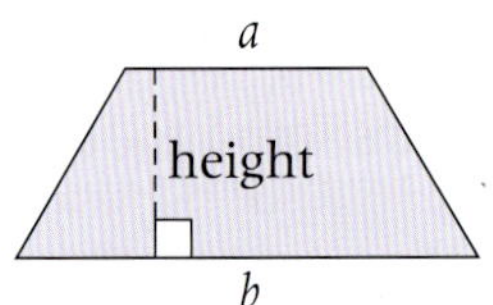

The height is the perpendicular distance between the parallel sides.

Example

Calculate the area of each shape.

a

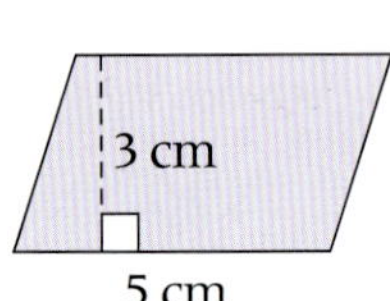

b

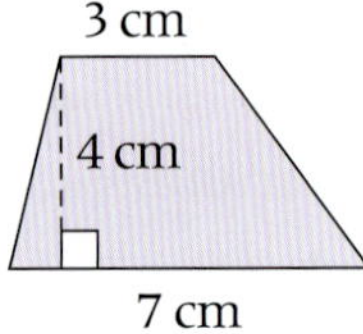

a Area of parallelogram = 5×3
$= 15\ \text{cm}^2$

b Area of trapezium = $\frac{1}{2}(3 + 7) \times 4$
$= 5 \times 4$
$= 20\ \text{cm}^2$

Exercise +11

1 Calculate the area of each parallelogram.

Give your answers in square units.

2 Calculate the area of each trapezium.

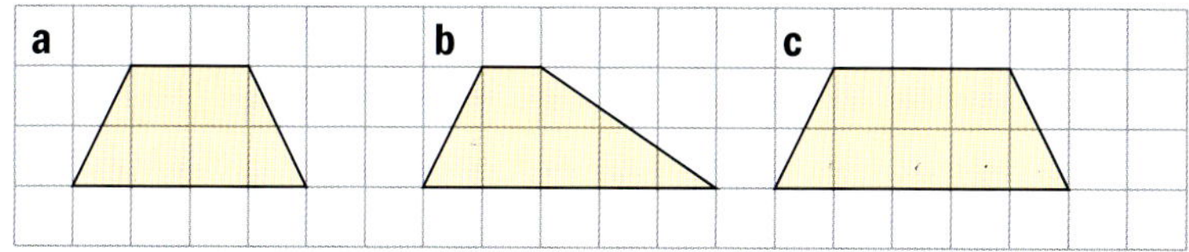

3 Calculate the area of each parallelogram. State the units of your answers.

a

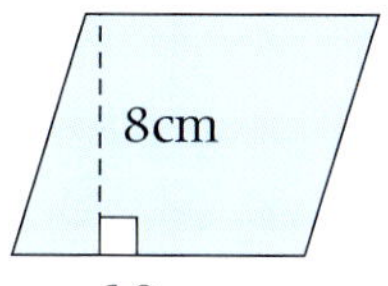

b

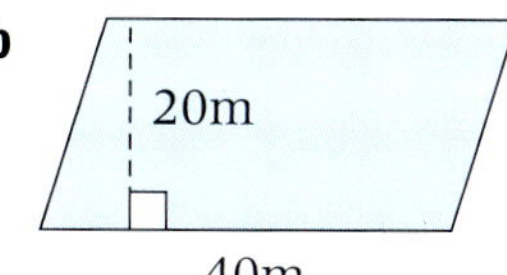

c

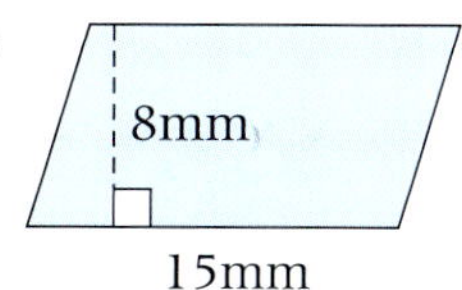

d

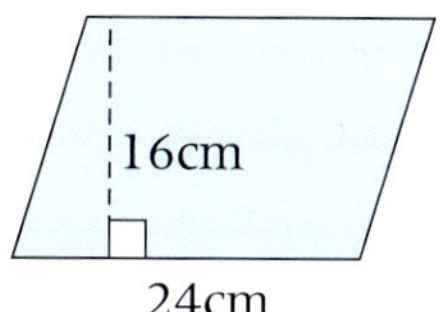

4 Calculate the area of each trapezium. State the units of your answers.

a

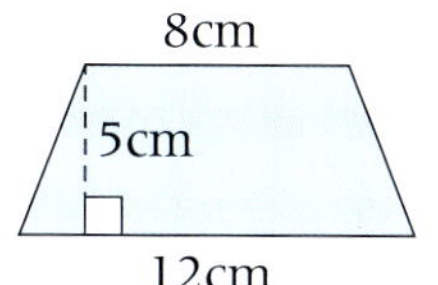

b

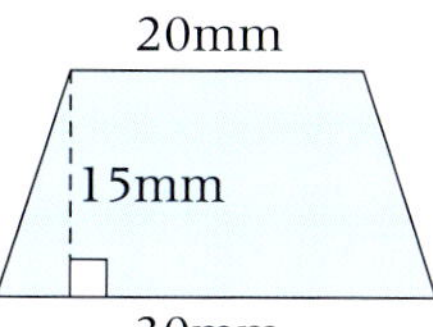

c

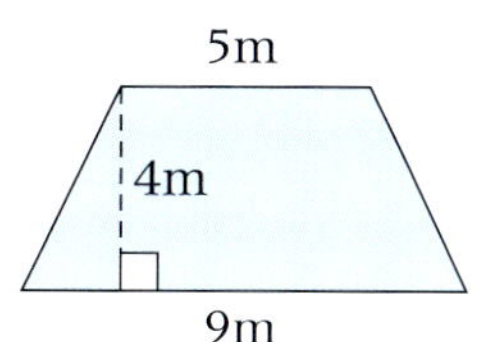

d

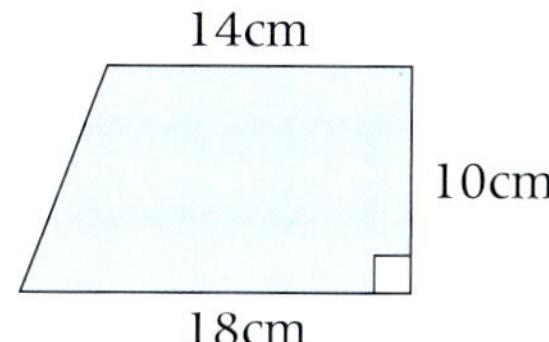

5 The area of these shapes is given. Calculate the unknown lengths.

a

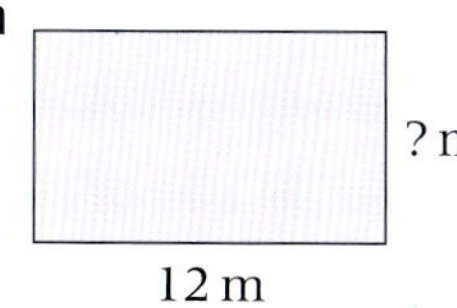

area = 72 m^2

b

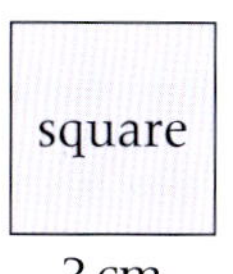

area = 196 cm^2

c

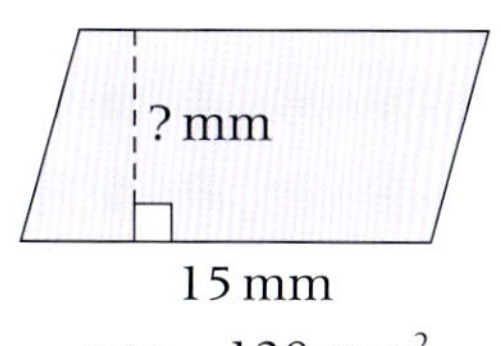

area = 120 mm^2

d

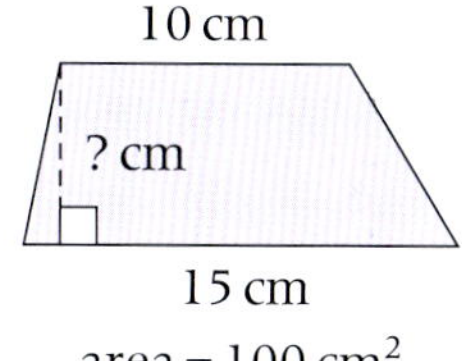

area = 100 cm^2

6 **a** Calculate the area of this shape using the formula for the area of a trapezium.

b Calculate the area by adding the areas of the triangles and the square.

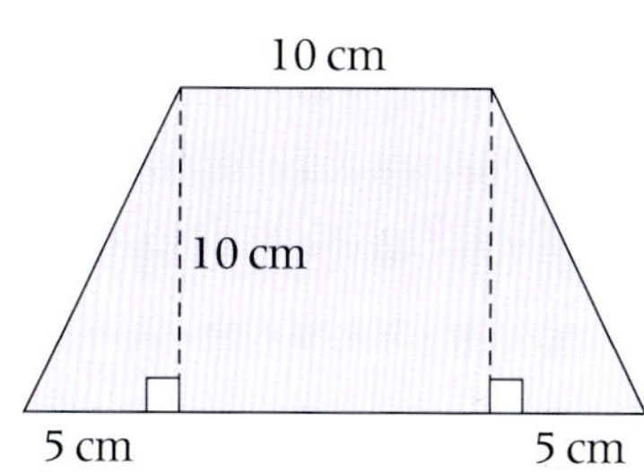

+12 Angles in parallel lines

This spread will show you how to:

- Use parallel lines and alternate angles

Keywords
Alternate
Corresponding
Parallel
Vertically opposite

When two lines cross, four angles are formed.

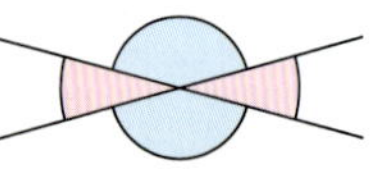

- **Vertically opposite** angles are equal.

When a line crosses two **parallel** lines, eight angles are formed.

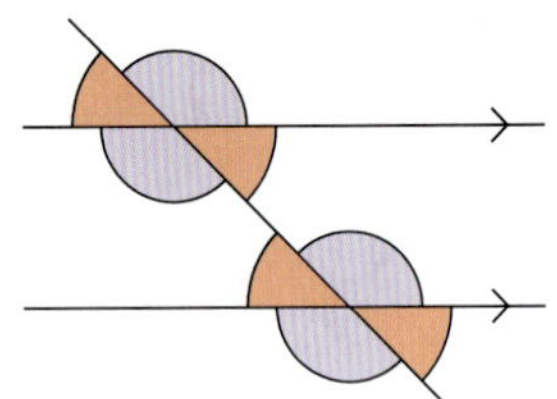

The four red **acute** angles are equal.

The four purple **obtuse** angles are equal.

Acute + obtuse = 180°

Parallel lines are always the same distance apart.

An acute angle is less than 90°.
An obtuse angle is more than 90° but less than 180°.

- **Alternate** angles are equal.

They are called Z angles.

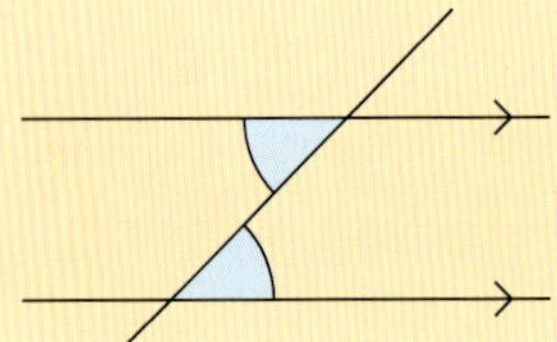

- **Corresponding** angles are equal.

They are called F angles.

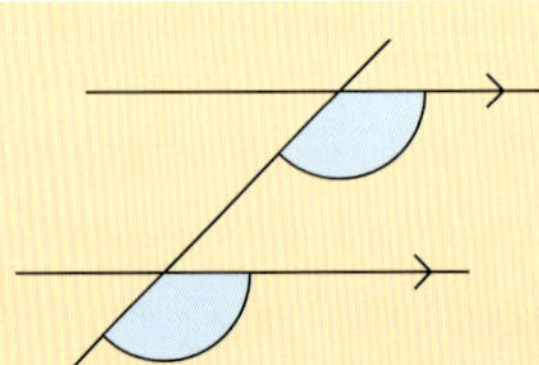

Example

Find the unknown angles in these diagrams.
Give reasons for your answers.

a

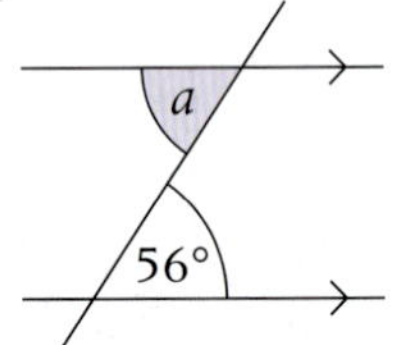

b

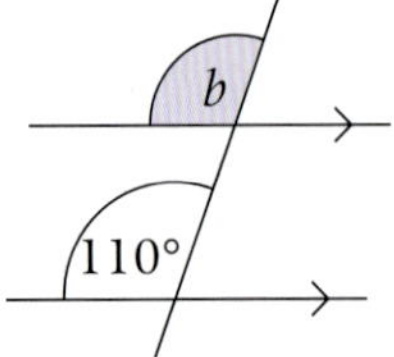

c

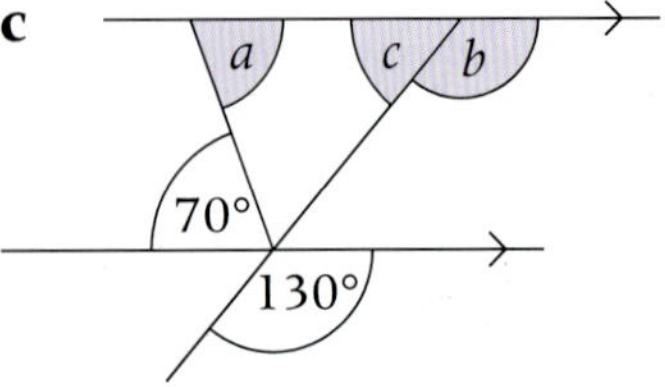

a $a = 56°$
(alternate angles)

b $b = 110°$
(corresponding angles)

c $a = 70°$ (alternate angles)
$b = 130°$ (corresponding angles)
$c = 180° - 130°$
$= 50°$ (angles on straight line add to 180°)

Exercise +12

The diagrams are not drawn to scale.

1 Calculate the size of the angles marked by a letter in each diagram. Give a reason for each answer.

a

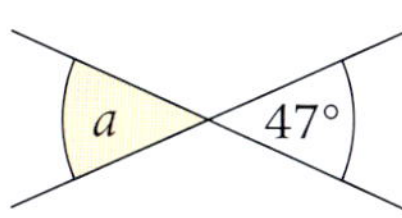

b

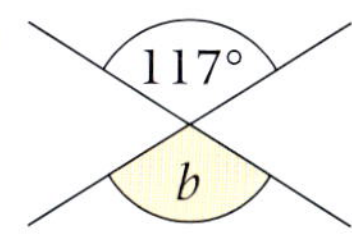

c

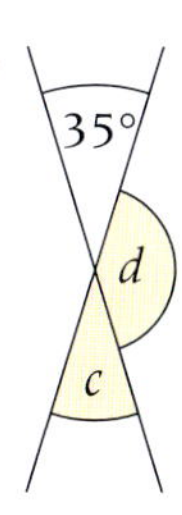

d

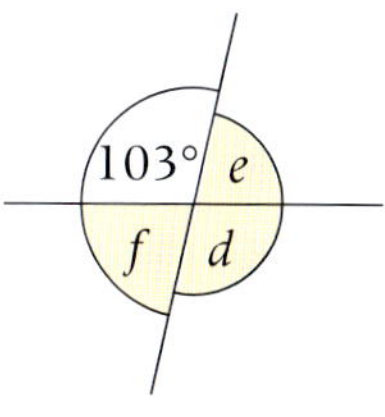

e

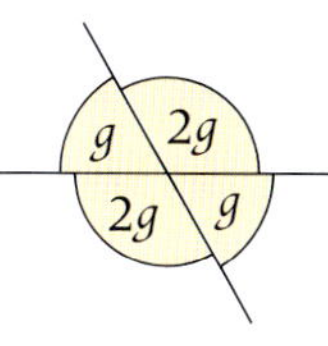

2 Find the value of each angle marked with a letter. Give a reason for each answer.

a

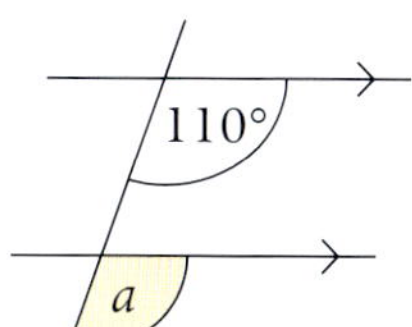

b

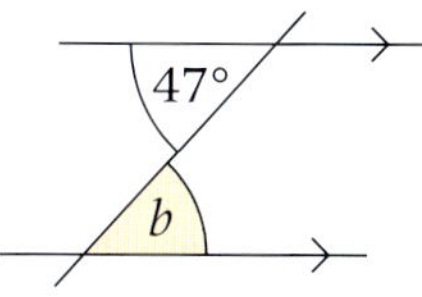

c

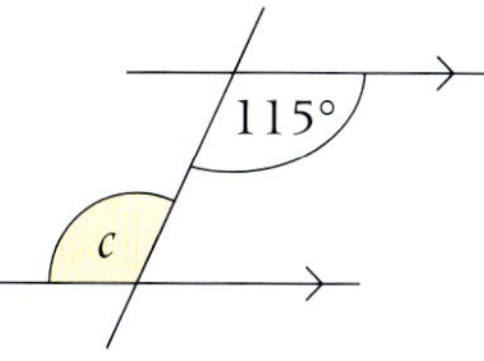

d

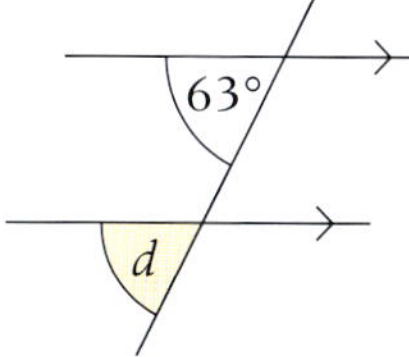

e

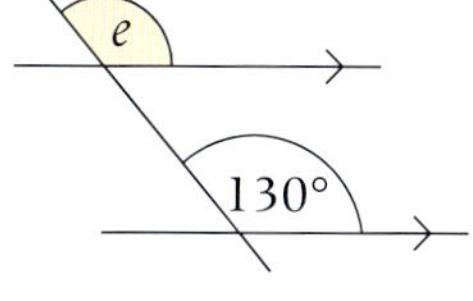

f

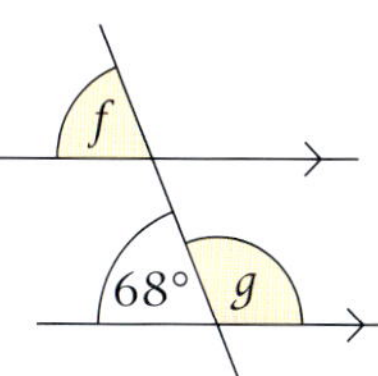

g

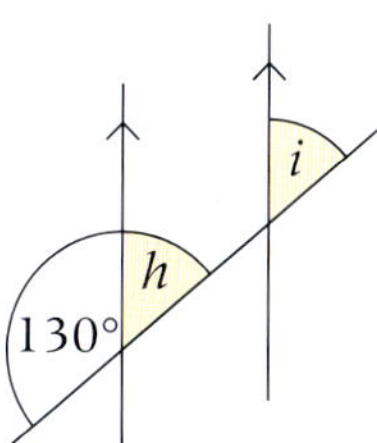

h

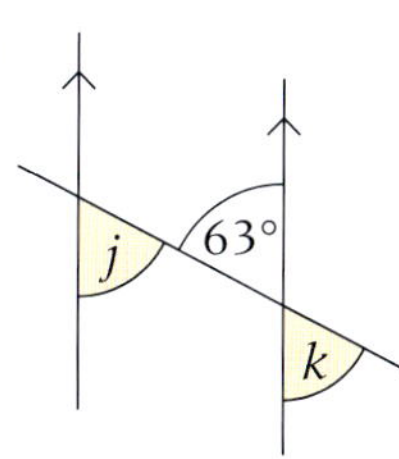

i

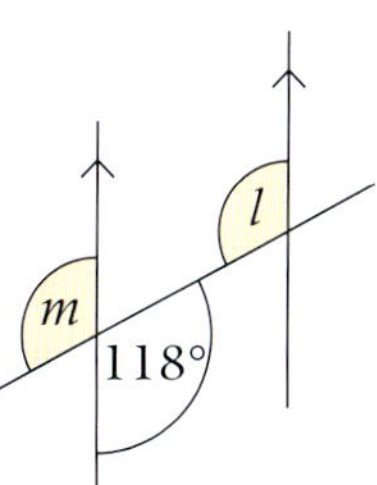

3 Find the value of each angle marked with a letter. Give a reason.

a

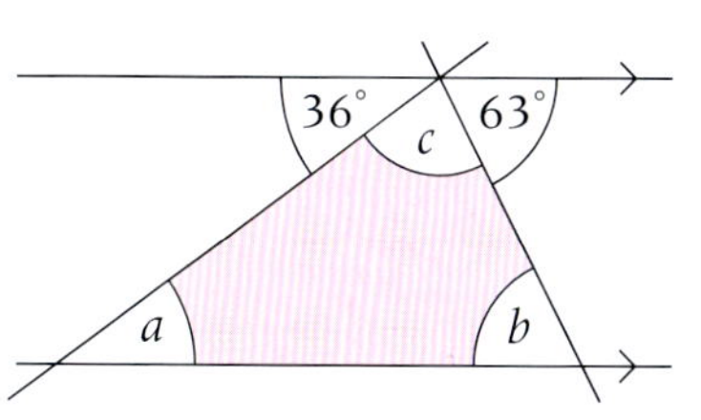

b

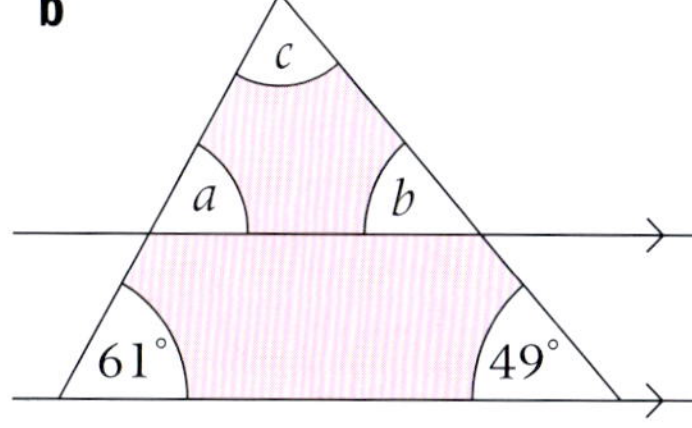

c

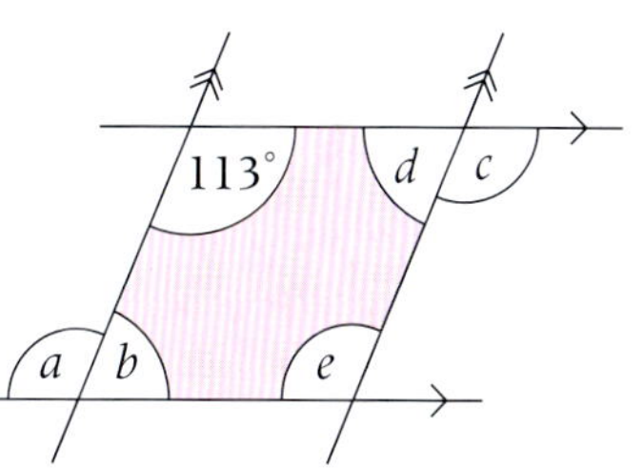

+13 Similar shapes

This spread will show you how to:

- Identify the scale factor of an enlargement as a ratio of the lengths

Keywords
Enlargement
Scale factor
Similar

- **In an enlargement, the object and the image are similar,**
  - **the angles stay the same**
  - **the lengths increase in proportion.**

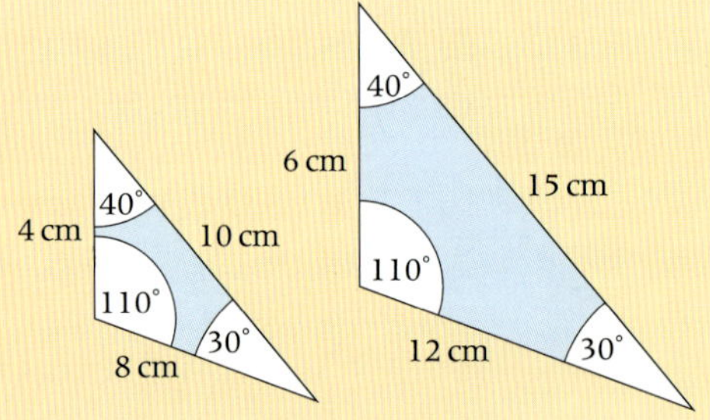

You use corresponding lengths to find the **scale factor**.

- **Scale factor** $= \dfrac{\text{length of image}}{\text{length of object}}$.

Example

These triangles are similar.
Find the length x.

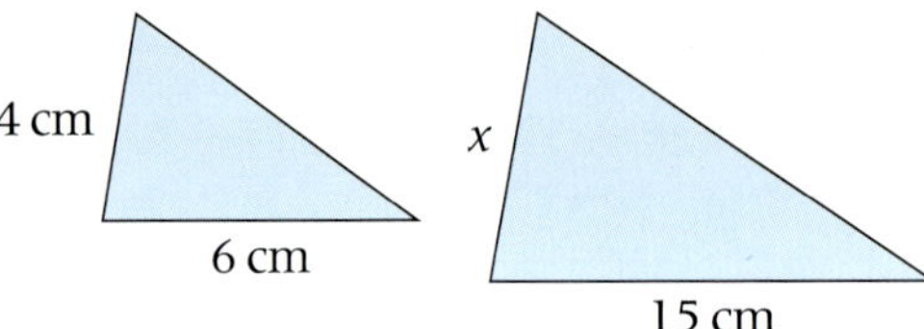

The scale factor is $\dfrac{15}{6} = 15 \div 6 = 2.5$

$x = 4 \text{ cm} \times 2.5 = 10 \text{ cm}$

Example

a Show that triangle ABE is similar to triangle ACD.
b Calculate the value of x.

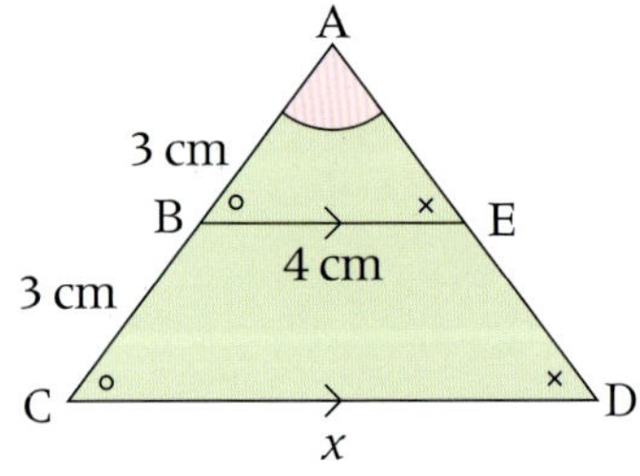

a Angle B = angle C (corresponding angles are equal)
Angle E = angle D (corresponding angles are equal)
Angle A is common to both triangles.
So △ABE and △ACD are similar.

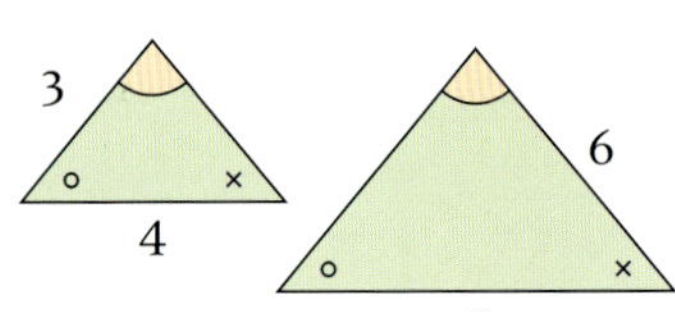

Similar shapes have the same angles but are different sizes.

b The scale factor is $6 \div 3 = 2$
$x = 4 \text{ cm} \times 2 = 8 \text{ cm}$

Exercise +13

1 In each question, the two triangles are similar.
Find the value of the unknown angles.

Angles in a triangle add to 180°.

a

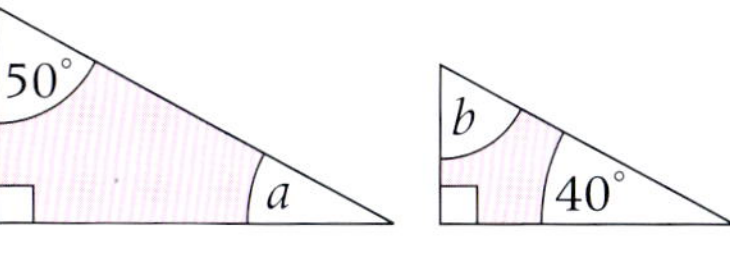

b

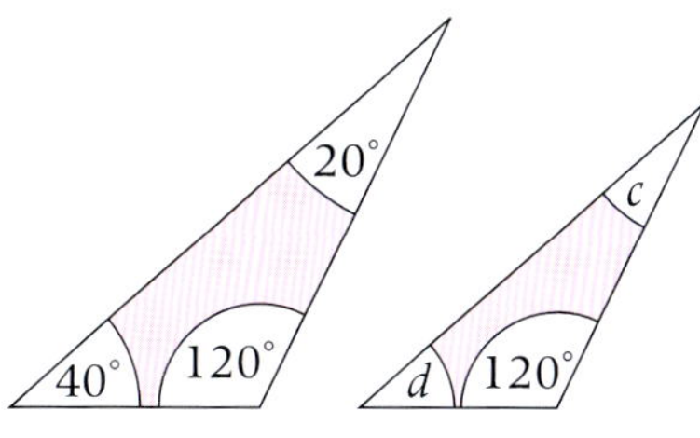

c

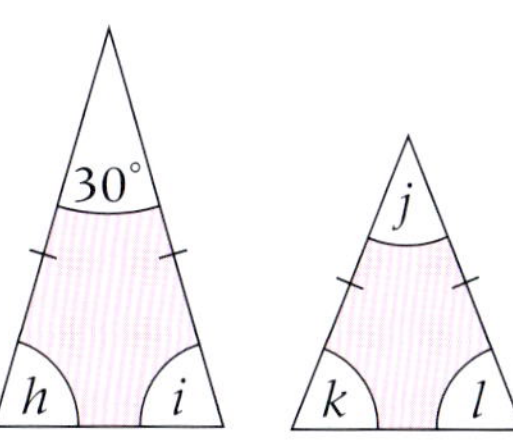

d

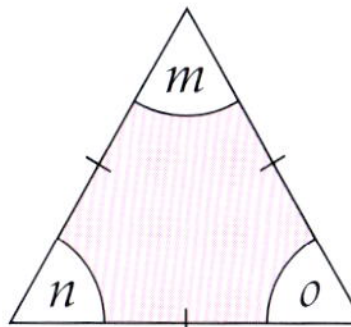

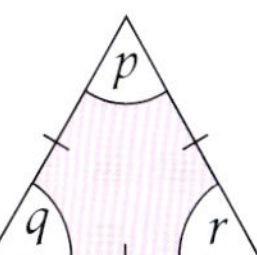

2 Which of these rectangles are similar to the green rectangle?
For the ones that are similar, give the scale factor of the enlargement.

a

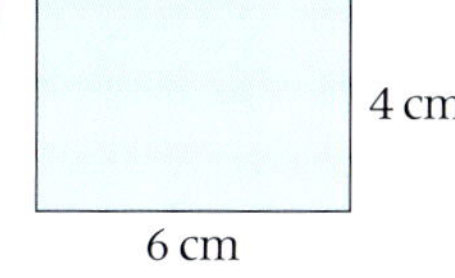

b

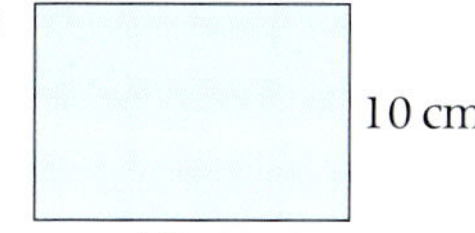

c

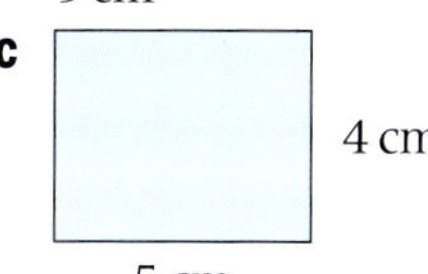

d

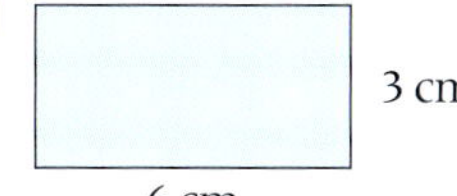

e

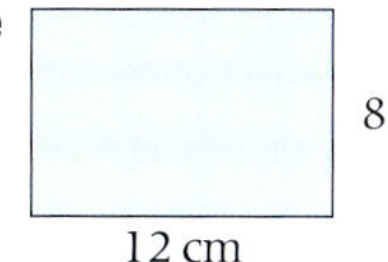

f

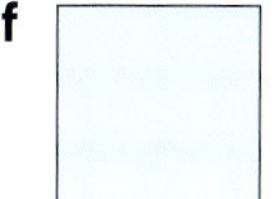

3 In each question, the two triangles are similar. Calculate the scale factor of the enlargement and the unknown length.

a

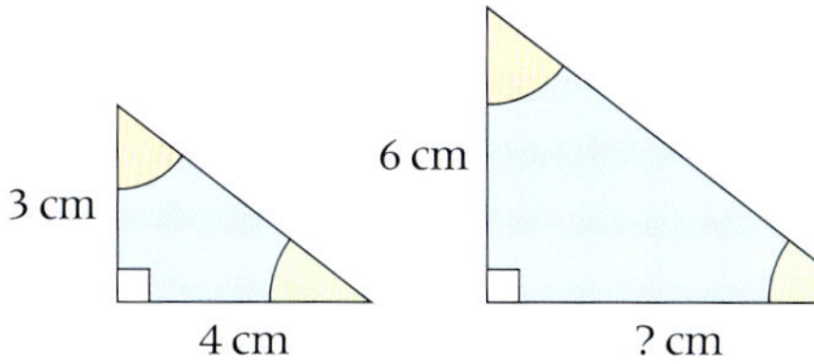

b

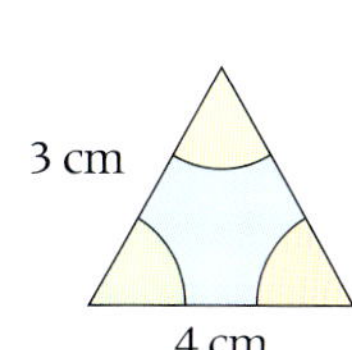

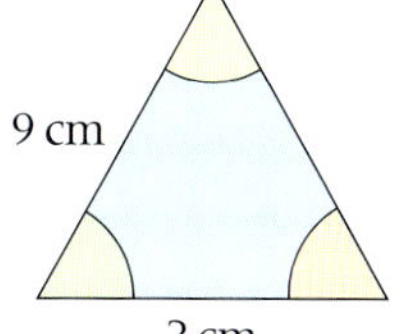

4 Calculate the value of each unknown length.

a

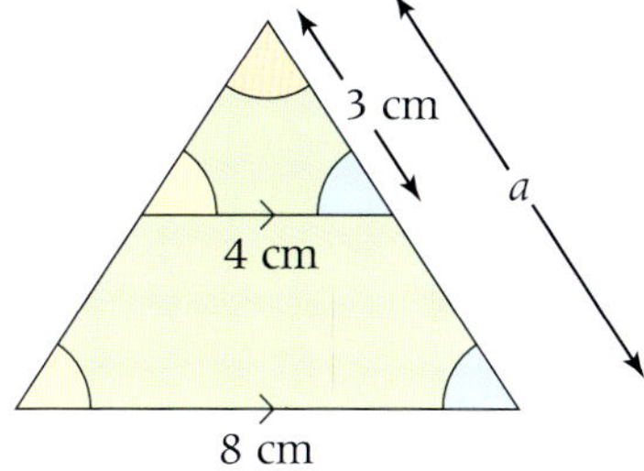

b

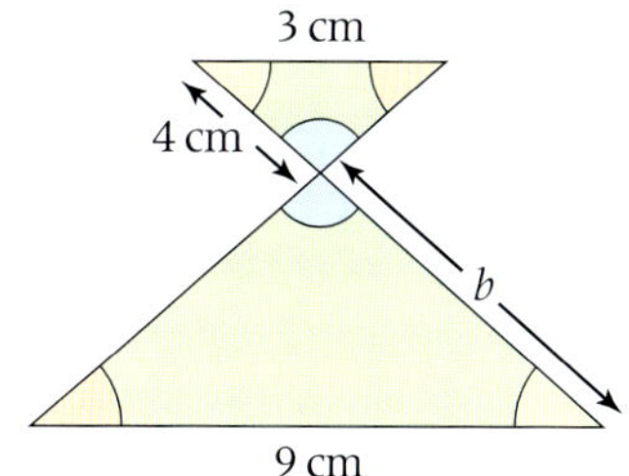

+14 Midpoint of a straight line

This spread will show you how to:

- Calculate the midpoint of the line AB

Keywords

Coordinates
Line
Midpoint

- **The midpoint M of a line AB is halfway along it.**

A M B

M is the midpoint.

If A = (−1, 3)
and B = (3, 3)
then M_1 = (1, 3)

If C = $(1, 1\frac{1}{2})$
and D = $(1, -1\frac{1}{2})$
then M_2 = (1, 0)

If E = (−2, 2)
and F = (0, −1)
then $M_3 = (-1, \frac{1}{2})$

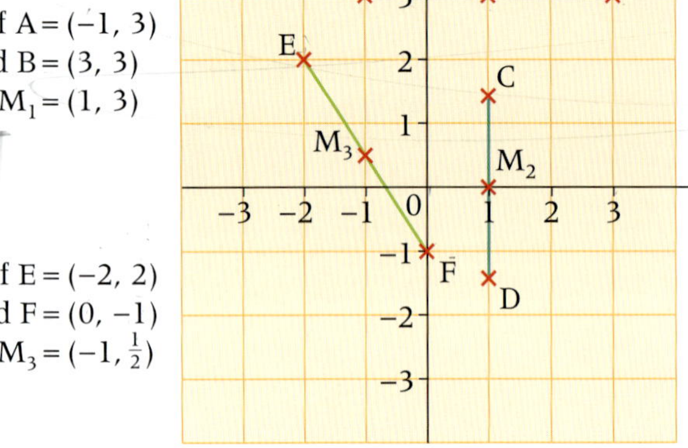

- **If A = (x_1, y_1) and B = (x_2, y_2) then M = $\left(\frac{x_1 + x_2}{2}, \frac{y_1 + y_2}{2}\right)$.**

The midpoint of AB is the mean of the **coordinates** of points A and B.

Example

Calculate the coordinates of the midpoint between the points

a (7, 1) and (−3, 5)

b (4, −1) and (2, −2)

a (7, 1) = (x_1, y_1) and (−3, 5) = (x_2, y_2)

$$\text{Midpoint} = \left(\frac{7+3}{2}, \frac{1+5}{2}\right)$$

$$= \left(\frac{4}{2}, \frac{6}{2}\right)$$

$$= (2, 3)$$

b (4, −1) = (x_1, y_1) and (2, −2) = (x_2, y_2)

$$\text{Midpoint} = \left(\frac{4+2}{2}, \frac{-1-2}{2}\right)$$

$$= \left(\frac{6}{2}, \frac{-3}{2}\right)$$

$$= (3, -1\tfrac{1}{2})$$

Exercise +14

1 **a** Draw the points A(1, 3) and B(5, 1) on a copy of the grid.

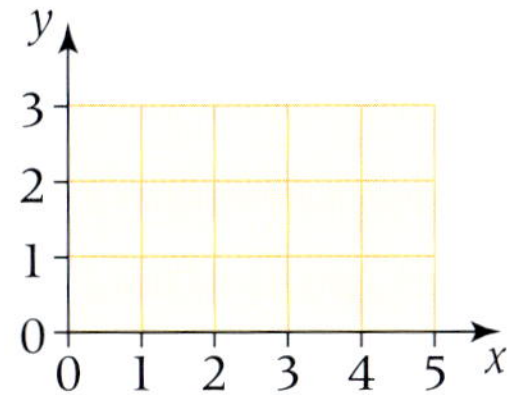

b M is the midpoint of the line AB.
Find the coordinates of the point M.

2 Calculate the coordinates of the midpoint between the points

a (3, 1) and (3, 5)

b (0, 4) and (4, 4)

c (2, −2) and (2, 4)

d (3, 1) and (7, 7)

e (−2, −1) and (6, 7)

Use a copy of the grid to help you.

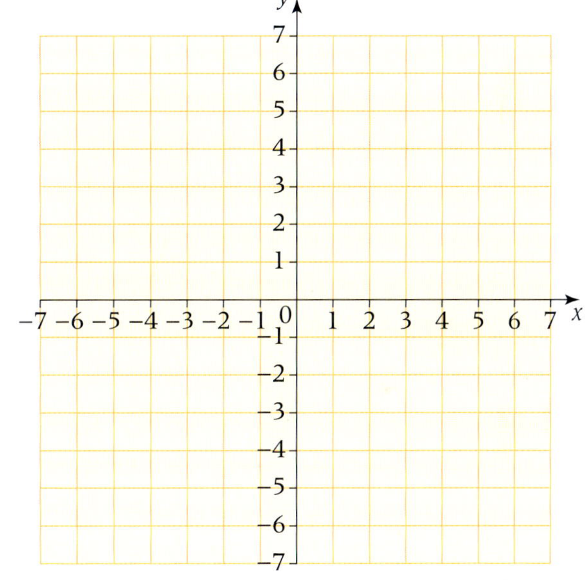

3 The point (3, 4) is the midpoint between (x, 6) and (1, y).
Find the values of x and y.

4 The point ($1\frac{1}{2}$, $5\frac{1}{2}$) is the midpoint between (x, 4) and (2, y).
Find the values of x and y.

5 The point (0, 2) is the midpoint between (x, 6) and (−4, y).
Find the values of x and y.

6 The coordinates (−1, 2), (−3, 0), (−1, −2) and (1, 0) are the midpoints of each side of a square.

a Find the coordinates of the vertices of the square.

b Find the area of the square.

+15 Pythagoras' theorem

This spread will show you how to:

- Understand, recall and use Pythagoras' theorem

Keywords
Hypotenuse
Pythagoras' theorem
Right-angled triangle
Square
Square root

The longest side of a **right-angled triangle** is called the **hypotenuse**.
The hypotenuse is always opposite the right angle.

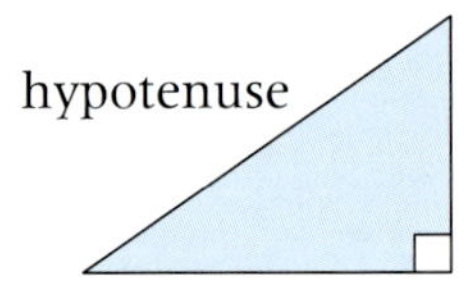

This is a right-angled triangle.

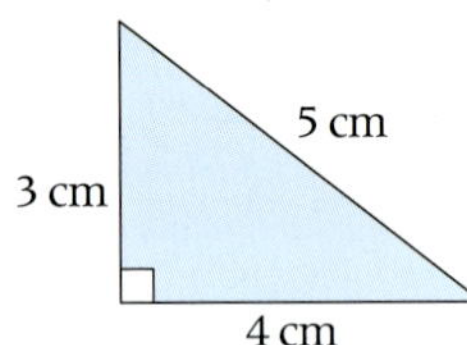

Draw the **squares** on each side.

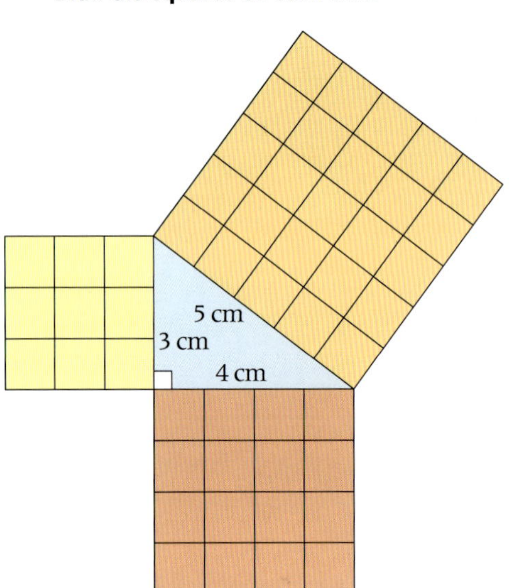

Area of yellow square $= 3 \times 3 = 9$ cm^2
Area of red square $= 4 \times 4 = 16$ cm^2
Area of orange square $= 5 \times 5 = 25$ cm^2

Area of orange square = area of yellow square + area of red square.

This is **Pythagoras' theorem**.

- **In a right-angled triangle, $c^2 = a^2 + b^2$ where c is the hypotenuse.**

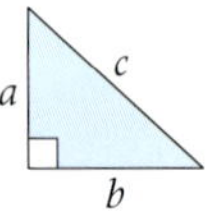

Example

Calculate the unknown lengths in these triangles.

a

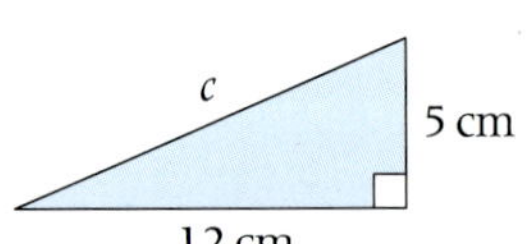

b

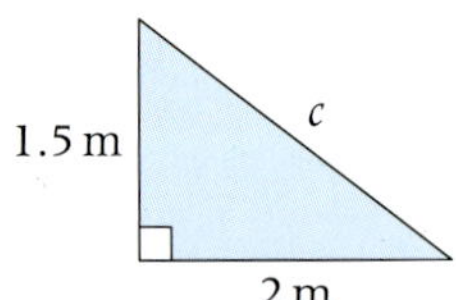

a Label the sides.

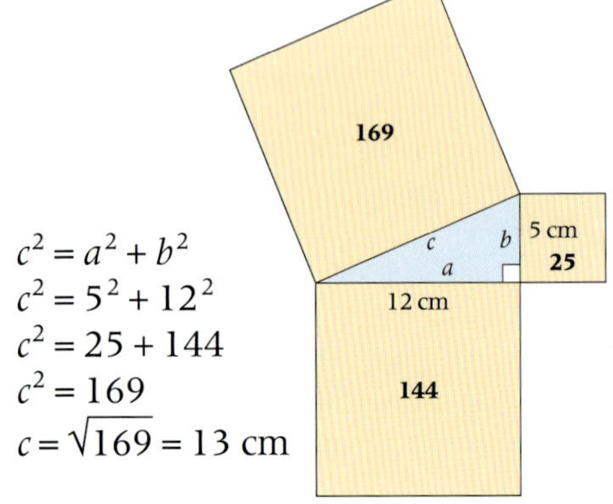

$c^2 = a^2 + b^2$
$c^2 = 5^2 + 12^2$
$c^2 = 25 + 144$
$c^2 = 169$
$c = \sqrt{169} = 13$ cm

b Label the sides.

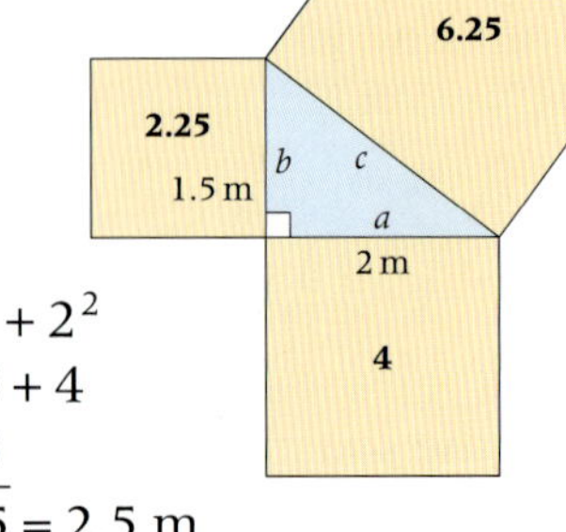

$c^2 = 1.5^2 + 2^2$
$c^2 = 2.25 + 4$
$c^2 = 6.25$
$c = \sqrt{6.25} = 2.5$ m

$\sqrt{\ }$ means **square root**.

$\sqrt{169} = 13$ because $13 \times 13 = 169$.

Exercise +15

1 Calculate the area of these squares. State the units of your answers.

a 8cm

b 10m

c 1.8m

d 36mm

e 4.5m

2 Calculate the length of a side of these squares.
State the units of your answers.

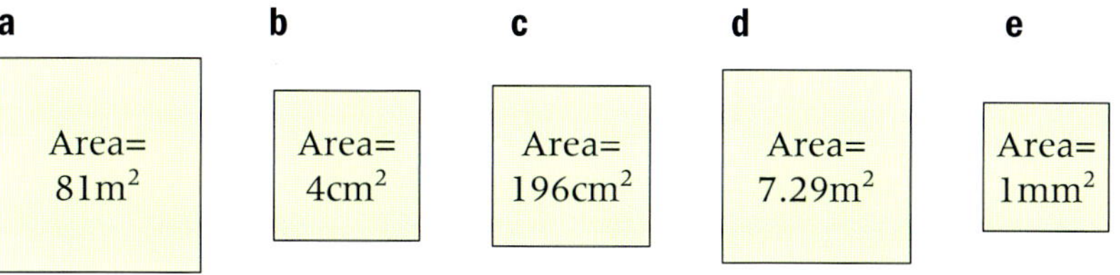

DID YOU KNOW?

Pythagoras was a Greek mathematician most famous for his theorem, who taught his students that 'all things are numbers'.

3 Calculate the unknown area for these right-angled triangles.

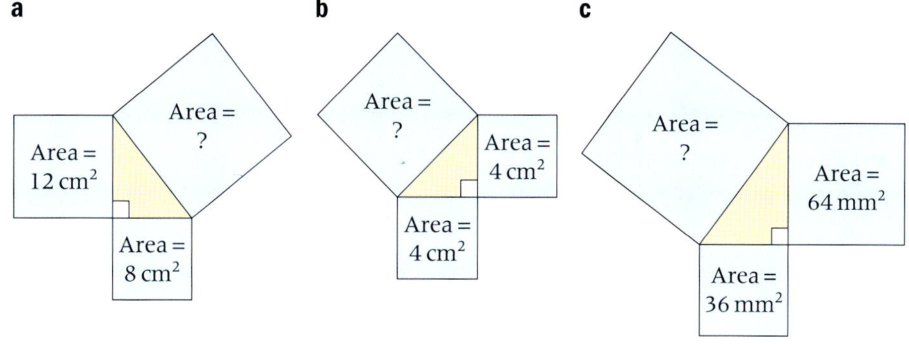

4 Calculate the length of the hypotenuse in these right-angled triangles. State the units of your answers.

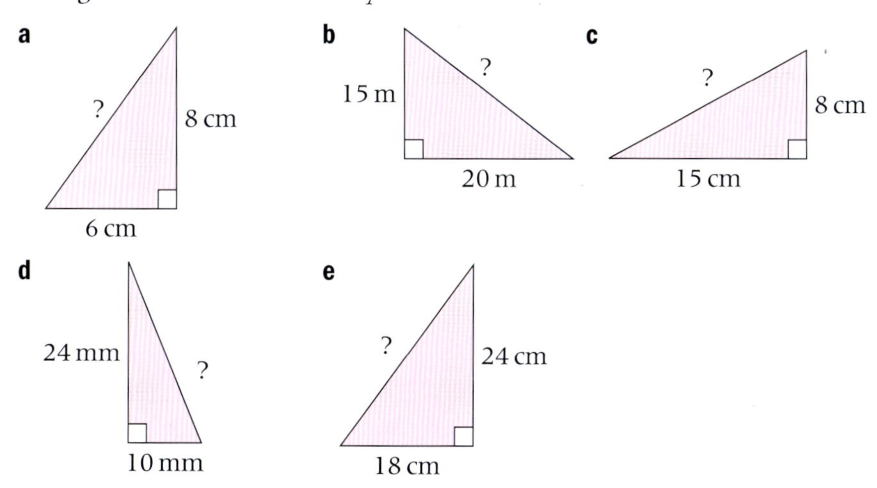

+16 Frequency tables and the mean

This spread will show you how to:

- Use a data collection sheet for discrete data
- Understand and use frequency tables

Keywords
Data
Data collection sheet
Frequency table
Tally chart

- **You can collect data using a data collection sheet.**

This one is a **tally chart**.

Coin face	Tally	Frequency
Head	~~||||~~ ~~||||~~ |||	13
Tail	~~||||~~ ~~||||~~ ||	12

~~||||~~ = 5

Data can also be shown using a **frequency table**.

Example

The number of televisions in each house in my street is shown in the frequency table.

Number of TVs	Number of houses
0	1
1	5
2	12
3	9
4	1

a Calculate the number of houses in my street.

b Calculate the total number of televisions in my street.

c Calculate the mean number of televisions per house.

The numbers in the table are:

0, 1, 1, 1, 1, 1, 2, 2, 2, 2, 2, 2, 2, 2,
2, 2, 2, 2, 3, 3, 3, 3, 3, 3, 3, 3, 3, 4

Add an extra column to the table

TVs	Houses	TVs × Houses
0	1	$0 \times 1 = 0$
1	5	$1 \times 5 = 5$
2	12	$2 \times 12 = 24$
3	9	$3 \times 9 = 27$
4	1	$4 \times 1 = 4$

a $1 + 5 + 12 + 9 + 1 = 28$ houses

b $0 + 5 + 24 + 27 + 4 = 60$ televisions

c $60 \div 28 = 2\frac{1}{7}$ televisions per house.

Exercise +16

1 **a** Copy and complete the tally chart to find the frequency of the vowels a, e, i, o, u in this sentence.

b Which vowel occurs the most often?

c Which vowel occurs the least often?

d Calculate the total number of vowels in the sentence.

e Find a paragraph of writing in a newspaper and complete a similar tally chart.

Vowel	Tally	Frequency
a		
e		
i		
o		
u		

2 Rainfall, measured in millimetres, is recorded daily for the month of April.

4 2 1 0 0 1 2 2 3 5
7 8 5 3 3 2 0 0 0 1
2 3 2 4 6 7 8 8 1 2

a Copy and complete the tally chart to show this information.

b State the number of completely dry days in April.

c Calculate the total amount of rain to fall throughout April. State the units of your answer.

Rainfall (mm)	Tally	Number of days
0		
1		
2		
3		
4		
5		
6		
7		
8		

3 Nails can be bought in bags.
There are approximately 20 nails in each bag.
The numbers of nails in 40 bags are recorded.

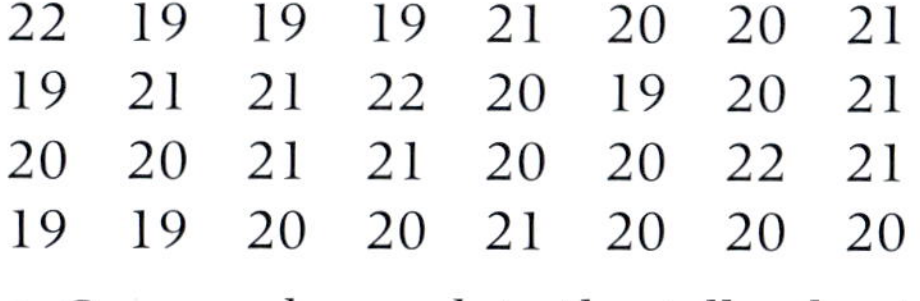

19 20 19 21 20 20 20 21
22 19 19 19 21 20 20 21
19 21 21 22 20 19 20 21
20 20 21 21 20 20 22 21
19 19 20 20 21 20 20 20

a Copy and complete the tally chart to show this information.

b Calculate the total number of nails in all 40 bags.

Number of nails	Tally	Number of bags
19		
20		
21		
22		

4 Sophie did a survey to find the number of CDs owned by the students in her class. The results are shown in the frequency table.

a Calculate the number of students in Sophie's class.

b Calculate the total number of CDs owned by the whole class.

c Calculate the mean number of CDs for each student.

Number of CDs	Number of students
0	1
1	8
2	6
3	8
4	2
5	3

+17 Scatter graphs

This spread will show you how to:

- Draw and use scatter graphs, and compare two data sets
- Understand correlation and use lines of best fit

Keywords
Correlation
Line of best fit
Relationship
Scatter graph
Variables

You can use a **scatter graph** to compare two sets of data, for example, height and weight.

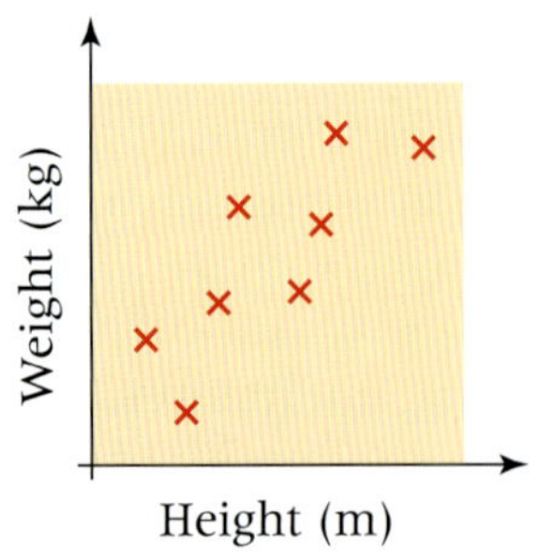

The data can be discrete or continuous.

- The data is collected in pairs and plotted as coordinates.
- If the points lie roughly in a straight line, there is a **relationship** or **correlation** between the two **variables**.

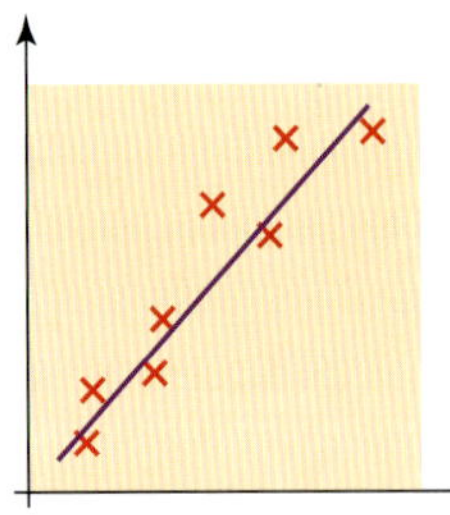
Positive correlation

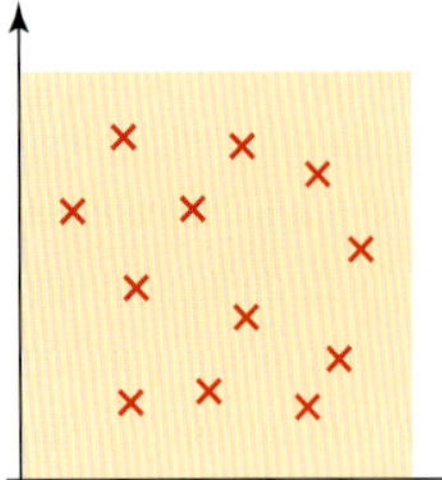
No correlation

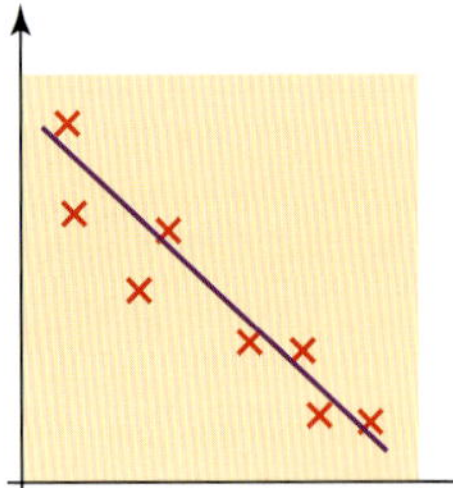
Negative correlation

Plotted points are not joined on a scatter diagram.

The straight line is the **line of best fit**.

Example

The exam results (%) for Paper 1 and Paper 2 for 10 students are shown.

Paper 1	56	72	50	24	44	80	68	48	60	36
Paper 2	44	64	40	20	36	64	56	36	50	24

a Draw a scatter graph and line of best fit.

b Describe the relationship between the Paper 1 results and Paper 2 results.

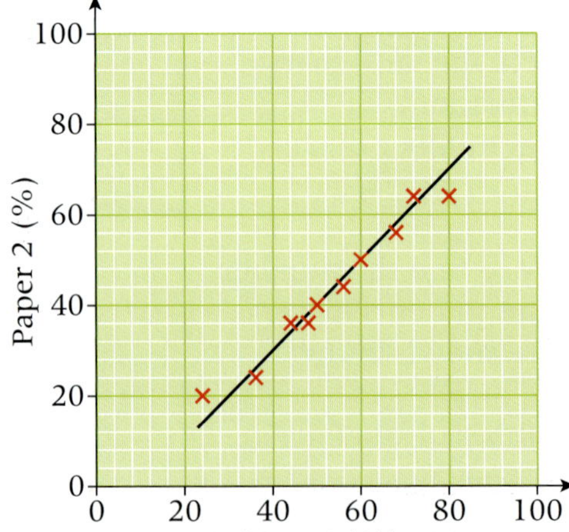

a Plot the exam marks as coordinates. The line of best fit should be close to all the points, with approximately the same number of crosses on either side of the line.

b Students who did well on Paper 1 did well on Paper 2. Students who did not do well on Paper 1 did not do well on Paper 2 either.

The line of best fit does not have to pass through (0, 0).

Exercise +17

1 Describe the type of correlation for each scatter graph.

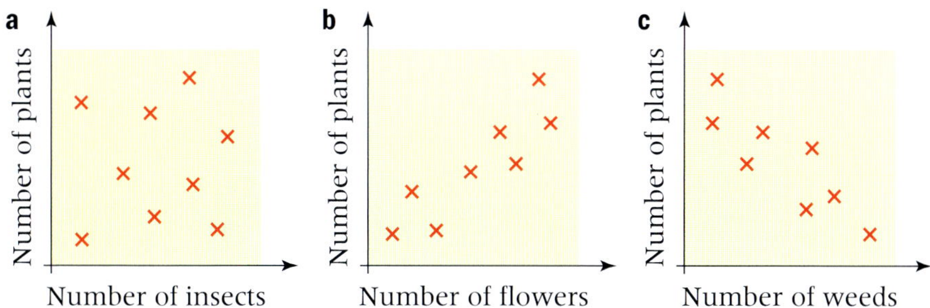

2 The table shows the amount of water used to water plants and the daily maximum temperature.

Water (litres)	25	26	31	24	45	40	5	13	18	28
Maximum temperature (°C)	24	21	25	19	30	28	15	18	20	27

a Copy and complete the scatter graph for this information.

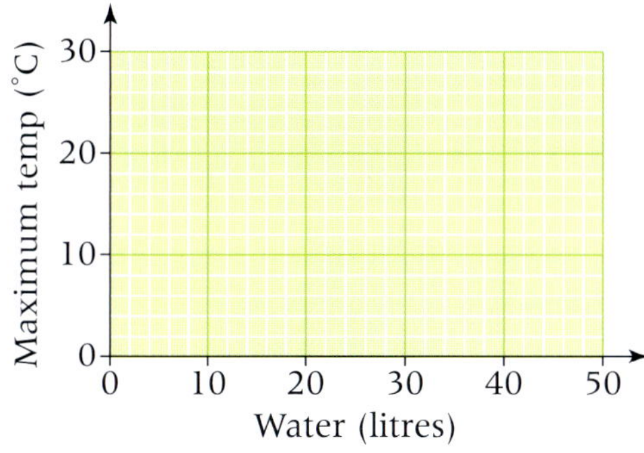

b State the type of correlation shown in the scatter graph.

c Copy and complete these sentences:

i As the temperatures increases, the amount of water used _____.

ii As the temperature decreases, the amount of water used _____.

3 The times taken, in minutes, to run a mile and the shoe sizes of ten athletes are shown in the table.

Shoe size	10	$7\frac{1}{2}$	5	9	6	$8\frac{1}{2}$	$7\frac{1}{2}$	$6\frac{1}{2}$	8	7
Time (mins)	9	8	8	7	5	13	15	12	5	6

a Draw a scatter graph to show this information.

Use 2 cm to represent 1 shoe size on the horizontal axis.
Use 2 cm to represent 5 minutes on the vertical axis.

b State the type of correlation shown in the scatter graph.

c Describe, in words, any relationship that the graph shows.

+18 Mutually exclusive outcomes

This spread will show you how to:

- Identify different mutually exclusive outcomes
- Know that the sum of the probabilities of all the outcomes is 1

Keywords
Mutually exclusive
Outcomes

The possible **outcomes** when a dice is rolled are 1, 2, 3, 4, 5, 6.

These outcomes are **mutually exclusive** because if you get one outcome, you cannot get another one.

The probability of rolling a 4 = $\frac{1}{6}$

The probability of rolling a 5 = $\frac{1}{6}$

The probability of rolling a 4 **or** a 5 = $\frac{2}{6}$

$P(4) = \frac{1}{6}$

$P(5) = \frac{1}{6}$

Or means either of the outcomes 4 or 5.

Notice that

$$P(4 \text{ or } 5) = P(4) + P(5)$$
$$\frac{2}{6} = \frac{1}{6} + \frac{1}{6}$$

- **In general, if outcomes A and B are mutually exclusive,**
  P(A or B) = P(A) + P(B)

Example

A spinner is made from a regular pentagon and numbered from 1 to 5. Calculate

a P(3) **b** P(3 or 4) **c** P(3 or an odd number)

a There is one 3. There are 5 possible outcomes. $P(3) = \frac{1}{5}$

b $P(3) = \frac{1}{5}$ $\quad P(4) = \frac{1}{5}$

The outcomes are mutually exclusive and so

$$\begin{aligned} P(3 \text{ or } 4) &= P(3) + P(4) \\ &= \tfrac{1}{5} + \tfrac{1}{5} \\ &= \tfrac{2}{5} \end{aligned}$$

c $P(3) = \frac{1}{5}$ $\quad P(\text{odd}) = \frac{3}{5}$

The outcomes are not mutually exclusive so P(A or B) = P(A) + P(B) cannot be used.
A list of the 5 possible outcomes is

$$\begin{pmatrix}1\\ \text{odd}\end{pmatrix} \begin{pmatrix}2\\ \text{even}\end{pmatrix} \begin{pmatrix}3\\ \text{odd}\end{pmatrix} \begin{pmatrix}4\\ \text{even}\end{pmatrix} \begin{pmatrix}5\\ \text{odd}\end{pmatrix}$$

There are 3 outcomes that are OK.

$P(3 \text{ or an odd number}) = \frac{3}{5}$

Exercise +18

1 Outcomes are mutually exclusive if they cannot occur at the same time. State if these outcomes are mutually exclusive.

a spinning a Head and spinning a Tail with a coin

b rolling a 2 and rolling a 3 with a dice

c rolling a 2 and rolling an even number with a dice

d rolling a 2 and rolling an odd number with a dice

e rolling a 2 and rolling a prime number with a dice

f winning and losing a game of chess

g sunny and rainy weather

h taking out a red ball and taking out a blue ball, when taking out one ball from a bag.

2 This is a net of a tetrahedral dice. The dice is made and rolled. What is the probability of rolling

a a 3 **b** a 2 **c** a 2 or a 3?

3 A bag contains 4 red discs, 5 blue discs and 1 white disc. One disc is taken out.
Calculate the probability that the disc is

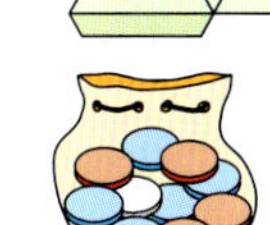

a red **b** blue **c** white

d red or white **e** blue or white **f** red or blue or white

4 Five names are written on cards and placed in a bag. One name is taken out of the bag at random.
Calculate the probability that

HENRY
EDWARD
JAMES
CHARLES
GEORGE

a the first letter on the card is H

b the first letter on the card is G

c the first letter on the card is H or G

d the card has 5 letters written on it

e the card has 5 or 6 letters written on it.

5 A survey of vehicles passing the school gate is taken.

C C C C V L C C V B
C L C C C C B C C V
V V C C L C C B C C
L B C C C V L C C C
B V C V C C V C C V

C = Car
L = Lorry
B = Bus
V = Van

a Copy and complete the frequency table.

Vehicle	Tally	Frequency
Car (C)		
Lorry (L)		
Bus (B)		
Van (V)		

b Calculate the estimated probability that the next vehicle that passes will be

i a car **ii** a lorry **iii** a bus or a van.

+19 Relative frequency

This spread will show you how to:

- Understand and use relative frequency
- Compare experimental data and theoretical probabilities

Keywords
Biased
Equally likely
Estimate
Experiment
Fair
Relative frequency
Trial

- **A dice is fair if the numbers are all equally likely to be rolled.**
  You would normally expect a dice to be fair, as each face is identical.

- **A spinner is biased if the colours are NOT all equally likely to happen.**
  This spinner is not fair as the size and shape of each colour are not identical.

It is not always possible to calculate the theoretical probability.

- **You can estimate the probability from experiments.**

Example

Sam knows the probability of a Head when spinning a coin should be $\frac{1}{2}$ (or 0.5).
She thinks the coin is biased and so she spins the coin 50 times. The results are shown in the frequency table.

	Frequency
Head	35
Tail	15

a Estimate the probability of getting a Head when spinning the coin.
b Do you think the coin is biased? Explain your answer.

a Sam got a Head on 35 out of 50 occasions.
Estimated probability of getting a Head $= \frac{35}{50} = \frac{7}{10} = 0.7$
b The spinner could be biased as 0.7 is significantly larger than 0.5.
However, Sam needs to spin the coin a lot more times before she can make the decision.

- **Estimated probability is called the relative frequency.**

Each spin of the coin is called a **trial**.

In the example, if Sam calculated the relative frequency of getting a Head after each spin, she could graph the results.

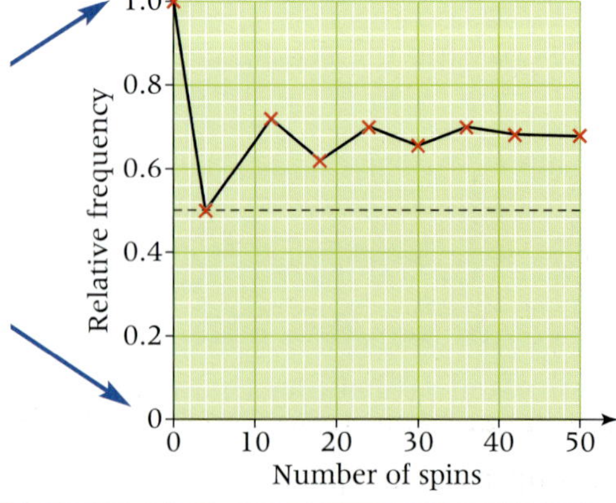

P(Head) $= \frac{1}{2} = 0.5$
This is the theoretical probability of spinning a Head.

- **The estimated probability becomes more reliable as you increase the number of trials.**

Exercise +19

1 The colours of 50 cars are recorded. The colours are shown

Blue	Red	Other	Silver	Blue	Red	Silver	Silver	Blue
Other	Red	Silver	Silver	Blue	Red	Red	Other	Silver
Silver	Red	Red	Silver	Silver	Blue	Blue	Silver	Silver
Red	Red	Other	Blue	Red	Red	Other	Red	Red
Red	Silver	Blue	Blue	Blue	Other	Silver	Other	Other
Other	Silver	Red	Red	Blue				

a Copy and complete the frequency chart to show the 50 colours.

Colour	Tally	Frequency
Blue		
Red		
Silver		
Other		

b State the modal colour.

c Give an estimate of the probability that the next car will be

i blue **ii** red **iii** silver.

2 A spinner is made from a regular pentagon. The scores are recorded

1	2	3	4	3	5	1	2
5	4	2	1	3	1	5	4
2	2	3	1	4	5	4	2
3	1	2	2	4	4	5	5
1	2	3	4	2	4	1	1

a Draw a frequency chart to show the scores.

b State the modal score.

c How many times is the spinner spun?

d Estimate the probability of scoring

i a 1 **ii** a 2 **iii** a 3 **iv** a 4 **v** a 5

e If the spinner is fair, how many times would you expect to spin a 3 from 100 spins?

3 There are 10 coloured balls in a bag.
One ball is taken out and then replaced in the bag.
The colours of the balls are shown in the frequency table.

Colour	Red	Green	Blue
Frequency	9	14	27

a How many times was a ball taken out of the bag?

b Estimate the probability of taking out

i a red ball **ii** a green ball **iii** a blue ball.

c How many balls of each colour do you think are in the bag?

d How could you improve this guess?

+20 Databases and random sampling

This spread will show you how to:

- Gather data from secondary sources
- Use effective methods for random sampling
- Identify sources of bias and plan to minimise it

Keywords
Biased
Database
Equally likely
RAN#
Random sample

- A **database** is an organised collection of data, especially in a form that can be used by a computer.

Examiner's tip
This topic will be useful for your Data coursework.

An example might be the records of the Year 11 students in a school.

You can sort a computer database

- alphabetically
- numerically.

You can sort using any or several of the columns (fields).

Number	Surname	Forename	Form	Gender
0001	Smith	Thomas	10E	Male
0002	Jones	Michaela	10B	Female
0003	Chapham	Leah	10A	Female
0004	Clark	Alan	10B	Male

This database has been sorted by number.

Sometimes there is too much data in a database to process all the data, and so a **random sample** is used.

- In a random sample, each person or item must be **equally likely** to be chosen.

If each person is not equally likely to be chosen, the sample is **biased**.

Example

Describe a method to choose a random sample of 30 students from a year group of 120 students.

- Number each student from 1 to 120.
- Generate 30 random numbers, by either
  - picking 30 numbers from a bag of 120, numbered 1 to 120, or
  - using the RAN# key on a calculator.

- You can generate random numbers on a calculator using the random function. RAN# generates random numbers from 0.000 to 0.999.

- You can generate a random number from, say, 1 to 120 by using 120 × RAN#.

Use the first three digits of the display.

Exercise +20

1 Sanjit wants to buy a laptop. He creates a database of the 10 laptops he is considering buying.

Laptop number	Speed of processor	RAM memory	Size of hard drive	Screen size	Warranty	Cost
1	1.2 GHz	128 MB	8 GB	14.1"	No	£250
2	1.3 GHz	256 MB	20 GB	14.1"	1 year	£300
3	1.8 GHz	256 MB	40 GB	15"	1 year	£450
4	1.2 GHz	512 MB	60 GB	12.1"	1 year	£1300
5	1.7 GHz	512 MB	80 GB	15.4"	3 years	£1400
6	2.4 GHz	1024 MB	100 GB	17"	2 years	£1500
7	2.0 GHz	512 MB	60 GB	14.1"	3 years	£1200
8	2.3 GHz	512 MB	80 GB	14.1"	3 years	£1400
9	1.6 GHz	512 MB	60 GB	14.1"	1 year	£700
10	1.6 GHz	256 MB	60 GB	15.1"	No	£600

a Write down the costs of the laptops in order of price, smallest first.

b Write down the speeds of the processors in order of size, smallest first.

c Which laptop has the smallest RAM memory?

d Which laptop has the largest RAM memory?

e List the laptops that have

i a screen size of 15" or more

ii at least 60 GB of hard drive memory

iii a 2-year or 3-year warranty.

f Sanjit wants a laptop with a screen size of 15" or more, at least 60 GB of hard drive memory and one with a 2- or 3-year warranty. Which laptop is the cheapest option?

2 A class of 30 students decide to elect a class representative by a random process. State whether these methods of selection are random or biased. Give a reason for each answer.

a Arrange the class list into alphabetical order and select the first name on the list.

b Arrange the class list into alphabetical order and select the last name on the list.

c Put the names of the students on cards of equal size. Put the cards into a bag and pick out one card.

d Arrange the students in order of height and select the smallest student.

e Hide a gold star in the classroom, and select the student who finds the star.

f Number the students from 1 to 30. Roll a dice and select the student with that number.

g Number the students from 1 to 30. Use a calculator to find 30 × RAN# and take the first two digits on the display.

GCSE formulae

In your Edexcel GCSE examination you will be given a formula sheet like the one on this page.

You should use it as an aid to memory, and it will be useful to become familiar with the information on the sheet.

Area of a trapezium = $\frac{1}{2}(a + b)h$

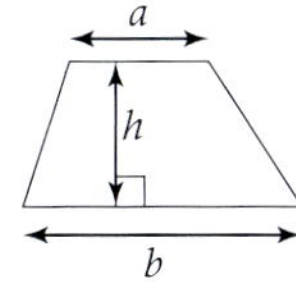

Volume of prism = area of cross section × length

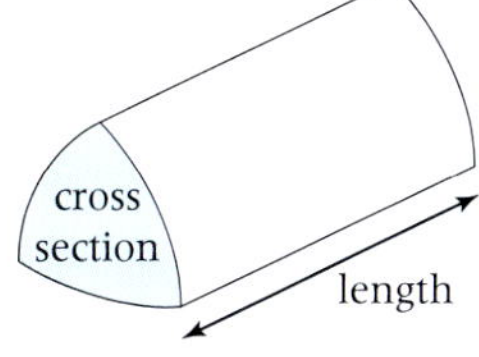

Here are some other formulae that you should learn.

Area of a rectangle = length × width

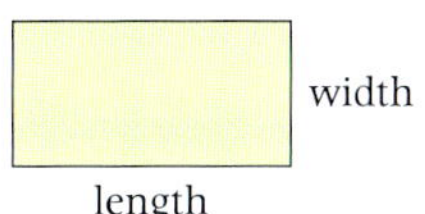

Area of a triangle = $\frac{1}{2}$ × base × height

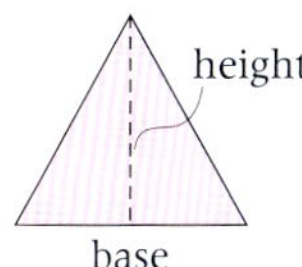

Area of a parallelogram = base × perpendicular height

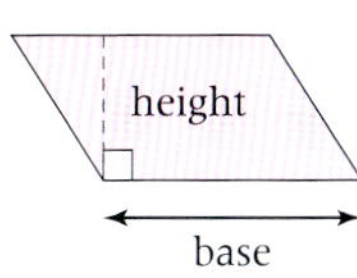

Area of a circle = πr^2

Circumference = $\pi d = 2\pi r$

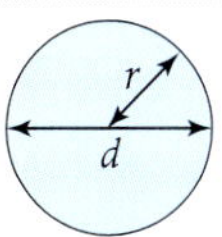

Volume of a cuboid = length × width × height

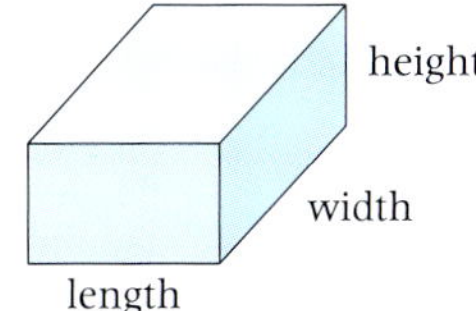

Volume of a cylinder = area of circle × length

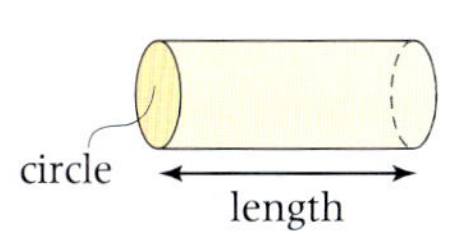

Pythagoras' theorem states

For any right-angled triangle, $c^2 = a^2 + b^2$ where c is the hypotenuse.

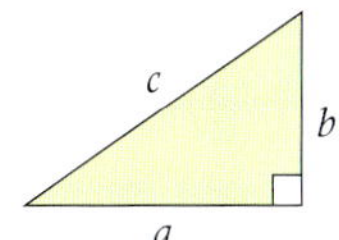

Answers

N1 Before you start ...

1 304
2 70
3 20 minutes
4 −7
5 −12
6 5

N1 Exam review

1 **a** 1.7 **bi** 88.5
bii

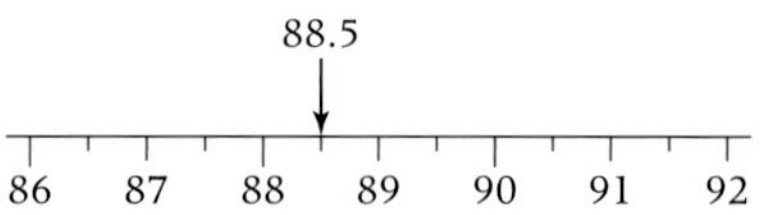

2 **ai** Edinburgh and Plymouth **aii** 12 °C
b Belfast and Cardiff, London and Plymouth

S1 Before you start ...

1 **a** 7.8 **b** 9.4 **c** 5.5
2 **a** 400 **b** 14 000 **c** 31 **d** 1340
e 6300 **f** 40 **g** 60 **h** 4.3
i 0.64 **j** 0.31
3 **a** 18 **b** 15 **c** 14 **d** 25 **e** 4.8
4 Perimeter = 12 units, Area = 5 square units
5 **a** 45 mm **b** 4.5 cm

S1 Exam review

1 **a** 24 cm **b** 11 cm^2
2 kilograms, litres, inches

A1 Before you start ...

1 **a** 5 **b** 9
2 **a** 9 **b** 4 **c** 16
3 15
4 **a** 7 **b** −1 **c** 2 **d** −3
5 **a** −6 **b** −8 **c** −2 **d** 4

A1 Exam review

1 **a** $3m - n$ **b** 7
2 $6x$

N2 Before you start ...

1 **a** 3000 **b** 2900 **c** 2920
2 **a** 48 **b** 105
3 **a** 815 **b** 229
4 **a** 30 **b** 6 **c** 72 **d** 141 **e** 168 **f** 13

N2 Exam review

1 **a** 79 **b** 407 **c** 75 **d** 13
2 **a** 54 000 **b** 50 000

A2 Before you start ...

1 **a** 6 **b** 15 **c** 11 **d** 3
2 **a** 4 **b** 6 **c** 8 **d** 5
3 **a** $12 \div 3 = 4$, $12 \div 4 = 3$
b $35 \div 5 = 7$, $35 \div 7 = 5$
4 **a** $18 \div 3 = 6$, $18 \div 6 = 3$
b $28 \div 4 = 7$, $28 \div 7 = 4$
c $36 \div 4 = 9$, $36 \div 9 = 4$
d $30 \div 6 = 5$, $30 \div 5 = 6$
5 **a** 7 **b** −1 **c** 12 **d** $2\frac{1}{4}$

A2 Exam review

1 **a** $y = 6$ **b** $x = 4$ **c** $z = 2$
2

Input	Output
1	1
2	3
3	5
5	9
8	15

D1 Before you start ...

1 **a** 6, 17, 19, 26, 29, 30, 37, 42
b 106, 115, 118, 121, 130, 135
c 144, 145, 154, 155, 156, 165, 166
2 **a** 121 **b** 144 **c** 252 **d** 413 **e** 68
f 23 **g** 49 **h** 82 **i** 189 **j** 266
3 **a** 04.58 **b** 20.45

D1 Exam review

1 **a** 0, 1, 1, 1, 2, 3, 2, 4, 1, 0, 2, 3 **b** August
2

	French	German	Spanish	Total
Female	15	11	13	39
Male	16	17	8	41
Total	31	28	21	80

N3 Before you start ...

1 $\frac{4}{6}$
2 e.g. $\frac{2}{4}, \frac{3}{6}$
3 **a** $\frac{1}{2}$ **b** $\frac{1}{4}$ **c** 0.2
4 80%
5 **a** $50\% = \frac{1}{2} = 0.5$ **b** $10\% = \frac{1}{10} = 0.1$
c $45\% = \frac{45}{100} = 0.45$

N3 Exam review

1 **a** $\frac{1}{5}$ **bi** 0.2 **bii** 20%
2 **a** 9, 37, 56, 59, 75
b 0.067, 0.56, 0.6, 0.605, 0.65
c −10, −6, −4, 2, 5
d $\frac{2}{5}, \frac{1}{2}, \frac{2}{3}, \frac{3}{4}$

S2 Before you start ...

1 **a** (1, 4) **b** (4, 2)
2 43
3 **a** 60 **b** 135 **c** 68 **d** 90 **e** 142

S2 Exam review

1 **a** Scalene **b** 38°
2 **a** Acute **b** Reflex
c Angles at a point should add to 360°, these only add to 350°.

A3 Before you start ...

1 **a** 3 **b** 9 **c** 4 **d** 3
2 1, 4, 9, 16, 25, 36, 49, 64, 81, 100
3 **a** 5 **b** 4 **c** 7 **d** 6
4 **a** 2 **b** 2 **c** 3 **d** 3
5 **a** 8 **b** 6 **c** 12 **d** 10

A3 Exam review

1 **a** 2, 5, 7 or 19 **b** 9 or 25 **c** 9 **d** 2, 5, 25
2 **a** 106, 102 **b** Subtract 4 each time **c** 46

D2 Before you start ...

1 **a** 5 **b** 5 **c** 10 **d** 4 **e** 8
2 **a** 360° **b** 110°
3 **a** 180 **b** 90 **c** 36 **d** 72 **e** 60
f 30 **g** 2 **h** 4 **i** 18 **j** 20
4 **a** 25, 29, 30, 31, 36, 41, 43, 49
b 234, 243, 324, 342, 423, 432
c 12.3, 13.2, 21.3, 23.1, 31.2, 32.1

D2 Exam review

1

70	6 7 9
80	0 1 4
90	1 2 2 3 7 8 8 9
100	2 8 9
110	2 7 8
120	0 2 3 5

Key: 120 | 3 means 123

2 **ai** 30 **aii** 25
b Wednesday: 4 videos, Thursday: $1\frac{1}{2}$ videos

A4 Before you start ...

1 **a** 7 **b** 17 **c** 19 **d** 1
2 **a** $p = 3$ **b** $q = 8$ **c** $r = 2$ **d** $s = 5$
3 **a** 8 **b** 5 **c** 8 **d** $3\frac{1}{2}$
4 **a** 5 **b** −4 **c** 1 **d** 11
5 **a** −1 **b** −5 **c** 1 **d** −7

A4 Exam review

1 **ai** (1, 4) **aii** (2, −3)
b

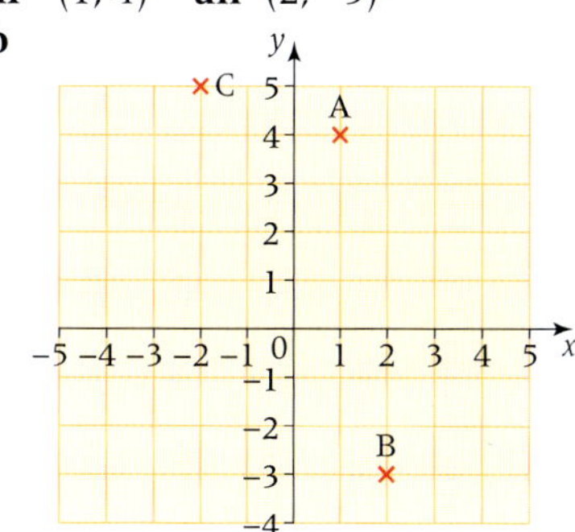

2 **a** $y = -4, -1, 2, 5, 8$
b

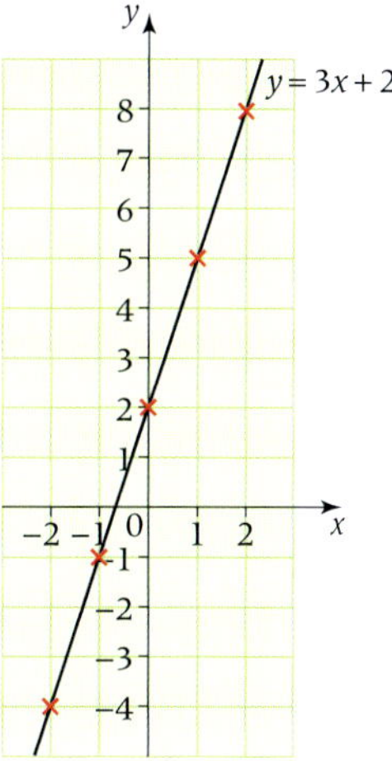

D3 Before you start ...

1 **a** $\frac{1}{10}$ **b** $\frac{1}{2}$ **c** $\frac{1}{5}$ **d** $\frac{7}{10}$ **e** $\frac{3}{5}$
2 **a** 0.2, 0.25, 0.3 **b** 0.7, 0.75, 0.8
c 0.8, 0.85, 1
3 **a** $\frac{3}{4}$ **b** $\frac{3}{8}$
4 **a** 1 **b** 1 **c** $\frac{1}{10}$ **d** $\frac{1}{5}$ **e** $\frac{1}{4}$
5 **a** 0.8 **b** 0.3 **c** 0.1

D3 Exam review

1 **a**

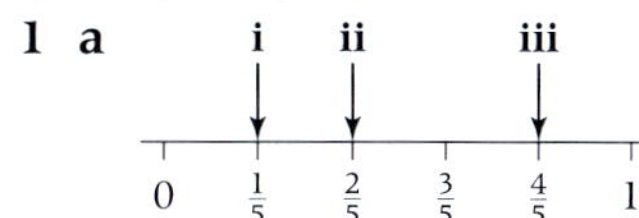

bi $\frac{1}{5}$ **bii** $\frac{2}{5}$ **biii** $\frac{4}{5}$
2 (1, 2), (1, 4), (1, 6), (1, 8), (5, 2), (5, 4), (5, 6), (5, 8), (7, 2), (7, 4), (7, 6), (7, 8)

N4 Before you start ...

1 $\frac{1}{6}$
2 £24
3 £5 per hour
4 AU$10

N4 Exam review

1 **a** 100 g **bi** $\frac{3}{20}$ **bii** 15% **c** 325 g
2 Ken £5, Susan £15

S3 Before you start ...

1 **a** 180° **b** 270° **c** 90° **d** 180°
2 **a** Clockwise **b** Anticlockwise
3 **a** (3, 1) **b** (−2, 1) **c** (2, −3) **d** (−3, −2)

S3 Exam review

1

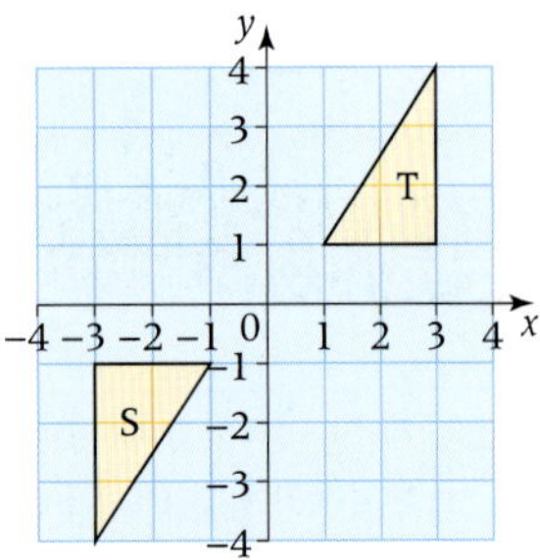

2 **a** 14 cm **b** 6 cm^2

c

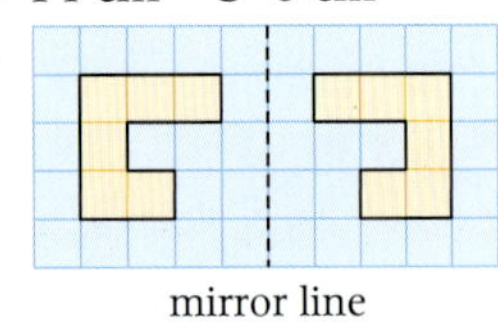

mirror line

d

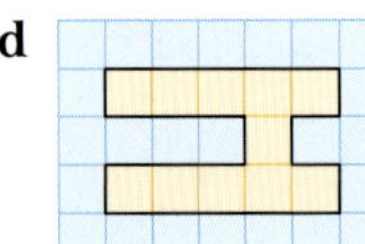

N5 Before you start ...

1 **a** 49 **b** 8 **c** 1000 **d** 0.32

2 5×5

3 1, 2, 3, 4, 6, 12

4 Yes, because half of 244 is even.

N5 Exam review

1 **ai** 1, 2, 13, 26 **aii** 1, 2, 17, 34 **b** 2

2 **ai** 16 **aii** 8 **b** 10^5

A5 Before you start ...

1 **a** 11 **b** 22 **c** 13 **d** 44

2 **a** $4x$ **b** m^2 **c** $4p$ **d** $6x$

3 **a** $5p + q$ **b** $5x + y$ **c** $-m + 5n - 4p$

4 **a** $t = 5$ **b** $x = 7$ **c** $m = 12$

A5 Exam review

1 **a** $3n$ **b** $-t$ **c** $x + y$

2 **a** 30 points **b** 65 points

S4 Before you start ...

1 180°

2 **a** (3, 2) **b** (1, −2) **c** (−2, −3) **d** (−2, 3)

3 **a** 57 mm **b** 5.7 cm

4 **a** 44° **b** 135°

S4 Exam review

1 **a** Line of length 5 cm

b Circle with diameter 5 cm

2 **a** F **b** B and D

N6 Before you start ...

1 17

2 $7 \times 4 = 28$

3 **a** 7.65 **b** 8.8

4 **a** 140 **b** 30 **c** $13\frac{3}{8}$ **d** 1104 **e** 0.23

5 **a** 10.24 **b** 3.4

N6 Exam review

1 **a** 6.438 095 238 **b** 6

2 **a** 855.4 kg **b** 14 crates

A6 Before you start ...

1 **a** −3 **b** +6 **c** ÷5 **d** ×4

2 **a** $2x + 2y$ **b** $5p + 2$ **c** $\frac{m}{8}$

3 **a** 54 **b** 32 **c** 9 **d** 9

4 **a** 12 **b** 11 **c** 4 **d** 13

5 **a** $x = 2$ **b** $x = 5$ **c** $y = 2$ **d** $y = 7$

A6 Exam review

1 **a** $x = 7$ **b** $y = 4$

1 **a** $3x$ **b** $x - 9$

D4 Before you start ...

1 **a** 23 **b** 150 **c** 23 **d** 42 **e** 64

2 **a** 26, 29, 31, 35, 36, 40, 41, 48

b 91, 92, 98, 101, 102

c 4, $4\frac{1}{2}$, 6, $7\frac{1}{2}$, 8

3 **a** 80, 80, 80, 82, 83, 83, 84, 84, 84, 84

b 45, 45, 46, 48, 48, 48, 48, 49, 49, 49

D4 Exam review

1 **a** 7 **b** 7 **c** 5 **d** 7

2 **a** 3, 7, 3, 5, 2 **b** 2 people **c** 7.4 rides

S5 Before you start ...

1 **a** 1 cm **b** 4 cm **c** 4.5 cm **d** 1 m

e 4 m **f** 1.5 m **g** 1 km **h** 6 km

i 0.5 km **j** 1.5 km

2 **a** 38° **b** 163° **c** 225°

3 360°

4 **a** 0.5 **b** 0.2 **c** 0.1

5 **a** 24 **b** cm^2

S5 Exam review

1 **a** 205° **b**

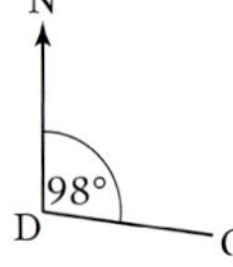

2 **b** Marked point 6 cm from each end of the line

c

N7 Before you start ...

1 £6

2 21

3 **a** £120 **b** $6

4 **a** £85 **b** £209

5 £175

N7 Exam review

1 a £7.50 b 37.5%

2 a 656 students b No, 25% is 200 students.
c 360 d 22%

D5 Before you start ...

1 a 360° b 295° c 140°

2 a 30 b 40 c 36 d 2 e 4

3 a Mean = 7, Mode = 3, Median = 6, Range = 13
b Mean = 5, Mode = 0, Median = 2, Range = 15
c Mean = 4, Mode = 3, Median = 3.5, Range = 3

D5 Exam review

1 a Chocolate
b Pie chart angles: Strawberry 72°, Chocolate 120°, Vanilla 72°, Mint 36°, Other 60°

2 a 26 b 22

S6 Before you start ...

1 a Diameter b Circumference c Radius

2 28 cm^2

3 a iii b i c ii

S6 Exam review

1 19.5 cm

2 ai 11 cm^2 aii 16 cm
b

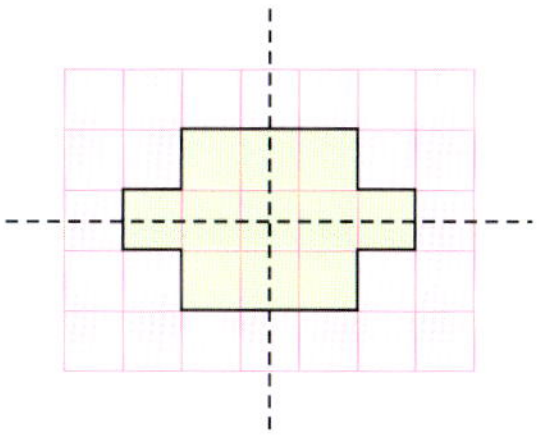

c 12 cm^3

A7 Before you start ...

1 a 2, 4, 6, 8, 10 b 4, 8, 12, 16, 20
c 3, 6, 9, 12, 15 d 5, 10, 15, 20, 25

2 19, 22

3 a 11 b 5 c 17 d 31

4 a $x = 6$ b $y = 3$

A7 Exam review

1 a 27, 32
b Add 5 to the previous term
c The units digit of all terms is either 2 or 7

2 a

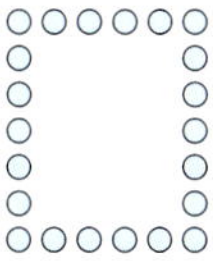

Pattern 4

b 22, 26 c 46

N8 Before you start ...

1 3 : 2

2 £6.20

3 a 24 b 12 kg

4 a 11.16 b 11.8336 c 15

N8 Exam review

1 a 3 : 4
b 90 g castor sugar, 37.5 g ground almonds, 1.5 tsp lemon zest, 120 g butter, 75 g flour

2 a 16 : 30 bi £8.39 bii £9

S7 Before you start ...

1 a $x = -2$ b $y = 1$

2 a 90° b 270° c 180°

3 a 6.4 b 7.6 c 9.4 d 11.4 e 18.4

S7 Exam review

1 2

2 a

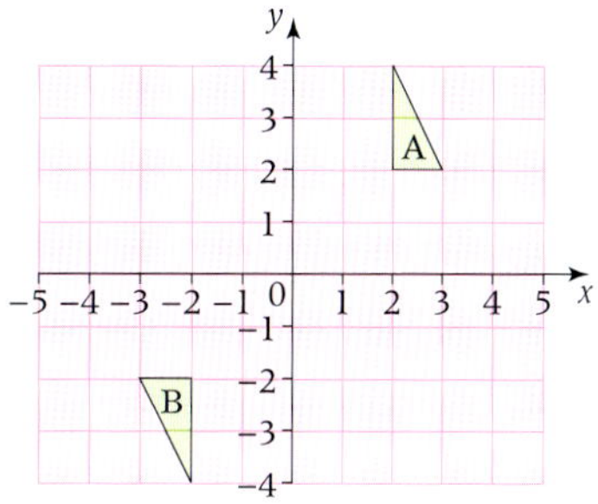

b Triangle with vertices (4, 4) (6, 4) (4, 8)

D6 Before you start ...

1 a 2, 3, 5 or 7 b 1, 4 or 9 c 1, 3, 6 or 10
d 4 or 8 e 1, 2, 5 or 10

2 a $\frac{2}{3}$ b $\frac{4}{5}$ c $\frac{1}{4}$ d $\frac{3}{5}$ e 1

3 a $\frac{3}{4}$ b $\frac{3}{10}$ c $\frac{2}{5}$

4 a 10 b 40 c 200

D6 Exam review

1 a 1T, 2T, 3T, 4T, 5T, 6T, 1H, 2H, 3H, 4H, 5H, 6H
b $\frac{1}{4}$

2 a $\frac{1}{2}$
b He should get Heads about half the time, but it is unlikely that it will be exactly half, especially with only 30 spins.

A8 Before you start ...

1 a, b Kite

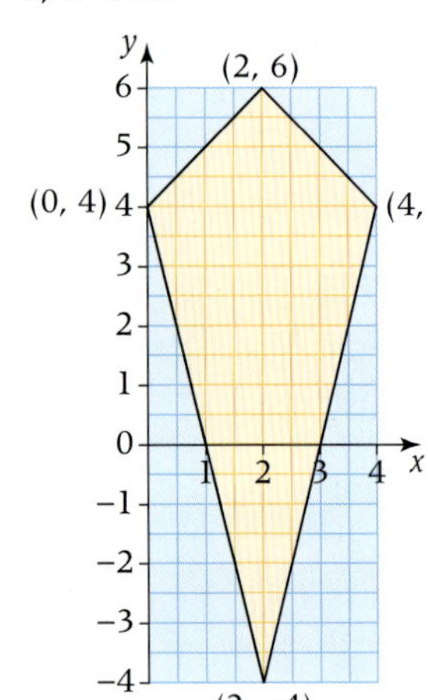

2 a 25
b 7
c 24
d 6.1

3 a 2000 m
b 60 000 m
c 500 m
d 3500 m

4 a €1.45
b €21.75
c €50.03

A8 Exam review

1 a 0 miles = 0 km, 5 miles = 8 km
b See answer to Exercise A8.3, question 2c–d
ci 6.4 km **cii** 7.5 miles

2 a 15 minutes **b** 15 km
c

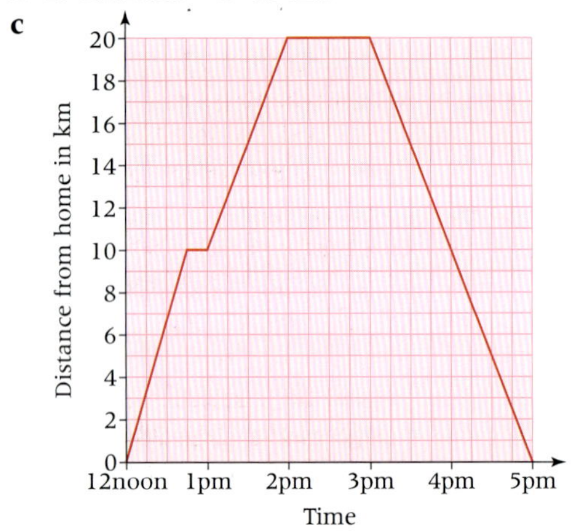

S8 Before you start ...

1 a 180° **b** 180° **c** 360°
2 a 50° **b** 70°
3 32 cm^2
4 a 4 cm^3 **b** 200 cm^3

S8 Exam review

1 a Octagon **b** B **c** Hexagon tessellation
2 ai 12 **aii** 8
b

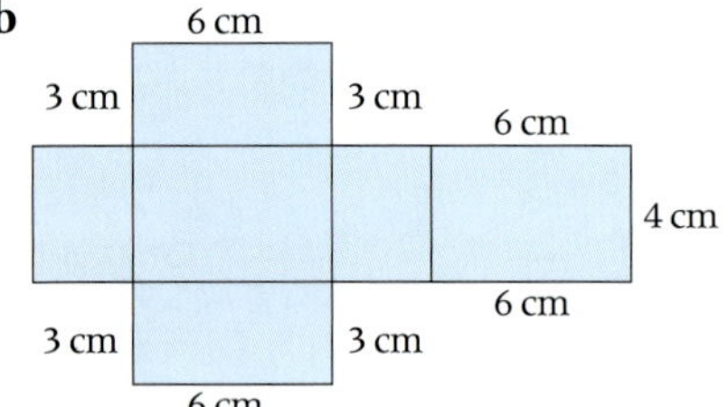

Index